Food Industrial Processes Handbook

Food Industrial Processes Handbook

Edited by **Hugh Brennan**

Callisto Reference

New York

Published by Callisto Reference,
106 Park Avenue, Suite 200,
New York, NY 10016, USA
www.callistoreference.com

Food Industrial Processes Handbook
Edited by Hugh Brennan

International Standard Book Number: 978-1-63239-337-1 (Hardback)

Printed in China.

Contents

Preface

This book has been a concerted effort by a group of academicians, researchers and scientists, who have contributed their research works for the realization of the book. This book has materialized in the wake of emerging advancements and innovations in this field. Therefore, the need of the hour was to compile all the required researches and disseminate the knowledge to a broad spectrum of people comprising of students, researchers and specialists of the field.

This book is based on the various food industrial processes. Nowadays, food industry consumers worldwide are becoming more cautious, knowledgeable and consequently demanding, since food is a basic necessity of every being. The consumers want the industries to maintain a food quality at par with the standards set by food industry organizations. Problems like global warming, climate change and greenhouse gas emissions cause disasters like floods, droughts, fires and storms which result in massive loss for the agricultural and husbandry sectors and thus, add to the already prevalent threat of food scarcity to mankind. This book is a collection of studies which take on the prevailing food industry related issues. These works were chosen to serve the purpose of investigation guides and documentations based on modern scientific and technical references. Therefore, this book can be used by students as well as industrialists or farmers. Not only are the issues related to control of food processing and production being discussed, but also the engineering, economic and financial avenues have been brought into light for the food industry managers.

At the end of the preface, I would like to thank the authors for their brilliant chapters and the publisher for guiding us all-through the making of the book till its final stage. Also, I would like to thank my family for providing the support and encouragement throughout my academic career and research projects.

Editor

Part 1

Physical and Chemical Features

1

Hydrocolloids in Food Industry

Jafar Milani and Gisoo Maleki
Sari Agricultural Sciences and Natural Resources University
Iran

1. Introduction

Hydrocolloids or gums are a diverse group of long chain polymers characterized by their property of forming viscous dispersions and/or gels when dispersed in water. These materials were first found in exudates from trees or bushes, extracts from plants or seaweeds, flours from seeds or grains, gummy slimes from fermentation processes, and many other natural products. Occurrence of a large number of hydroxyl groups noticeably increases their affinity for binding water molecules rendering them hydrophilic compounds. Further, they produce a dispersion, which is intermediate between a true solution and a suspension, and exhibits the properties of a colloid. Considering these two properties, they are appropriately termed as 'hydrophilic colloids' or 'hydrocolloids'.

Hydrocolloids have a wide array of functional properties in foods including; thickening, gelling, emulsifying, stabilization, coating and etc. Hydrocolloids have a profound impact on food properties when used at levels ranging from a few parts per million for carrageenan in heat-treated dairy products to high levels of acacia gum, starch or gelatin in jelly confectionery. The primary reason behind the ample use of hydrocolloids in foods is their ability to modify the rheology of food systems. This includes two basic properties of food systems that is, flow behavior (viscosity) and mechanical solid property (texture). The modification of texture and/or viscosity of food systems helps modify its sensory properties, therefore hydrocolloids are used as significant food additives to perform specific purposes. It is evident that several hydrocolloids belong to the category of permitted food additive in many countries throughout the world. Various food formulations such as soups, gravies, salad dressings, sauces and toppings use hydrocolloids as additives to achieve the preferred viscosity and mouth feel. They are also used in many food products like ice-creams, jams, jellies, gelled desserts, cakes and candies, to create the desired texture.

In addition to the functional attributes, future acceptance and, possibly, positive endorsement may derive from the recognition that fibers contribute many physiological benefits to the natural function and well-being of the body.

The aim of this chapter of the book is to highlight the importance of the hydrocolloids in food industry.

2. Functional properties

2.1 Viscosity enhancing or thickening properties

The foremost reason behind the ample use of hydrocolloids in foods is their ability to modify the rheology of food system. The modification of texture and/or viscosity of food system helps to modify its sensory properties, and hence, hydrocolloids are used as important food additives to perform specific purposes. The process of thickening involves nonspecific entanglement of conformationally disordered polymer chains; it is essentially polymer-solvent interaction (Philips et al., 1986). Hydrocolloids that have been used as thickening agents are shown in Table 1. The thickening effect of produced by the hydrocolloids depends on the type of hydrocolloid used, its concentration, the food system in which it is used and also the pH of the food system and temperature. Ketchup is one of the most common food items where the hydrocolloid thickeners are used to influence its viscosity (Sahin & Ozdemir, 2004).

The question that arises is how hydrocolloids thicken solution. In dilute dispersion, the individual molecules of hydrocolloids can move freely and do not exhibit thickening. In concentrated system, these molecules begin to come into contact with one another; thus, the movement of molecules becomes restricted. The transition from free moving molecules to an entangled network is the process of thickening.

The viscosity of polymer solutions is influenced significantly by the polymer molecular mass. In addition to molecular mass effects, the hydrodynamic size of polymer molecules in solution is significantly influenced by molecular structure. Linear, stiff molecules have a larger hydrodynamic size than highly branched, highly flexible polymers of the same molecular mass and hence give rise to a much higher viscosity.

2.2 Gelling properties

Swollen particulate forms of gelled hydrocolloids are particularly useful as they combine macroscopic structure formation with an ability to flow and often have an attractive soft solid texture, which is especially sought in food applications, all at high water contents (>95%). There is a potential opportunity for particulate hydrocolloid systems to replace chemically cross-linked starches based on appropriate structuring, processing, and molecular release properties without the need for chemical treatment.

The characteristics of gel particles, and the application for which they are used, will depend on the type of hydrocolloid, the network formation mechanism and the processing method used for particle formation (Burey et al., 2008).

Hydrocolloid gel networks form through entwining and cross-linking of polymer chains to form a three-dimensional network. The mechanism by which this interchain linking occurs can vary (Djabourov, 1991).

Hydrocolloid gelation can involve a hierarchy of structures, the most common of which is the aggregation of primary interchain linkages into "junction zones" which form the basis for the three-dimensional network characteristic of a gel.

Various parameters such as temperature, the presence of ions, and the inherent structure of the hydrocolloid can affect the physical arrangement of junction zones within the network.

2.3 Surface activity and emulsifying properties

The functionality of hydrocolloids as emulsifiers and/or emulsion stabilizers correlates to phenomena such as: retardation of precipitation of dispersed solid particles, decreased creaming rates of oil droplets and foams, prevention of aggregation of dispersed particles, prevention of syneresis of gelled systems containing oils and retardation of coalescence of oil droplets. It is believed that gums will adsorb (onto solid or liquid surfaces) very slowly, weakly and with very limited surface load if at all. The hydrocolloids were classified according to their activity at the interface. Gum Arabic is probably the most studied hydrocolloid that proved significant surface activity. Gum arabic is the only gum adsorbing onto oil-water interfaces and imparting steric stabilization. Other gums such as galactomannans, xanthan, pectin, etc. have been known to reduce surface and interfacial tensions, to adsorb onto solid surfaces and to improve stability of oil-in-water emulsions. Micro crystalline cellulose (MCC) is an example of a hydrocolloid with no solubility in water that adsorbs mechanically at the interface.

It well documented that gum arabic, a natural polysaccharide, has excellent emulsification properties for oil-in-water emulsions. An excellent example of its use is in cloudy emulsions, as opacity builders for citrus beverages (Connolly et al. 1988). Related the significance of protein components presenting in gum arabic to its emulsifying properties. It seems that the protein-hybrid in gum arabic meets all the necessary requirements in a capacity similar to emulsifying proteins (such as casein, or soy protein) via its numerous adsorption sites, flexibility, conformational change at the interface and the entropy gain (solvent depletion). Gum arabic works by reducing the oil-water interfacial tension, thereby facilitating the disruption of emulsion droplets during homogenization. The peptides are hydrophobic and strongly adsorb on to the surface of oil droplets, whilst the polysaccharide chains are hydrophilic and extend out into the solution, preventing droplet flocculation and coalescence through electrostatic and steric repulsion forces.

Microcrystalline cellulose is also able to stabilize the oil-in-water emulsions. Its strong affinity for both the oil and the water results in precipitation and some orientation of the solid particles at the oil-in-water interface (Philips et al., 1984). It was proposed that the colloidal network of the free MCC thickens the water phase between the oil globules preventing their close approach and subsequent coalescence. Therefore, the MCC provides long term stability (Philips et al., 1990).

As the galactomannans structure gives no suggestion of the presence of any significant proportion of hydrophobic groups, it is generally assumed that this type of hydrocolloid functions by modifying the rheological properties of the aqueous phase between the dispersed particles or droplets. It has been suggested that these gums stabilize emulsions by forming liquid crystalline layers around the droplets. It should be noted that this putative adsorption of the gum to the oil-water interface is reportedly rather weak and reversible. That is, the associated emulsion stability is lost on diluting the aqueous phase of the emulsion with water (Dickinson, 2003).

Pectin is another class of hydrocolloid whose emulsifying character has attracted attention in recent years. While citrus and apple pectin is normally used as low-pH gelling or thickening agents (are not effective as emulsifying agents), sugar beet pectin does not form gels with calcium ions or at high sugar concentrations. Due to its higher protein content, sugar beet pectin is considerably more surface-active than gum arabic, and it is very effective in stabilizing fine emulsions based on orange oil or triglyceride oil at a pectin/oil ratio of 1:10 (Williams et al., 2005).

2.4 Hydrocolloids as edible films and coatings

An edible film is defined as a thin layer, which can be consumed, coated on a food or placed as barrier between the food and the surrounding environment. The most familiar example of edible packaging is sausage meat in casing that is not removed for cooking and eating. Hydrocolloids are used to produce edible films on food surfaces and between food components. Such films serve as inhibitors of moisture, gas, aroma and lipid migration. Many gums and derivatives have been used for coating proposes. They include alginate, carrageenan, cellulose and its derivatives, pectin, starch and its derivatives, among others. Since these hydrocolloids are hydrophilic, the coatings they produce have nature limited moisture barrier properties. However, if they are used in a gel form, they can retard moisture loss during short term storage when the gel acts as sacrificing agent rather than a barrier to moisture transmission. In addition, since in some cases an inverse relationship between water vapor and oxygen permeability has been observed, such films can provide effective protection against the oxidation of lipid and other susceptible food ingredient. The hydrocolloid edible films are classified into two categories taking into account the nature of their components: proteins, polysaccharides or alginates. Hydrocolloidal materials, i.e. proteins and polysaccharides, used extensively for the formation of edible films and coatings are presented in Table 1 (Hollingworth, 2010).

	Film forming material		Principal function
Polysaccharide	Agar		Gelling agent
	Alginate		Gelling agent
	Carrageenan		Gelling agent
	Cellulose derivatives	CMC	Thickener
		HPC	Thickener and emulsifier
		HPMC	Thickener
		MC	Thickener, emulsifier and gelling agent
	Chitosan		Gelling agent Antimicrobials
	Gum	Arabic gum	Emulsifier
		Guar gum	Thickener
		Xanthan gum	Thickener
	Pectin		Gelling agent
	Starches		Thickener and gelling agent
protein	Gelatin	Bovine gelatin	Gelling agent
		Fish gelatin	Gelling agent
		Pig gelatin	Gelling agent

Table 1. Hydrocolloidal materials that have been studied extensively for the formation of edible films and coatings in foods.

2.5 Hydrocolloids as fat replacers

The changes in modern lifestyle, the growing awareness of the link between diet and health and new processing technologies have led to a rapid rise in the consumption of ready-made meals, novelty foods and the development of high fiber and low-fat food products. Calorie-dense materials such as fats and oils may be replaced with 'structured water' to give healthy, reduced-calorie foods with excellent eating quality. In particular, numerous hydrocolloid products have been developed specifically for use as fat replacers in food. This has consequently led to an increased demand for hydrocolloids. As an example, the Italian dressing includes xanthan gum as a thickener and the 'Light' mayonnaise contains guar gum and xanthan gum as fat replacers to enhance viscosity.

The traditional approach is the partial replacement of fat using starches which, when dissolved in water, create stable thermo-reversible gels. Soft, fat-like gels can be created by conversion modifications to the degree necessary to produce thermo-reversible, spreadable gels. Typically, 25–30% solids, i.e. starch in water, form an optimal stable structure for fat replacement. New generation fat replacers are tailored to mimic more closely the many and complex properties of fats or oils in a particular application. These are referred to as fat mimetics. Maximising the synergies of functional ingredients such as hydrocolloids generally in combination with specific starch fat mimetics can mean that 100% fat reduction is achievable (Phillips & Willians, 2000).
Based on the particle gel characteristics of inulin, it can be concluded that inulin functions as a fat replacer but only in water-based systems. When concentrations exceed 15%, insulin has the ability to form a gel or cream, showing an excellent fat-like texture. This inulin gel is a perfect fat replacer offering various opportunities in a wide range of foods. Each inulin particle dispersed in the water phase of any food system will contribute to the creaminess of the finished food. Inulin is also destined to be used as a fat replacer in frozen desserts, as it processes easily to provide a fatty mouth-feel, excellent melting properties, as well as freeze–thaw stability, without any unwanted off-flavor.

3. Origins and structures of hydrocolloids

3.1 Plant hydrocolloids

3.1.1 Cellulose and derivatives

Cellulose is the most abundant naturally occurring polysaccharide on earth. It is the major structural polysaccharide in the cell walls of higher plants. It is also the major component of cotton boll (100%), flax (80%), jute (60 to 70%), and wood (40 to 50%). Cellulose can be found in the cell walls of green algae and membranes of fungi. Acetobacter xylinum and related species can synthesize cellulose. Cellulose can also be obtained from many agricultural by-products such as rye, barley, wheat, oat straw, corn stalks, and sugarcane. Cellulose is a high molecular weight polymer of (1→4)-linked β-D-glucopyranose residues. The β-(1→4) linkages give this polymer an extended ribbon-like conformation. The tertiary structure of cellulose, stabilized by numerous intermolecular H-bonds and van der Waals forces, produces three-dimensional fibrous crystalline bundles. Cellulose is highly insoluble and impermeable to water. Only physically and chemically modified cellulose finds applications in various foodstuffs (Cui, 2005).

3.1.1.1 Microcrystalline cellulose

Microcrystalline cellulose (MCC) is purified cellulose, produced by converting fibrous cellulose to a redispersible gel or aggregate of crystalline cellulose using acid hydrolysis. Microcrystalline cellulose is prepared by treating natural cellulose with hydrochloric acid to partially dissolve and remove the less organized amorphous regions of this polysaccharide. The end product consists primarily of crystallite aggregates. MCC is available in powder form after drying the acid hydrolysates. Dispersible MCC is produced by mixing a hydrophilic carrier (e.g., guar or xanthan gum) with microcrystals obtained through wet mechanical disintegration of the crystallite aggregates (Cui, 2005). These colloidal dispersions are unique when compared to other soluble food hydrocolloids. They exhibit a variety of desirable characteristics including suspension of solids, heat stability, ice crystal control, emulsion stabilization, foam stability, texture modification and fat replacement (Imeson, 2010).

3.1.1.2 Carboxymethylcellulose

Carboxymethylcellulose (CMC) is an anionic, water-soluble polymer capable of forming very viscous solutions. CMC is prepared by first treating cellulose with alkali (alkali cellulose), and then by reacting with monochloroacetic acid. The degree of substitution (DS) with the carboxyl groups is generally between 0.6 to 0.95 per monomeric unit (maximum DS is 3), and occurs at *O*-2 and *O*-6, and occasionally at *O*-3 positions.

3.1.1.3 Methylcellulose and its derivatives

Methylcellulose (MC) has thickening, surface activity (due to hydrophobic groups), and film forming properties. MC is prepared by treating alkali cellulose with methyl chloride. Other MC derivatives are also available, of which hydroxypropylmethylcellulose (HPMC) has been widely used. The reagent for HPMC is mixture of methyl chloride and propylene oxide. These two cellulose derivatives are non-ionic.

3.1.2 Hemicelluloses

Hemicelluloses are a heterogeneous group of polysaccharides constituting the cell walls of higher plants; these polysaccharides are often physically entangled, covalently and/or noncavalently bonded to cellulose and lignins. The structure of hemicelluloses may vary depending on their origin, but they can be divided into four groups based on composition of their main backbone chain:D-xylans with (1→4)-linked β-D-xylose; D-mannans, with (1→4)-linked β-D mannose; D-xyloglucans with D-xylopyranose residues attached to the cellulose chain; and D-galactans with (1→3)-linked β-D-galactose. The first three groups are very similar to cellulose in having the main chain backbone linked via (1→4) diequatorial linkages and capable of adopting extended ribbon conformations. Most of the hemicelluloses, however, are substituted with various other carbohydrate and noncarbohydrate residues, and unlike cellulose, they are heteropolysaccharides. This departure from uniformity because of various side branches renders them at least partially soluble in water (Cui, 2005).

3.1.2.1 Mannans and galactomannans

The cell walls of seeds are especially rich in mannans and galactomannans. D-Mannans, found in tagua palm seeds, have a backbone composed of linear (1→4)-linked β-D-mannose chains. The best known D-galactomannans, locust bean, guar, and tara gums have the same linear mannan backbone but they are substituted with α-D-Gal*p* side units linked to O-6. The degree of substitution in galactomannans, which profoundly affects their solution properties, differs in galactomannans extracted from various plants (Table 2) (Cui, 2005).

Widely used galactomannans are from the carob tree (*Ceratonia siliqua*), named as locust bean gum (LBG) or carob bean gum (CBG), and the guar plant (*Cyamopsis tetragonoloba*), namely guar gum (Philips & Williams, 2000). The ratio of D-mannosyl to D-galactosyl units is about 1.8:1 in guar gum and 3.9:1 in LBG. Guar gum containing galactose content of 33-40% (w/w) is soluble in water of 25°C. The rate of dissolution of guar gum increases with decreasing particle size and with increasing temperature. In the case of LBG, with the larger part containing galactose contents of about 17-21% (w/w), it needs a heat treatment during 10 min at 86-89°C under stirring to dissolve in water. Like most hydrocolloids, both guar gum and LBG shows pseudoplastic, or shear-thinning, behavior in solution. The degree of pseudoplasticity increases with both concentration and molecules weight.

Galactomannan	Species of Origin	Man:Gal Ratio
Locust bean gum	Ceratonia siliqua	3.5
Senna gum	Senna occidentalis	3.5
Guar gum	Cyamopsis tetragonolobus	1.6
Tara gum	Caesalpinia spinosa	1.3
Fenugreek gum	Trigonella foenum graecum	1

Table 2. Botanical Origin and Main Structure Features of Galactomannans

3.1.2.2 Xyloglucans

Xyloglucans, like cellulose, have linear backbones of (1→4)-linked β-D glucopyranoses. Numerous xylopyranosyl units are attached along the main backbone. In many plant xyloglucans, the repeating unit is a heptasaccharide, consisting of a cellotetraose with three subtending xylose residues (Phillips & Williams, 2000). Some xylose residues may carry additional galactosyl and fucosyl units. A few plants may have arabino- instead of fucogalactosylgroups attached to the xylose residues. One of the best characterized is the xyloglucan from the cotyledons of the tamarind seed (*Tamarindus indica*) (Shirakawa et al., 1998).

3.1.2.3 Glucomannans

Glucomannans are linear polymers of both (1→4)-linked β-D-mannose and (1→4)-linked β-D-glucose residues. Glucomannans are obtained from dried and pulverized root of the perennial herb *Amorphophallus konjac*. Acetyl groups scattered randomly along the glucomannan backbone promote water solubility. Konjac glucomannan is a high molecular weight polymer (>300 kDa) which can form viscous pseudoplastic solutions. It can form a gel in the presence of alkali.

3.1.2.4 Arabinoxylans

D-Xylans are composed of (1→4)-linked β-D-xylopyranoses with various kinds of side branches, the most common being 4-*O*-methyl-D-glucopyranosyl uronic acid linked mostly to *O*-2 of β-Xyl*p* units and α-L-Araf linked to *O*-3 of β-Xyl*p* units. The amount of arabinose and glucuronic acid in glucuronoarabinoxylans may vary substantially, ranging from substitution at almost all Xyl*p* to polymers having more than 90% of unsubstituted β-Xyl*p* units. Many cereal (wheat, barley, rye, oats) arabinoxylans do not carry glucuronic acid units.

3.1.2.5 β-D-Glucans

β-D- Glucans are high molecular weight, viscous polysaccharides. Mixed linkage (1→3), (1→4) β-D-glucans are present in the grass species, cereals, and in some lichens (e.g., *Cetraria islandica*). Cereal β-D-glucans contain predominantly (1→4) linked β-D-Glc*p* units (~70%) interrupted by single (1→3)-linked β-D-Glc*p* units (~30%). The distribution of β-(1→4) and β-(1→3) linkages is not random; this leads to a structure of predominantly β-(1→3)-linked cellotriosyl and cellotetraosyl units. There are also longer fragments of contiguously β-(1→4)-linked glucose units (cellulose fragments) in the polymer chain. The main source of food β-D-glucans are the kernels of oats, barley, wheat, and rye. β-D- glucans have been ascribed cholesterol and blood glucose lowering properties.

3.1.2.6 Arabinogalactan

Arabinogalactan is a major D-galactan obtained from soft-woods such as pine, larch, cedar, and spruce. This polymer has a main backbone of (1→3)-linked β-D-galactopyranosyl residues with β-(1→6)-linked disaccharides of β-D-Gal*p*-(1→6)-β-D-Gal*p* and α-(1→6)-linked disaccharides of β-L-Ara*f*-(1→3)-α-L-Ara*f*. Arabinogalactan is generally a highly branched polymer with arabinose and galactose ratio of 1:6. Commercially available arabinogalactan, obtained from the butt wood of Western larch, has a relatively low molecular weight of 15,000 to 25,000, little impact on viscosity, color, and taste. It is used as a low-calorie additive in beverages to increase the fiber content.

3.1.3 Pectins

Pectins are polysaccharide and are the major components of most higher plant cell walls; they are particularly prevalent in fruits and vegetables. Commercial pectins are prepared mostly from some by-products of the food industry, such as apple pulp, citrus peels, and sugarbeet pulp. Pectins are the most complex class of plant cell wall polysaccharides. They comprise of two families of covalently linked polymers, galacturonans and rhamnogalacturonans.

Galacturonans are segments of pectins with (1→4)-linked α-D-galactosyluronic acid residues in the backbone, such as those in the linear homogalacturonans, in the substituted xylogalacturonans and in rhamnogalacturonans type II (RG II) (Cui, 2005). The carboxylic acid groups in galacturonans may be methyl esterified; the degree of esterification has an important effect on the conformation and solution properties of these polymers. Based on the degree of esterification, pectins are divided into two categories: low methyl (LM) pectin that contains less than 50% methyl esters, and high methyl (HM) pectin with more that 50% methyl esters. Xylogalacturonans are relatively recently discovered subunits of pectic polysaccharides, present in storage tissue of reproductive organs of peas, soybeans, apple fruit, pear fruit, onions, cotton seeds, and watermelon. The rhamnogalacturonans type II have been found in the cell walls of many tissues of edible plants, such as apple (juice), kiwi, grape (wine), carrot, tomato, onion, pea, and radish. Rhamnogalacturonans type I (RG I) have a backbone composed of alternating (1→2)-linked α-L rhamnosyl and (1→4)-linked α-D-galacturonic acid residues. Depending on the source of pectins, 20 to 80% of rhamnose residues may be branched at *O*-4 with side chains which vary in length and composition. The side branches may be composed of arabinans, galactans and type I arabinogalactans. Pectins with type I arabinogalactans have been found in potato, soybean, onion, kiwi, tomato, and cabbage.

3.1.4 Exudate gums

Exudate gums are polysaccharides produced by plants as a result of stress, including physical injury and/or fungal attack. Gum arabic, gum tragacanth, gum karaya, and gum ghatti have been used by humans for many thousands of years in various food and pharmaceutical applications (Table 3). Generally, these gums are structurally related to arabinogalactans, galacturonans, or glucuronomannans. They all contain a high proportion of glucuronic or galacturonic acid residues (up to 40%).

Gum	Species of Origin	General structure	Viscosity Concentration %	(Pas×10^-3)[a]
Gum arbic	Acacia Senegal	Substituted acidic arabinogalactan	5.01	717
Gum tragacanth	Astragalus gummifer	Mixture of arabinogalactan and glycano-rhamnogalacturonan	1.0 3.0	1,000 >10,000
Gum karaya	Sterculia urens	glycano-rhamnogalacturonan	1.0 5.0	300,045,000
Gum ghatti	Anogeissus latifolia	glycano-glucuronomannoglycan	5.01	2,882,440

[a] Viscosity obtained at shear rates 10-S

Table 3. Main botanical and structural characteristics exudates gums

3.1.4.1 Gum arabic

Acacia gum, also known as gum arabic, is a natural, vegetable exudate from acacia trees (primarily in Africa) known since antiquity and used for thousands of years in foods as an additive and ingredient, in the pharmaceutical industry and for technical purposes (Imenson, 2010). The thorny trees grow to a height of 7 to 8 meters, and the gum is obtained by cutting sections of the bark from the tree. The structure of gum arabic is relatively complex. The main chain of this polysaccharide is built from (1→3) and (1→6)-linked β-D-galactopyranosyl units along with (1→6)-linked β-D-glucopyranosyl uronic acid units. Side branches may contain α-Lrhamnopyranose, β-D-glucuronic acid, β-D-galactopyranose, and α-L-arabinofuranosyl units with (1→3), (1→4), and (1→6) glycosidic linkages. Gum arabic has a high water solubility (up to 50% w/v) and relatively low viscosity compared to other exudate gums. The highly branched molecular structure and relatively low molecular weight of this polymer are responsible for these properties. Another unique feature of gum arabic is its covalent association with a protein moiety. It is thought that the protein moiety rich in hydroxyproline (Hyp), serine (Ser), and proline (Pro) constitutes a core to which polysaccharide subunits are attached via Ara-Hyp linkages (the wattle blossom model). The protein moiety of gum arabic is responsible for the surface activity, foaming, and emulsifying properties of this polymer (Phillips & Williams, 2000).

3.1.4.2 Tragacanth gum

In the *European Pharmacopoeia* (6th edition, 2007), *gum tragacanth* is defined as 'the air-hardened gummy exudates, flowing naturally or obtained by incision from the trunk and branches of *Astragalus gummifer* Labillardiere and certain other species of *Astragalus* from western Asia (mostly in Iran, some in Turkey)'. Tragacanth gum contains a water-soluble fraction and a water-insoluble fraction and the water-soluble fraction is accounted for 30 to 40% of the total gum. The water soluble fraction (tragacanthin) is a highly branched neutral polysaccharide composed of 1→6-linked D-galactosyl backbones with L-arabinose side chains joined by 1→2-, 1→3- and/or 1→5-linkages. The water-insoluble fraction (~60 to 70%), is tragacanthic acid (bassorin) which is a water-swellable polymer and is consisted of D-galacturonic acid, D-galactose, L-fucose, D-xylose, L-arabinose and L-rhamnose. It has a

(1→4)-linked α-D-galacturonopyranosyl backbone chain with randomly substituted xylosyl branches linked at the 3 position of the galacturonic acid residues. In spite of the availability of alternative materials, the continued use of the gum is the result of its unique functional properties combined with a high degree of stability in a range of conditions.

3.1.4.3 Gum karaya

Gum karaya, also known as sterculia gum, is a branched acidic polysaccharide obtained from the exudates of the *Sterculia urens* tree of the Sterculiaceae family grown in India. The backbone chain is a rhamnogalacturonan consisting of α-(1→4)-linked D-galacturonic acid and α-(1→2)-linked-L-rhamnosyl residues. The side chain is made of (1→3)-linked β-D-glucuronic acid, or (1→2)-linked β-D-galactose on the galacturonic acid unit where one half of the rhamnose is substituted by (1→4) linked β-D-galactose.

3.1.4.4 Gum ghatti

Gum ghatti is an amorphous translucent exudate of the *Anogeissus latifolia* tree of the Combretaceae family grown in India. The monosaccharide constituents of gum ghatti are L-arabinose, D-galactose, D-mannose, D-xylose, and D-glucuronic acid in the ratio of 10:6:2:1:2, with traces of 6-deoxyhexose.

3.1.5 Mucilage gums

Mucilage gums are very viscous polysaccharides extracted from seeds or soft stems of plants; examples are psyllium (from *Plantago* species), yellow mustard (from *Sinapis alba*), and flax mucilage (from *Linum usitatissimum*). All of them are acidic polysaccharides with structures somewhat related to some of the exudate gums. Their utilization in certain food products is increasing due to their functional properties (viscosity, gelation, water binding) as well as to their bio-active role in prevention and/or treatment of certain diseases (Cui, 2005).

3.1.5.1 Psyllium gum

Psyllium gum can be extracted from seeds of the *Plantago* species. The gum is deposited in the seed coat; it is, therefore, advantageous to mechanically separate the outer layers from the rest of the seed before extraction. Psyllium gum can be extracted with hot water or mild alkaline solutions. The molecular structure of the gum is a highly branched acidic arabinoxylan. D-Glucuronic acid residues have also been found in this gum. Psyllium gum has a very high molecular weight (~1500 kDa) and does not completely dissolve in water. When dispersed in water, it swells and forms a mucilageous dispersion with gel-like properties. It is used primarily as a laxative and dietary fiber supplement in pharmaceutical and food industries.

3.1.5.2 Yellow mustard mucilage

Yellow mustard mucilage can be extracted from whole mustard seeds or from the bran. The mucilage contains a mixture of a neutral polysaccharide, composed mainly of glucose, and an acidic polysaccharide, containing galacturonic and glucuronic acids, galactose, and rhamnose residues. Detailed analysis of the neutral fraction of yellow mustard mucilage showed that it contains mainly (1→4)-linked β-D-glucose residues. The *O*-2, *O*-3, and *O*-6 atoms of the (1→4)-β-D-glucan backbone may carry ether groups (ethyl or propyl). Depending on the polymer concentration, yellow mustard mucilage can form either viscous solution of weak gels. When it is mixed with locust bean gum, however, the gel rigidity can be increased substantially. It has been shown that the neutral (1→4)-β-D-glucan fraction of yellow mustard mucilage synergistically interacts with galactomannans. Yellow mustard is used in processed meat formulations and salad dressing as a stabilizer and bulking agent.

3.1.5.3 Flaxseed mucilage

Flaxseed mucilage can be easily extracted from the seeds by soaking them in warm water. The mucilage constitutes the secondary wall material in the outermost layer of the seed. Upon hydration of the seeds, it expands, breaks the mucilage cells, and exudes on the surface of the seeds. Flaxseed mucilage contains 50 to 80% carbohydrates and 4 to 20% proteins and ash. Flaxseed mucilage contains a mixture of neutral polysaccharides, composed mainly of xylose, arabinose and galactose residues, and acidic polysaccharides, containing galactose, rhamnose, and galacturonic acid residues. The neutral fraction of flaxseed mucilage has a backbone of (1→4)-linked β-D-xylopyranosyl residues, to which arabinose and galactosecontaining side chains are linked at *O*-2 and/or *O*-3.

The acidic fraction of flaxseed mucilage has a rhamnogalacturonan backbone with (1→4)-linked α-D-galacturonopyranosyl and (1→2)-linked α-L-rhamnopyranosyl residues. The ratio of neutral to acidic polysaccharides in flaxseed may vary substantially with their origin. Unfractionated flaxseed mucilage forms a viscous solution, but it is the neutral fraction that mainly contributes to the high viscosity and weak gel-like properties of this gum. Flaxseed mucilage has not yet been widely utilized mostly because of limited information about the structure and functional properties of this gum. Similar to other gums, flaxseed mucilage can be used as a thickener, stabilizer, and water-holding agent.

3.1.6 Fructans

Fructans are reserve polysaccharides in certain plants, either complementing or replacing starch. They can also be produced by certain species of bacteria. A main kind of fructans is inulin. Inulins are found in roots or tubers of the family of plants known as Compositae, including dandelions, chicory, lettuce, and Jerusalem artichoke. They can also be extracted from the Liliacae family, including lily bulbs, onion, tulips, and hyacinth. Inulin is a low molecular weight polysaccharide containing (2→1) linked β-D-Fru*p* residues.

3.2 Seaweed hydrocolloids

3.2.1 Alginates

Alginates constitute the primary structural polysaccharides of brown seaweeds (*Phaeophyceae*). The alginate molecules provide both flexibility and strength to the plants and these properties are adapted as necessary for growth conditions in the sea. The major species of seaweeds that produce alginates are *Macrocystis pyrifera*, grown primarily along the California coast of the USA, south- and north-western coasts of South America, and coasts of Australia and New Zealand. Other good sources of alginates are *Laminaria hyperborea*, *Laminaria digitata*, and *Laminaria japonica*, grown along the north Atlantic coast of the USA, Canada, France, and Norway. Alginates can also be synthesized by bacteria, *Pseudomonas aeruginosa* and *Azobacter vinelandii*. Alginates are unbranched copolymers of (1→4)-linked β-D-mannuronic acid (M) and α-L-guluronic acid (G) residues. If the uronic acid groups are in the acid form (-COOH), the polysaccharide, called alginic acid, is water insoluble. The sodium salts of alginic acid (-COONa), sodium alginates, are water soluble. The sequence of mannuronic and guluronic residues significantly affects the physicochemical properties of alginates. The ratio of β-D-mannuronic acid to α-L-guluronic acid residues is usually 2:1, although it may vary with the algal species, the age of the plant as well as the type of tissue the alginates are extracted from (Cui, 2005). The main advantage of alginate as a gel former is its ability to form heat-stable gels which can set at room temperatures. In food applications, it is primarily gel formation with calcium ions which is of interest.

3.2.2 Carrageenans

Carrageenans are structural polysaccharides of marine red algae of the Rhodophyceae class. They are extracted mainly from *Chondrus crispus*, *Euchema cottoni*, *Euchema spinosum*, *Gigartina skottsbergi*, and *Iradaea laminarioides*. These red seaweeds grow mostly along the Atlantic coasts of North America, Europe, and the western Pacific coasts of Korea and Japan. Carrageenan extracted from seaweed is not assimilated by the human body, providing only fibre with no nutritional value, but it does provide unique functional characteristics that can be used to gel, thicken and stabilise food products and food systems. κ-carrageenans, ι-carrageenans, and furcellarans are linear polysaccharides whose backbone structure is based on a repeating disaccharide sequence of sulphate esters of (1→3) linked β-D-galactose and (1→4) linked 3,6-anhydro-α-D-galactose. They differ from each other in the number and position of sulphate groups. κ-carrageenans have one sulphate group per repeating disaccharide unit, positioned at C-4 of the β-D-galactopyranosyl residue, whereas ι-carrageenans have two sulphate groups, positioned at C-4 of the β-D-galactopyranosyl residue and C-2 of the 3,6-anhydro-α-D-galactopyranosyl residue. Furcellaran has a similar structure to κ-carrageenan, but it is less sulphated; only 40% of the β-D-galactopyranosyl residues carry the sulphate group at C-4. These two types of monosaccharide conformations, along with the presence of axial and equatorial glycosidic linkages, allow κ- and ι-carrageenans to assume a helical conformation. In solution, in the presence of some cations (K^+, Rb^+, Ca^{++}), the double helices of furcellaran, κ- and ι-carrageenans can aggregate and form gel. κ-carrageenan and ι-carrageenan form thermally reversible gels, which range in texture from firm and brittle to soft and elastic.

The functional properties of carrageenan gels, such as rigidity, turbidity, and tendency to syneresis (separation of water from gel upon aging), generally decrease with the increasing degree of sulphation in these polymers. λ-carrageenans constitute another group of the red seaweed polysaccharides. The repeating disacharide unit in λ-carrageenans consists of β-D-galactopyranosyl residue sulphated at C-2 (instead of C-4 as in ι- and κ-carrageenans) and 2, 6-di-*O*-sulfato-α-D-galactopyranosyl units (instead of 3, 6-anhydro-α-D-galactopyranosyl residue). λ-carrageenans are nongelling polysaccharides used as cold soluble thickeners in syrups, fruit drinks, pizza sauces, and salad dressings.

3.2.3 Agar

Agar constitutes another group of polysaccharides from red-purple algae of the *Rhodophyceae* class. The agar-yielding species of *Gracilaria* and *Gelidium* grow in the waters along the coast of Japan, New Zealand, South Africa, Southern California, Mexico, Chile, Morocco, and Portugal. Agar is a linear polysaccharide built up of the repeating disaccharide unit of (1→3)-linked β-D-galactose and (1→4)-linked 3,6-anhydro-α-L-galactose residues. In contrast to carrageenans, agar is only lightly sulphated and may contain methyl groups. Methyl groups, when present, occur at C-6 of the (1→3)-linked β-D-galactose or C-2 of (1→4)-linked 3, 6-anhydro-α-L-galactose residues. Agar, containing

3,6-anhydro-α-L-galactose residues, forms three-fold left-handed helices. These left-handed threefold helices are stabilized by the presence of water molecules bound inside the double helical cavity (Labropoulos *et al.*, 2002) and exterior hydroxyl groups allow aggregation of up to 10 000 of these helices to form microdomains of spherical microgels (Boral *et al.*, 2008).

The agar helix is more compact due to the smaller amount of sulphate groups. Agar is a well known thermo-reversible gelling polysaccharide, which sets at 30 to 40°C. Being less sulphated than furcellaran, and κ- and ι-carrageenans, agar can form strong gels, which are, subject to pronounced syneresis, attributed to strong aggregation of double helices (not weakened by the sulphate groups). The ability to form reversible gels by simply cooling hot, aqueous solutions is the most important property of agar. Gelation depends exclusively on the formation of hydrogen bonds, where the random coils associate to form single helices (Foord & Atkins, 1989) and double helices (Rees & Welsh, 1977).

3.3 Microbial hydrocolloids

3.3.1 Xanthan gum

Xanthan gum is an extracellular polysaccharide produced by the bacterium *Xanthomonas campestris*. The primary structure of xanthan gum consists of a cellulosic backbone of *β*-(1→4) linked D-glucose units substituted on alternate glucose residues with a trisaccharide side chain. The trisaccharide side chain is composed of two mannose units separated by a glucuronic acid (Melton *et al.*, 1976). Approximately half the terminal mannose units are linked to a pyruvate group and the non-terminal residue usually carries an acetyl group. The carboxyl groups on the side chains render the gum molecules anionic. The pyruvic acid content of xanthan can vary substantially depending on the strain of *X. campestris*, resulting in different viscosities of xanthan solutions.

Molecular modelling studies suggest that xanthan gum can assume a helical structure, with the side branches positioned almost parallel to the helix axis and stabilizing the structure. Xanthan gum forms very viscous solutions, and, at sufficiently high polymer concentration, it exhibits weak gel-like properties. It can form thermo-reversible gels when mixed with certain galactomannans (e.g., locust bean gum) or konjac glucomannan. Xanthan is widely used in foods because of its good solubility in either hot or cold solutions, high viscosity even at very low concentrations, and excellent thermal stability.

3.3.2 Pullulan

Pullulan is an extracellular homopolysaccharide of glucose produced by many species of the fungus *Aureobasidium*, specifically *A. pullulans*. Pullulan contains (1→4) and (1→6)-linked α-D-glucopyranosyl residues. The ratio of (1→4) to (1→6) linkages is 2:1. Pullulan is generally built up of maltotriose units linked by (1→6) with much smaller amount of maltotetraose units. The presence of (1→6) glycosidic linkages increases flexibility of pullulan chains and resulted in their good solubility in water compared with other linear polysaccharides (e.g., amylose) (Cui, 2005). Pullulan easily dissolves in cold or hot water to form a stable, viscous solution that does not gel. A pullulan solution is stable over a wide range of pH and is also relatively stable to heat (Imenson, 2010).

3.3.3 Gellan gum

Gellan gum is a fermentation polysaccharide produced by the microorganism *Sphingomonas elodea* (previously identified as *Pseudomonas elodea,* but later reclassified). Gellan gum is now

approved for food use in many countries including Australia, Canada, United States, Mexico, Chile, Japan, South Korea, and Philippines.
The molecular structure of gellan gum is a straight chain based on repeating glucose, rhamnose and glucuronic acid units. In its native or high-acyl form, two acyl substituents - acetate and glycerate - are present. Both substituents are located on the same glucose residue and, on average, there is one glycerate per repeat and one acetate per every two repeating unit (Kuo et al., 1986). In low-acyl gellan gum, the acyl groups are absent. Upon cooling of gellan solutions, the polysaccharide chains can assume double helices, which aggregate into weak gel structures (supported by van der Waals attractions). In the presence of appropriate cations (Na^+ or Ca^{++}), the double helices form cation-mediated aggregates, which leads to formation of strong gel networks. Acyl substituents present in native gellan interfere with the aggregation process, giving much weaker gels. In the branched variants of gellan, the side chains also interfere with the cation-induced aggregation, allowing only 'weak gel' formation.

3.4 Animal hydrocolloids

3.4.1 Chitin and chitosan

Chitin is a structural polysaccharide that replaces cellulose in many species of lower plants, e.g., fungi, yeast, green, brown, and red algae. It is also the main component of the exoskeleton of insects and shells of crustaceans (shrimp, lobster, and crab). The molecular structure of chitin is similar to that of cellulose, except that the hydroxyl groups at *O*-2 of the β-D-Glc*p* residues are substituted with *N*-acetylamino groups. Chitin forms a highly ordered, crystalline structure, stabilized by numerous intermolecular H-bonds. It is insoluble in water. However, when chitin is treated with strong alkali, the *N*-acetyl groups are removed and replaced by amino groups. This new water-soluble polysaccharide, called chitosan, contains, therefore, (1→4)-linked 2-amino-2-deoxy-β-D-glucopyranosyl residues. Chitosan is the only polysaccharide carrying a positive charge. It is not digested by humans and can be used as a dietary fiber.

3.4.2 Gelatin

Gelatin is a proteinaceous material obtained from animal connective tissue (collagen) using hydrolysis in acidic (type A) or basic (type B) solution followed by hot water extraction. Commercially, skins or bones of different animal species, such as beef, pork, fish and poultry, form the main raw material for gelatin production. The extracted gelatin is a group of molecules of different molecular weight (Imenson, 2010). The molecular weight profile depends on the process. The amino acid profile determines hydrogen bond formation and reactivity via side groups such as amine, imidazole, alcohol, amide and carboxylic acid. It hydrates readily in warm or hot water to give low-viscosity solutions that have good whipping and foaming properties. After cooling, the network of polypeptide chains associates slowly to form clear, elastic gels that are syneresis free.

3.5 Chemically modified hydrocolloids

Although all natural gums have inherently useful and uniquely functional properties, they also have inherent limitations and deficiencies which restrict their overall utilization. In many cases, these limitations can be removed by selective chemical modification and derivatization of the gum. In other cases, the overall functional properties can be improved by the chemical modification of the natural hydrocolloid.

Thus, while sodium alginate is quite soluble, it does not have good stability at low pHs. By treating alginates with propylene oxide to form propylene glycol alginate ester, a modified soluble alginate is formed that has exceptional stability under acidic conditions.

In a similar fashion, while normal guar gum is quite soluble in cold water, solubility can be greatly increased by forming the hydroxypropyl guar derivative, while simultaneously giving a greatly increased viscosity.

Pure cellulose is completely in soluble in water as well as being poorly absorptive in its native form. By chemical treatment to form cellulose ether compounds, such as methyl cellulose and hydroxypropyl cellulose, water solubility can be imparted, thus making a useful series of water soluble functional hydrocolloid polymers.

4. Hydrocolloids in the production of special products

4.1 Soft gelatin capsules

Liquid foods, as well as instant (soluble) coffee and other food powders, can be conveniently contained in a gelatin capsule (Maddox, 1971). The interior of the capsule contains a suitable instant food which dissolves or disperses promptly upon addition of water. The capsule is maintained in a dry form in a suitable enclosure, such as a hermetically sealed bottle, blister-pack packaging or the like, until use. Soft gelatin capsules are commonly used in food supplements. Gelatin is the basic capsule shell component and it is formulated with suitable ingredients to encapsulate a wide variety of materials. Gelatin's special properties are of particular interest in foods since it acts as a barrier and protects liquid capsule contents from the outside environment. On the one hand, gelatin acts as a physical barrier to bacteria, yeasts and molds. On the other, it provides a low-permeability membrane to gases. The gelatin shell is transparent, can be formed in a wide range of sizes and shapes and dissolves quickly in hot water, releasing its encapsulated liquid. The advantages of encapsulation are: portion control, easy use and storage, extended shelf-life, improved aesthetic appeal, the variety of sizes available, disposability and edibility, improved product aromatics versus time, and biodegradability. A wide range of filler materials can be encapsulated within these capsules, such as most vegetable oils, essential oils and fish oils, as well as suspensions of crystalline materials milled with oils. A few food applications are: real chicken broth capsules which retain and deliver flavor more effectively than the powder system, encapsulated lemon oil for meringue pie mix, mint essence capsules for the tinned goods market (Moorhouse & Grundon, 1994).

4.2 Liquid-core capsules

Liquid-core hydrocolloid capsules are liquids encapsulated in a spherical polymer membrane (Vergnaud, 1992). Production of these capsules included suspending cells in a sodium alginate solution, forming small spherical calcium alginate beads by cross-linking with calcium salt, and reacting with polylysine to create a polylysine alginate membrane around the bead. In the final stage, the bead's core, composed of calcium alginate gel, was solubilized, thus forming a liquid-core micro-capsule containing cells (Lim & Sun, 1980). With this procedure, cells could also be found in the membrane matrix, leading to the proposal of an approach to eliminate this possibility (Wong & Chang, 1991). In the latter approach, cells were entrapped in alginate-gel micro-spheres, which in turn were contained within larger beads, resulting in a greater distance between the cells and the surface of the larger alginate bead. Similar to (Lim and Sun's 1980) procedure, the surface of the larger micro-sphere was reacted with poly-L-lysine and then with alginate to form a coating membrane. The contents of the micro-capsule were then liquefied with sodium citrate to remove the calcium from the array. The cells in the smaller entrapped gel micro-sphere were released and allowed to float freely in the liquid core of the resultant beads (Wong & Chang, 1991).

The contents of the capsule were either distilled water or sucrose solutions (2.5 and 30%, w/w), although other viscous liquids can be used. Beads with 0, 2 and 5% sucrose were produced by diffusion of sucrose out of liquid-core capsules containing 30% sucrose. The spherical shape of the capsule was retained after diffusion. Capsules with a higher hydrocolloid concentration within their membrane displayed more stress at failure (strength) and less brittleness than those with lesser solid membrane content. Following diffusion, capsules with 2 and 5% sucrose were weak compared to those with 30% sucrose; however, no membrane rupture was observed after incubation.

4.3 Jelly-like foods

Natural gums are used in the confectionary industry. At one time, guar was used for production of jellies (candies) and marshmallows, and gum arabic was used gum drops. The gum within the formulation served to form *jelly*, but an additional function was to prevent sugar crystallization and to emulsify fat, keeping it evenly distributed within the product (Furia, 1980). The gum powder swells and gels when added to water and heated. Its gels are thermally irreversible and unaffected by further addition of water and can be produced over a pH range of 2.0-9.5 in the presence of many food additives. The gels may be used to make novel food products consisting of a jelly-like skin with a liquid core, and canned jellies. The concentration of the polysaccharide in water must be greater than 1.5% for gel stability and less than 0.6% for taste acceptability. The gels are freeze-thaw stable and may be used to make an ice confection contained in an elastic gel skin (Anon, 1977).

4.4 Fruit products

A combination of compression and shearing forces is used to extract juice from fruits or vegetables. For pulp production, and in the case of grapes, tomatoes or other soft fruits, are heated, if necessary, to soften their tissues and pulp is forced through the perforations of the pulping equipment's screen, the size of which determines the consistency of the resultant product (Fellows, 2000). Unique uses of such fruit products (i.e. juice, pulp or puree) for production of soft viscous, fruit-based, membrane-coated items by a membrane were described decades ago. For example fruit juice, pulp or puree containing soluble Ca salt is extruded to form drops which are coated with a thin skin of alginate or pectate sol. The coated drops are exposed to an aqueous setting bath containing a soluble Ca salt (Sneath, 1975). Drops of aqueous fruit material are coated with an aqueous alginate or pectate solution and applied in a solution containing Ca or Al ions to gel the surface.

4.5 Frozen product

Frozen desserts are mixtures of ice crystals in flavored liquid syrup. The most common frozen dessert is ice cream. During the last 50 years, a huge change in the texture of ice-cream products has occurred.

Gum karaya can be used as a stabilizer in ice cream, ice milk, mellorine and related products. In ice pops and sherbets, formation of large ice crystals and syneresis can be prevented by including 0.2-0.4% gum karaya. Combinations of 0.15% gum karaya and 0.15%

LBG can be used successfully for ice pop stabilization. Karaya, as well as carrageenan, can be used as a binder and emulsifying agent in quantities of less than ~1%. The binders are used to absorb the water resulting from the ice during chopping. LBG is used in the food industry for its ability to bind and immobilize large amounts of water. This property helps inhibit ice crystal formation in frozen products, produce viscosity, modify texture and stabilize product consistency in the face of temperature changes (Glicksman, 1969). Sodium CMC in its highest purified form is used in many food applications. In frozen desserts (such as ice cream), cellulose gum inhibits the formation of ice crystals (Davidson, 1980). Gum arabic, because of its water-absorbing properties, gum inclusion inhibits the formation and growth of ice crystals. Other stabilizers, such as carrageenan and LBG, can be used for the same purposes (Glicksman, 1969). Guar gum is used in the food industry for its ability to bond and immobilize large amounts of water. This property contributes to inhibition of ice crystal formation, product texture, stabilization of product consistency to changes in temperature, and viscosity (Davidson, 1980).

Freezing often causes undesirable changes in foods, and hydrocolloids are used to improve their quality. To produce a high-quality ice cream, a blend of guar gum or CMC with a smaller amount of carrageenan may be used. If xanthan and guar gum are used instead, viscosity is lower and faster processing is obtained. Karaya gum has been used in the past as a stabilizer for frozen desserts, but has been replaced almost completely by other gums. Carrageenan, guar gum and CMC have also been used as stabilizers in other frozen products. In foods containing starch as the main ingredient, there is a tendency for water to exude from the gel. Thus starch-based products curdle and undergo syneresis (loss of water) after freezing and thawing. Modified starches have been developed to deal with the problem (Nussinovitch, 1997).

Frozen doughs are widely used in industrial bakeries to make baking more profitable. However, loaf volumes are usually smaller and quality poorer for breads baked from frozen doughs, especially in those with low fat content (Williams & Philips, 1998). Addition of hydrocolloids such as CMC, alginate, and different blends in quantities of up to 1.5% yielded higher total dough water content without changing baking properties. There were no obvious differences between analyzed samples or added hydrocolloid levels.

4.6 Candies

Candies are popular products among children and adults and their versatility is visually alluring as well as pleasing to the consumer. The confectionery industry uses gum arabic to a great extent, for crystallization prevention, as an emulsification agent of fat and as a glaze in candies, chewing gum. Gum arabic serves to coat the center of sugar-coated tablets. It is the main ingredient in gumdrops (regular and dietetic) and other chewy-type gums, where pectin or modified starches can also play a major role (Davidson, 1980). The incorporation of sorbitol, mannitol and gum arabic can produce dietetic candies. The higher the gum arabic content, the softer and chewier the candy.

4.7 Fabricated foods

Using fish, meat, fruit or vegetables as main ingredients within a matrix, which is usually produced from a gum, can create fabricated foods. Gums were incorporated into meat products to achieve better control of their texture, improve sliceability and increase yield. In some meat products, hydrocolloids are responsible for the undesirable broad dark striations (called tiger stripping), running parallel to the meat fibers (Williams & Philips, 1998). The swelling ability of the type of carrageenan used influences its activity within the product.

Semi-refined carrageenan (less swelling) improved performance in injected poultry by reducing the incidence of tiger stripping without reducing purge controls (Williams & Philips, 1998).

Fabricated fruit is easily manufactured with alginates. A gel is readily formed when a soluble calcium salt is added to a sodium alginate solution. This gel is stable over a wide range of temperatures, has excellent syneresis control, and is irreversible to heat. Possible uses for this fabrication concept include imitation cocktail cherries; imitation glazed fruit pieces for cakes, breads, cookies, ice cream and candy products; icing; and gelled products containing pureed fruit. Other hydrocolloids - carrageenan, gelatin (and recently gellan), and combinations of gums such as carrageenan and locust bean gum (LBG) - have been used to fabricate food products. Examples include reconstituted pimento strips (based on alginate and gum arabic), the aforementioned imitation caviar, and restructured fish and shellfish (Nussinovitch, 1997).

Use of sodium carboxymethyl cellulose (CMC) in food applications is on the rise, especially in developed countries where the popularity and convenience of fabricated foods has grown rapidly since the early 1950s. Hydroxypropylcellulose can also be used in fabricated foods to a large extent. Its useful properties are its ability to form solvent-soluble films and its surface-active stabilization (Davidson, 1980).

5. Health benefits of hydrocolloids

5.1 Hydrocolloid and the risk of cardiovascular disease (CVD)

Dietary fiber was briefly mentioned in the WHO report to reduce total and LDL cholesterol, and probably also to decrease the risk for cardiovascular diseases. Several studies have dealt with the association between dietary fiber intake and risk for cardiovascular disease. The main interest has been focused on effects of soluble fibers, such as different hydrocolloids, and thus on foods rich in soluble fibers. Both mixture of hydrocolloids and only one hydrocolloid were investigated. In one study, subjects with increased plasma cholesterol values were given a daily supplement of 15 g of psyllium, pectin, guar gum, and locust bean gum during 6 months (Jensen et al., 1997). The fibers were mixed in water and consumed with each of three major daily meals. In comparison with the control group given acacia gum, the total and LDL cholesterol values were significantly lower in the test group. After 8 weeks, the reductions in comparison with baseline were 6.4 and 10.5%, respectively, and about the same reductions were found at weeks 16 and 24. In another study, a combination of soluble fibers from psyllium, oats, and barley was given to men with hypercholesterolemia (Roberts et al., 1994). They consumed the fibers as a breakfast cereal (50 g containing 12 g of soluble fiber) for 6 weeks. In comparison with a control group given a breakfast cereal based on wheat, the total cholesterol and LDL cholesterol levels fell significantly in the test group, with 3.2 and 4.4%, respectively.

Psyllium has been extensively investigated in relation to its effects on CVD. As in the study of Anderson et al. (2000), the fiber preparations were mixed in water and taken before regular meals three times per day. A dose–response study was made by Davidson et al. (1998) using psyllium seed husk given in daily doses of 0, 3.4, 6.8, or 10.2 g for 24 weeks. The fibers were included in different foods like ready-to-eat cereals, bread, pasta, and snack bars. A change in LDL cholesterol (–5.3% in comparison to control) after 24 weeks consumption was only shown for the group that took the highest dose of psyllium husk-10.2 g/day. The reduction in LDL cholesterol was more pronounced in the beginning of the intervention (week 4) for all groups. Davidson et al. (1998) also investigated the lipid-lowering effect of psyllium in hypercholesterolemic children (6 to 18 years). They were given psyllium for 6 weeks and, after a 6-week washout period, a control cereal. Consumption of psyllium gave a 7% reduction in LDL cholesterol compared with the control cereal.

Many human studies on guar effect on lipid metabolism have been conducted. The dose of guar gum was 10 g and it was taken three times a day for 6 weeks. In comparison with a placebo, the guar gum decreased the blood cholesterol and triglyceride levels and blood pressure significantly (Landin et al., 1992). The effect of guar gum on LDL metabolism seemed to be related to an increased LDL a polipoprotein B fractional catabolism. Modified guar gum has also been studied, and in one study partially depolymerized guar gum decreased the total cholesterol levels by 10%, which is a reduction similar to that found earlier for high molecular weight guar gum (Blake et al., 1997). The effects of solid or liquid guar gum and preparations with high or medium viscosity on lipid metabolism were followed in hypercholesterolemic subjects (Superko et al., 1988). Both solid and liquid guar gum preparations lowered the total and LDL cholesterol, but the high-viscosity preparation gave a larger reduction in blood lipid levels than the medium-viscosity preparation.

Pectin can be included in the diet as a supplement, but also as fruits, which often contain much pectin. In a study, subjects with hypertension were given guava fruits before meals during 12 weeks, and the effect on the blood lipids and blood pressure was followed (Singh et al., 1992). In comparison with a group that was not given guava, the total cholesterol, HDL cholesterol, triglycerides, and blood pressure decreased significantly. Several studies have also been done with different kinds of pectins. Dongowski & Lorentz (2004) gave diets containing pectin with different degrees of methylation (34.5, 70.8, and 92.6%) to rats for 3 weeks. The concentration of bile acids in the plasma decreased when pectin was given, and with increasing degree of methylation more bile acids were excreted with the feces.

5.2 Hydrocolloids and type 2 diabetes

It has been reported that dietary soluble fiber such as β-glucan, psyllium, and guar gum decreases glucose and insulin responses to carbohydrates if taken in sufficient amounts. A study comparing the effects over 6 months of barley bread, high in β-glucan, to white bread found that barley bread improved glycemic control compared with white wheat bread in 11 men with type 2 diabetes.97 Insulin responses were increased, which hypothetically could reflect recovered β-cell function. In men with diabetes and hypercholesterolemia participating in a crossover trial, 8 weeks of psyllium (15 g/day) decreased hemoglobin A_{1c} 6.1% (absolute change, 0.8%), with similar 6% decreases in fasting postprandial glucose (Anderson et al., 1999). Improved fasting and postprandial glycemic control was found in 11 type 2 diabetic patients taking 21 g/day of guar gum or placebo in a randomized double-blinded crossover trial (Aro et al., 1981). Small improvements in overall glycemic control and sizable improvements in postprandial glycemia after 4 weeks of treatment in a randomized controlled crossover trial were reported by Fuessl et al. (1987) Guar gum decreased fasting blood glucose from 11.4 to 9.5 mmol/l in 19 obese patients with type 2 diabetes who were enrolled in a randomized double-blind crossover trial (Lalor et al., 1990). Guar gum (15 g/day) has also improved long-term glycemic control and postprandial glucose tolerance in 15 type 2 diabetic patients treated with guar gum over an 8-week period (Groop et al., 1993). Viscosity is an important determinant of soluble hydrocolloid in retarding glycemic responses (Wood et al., 2000).

5.3 Hydrocolloids as laxative and antidiarrhea

The laxative activity of bulk forming substances has been known since the time of Hippocrates. Hydrocolloid fractions of psyllium and ispaghula are common bulk forming laxative (Fingal & Feston, 1979). The fiber from foods, such as carrots, cabbage, apple, and bran compared to guar, produces very different responses in colon function 104. Fecal weight is increased more by a bran supplement than by guar supplement. The results from a study of digestion of hemicelluloses in humans suggest that arabinoxylan is not digested and perhaps may be the active component in laxation (Holloway et al., 1980). One hypothesis for judging the value of a bulk former as a laxative concerns its ability to hold water. However, some new evidence appears to contradict this. Since the greater the water holding capacity of a fiber source, the less the effect on fecal bulk (Stephen & Cummings, 1979).

6. References

Anderson, J.W., Allgood, L.D., Lawrence, A., Altringer, L.A., Jerdack, G. R., Hengehold, D.A. & Morel, J.G. (1999). Cholesterol-lowering effects of psyllium intake adjunctive to diet therapy in men and women with hypercholesterolemia: meta-analysis of 8 controlled trials, *American Journal of Clinical Nutrition*, vol. 71, No. 2, pp. 472-479

Anderson, J.W., Davidson, M.H., Blonde, L., Brown, W.V., Howard, W.J., Ginsberg, H., Allgood, L.D. & Weingand, K.W. (2000). Long-term cholesterol lowering effects of psyllium as an adjunct to diet therapy in the treatment of hypercholesterolemia, *American Journal of Clinical Nutrition*, vol. 71, No. 6, pp. 1433-1438

Anon, (1977). Facinating jelly-like foods. *Food Engineering International*, vol. 2, No. 4, pp. 38-39

Aro, A. Uusitupa, M., Voutilainen, E., Hersio, K., Korhonen, T. & Siitonen, O. (1981). Improved diabetic control and hypocholesterolaemic effect induced by long-term dietary supplementation with guar gum in type 2 (insulin-independent) diabetes. *Diabetologia*, vol. 21, No. 1, pp. 29-33

Blake, D.E., Hamblett, C.J., Frost, P.J., Judd, P.A. & Ellis, P.R. (1997). Wheat bread supplemented with depolymerized guar gum reduces the plasma cholesterol concentration in hypercholesterolemic human subjects. *American Journal of Clinical Nutrition*, Vol. 65, pp. 107-113

Boral, S., Saxena, A. & Bohidar, H.B. (2008). Universal growth of microdomains and gelation transition in Agar hydrogels. *Journal of Physical Chemistry B*, vol. 112, No. 12, pp. 3625-3632.

Burey, P., Bhandari, B. R., Howes, T. & Gidley, M. J. (2008). Hydrocolloid Gel Particles: Formation, Characterization, and Application. *Critical Reviews in Food Science and Nutrition*, vol. 48, No. 5, pp. 361–377

Connolly, S., Fenyo, J.C. & Vandevelde, M.C. (1988). Effect of proteinase on the macromolecular distribution of Acacia Senegal gum. *Carbohydrate Polymers*, Vol. 8, No. 1 pp. 23-32

Cui, S.W. (2005). *Food carbohydrates: Chemistry, Phisical properties and application*, CRC/ Taylor and Francis, ISBN 0-8493-1574-3, USA

Davidson, R.L. (1980). *Handbook of Water-soluble Gums and Resins*, McGraw-Hill, New York

Davidson, M.H., Maki, K.C., Kong, J.C., Dugan, L.D., Torri, S.A., Hall, H.A., Drennan, K.B., Anderson, S.M., Fulgoni, V.L., Saldanha, L.G. & Olson, B.H. (1998). Long-term effects of consuming foods containing psyllium seed husk on serum lipids in subjects wih hypercholesterolemia, *American Journal of Clinical Nutrition*, vol. 67, No. 6, pp. 367-376

Dickinson, E. (2003). Hydrocolloids as interfaces and the influence on the properties of dispersed systems. *Food hydrocolloids*, vol. 17, No. 1, pp. 25-39

Djabourov, M. (1991). Gelation-A Review. *Polymer International*, vol. 25, No. 3, pp. 135–143

Dongowski, G. & Lorentz, A. (2004). Intestinal steroids in rats are influenced by the structural parameters of pectin, *Journal of Nutritional Biochemistry*, vol. 15, No. 4, pp. 196-205

Fellows, P. (2000). *Food processing technology: principles and practice*, CRC press. Boca Raton, Florida

Fingal, E. & Feston, J.W. (1979). Antidiarrheal agents and laxatives: changing concepts. *Clinical Gastroenterology*, vol. 8, No. 1, pp. 161-185

Foord, S.A. & Atkins, E.D.T. (1989). New x-ray diffraction result from agarose: extended single helix structure and implications for gelation mechanism. *Biopolymers*, vol. 28, No. 8, pp. 1345–1365

Fuessl, H.S., Williams, G., Adrian, T.E. & Bloom, S.R. (1987). Guar sprinkled on food: effect on glycaemic control, plasma lipids and gut hormones in non-insulin dependent diabetic patients. *Diabetic Medicine*, vol. 4, No. 5, pp. 463-468

Furia, T.E. (Ed.). (1980). *Handbook of food additives*, CRC press, ISBN 0849305438, Boca Raton, Florida

Glicksman, M. (1969). *Gum Technology in the Food and Other Industries*, Academic Press, ISBN 12286350X, New York

Groop, P.H., Aro, A., Stenman, S. & Groop, L. (1993). Long-term effects of guar gum in subjects with non-insulin dependent diabetes mellitus. *American Journal of Clinical Nutrition*, vol. 58, pp. 513-518

Hollingworth, C. S. (ed.). (2010). Food Hydrocolloids: Characteristics, Properties and Structures. Nova Science Publishers, ISBN 978-1-60876-222-4, New York

Holloway, W.D., Tamsan-Joens, C. & Bell, E. (1980). The hemicelluloses component of dietary fiber. *American Journal of Clinical Nutrition*, vol. 33, No. 2, pp. 260-263

Imeson, A. (2010). *Food stabilisers, thickeners and gelling agents*, Blackwell Publishing Ltd, ISBN 978-1-4051-3267-1, Oxford, England

Jensen, C.D., Haskell, W. & Whittam, J.H. (1997). Long-term effects of water-soluble dietary fiber in the management of hypercholesterolemia in healthy men and women. *American Journal of Cardiology*, vol. 79, No. 1, pp. 34-37

Kuo, M.-S., Mort, A.J. & Dell, A. (1986). Identification and location of l-glycerate, an unusual acyl substituent in gellan gum. *Carbohydrate Research*, vol. 156, pp. 173–187

Labropoulos, K.C., Niesz, D.E., Danforth, S.C. & Kevrekidis, P.G. (2002). Dynamic rheology of agar gels: theory and experiment. Part II. gelation behavior of agar sols and fitting of a theoretical rheological model. *Carbohydrate Polymeres*, vol. 50, No 4, pp. 407–415

Lalor, B.C, Bhatnagar, D., Winocour, P.H., Ishola, M., Arrol, S., Brading, M. & Durrington, P.N. (1990). Placebo-controlled trial of the effects of guar gum and metformin on fasting blood glucose and serum lipids in obese, type 2 diabetic patients. *Diabetic Medicine*, vol. 7, No. 3, pp. 242-245

Landin, K., Holm, G., Tengborn, L. & Smith, U. (1992). Guar gum improves insulin sensitivity, blood lipids, blood pressure, and fibrinolysis in healthy men. *American Journal of Clinical Nutrition*, vol. 56, No. 6, pp. 1061-1065

Lim, F. & Sun, A.M (1980). Microencapsuled islets as bioartificial endocrine pancreas. *Science*, vol. 210, No. 4472, pp. 908-910

Maddox, V.H. (1971). Gelatin capsule. *United States Patent 3,620,759*

Melton, L.D., Mindt, L., Rees, D.A. & Sanderson G.R. (1976). Covalent structure of the polysaccharide from *Xanthomonas campestris*: evidence from partial hydrolysis studies. *Carbohydrate Research*, vol. 46, No. 2, pp. 245–257

Moorhous, S. & Grudon, V. (1994). Encapsulating: A new concept for the food industry. *Nutrition and Food science*, vol. 94, No. 2, PP. 17-19, ISSN 0034-6659

Nussinovitch, A. (1997). *Hydrocolloid Applications Gum Technology in the Food and Other Industries*, Blackie Academic & Professional, London, UK

Philips, G.O., Wedlock, D.J. & Williams, P.A. (eds). (1984). *Gums and stabilizers for the food industry*, vol. 2, Pergamon Press, Oxford, UK

Philips, G.O., Wedlock, D.J. & Williams, P.A. (eds.). (1990). *Application of microcrystalline cellulose. In: Gums and stabilizers for the food industry*, Vol. 5, IRL Press, Oxford, UK

Philips, G.O., Williams, P.A. & Wedlock, D.J. (eds). (1986). *Gums and stabilizers for the food industry*, vol. 3, IRL Press, Oxford, UK

Phillips, G.O. & Williams, P.A. (eds.). (2000). *Handbook of Hydrocolloids*, CRC Press, ISBN-978 1 84569 414 2, Boca Raton, Florida

Rees, D.A. & Welsh, E.J. (1977). Secondary and tertiary structure of polysaccharides in solution and gels. *Angewandte Chemie International Edition in English*, vol. 16, No. 4, pp. 214–224

Roberts, D.C., Truswell, A.S., Bencke, A., Dewar, H.M. & Farmakalidis, E. (1994). The cholesterol lowering effect of a breakfast cereal containing psyllium fibre. *Medical Journal of Australia*, vol. 161, No. 11-12, pp. 660-664

Sahin, H. & Ozdemir, F. (2004). Effect of some hydrocolloids on the rheological properties of different formulated ketchups. *Food Hydrocolloids*, vol. 18, No. 6, pp. 1015-1022

Singh, R.B., Rastogi, S.S., Singh, R., Ghosh, S. & Niaz, MA. (1992). Effects of guava intake on serum total and high density lipoprotein cholesterol levels and on systemic blood pressure. *American Journal of Cardiology*, vol. 70, No. 15, pp. 1287-1291

Shirakawa, M., Yamatoya, K. & Nishinari, K. (1998). Tailoring of xyloglucan properties using an enzyme. *Food hydrocolloids*, vol. 12, No. 1, pp. 25-28

Sneath, M.E. (1975). Simulated soft fruits. *U.S. Patent, 3922360*

Stephen, A.M. & Cummings, J.H. (1979). Water holding by diatary fiber in vitro and its relationship faecal output in man. *Gut*, vol. 20, No. 8, pp. 722-729

Superko, H.R., Haskell, W.L., Sawrey-Kubicek, L. & Farquhar, J.W. Effects of solid and liquid guar gum on plasma cholesterol and triglyceride concentrations in moderate hypercholesterolemia, *American Journal of Cardiology*, vol. 62, No. 1, pp. 51-55, 1988.

Vergnaud, J.M. (1992). *Drying of polymeric and solid materials*, Springer Verlag, London, Berlin

Williams, P.A., & Philips, G.O. (1998). *Gums and Stabilisers for the food industry*, vol. 9. IRL Press at Oxford University Press, Oxford, UK

Williams, P.A., Sayers, C., Viebke, C., Senan, C., Maoyer, J. & Boulenguer, P. (2005). Elucidation of the emulsification properties of sugar beet pectin. *Journal of Agricultural Food Chememistry*, Vol. 53, No. 9, pp. 3592-3597

Wong, H. & Chang, T.M.S. (1991). A novel two step procedure for immobilizing living cells in microcapsules for improving xenograph survival. *Journal of the International Society for Artificial Cells and Immobilization Biotechnology*, vol. 19, No. 4, pp. 687-690

Wood, P.J., Beer, M.U. & Butler, G. (2000). Evaluation of the role of concentration and molecular weight of oat β-glucan in determining the viscosity on plasma glucose and insulin following an oral glucose load. *British Journal of Nutrition*, vol. 84, No. 1, pp. 19-23

2

Physical and Chemical Characteristics of Tropical and Non-Conventional Fruits

Ítalo Herbert Lucena Cavalcante[1], Lourival Ferreira Cavalcante[2], João Marcos de Sousa Miranda[1] and Antonio Baldo Geraldo Martins[3]

[1]*Federal University of Piauí*
[2]*Federal University of Paraíba*
[3]*São Paulo State University*
Brazil

1. Introduction

Fruit quality is one of the most important themes of fruit industry, especially when concerning juice and pulp ones, since they have a direct impact on the use of synthetic products such as acidifiers, colorants and sugars, for instance, i.e., fruits with adequate physical and chemical properties have the use of synthetic composts reduced on their processed products.

The physical and chemical parameters of fruits are important indicators of their maturation and internal and external quality, decisive factors for accomplishment of market demands, that have encouraged a lot of researches under different conditions overseas.

In this sense, this book chapter includes three themes related to fruit quality for industry: i) the importance of physical and chemical characteristics of cultivated tropical fruits; ii) the effects of management techniques on fruit quality; and iii) the physical and chemical characteristics of non-conventional fruits.

2. Importance of physical and chemical characteristics of cultivated tropical fruits

Quality is defined as the absence of defects or degree of excellence and it includes appearance, color, shape, injuries, flavor, taste, aroma, nutritional value and being safe for the consumer (Abbott, 1999). Due to a higher market exigency as for high quality products, the juice and pulp industries have been looking for fruits with better internal and external features, including fruit length and width; fruit weight; pulp, seed and peel percentages per fruit; number of seeds per fruit; seed size and peel diameter; soluble solids (°Brix); titratable acidity (%); vitamin C content (mg/100g of fresh fruit); pulp pH and soluble solids/titratable acidity ratio.

The maintenance of fruit quality characteristics (internal and external features above cited) demands postharvest handlings, such as: preventing mechanical injury, water loss and disease development, limiting unwanted physiological changes and preventing chemical and microbial contamination (Cook, 1999). It is important to mention that farmers, packers, shippers, wholesalers, retailers and consumers frequently have different perspectives regarding to quality and often place different emphasis on the different components of quality. In addition to that, Paull & Duarte (2011) reported that fruit quality is related to some intrinsic characters (appearance, colour, acids, sugars, etc.) and since they change during handling research data can give us information on the waya product should be handled postharvest.

Inherently, the demand of fruit quality, physically and chemically talking, by industries, depends on fruit species and the product processed by each one of them.

Orange juice industry requires fruits for processing during all months of the year, what can be considered a great problem since the most cultivated orange cultivars in São Paulo State are, for instance, Hamlin, Natal, Pera and Valencia (Cavalcante et al., 2009a), whose production is concentrated in few months, fact that makes the commercial activity vulnerable to incidence of harmful diseases and concentrated juice price fluctuations. This way, lots of research work have been developed aiming the expansion of harvest time through new cultivars, with different fruit maturation curves as for the most recently cultivated ones, such as Cavalcante et al. (2006b), which has established a physical and chemical characterization of eighteen orange cultivar fruits and has also concluded that some of them presented potential for consumption as fresh fruit while others are much more appropriate for industry, indicating the importance of diversification.

Cavalcante et al. (2009a) evaluated, under São Paulo State (Brazil) conditions, fruit maturation of 18 sweet orange cultivars and have found that, according to Figure 1, four groups are formed considering its maturation: Orange cultivars João Nunes and Hamlin are the earliest ones, followed by Kawatta, Mayorca, Rubi and Westin; Pineapple and Tarocco A and Oliverlands, Cadenera and Homosassa, respectively; while Torregrosa, Jaffa, Biondo, Finike, Sanguinea, Moro and Early Oblong are the cultivars whose maturation is later than the other ones studied.

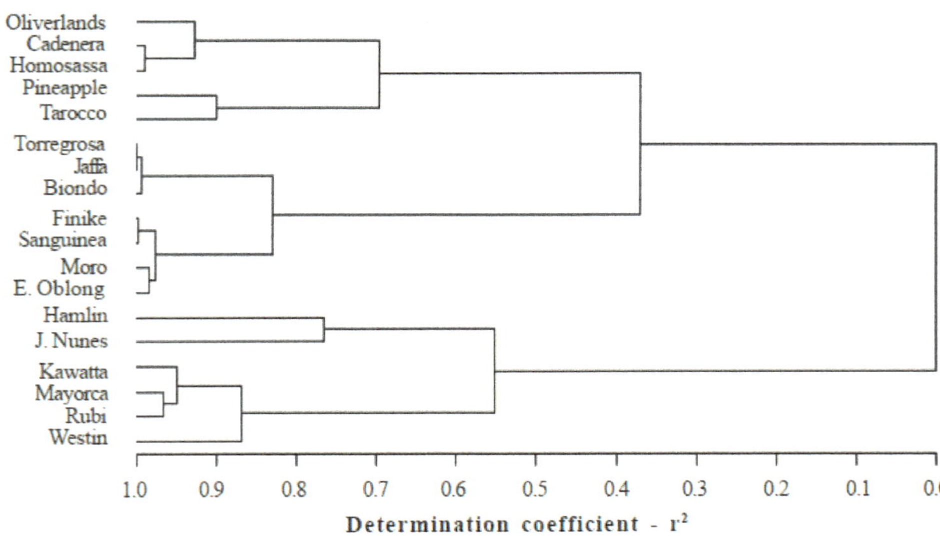

Fig. 1. Cluster diagram of sweet orange cultivars, obtained with SS/TA ratio, technological index and fruit mass. Cavalcante et al. (2009a)

Underutilized citric fruits also have been studied in Brazil aiming to provide, to orange industry, new options. One of these species is sour orange (*Citrius aurantium* L.; Figure 2), generally more consumed by people in Piaui State (Northeasthern Brazil) as for juice or fresh fruit than common orange (*Citrus sinensis* L.) cultivars. Silva Júnior et al. (2010) evaluated the physical and chemical properties of "laranja da terra" (in English, "native orange") fruits and concluded that its fruit presents good physico-chemical properties, with an average of 125.76 mg/100g of fresh fruit, indicating that it is a good natural source of vitamin-C; "laranja da terra" fruits present low titratable acidity and high soluble solids, so they have a great potential for consumption as fresh fruit. Further studies are necessary to determine the technological parameters for its consumption as fresh or processed fruit.

Fig. 2. Sour orange (*Citrus aurantium* L.) fruit.

It is important to detach that, orange fruit maturation considers, mainly, the concentration of soluble solids (°Brix) and titratable acidity (%), while soluble solids: titratable acidity ratio (SS/TA ratio) of 12 (Di Giorgi, 1990), in a general way, is required for orange harvest aiming pasteurized juice industry and focusing the more profitable harvest time for farmers.

Another important fruit for industry is acerola (*Malpighia glabra*) also known as Barbados cherry. The importance of this species is directly related to the chemical and nutritional status of its fruits, especially vitamin C content, which has motivated the expansion of the cultivated area and has increased the demand for acerola in the world market. On the other hand, vitamin C content of acerola fruits depends on plant genetics since, according to Cavalcante et al. (2007a), high variability in fruit quality is observed in Brazilian acerola crops, especially those propagated by seeds. In this area, Cavalcante et al. (2007a) developed a research work with the objective of evaluating physical and chemical characteristics of fruits of sixteen acerola genotypes in Jaboticabal, São Paulo State, Brazil, and they have observed that eleven groups were formed, what indicates high variability among the acerola genotypes studied in relation to fruit quality parameters. This way, some genotypes presented potential for consumption as fresh fruit and others for industry. Figure 3 shows acerola fruits of genotype ACER-3, which presented almost 1.200 mg of vitamin C in 100g of fresh fruit, fruits with 6.9 °Brix, 0.73% of titratable acidity, pulp percentage of 47% and fruit weight 6.82g.

Fig. 3. Acerola fruits of genotype ACER-3 from the Active Germplasm Bank of São Paulo State University, Brazil.

Among the other members of *Caricaceae* family, the cultivated papaya (*Carica papaya* L.) is the only one that belongs to *carica* genus. Papaya fruit is native of tropical America, where it is popular and grows for its small to large melon-like fruit.

In consonance with Paull & Duarte (2011), papaya fruit is a fleshy berry, from 50 g to well over 10 kg, and it superficially resembles a melon, being spherical, pyriform, oval or elongated in shape. Fruit shape is a sex-linked character and ranges from spherical to ovoid in female flowers to long, cylindrical or pyriform (pear-shaped) in hermaphrodite flowers. The skin of the fruit is thin and it is usually green when unripe, turning to yellow or orange when ripe, with flesh total soluble solids ranging from 5% to 19%.

In Brazil, the main papaya production country worldwide (Fao, 2011), papaya has been widely produced aiming its consumption as fresh (mainly) or processed fruit into various products, such as chunks and slices for tropical fruit salads, dehydrated slices, cocktails or processed into puree for juices and nectar base, usually frozen; and as canned nectar, mixed drinks and jams, because the puree of papaya fruit is the basis for the remanufacturing of many products. For papaya fruit evaluation, many studies have been conducted in order to understand the postharvest factors that influence papaya quality just as Bron & Jacomino (2006) and Mesquita et al. (2007), and others have been developed on the importance of an extra variable and on fruit firmness. Better firmness of papaya fruit delays membrane lipid catabolism thus extending storage life of fresh fruits, what could be improved by fruit treatments before its ripening, likeextra calcium supply, the same manner Mahmud et al. (2008) has studied and verified that the infiltration treatment at 2.5% demonstrated the best effect on maintaining fruit firmness. Additionally, the desired effect of calcium infiltration at 2.5% on maintaining fruit firmness may be due to the calcium binding to free carboxyl groups of polygalacturonate polymer, stabilizing and strengthening the cell walls, as explained by Conway & Sams (1983).

Another very appreciated tropical fruit specie in North Eastern of Brazil is yellow mombin (*Spondias mombin* L.), a native fruit with expressive potential for food industry in this region, concerning the sensorial quality of its fruit, being a species that should be used for more extensive commercial crops. During the last years, yellow mombin have been commercialized in higher amounts on different commercial sectors and Brazilian regions, mainly due to the possibility of fruit consumption as *in natura* or processed.

Cavalcante et al. (2009b) evaluated physical and chemical characteristics of yellow mombin fruits [fruit mass, width and length of fruits and seeds, percentages of pulp, skin and seed, water (%), soluble solids, industrial index and pulp pH] for consumption as fresh or processed fruit from natural plants native of North Eastern of Brazil. These authors revealed that yellow mombin fruit mass evaluated varies from 8.36 to 20.4 g; soluble solids and pH are in agreement to the standard market parameter (i.e. SS>9.0 °Brix and pH>2.2) but pulp percentage of most evaluated counties was below food industry exigency, which standardized the minimum of 60% for pulp percentage. The industrial index for yellow mombin fruits reached 9.0%, pointing out that, for fruit industry, higher industrial indexes are required focusing a prominent possibility of higher soluble solids concentrations. This way, 9.0% is considered a satisfactory index for yellow mombin fruits.

Custard apple (*Annona squamosa* L.), also known as sweetsop (English); anon, riñón (Spanish); noina (Thai); nona seri kaya (Malay); custard apple (Indian); and pinha, ata or fruta-do-conde (Portuguese), is a fruit species native of tropical America, occurring spontaneously in Northeastern Brazil, where it is exploited mainly as subsistence without adequate management and without genetic material selection. Custard apple fruit has been usually consumed as dessert, i.e., several custards and fine desserts, which can include combinations with whipped cream and meringues. Adversely, the perishable nature of the fruit and often short supply limits availability to local markets or air shipment to more-distant markets. On the other way, lately, custard apple fruit has been studied for being included among tropical fruits widely used by industry due to its flavor and high concentrations of titratable acidity, which is particularly important for fruit processing, reducing the addiction of artificial acid components.

In a research work on custard apple fruits, Cavalcante et al. (2011b) evaluated yield, physical and chemical characteristics of the fruits of ten sugar apple genotypes in Bom Jesus, Piauí State, Brazil. The best results of the mentioned study have revealed that custard apple fruits presented a relation between longitudinal diameter and transversal diameter of 1.09, what classifies them as good quality fruits, according to the format variable, with an average fruit weight of 203.69 g, 11 seeds per fruit, soluble solids of 22.8°Brix, titratable acidity of 0.16% and vitamin C of 138.55 mg/100g of fresh fruit; results that characterize custard apple fruits as a natural source of vitamin C.

Tropical fruits are usually sold fresh, and off-grade fruit is processed, exception done for coconut, which is grown mainly for the production of other products, such as copra, oil and coir with a small acreage, often of special varieties that are grown for fresh consumption of its water of albumen, fact that highlights that coconut fruit, among tropical fruits, is classified as a low water moisture loss rate. In Brazil, coconut has been grown preferentially for food industry purposes as for its water of albumen, although fresh water consumption is very appreciated by people.

3. The effects of management techniques on fruit quality

In general, the most important quality factors for tropical fruit growers, production managers, processors, and packers are fruit juice content, soluble solids and acid concentrations, soluble solids-acid ratio, fruit size, and color, showing that these characteristics may change regarding fruit species.

Fruit qualities, when consumed, are decided largely before harvest and depend on the variety grown, crop management (fertilization, irrigation, etc.), environment [climate (excessive rainfall causes major problems with flowering, pests, diseases and fruit quality) and soil] and other preharvest factors. Specifically, fruit quality, depends on several factors including cultivar, rootstock, climate, soil, pests, irrigation and nutrition, although standardized foliar levels have not been established for some tropical fruit species yet, such as yellow passion fruit, coconut and papaya.

In this sense, some of other research works have been developed aiming to measure the contribution of each factor on fruit quality for several tropical fruits and, additionally, trying to reduce the use of synthetics for the production of fruits with less inorganic products and they have also have as their objective the production of adequate amounts of high-quality food, protecting resources and being both environmentally safe and profitable.

Since excessive irrigation and fertilization reduce fruit quality, supplying sufficient nutrition and using sound irrigation scheduling techniques should be a high-priority management practice for every tropical fruit grower. Each fruit species requires a properly designed, operated, and maintained water management system and a balanced nutrition program formulated to provide specific needs for maintenance, expected yield and fruit quality. Such information depends on scientific studies to make fruit production reach high incomes and to manage adequately environmental factor.

Accordingly, irrigation contributes to the efficiency of nutrient programs because fruit trees with sufficient water and nutrients grow stronger, tolerate pests and stresses in a better way, yield more consistently, and produce good quality of fruit.

Adversely, deficient or excessive irrigation or fertilization may result in poor fruit quality, considering that the most important management practices that influence fruit quality are irrigation and an adequate nutrient management, mainly in relation to nitrogen, phosphorus, potassium, and magnesium. In addition to that, some micronutrients like boron and copper, in deficient plants, can also affect fruit quality, as long as when any nutrient is severely deficient, fruit yield and fruit quality will be negatively affected.

Among many economically important tropical fruits, in this book chapter, yellow passion fruit (*Passiflora edulis* f. flavicarpa Deg), coconut (*Cocos nucifera* L.) and papaya (*Carica papaya*) are going to be focused.

Yellow passion fruit is one of the most potentially fruitful plants for tropical regions, where climatic and soil conditions are favorable for its cultivation. Yellow passion fruits produced in Brazil present good flavor, high nutritional value, pharmacological applications and they are especially consumed as juice, although being also very appreciated as cold drinks, yogurt, sauce, gelatin desserts, candy, ice cream, sherbet, cake icing, cake filling, meringue or chiffon pie, cold fruit soup, or cocktails.

Fruit quality of yellow passion fruit has been widely studied since it is affected by several agronomic treats during crop growth, such as fertilizing, mulching, biofertilizing, irrigation water quality in relation to salinity and soil water losses.

In a study about potassium fertilizing of yellow passion fruit, Campos et al. (2007) registered weightier and sweeter fruits in plants submitted to larger amounts of potassium, agreeing with Marschner (2005), who informs us about potassium influence on increasing soluble solids of fruits.

The effect of mulching is also confirmed because it reduces soil water loses for atmosphere trough evaporation and, consequently, soil moisture becomes higher on root zone making it possible a better nutrient solubility and availability for plants. Campos et al. (2007) verified increase in fruit soluble solids (°brix) from plants grown with soil mulching, with a quantitative difference of 0.2 °brix; additionally, Freire et al. (2010) found that soil mulch promoted enhancement on fruit weight, fruit firmness, pulp percentage, pulp pH and ascorbic acid, important variables for fruit industry, specially for yellow passion fruit which is at most consumed in processed forms.

Another important theme on yellow passion fruit quality is the use of saline water for growth and production of this tropical fruit. In tropical regions, where water availability is restricted to low quality in relation to salt contents but climatic and soil conditions are adequate for yellow passion fruit growth, researchers have tried to find alternative managements to make its cultivation viable. North-Eastern Brazil is one of these regions, in which, nowadays, there is an important production of yellow passion fruit, in many cases, under irrigation with saline water.

Costa et al. (2001) characterized yellow passion fruit from plants irrigated with different saline levels and concluded that water salinity higher than 3.0 dS m^{-1} does not affect the external and internal quality of fruits. In the same study, it was reported that fruit quality related to pulp percentage, soluble solids, titratable acidity and vitamin C contents were similar to standardized values for plants irrigated with good quality water. These successful results were possible because, as forthis study, planting pits without coating (R0) and with side coatings (R1; R2; R3; R4; respectively one, two, three and four lateral sides) were used, so plating pits with coating, promoted a better water use and, consequently, better fruits. Accordingly, Cavalcante et al. (2005) found a positive correlation between pit coating and number of fruits and fruit weight, under irrigation with saline water, i.e., coated pits increased fruit production under irrigation with saline water with results compatible to fruit production by plants irrigated with good quality water. Cavalcante et al. (2003) associated soil mulch with saline water levels for yellow passion fruit production and identified that increasing salinity of irrigation water has no effect on fruit quality when soil mulch also was used, it was also noticed that soluble solids (°brix), titratable acidity and pulp pH produced under irrigation with saline water and soil mulching are compatible to low demanding markets.

Yellow passion fruit is also positively affected by organic fertilizers, such as biofertilizers, which are obtained by anaerobic fermentation and could be simple [water + fresh bovine manure] at a ratio of [1:1] (in volume) or enriched [water + fresh bovine manure + a protein mix + nutrient sources], as can be seen in figure 4, adapted from Cavalcante et al. (2011c).

During the last years, biofertilizers have emerged as an important component of the integrated nutrient supply system and they have been tested on yellow passion fruit production and nutrition through environmentally better nutrient supplies. It is possible to find in the scientific literature some research work that has studied biofertilizers effects on fruit quality of yellow passion fruit, including the ones under irrigation with saline water. According to Cavalcante et al. (2011a), fruit quality and nutritional status of yellow passion fruit are affected by biofertilizer doses applied; fruit length, width, pulp percentage, skin diameter, mass, soluble solids and titratable acidity were improved with biofertilizer application, independently of type; simple biofertilizer promotes optimum supplies of potassium, calcium and sulphur, while enriched one promotes optimum supplies of nitrogen, phosphorus, potassium and calcium; and bovine biofertilizer is an important key to the production of yellow passion fruit with less use of chemical fertilizers, maintaining fruit quality.

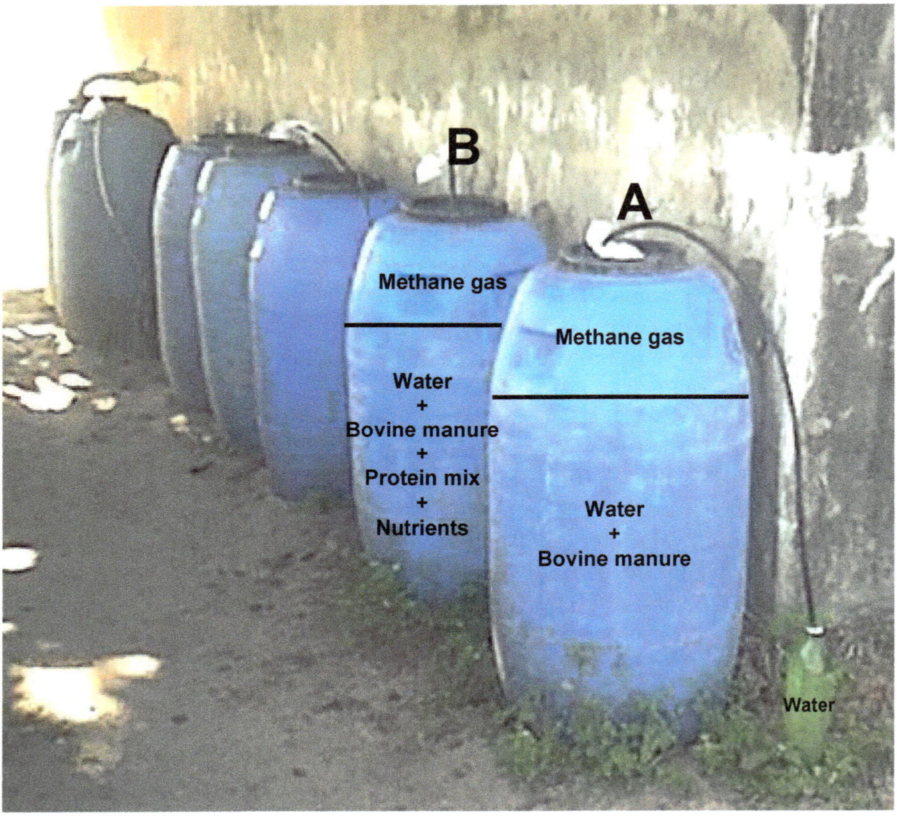

Fig. 4. Preparation system of both biofertilizers (simple - A; enriched - B) by anaerobic fermentation.

Cavalcante et al. (2007b) reported that enriched biofertilizer inhibited the mean weight of fruits and yield of yellow passion fruit in relation to simple biofertilizer, and fruits produced under simple fertilizer use were compatible to the average standardized fruit weight of 200g quoted by Meletti et al. (2002). Despite, different results can be found: Rodrigues et al. (2008), adversely, concluded that yellow passion plants submitted to enriched biofertilizer produced adequate titratable acidity (citric acid) and low pulp percentage and vitamin C content, including that soluble solids (°brix) ranged from 12.9 to 13.9°brix, thus above the limit of 11°brix required by the industry (Anonymous, 1999).

Simple biofertilizer promoted enhancement on titratable acidity of yellow passion fruits (Dias et al., 2011), what is pretty much important for juice industry, since it reduces the addition of artificial acid components. Furthermore, all the yellow passion fruit plants biofertilized in the work cited above presented the minimum titratable acidity required by juice industry, which is 2.5% (Anonymous, 1999). Dias et al. (2011) also concluded that higher biofertilizing frequency stimulates pulp percentage, titratable acidity, vitamin C contents and juice electrical conductivity. Enriched biofertilizer affected yellow passion fruit mass, promoting production of fruits with average mass compatible to the fresh fruit market (Rodrigues et al., 2009).

Coconut is another tropical fruit which is influenced by management techniques. According to Ferreira Neto et al. (2007), the mean weight of fruit is not influenced by the application of N and K through fertigation, but increasing N has decreased water volume and °Brix and increased pH, while increasing K dose has decreased salinity and increased °brix of coconut water. On another paper, Silva et al. (2006) verified that coconut water volume is affected by nitrogen and potassium, since the maximum coconut water volume observed (417.75 mL) was found when 818 g of nitrogen plant^{-1}year^{-1} and 1487 g of potassium plant^{-1}year^{-1} were tested. Nitrogen and potassium levels also demonstrated a linear effect on the soluble solids content of coconut water, on which nitrogen had a negative and potassium a positive effect. A negative linear effect was observed between nitrogen concentration and electrical conductivity of coconut water, while it was observed that potassium levels showed a quadratic effect on this same parameter. Coconut water quality is directly and mainly affected by coconut plant genotype, maturation, plant nutrition, irrigation and climatic conditions, thus it is important monitoring all fruit growth and maturation circle to obtain fruits with high amounts of water, what is very important for coconut water industry.

As for papaya fruit processing it is necessary good quality fruits, produced under low costs, as suggested by Mesquita et al. (2007), who have evaluated the fruit quality and yield of papaya fertilized with bovine biofertilizers. The results of their study, except for fruit firmness, registered positive effects of the biofertilizers on 'Baixinho de Santa Amália' papaya cultivar in relation to yield, physical and chemical fruit quality.

4. Physical and chemical characteristics of non-conventional fruits

The higher and increasing demand, during the last decades, for exotic fruits has offered greater variety in the production market, the same way expanded marketing opportunities have been of fruit producers' interest, especially in Brazil where climatic and soil conditions are favorable for the production of non-conventional fruits. Equally, scientific information that characterizes these species in relation to quality and maturation of fruits is important to the development of this industry. In this sense, the determination of physical and chemical fruit characteristics constitutes an important reference for studies about the maturation and quality of fruits, with the ultimate aim of determining consumer acceptance requirements.

Additionally, nowadays, fruit consumers are becoming increasingly aware of health and nutritional aspects of their food, demonstrating a tendency of avoiding synthetic products, such as additives, in their food, since they are obtaining therapeutic effects and nutrition from natural resources. This way, non-conventional fruits could satisfy these demands because they are also produced, in most cases, without agronomic techniques, as mentioned chemical fertilizing and pesticides, especially in Brazil where the Agriculture Ministry regulates the pesticides, dividing them into two groups: the ones which have been scientifically tested and proved their efficiency and, the ones which have not been through this process concerning non-conventional fruits. On the other hand, many of these fruits are highly perishable and difficult to store in their fresh form, some of them are not acceptable as fresh fruits due to its high acidity and/or a strong astringent taste, although most of non-conventional fruit have unlimited potential in the world trade in their processed form.

Reliable information about physical and chemical characteristics of non-conventional fruits is poorly found in scientific literature, including books and manuscripts around the world. Among the non-conventional fruits dovyalis (*Dovyalis* sp.), yellow mangosteen (*Garcinia* sp.) and mamey (*Pouteria sapota*) have particular importance and will also be focused in this chapter.

Dovyalis belongs to the Flacourtiaceae family native to India or Sri Lanka (Ferrão, 1999), which have fleshy, yellow, spherical in form, succulent and extremely acidic fruits (Figure 5). Cavalcante & Martins (2005) established the physical and chemical characterizations of dovyalis hybrid fruits (*Dovyalis abyssinica* and *D. hebecarpa*) and observed that dovyalis fruits have good physical qualities for the fresh market and juice industry averaging 75% pulp; soluble solids (SS) results ranged between 14.9 and 14.0%; titratable acidity (TA) was considered high, i.e., from 2.9 to 3.6 mg/100 g of fresh fruit. Consorting to these results, the SS/TA rate ranged between 4.1 and 5.4, numbers that are considered low, since the higher this ratio, the better fruit quality for fresh fruit consumption is. On the contrary, the 120.3 mg/100 g of fresh fruit content demonstrates that dovyalis fruit is a good natural source of vitamin C. However, further studies are needed to determine technological parameters for consumption as fresh or processed fruit.

Fig. 5. Dovyalis fruits from the Active Germplasm Bank of São Paulo State University, Brazil.

Another non-conventional and potential fruit is yellow mangosteen (*Garcinia xanthochymus* Hook), which is also known as false mangosteen and belongs to Clusiaceae family, which also includes 35 genus and more than 800 species (see Figure 6). According to Cavalcante et al. (2006a), yellow mangosteen fruits present 76.03-95.04g, 5.09-5.50cm in length, 5.54-5.72cm in width, 1.45-1.95 seeds/fruit, 71.13-76.61% of pulp percentage, 10.8-12.6°Brix of soluble solids, 3.85-4.42% of titratable acidity and vitamin C content varying from 31.21 to 46.82 mg/100 of fresh pulp. These results indicate that yellow mangosteen fruit has a good pulp percentage, what is important for fruit industry; this is a very acid fruit with a TA average 3.51% and a good natural source of vitamin C, although this last variable is lower than that registered as for dovyalis fruits.

Fig. 6. Yellow mangosteen fruit from the Active Germplasm Bank of São Paulo State University, Brazil.

Mamey (*Pouteria sapota*) is native to Mexico and Central America and, after its introduction to Florida (USA), it has become much more known and sought especially by Latin Americans. Mamey is also known as mammee, mammee apple, St. Domingo apricot and South American apricot (English); mamey de Santo Domingo, mamey amarillo, mamey de Cartagena, mata serrano, zapote mamey, or zapote de Santo Domingo (Spanish); abricote, abrico do Pará or abrico selvagem (Portuguese); and abricot d' Amerique, abricot des Antilles, abricot pays, abricot de Saint-Dominque or abricotier sauvage (French).
Fruits of fresh mamey are consumed in fruit salads or served with cream and sugar or wine; they are also minimally processed and sliced to be used in pies or tarts, and may be seasoned with cinnamon or ginger. The mamey is widely turned into preserves such as spiced marmalade and pastes (resembling guava paste) and used as filler for products made of other fruits. Slightly under-ripe fruits, rich in pectin, are turned into jelly. Wine is made from the fruit and fermented "toddy" from the sap of the tree in Brazil.
Instead of the large options of mamey fruit consumption, it is still poorly studied in relation to fruit quality. This way, Nascimento et al. (2008) found large ranges of fruit quality parameters of mamey fruits in Brazil, i.e., average values ranged from 11.9-17.31cm in fruit length, 5.77-9.50cm in fruit width, 317.71-765.82g in fruit mass, 57.57-82.49% in pulp percentage, 20.60-26.40°Brix of soluble solids, 0.51-0.07% of titratable acidity, 47.89-387.49 in SS/AT ratio and 12.29-36.98 mg/100 of fresh pulp, indicating that mamey fruit presents high pulp percentage, low titratable acidity and high soluble solids, demonstrating to be a good flavor fruit, but not a vitamin C source. In Figure 7 it is possible to identify mamey fruits.

Fig. 7. Yellow mangosteen fruit from the Active Germplasm Bank of São Paulo State University, Brazil.

In fact, there are many unconventional fruits also underutilized around the world, a lot to study, especially those which present potential for consumption as fresh or processed fruit, and much more have to be done and published to make good flavored but unconventional fruits, also traditional ones as for the main world fruit market.

5. References

Abbott, J.A. (1999). Quality Measurement of Fruits and Vegetables. *Postharvest Biology and Technology*, Vol. 15, pp. 201–206, ISSN 09251929

Anonymous. (1999). *Instrução Normativa nº12, de 10 de setembro de 1999, Seção1*. Brasília, Brazil: Ministério da Agricultura e do Abastecimento.

Bron, I.U. & Jacomino, A.P. (2006). Ripening and Quality of 'Golden' Papaya Fruit Harvested at Different Maturity Stages. *Brazilian Journal Plant Physiology*, Vol. 18, N°. 3, pp. 389-396, ISSN 1677-0420

Campos, V.B.; Cavalcante, L.F.; Dantas, T.A.G.; Mota, J.K.M.; Rodrigues, A.C. & Diniz, A.A. (2007). Caracterização Física e Química de Frutos de Maracujazeiro Amarelo sob Adubação Potássica, Biofertilizante e Cobertura Morta. *Revista Brasileira de Produtos Agroindustriais*, Vol. 9, N°. 1, pp. 59-7, ISSN 1517-8595

Cavalcante, I.H.L. & Martins, A.B.G. (2005). Physical and Chemical Characterization of Dovyalis Fruits. *International Journal of Fruit Science*, Vol. 5, N°. 4, pp. 39-46, ISSN 1553-8621

Cavalcante, I.H.L.; Beckmann, M.Z.; Martins, A.B.G. & Campos, C.C.M. (2007a). Preliminary Selection of Acerola genotypes in Brazil. *Fruits*, Vol. 62, pp. 1-8, ISSN 0248-1294

Cavalcante, I.H.L.; Cavalcante, L.F.; Santos, G.D.; Cavalcante, M.Z.B. & Silva, S.M. (2011a). Impacto f Biofertilizers on Mineral Status and Fruit Quality in Yellow Passion Fruit in Brazil. *Communications in Soil Science and Plant Analysis*, In Press, ISSN 0010-3624

Cavalcante, I.H.L.; Jesus, N. & Martins, A.B.G. (2006a). Physical and Chemical Characterization of Yellow Mangosteen Fruits. *Revista Brasileira de Fruticultura*, Vol. 28, N°. 2, pp. 325-327, ISSN 0100-2945

Cavalcante, I.H.L.; Martins, A.B.G. & Stuchi, E.S. (2006b). Fruit Characteristics of Eighteen Orange Cultivars. *Revista de Biologia e Ciências da Terra*, Vol. 6, Nº. 2, pp. 72-77, ISSN 1519-5228

Cavalcante, I.H.L.; Martins, A.B.G.; Stuchi, E.S. & Campos, C.C.M. (2009a). Fruit Maturation as a Parameter for Selection of Sweet Orange Cultivars in Brasil. *Journal of Food Agriculture & Environment*, Vol. 7 (3 & 4), pp. 316-319, ISSN 1459-0255

Cavalcante, I.H.L.; Moura, M.C.S.; Rocha, L.F.; Silva Júnior, G.B.; Martins, L.V. & Silva, R.R.S. (2011b). Seleção Preliminar de Genótipos de Pinheira em Bom Jesus-PI. *Revista de Ciências Agrárias*. Vol. 34, pp. 171-179, ISSN 0871-018X

Cavalcante, L.F.; Andrade, R.; Mendonça, R.M.N.; Silva, S.M.; Oliveira, M.R.T.; Araújo, F.A.R. & Cavalcante, I.H.L. (2003). Caracterização Qualitativa de Frutos de Maracujá-Amarelo (*Passiflora edulis* f. *flavicarpa* Deg.) em Função da Salinidade da Água de Irrigação. *Agropecuária Técnica*, Vol. 24, N°. 24, pp. 39-45, ISSN 0100-7467

Cavalcante, L.F.; Cavalcante, I.H.L.; Júnior, F.R.; Cavalcante, M.Z.B. & Santos, G.P. (2011c). Leaf-Macronutrient Status and Fruit Yield of Biofertilized Yellow Passion Plants. *Journal of Plant Nutrition*, In Press, ISSN 0190-4167

Cavalcante, L.F.; Costa, J.R.M.; Oliveira, F.K.D.; Cavalcante, I.H.L. & Araújo, F.A.R. (2005). Produção do Maracujazeiro-Amarelo Irrigado com Água Salina em Covas Protegidas Contra Perdas Hídricas. *Irriga*, Vol. 10, N°. 3, pp. 229-240, ISSN 1808-3765

Cavalcante, L.F.; Lima, E.M.; Freire, J.L.O.; Pereira, W.E.; Costa, A.P.M. & Cavalcante, I.H.L. (2009b). Componentes Qualitativos do Cajá em Sete Municípios do Brejo Paraibano. *Acta Scientiarum Agronomy*. Vol. 31, N°. 4, pp. 627-632, ISSN 1807-8621

Cavalcante, L.F.; Santos, G.D. dos; Oliveira, F.A.; Cavalcante, I.H.L.; Gondim, S.C & Cavalcante, M.Z.B. (2007b). Crescimento e Produção do Maracujazeiro-Amarelo em Solo de Baixa Fertilidade Tratado com Biofertilizantes Líquidos. *Revista Brasileira de Ciências Agrárias*, Vol. 2, N°. 1, pp. 15-19, ISSN 1981-0997

Conway, W.S. & Sams, C.E. (1983). Calcium Infiltration of Golden Delicious Apples and its Effect on Decay. *Phytopathology*, Vol. 73, pp. 1068-1071, ISSN 0031-949X

Cook, R. (1999). Anoverview of Key Food Industry Drivers: Implications for Fresh Produce Industry. *Journal of Food Distribution Research*, Vol. 30, pp. 1–4, ISSN 0047-245X

Costa, J.R.M.; Lima, C.A.A.; Lima, E.D.A; Cavalcante, L.F. & Oliveira, F.K.D. (2001). Caracterização dos Frutos de Maracujá Amarelo Irrigados com Água Salina. *Revista Brasileira de Engenharia Agrícola e Ambiental*, Vol. 5, N°. 1, pp. 143-146, ISSN 1807-1929

Di Giorgi, F.; Ide, B.Y.; Dib, K.; Marchi, R.J.; Triboni, H.R. & Wagner, R.L. (1990). Contribuição ao Estudo do Comportamento de Algumas Variedades de Citros e suas Implicações Agroindustriais. *Laranja*, Vol. 11, N°. 2, pp. 567-612, ISSN 0102-1907

Dias, T.J.; Cavalcante, L.F.; Freire, J.L.O.; Nascimento, J.A.M.; Cavalcante, M.Z.B. & Santos, G.P. (2011). Qualidade Química de Frutos do Maracujazeiro-Amarelo em Solo com Biofertilizante Irrigado com Águas Salinas. *Revista Brasileira de Engenharia Agrícola e Ambiental*, Vol. 15, N°. 3, pp. 229-236, ISSN 807-1929

FAO (February 2011). Statistics Database, In: *FAOSTAT*, 20.02.2011, Available from: http://apps.fao.org/

Ferrão, J.E.M. (1999). *Fruticultura Tropical*: Espécies com Frutos Comestíveis. Instituto de Investigação Científica Tropical, ISBN: 972-672-879-7, Lisboa, Portugal

Ferreira Neto, M.; Holanda, J.S.; Folegatti, M.V.; Gheyi, H.R.; Pereira, W.E. & Cavalcante, L.F. (2007). Qualidade do Fruto do Coqueiro Anão Verde em Função de Nitrogênio e Potássio na Fertirrigação. *Revista Brasileira de Engenharia Agrícola e Ambiental*, Vol. 11, N°. 5, pp. 453-458, ISSN 1807-1929

Freire, J.L.O.; Cavalcante, L.F.; Rebequi, A.M.; Dias, T.J.; Nunes, J.C. & Cavalcante, I.L.C. (2010). Atributos Qualitativos do Maracujá Amarelo Produzido com Água Salina, Biofertilizante e Cobertura Morta no Solo. *Revista Brasileira de Ciências Agrárias*, Vol. 5, N°. 1, pp. 102-110, ISSN 1981-0997

Mahmud, T.M.M.; Al Eryani-Raqeeb, A.; Syed Omar, S.R.; Mohamed Zaki, A.R. & Al Eryani, Abdul-R. (2008). Effects of Different Concentrations and Applications of Calcium on Storage Life and Physicochemical Characteristics of Papaya (*Carica Papaya* L.). *American Journal of Agricultural and Biological Sciences*, Vol. 3, N°. 3, pp. 526-533, ISSN 1557-4989

Marschner, H. (2005). *Mineral Nutrition of Higher Plants*. (6th edition). UK: Academic Press, ISBN: 978-0-12-384905-2, London, Great Britain

Meletti, L.M.M.; Soares Scott, M.D.; Bernacci, L.C. & Azevedo, F.J.A. (2002). Desempenho das Cultivares IAC-273 e IAC-277 de Maracujazeiro-Amarelo (*Passiflora edulis* f. *flavicarpa* Deg.) em Pomares Comerciais. In: *Reunião Técnica de Pesquisa em Maracujazeiro*, Vol. 3, pp. 196-197

Mesquita, E.F.; Cavalcante, L.F.; Goldim, A.C.; Cavalcante, I.H.L.; Araújo, F.A.R. & Cavalcante, M.Z.B. (2007). Produtividade e Qualidade de Frutos do Mamoeiro em Função de Tipos e Doses de Biofertilizantes. *Semina: Ciência Agrárias*, Vol. 28, N°. 4, pp. 589-596, ISSN 1679-0359

Nascimento, V.E.; Martins, A.B. & Hojo, R.H. (2008). Caracterização Física e Química de Frutos de Mamey. *Revista Brasileira de Fruticultura Brasileira*, Vol. 30, N°. 4, pp. 953-957, ISSN 0100-2945

Paull, R.E. & Duarte O. (2011). *Tropical Fruits*, (2nd edition), ISBN 978-1-84593-672-3, London, Great Britain

Rodrigues, A.C.; Cavalcante, L.F.; Dantas, T.A.G.; Campos, V.B. & Diniz, A.A. (2008). Caracterização de Frutos de Maracujazeiro-Amarelo em Solo Tratado com Biofertilizante Supermagro e Potássio. *Magistra*, Vol. 20, N°. 3, pp. 264-272, ISSN 0102-5333

Rodrigues, A.C.; Cavalcante, L.F.; Oliveira, A.P.; Sousa, J.T. & Mesquita, F.O. (2009). Produção e Nutrição Mineral do Maracujazeiro-Amarelo em Solo com Biofertilizantes Supermagro e Potássio. *Revista Brasileira de Engenharia Agrícola e Ambiental*, Vol. 13, N°. 2, pp. 117-123, ISSN 1807-1929

Silva Júnior, G.B.; Rocha, L.F.; Amaral, H.C.; Andrade, M.L.; Neto, R.F. & Cavalcante, I.H.L. (2010). Laranja-da-Terra: Fruta Cítrica Potencial Para o Piauí. *Semina: Ciência Agrária*, Vol. 31, N°. 3, pp. 557-562, ISSN 1679-0359

Silva, R.A.da; Cavalcante, L.F.; Holanda, J.S.de; Pereira, W.E.; Moura, M.F. & Ferreira Neto, M. (2006). Qualidade de Frutos do Coqueiro-Anão Verde Fertirrigado com Nitrogênio e Potássio. *Revista Brasileira de Fruticultura, Jaboticabal*, Vol. 28, N°. 2, pp. 310-313, ISSN 0100-2945

3

Photodecomposition Behaviors of Pesticides in the Source for Water Supply Using an Alumina Carrier-Titanium Dioxide Photocatalyst

Seiichi Ishikawa, Bunko Cho,
Shin Li, Yuji Okumura, Yoshikazu Iida,
Teiji Tanizaki and Masayuki Higuchi
The University of Kitakyushu
Japan

1. Introduction

The countermeasures against toxic substances or musty odor cause substances in the source for water supply have become a serious subject with the pollution of environmental waters (Coleman et al., 1980; Nakasugi, 1993; Tanada et al., 1995; U. S. EPA, 1993). For the purpose of the utilization of the photodecomposition method for the water purification at a water purification plant or a water purifier, we have investigated on the photodecomposition behaviors of organic pollutants (Matsuda et al., 2002; Okumura et al., 2003), total organic compounds (Tanizaki et al., 2005), formaldehyde (Tanizaki et al., 2005), total organic halides (Tanizaki et al., 2005), trihalomethanes (Tanizaki et al., 1997, 2005), geosmin (Okumura et al., 2004) and 2-methyl-iso-borneol (Okumura et al., 2004) using a titanium dioxide (TiO_2) photocatalyst. Relatively high photodecomposition efficiencies were obtained for these substances.

Pesticide is one of the toxic chemical substances polluting the source for water supply (Tanada et al., 1995), vegetables, fruits, etc (Ishikawa et al., 2004). Then, the residual pesticide in many kinds of food have already been regulated and the aiming standard values of 101 types of pesticide for tap water quality control were set up in 2003 (Ando, 2004). However, only 21 in 117 types of pesticide showed over 80% of removal ratio in the coagulation and sedimentation usually performed at a water purification plant (Ishikawa et al., 2006). Then, we investigated on the photodecomposition behaviors of the 5 types of pesticide which were largely used in Kitakyushu district.

We have ever used silica gel having high decomposition efficiency as a carrier of TiO_2 photocatalyst. However, a silica gel carrier was fragile. In addition, sufficient endurance of the carrier is demanded when the decomposition of chemicals in water and food or that disinfection using supersonic wave together is performed (Ishikawa et al., 2008). Then, we also investigated on the photodecomposition ability with an alumina carrier, which could stand against ultra violet (UV) light and water, instead of silica gel carrier.

2. Material and methods

2.1 Instruments

Gas chromatograph/mass spectrometer (GC/MS) was a Hewlett Packard HP-5890 series and a JEOL auto mass system. Total nitrogen (T-N) analysis was performed using a Tokyo Kasei Ind. Co. TCI-NOX 1000, GASTORR GT-102, VISIBLE DETECTOR S-3250 and AUTO SAMPLER SS-3600, while other determinations Hitachi Co. U-2000A a spectrophotometer was used.

2.2 Objective pesticides and other reagents

Five pesticides (Table 1) were selected for an objective pesticide. These pesticides were used in the Onga River Basin and their aiming standard values in tap water quality were established (Ando, 2004). Their structural formulas were shown in Fig. 1. Dinitrophenol (DNP) was used as a chemical substance for the capability evaluation of alumina carrier (Okumura et al., 2003). These chemical substances were obtained from Hayashi Pure Chemical Industries, Wako Pure Chemical Industries and Tokyo Kasei Kogyo Co.. Each standard solution was prepared by the dilution with acetone. An alumina carrier was obtained from Sumitomo Chemical Co.. The properties of the alumina and silica gel carriers, comparative carriers, were shown in Table 2. Two types of alumina, NK124 and NKHO24
(Fig. 2), and silica gel carrier-TiO_2 photocatalysts were produced by the Sol-Gel Method (Okumura et al., 2004; Tanizaki et al., 1997). All solvents were the grade reagents for pesticide residue analysis, which were obtained from Kanto Chemical Co. and Wako Pure Chemical Industries. Other reagents were special grade reagents, which were obtained from Wako Pure Chemical Industries. Anhydrous Na_2SO_4 and NaCl were heated at 800 °C for 3 h after acetone-washing. The water was purified using a Millipore Milli-Q Ultra-pure Water System.

Pesticide	Use	Shipped amount [a] (kg)	Aiming standard value (mg l^{-1})
Chlorthalonil	Disinfectant	185	0.05
Pencycuron	Disinfectant	686	0.04
Cafenstrole	Herbicide	1,946	0.008
Thiobencarb	Herbicide	155	0.02
Trifluralin	Herbicide	5,238	0.06

[a] The values in 2003

Table 1. The shipped amounts and the aiming water qualities of the objective pesticides in the Onga River Basin

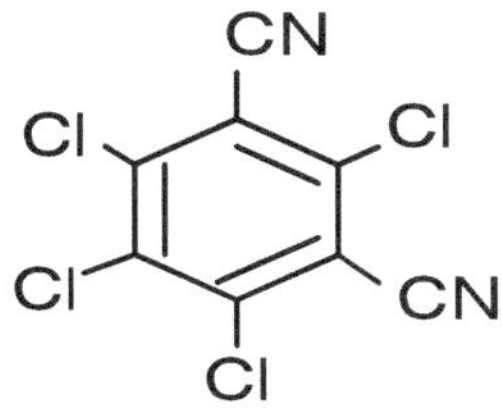

Chlorthalonil

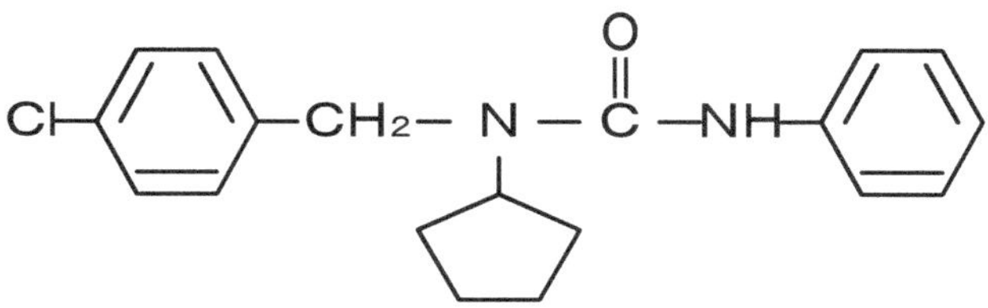

Pencycuron

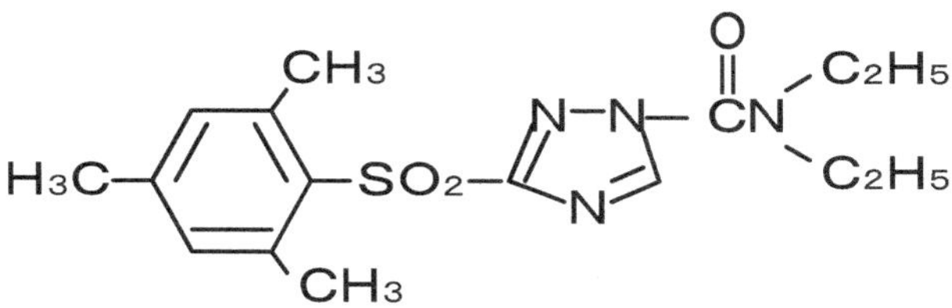

Cafenstrole

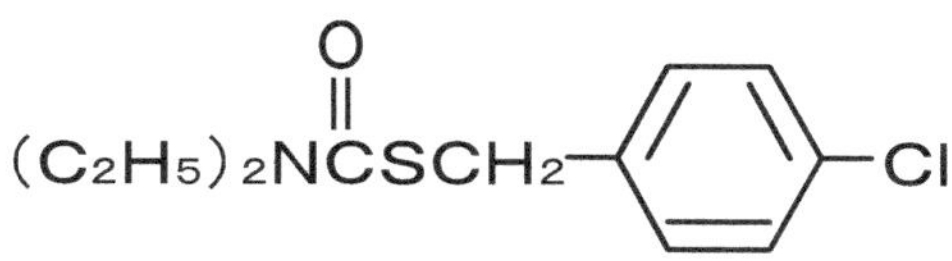

Thiobencarb

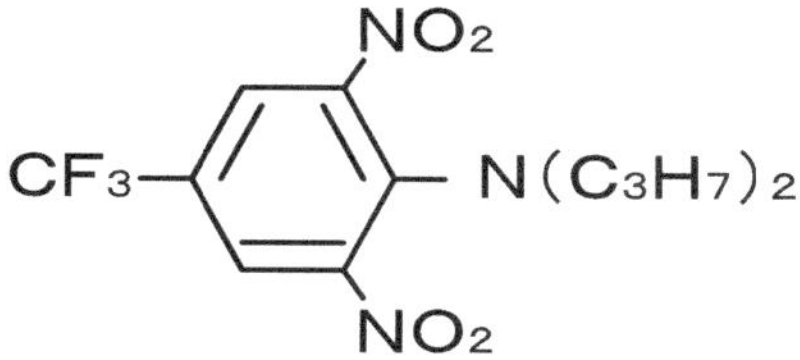

Trifluralin

Fig. 1. Structural formulas of the objective pesticides

Item	Carrier		
	NK124	NKHO24	Silica gel
Particle size (mm)	2-4	2-4	1.7-4
Al_2O_3 contents (%)	99.9	99.7	—
Micropore volume ($cm^3\ g^{-1}$)	0.77	0.58	—
Relative surface area ($m^2\ g^{-1}$)	130	160	—
Compacting strength (kg)	2	6	—
Supporting ratio (%)	20	17	25

Table 2. Properties of alumina and silica gel carriers

NK124 NKHO24

Fig. 2. Prepared alumina carrier-TiO_2 photocatalysts

2.3 Apparatus for the photodecomposition of the pesticides

The apparatus for the photodecomposition experiments of the pesticides and DNP was shown in Fig. 3. The photoreactor (Fig, 4) was made of stainless steel and equipped with a 6 W low pressure mercury lamp (a Matsushita Electric Ind. Co. GL6/Q) and a stabilizer (Nihon Fluorescence Electric Co.). Eighty five milliliters of each photocatalyst was packed in a thickness of about 5 mm. The UV illumination intensity on the surface of catalyst was 10 mW cm^{-2}. Sample water was circulated with a roller pump (Furue Science Co.).

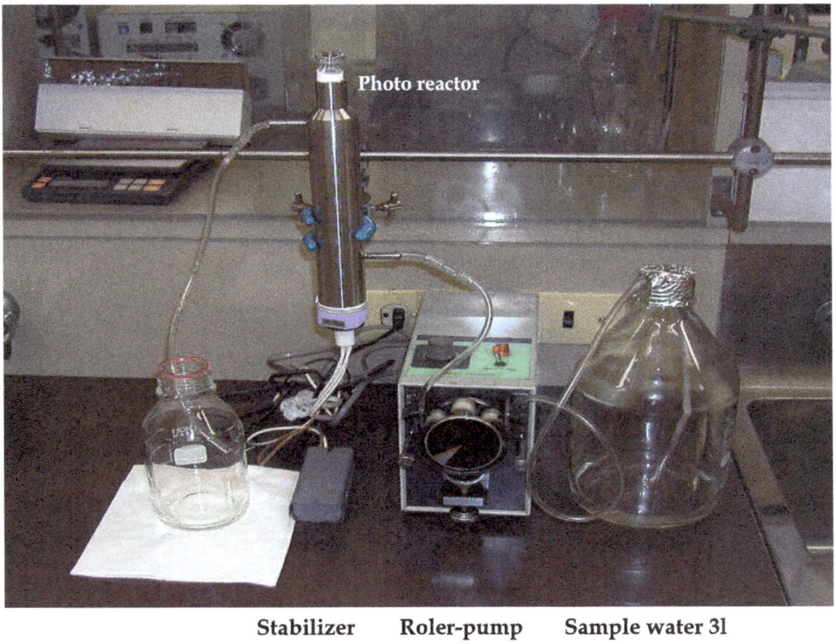

Fig. 3. Apparatus for the photodecomposition of the pesticides

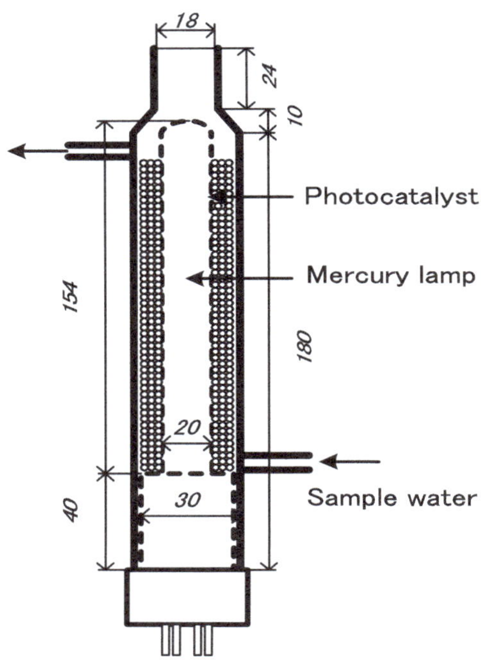

Fig. 4. Structure of photoreactor

2.4 Analyses of pesticides, DNP and other items

The quantification of the pesticides and DNP in UV irradiated solution was performed in the next procedure. About 2 g of NaCl was added in 40 ml of UV irradiated solution and each pesticide or DNP was extracted with 4 ml of dichloromethane. The dichloromethane layer was separated from aqueous layer, dehydrated with anhydrous Na_2SO_4 and analyzed by the GC/MS method (Nakano et al., 2004; Yamaguchi et al., 1997). The GC/MS conditions were shown in Table 3. Each calibration curve showed good linearity in the quantification range. Their recoveries by the method were over 85%.

pH, suspended solid matter (SS), BOD, $KMnO_4$ consumption, total nitrogen (T-N), total phosphorus (T-P) and electric conductivity (EC) were measured by the method of Japanese Industrial Standard K0102 (Japan Industrial Standards Committee, 1995).

Column	DB-5MS (5%Diphenyl 95%dimethyl polysiloxane) 0.25 mm x 30 m x 0.25 μm
Column temperature	60 °C (1 min) – 30 °C min^{-1} – 130 °C – 5 °C min^{-1} – 240 °C – 10 °C min^{-1} – 300 °C (10 min)
Injector temperature	250 °C
Carrier gas	He 1.5 ml min^{-1}
Transfer line temperature	260 °C
Mode	EI

Table 3. GC/MS conditions

2.5 The photodecomposition capability experiments of alumina carrier-TiO_2 photocatalysts and photodecomposition experiments of pesticides

Each alumina or silica gel carrier-TiO_2 photocatalyst was packed in the photoreactor. The water samples for DNP and pesticides experiments were prepared by adding 3 ml of each 1,000 mg l^{-1} DNP or pesticide acetone solution in 3 l of purified water (for the photodecomposition capability experiment of photocatalyst) or the river water (Table 4) (for the photodecomposition experiment of pesticide). The water sample was vigorously shaken for 30 min using a separatory funnel and placed in 5 l glass bottle. The water sample was firstly circulated for 30 min at 1 min^{-1} of flow rate by stirring and the system was allowed to reach equilibrium. Then the mercury lamp was switched on. The UV irradiated solution was periodically withdrawn during irradiation and DNP or each pesticide was quantified by the GC/MS method. The photodecomposition experiments of pesticides without a photocatalyst were also performed.

Item	Concentration
pH	7.6
SS (mg l^{-1})	6
BOD (mg l^{-1})	3.5
$KMnO_4$ consumption (mg l^{-1})	5.8
T-N (mg l^{-1})	1.52
T-P (mg l^{-1})	0.08
EC (μS cm^{-1})	264

Table 4. Quality of the river water used for this experiment

3. Results and discussion

3.1 Comparison of the photodecomposition capability of alumina carrier-TiO_2 photocatalysts

Figure 5 shows the photodecomposition rates of DNP using the alumina and silica gel carrier-TiO_2 photocatalysts. DNP was decreased exponentially with reaction time (t) and the rate of DNP disappearance was nearly represented by a first-order process. The values of pseudo-first-order rate constant (k: $C=C_0e^{-kt}$) of NK124, NKHO24 and silica gel carrier-TiO_2 photocatalysts determined from the plot of data points (C/C_0 vs. t) were 0.027, 0.016 and 0.030 min^{-1}, respectively. The rate constant of NK124 carrier-TiO_2 photocatalyst was near that of silica gel-TiO_2 photocatalyst.The micropore volume of NK124 is lager than that of NKHO24 but its relative surface area is smaller than that of NKHO24. The supporting ratio of NK124 was higher than that of NKHO24. It was supposed that the deference of DNP photodecomposition rate was caused by the deference of supporting ratio. Then, the photodecomposition experiments of the pesticides were performed using a NK124 carrier-TiO_2 photocatalyst.

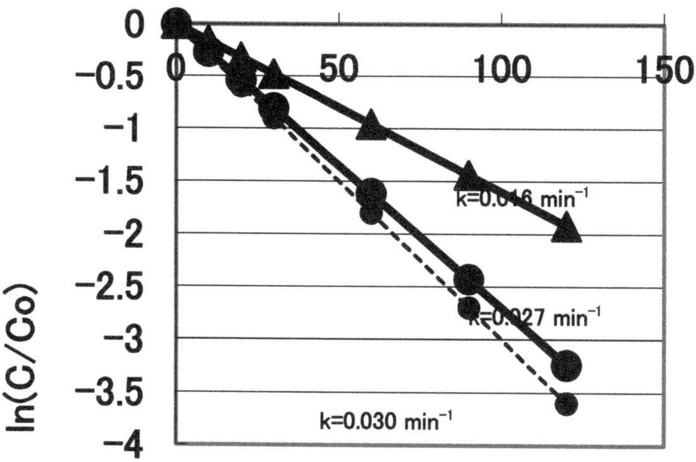

Fig. 5. Photodecomposition rates of DNP

3.2 Photodecomposition behaviors of pesticides

Figure 6-10 show the photodecomposition ratios of the pesticides. These pesticides were decomposed faster than DNP and the removal efficiencies after 3 min UV irradiation were 94% (Cafenstrole), 92% (Chlorthalonil), 75% (Thiobencarb), 67% (Pencycuron), 58% (Trifluralin) and 8% (DNP). After 30 min UV irradiation, the removal efficiencies of Cafenstrole and Chlorthalonil, and Thiobencarb, Pencycuron and Trifluralin, were 98 and 94%, respectively. The UV illumination intensity in the experiments was so strong that the same removal efficiencies as with the photocatalyst were obtained without the

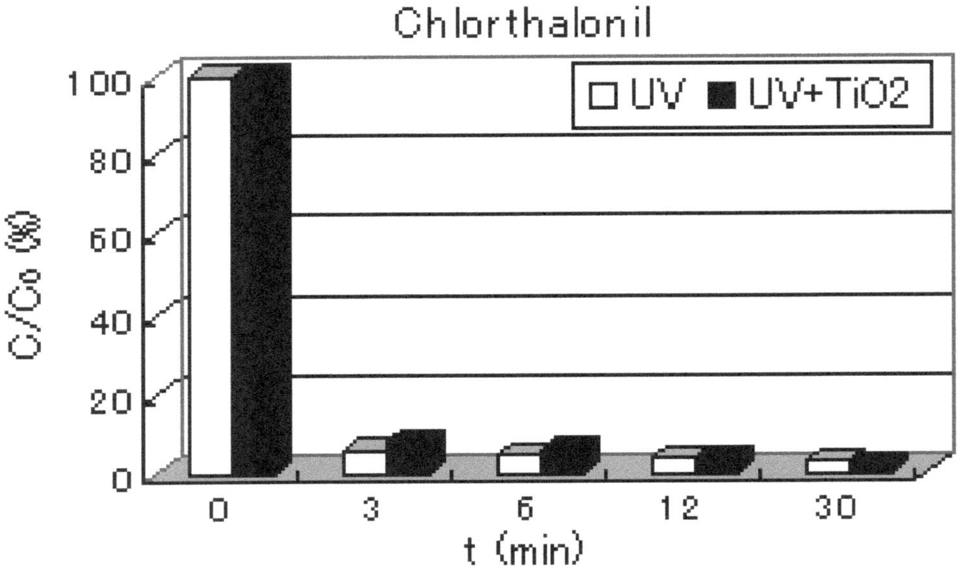

Fig. 6. Photodecomposition of Chlorthalonil

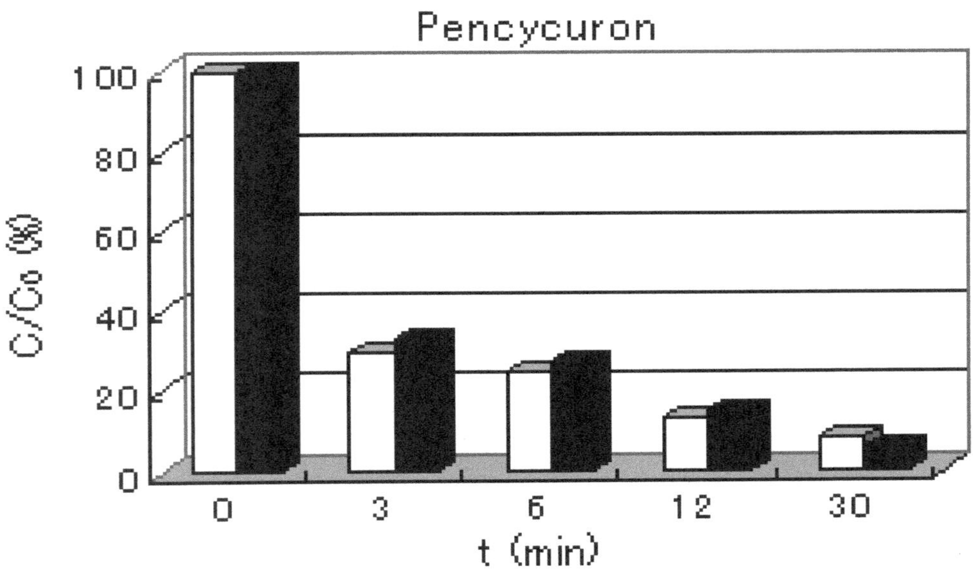

Fig. 7. Photodecomposition of Pencycuron

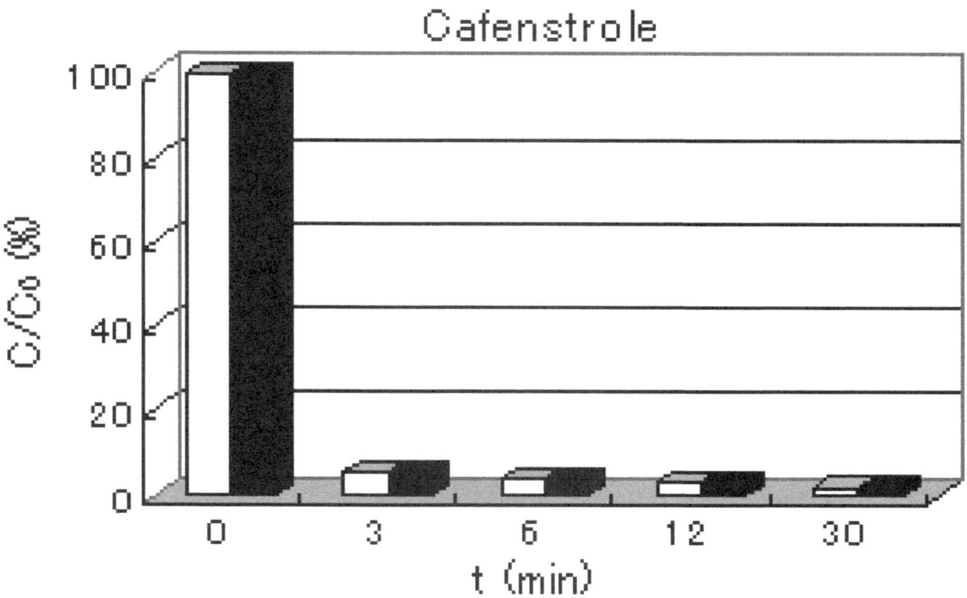

Fig. 8. Photodecomposition of Cafenstrole

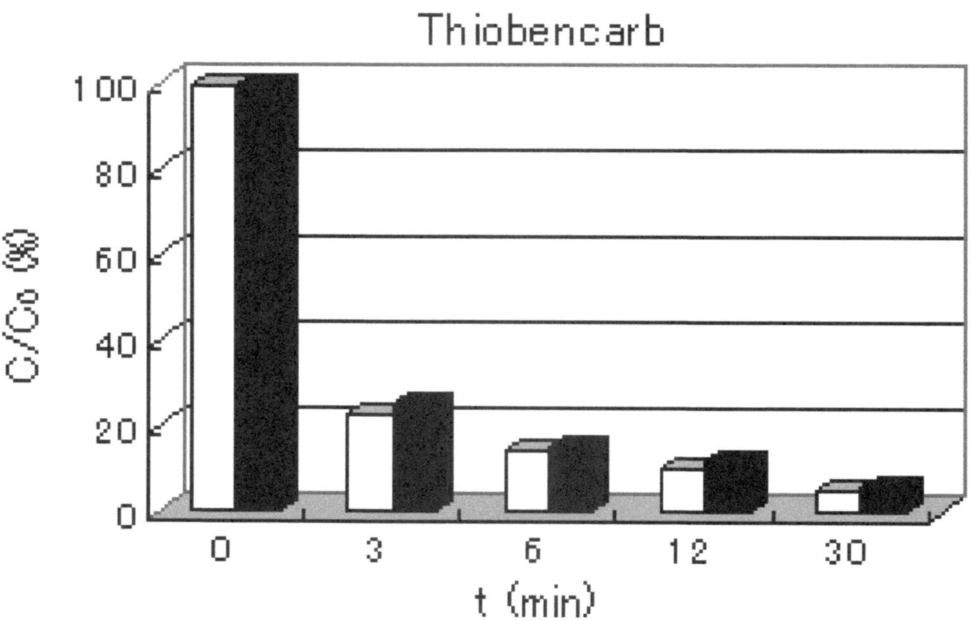

Fig. 9. Photodecomposition of Thiobencarb

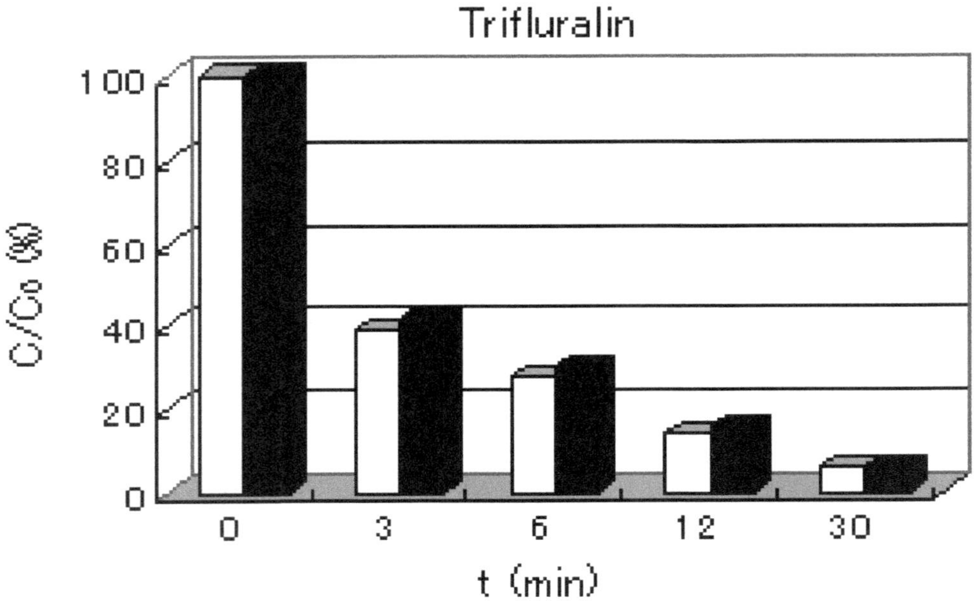

Fig. 10. Photodecomposition of Trifluralin

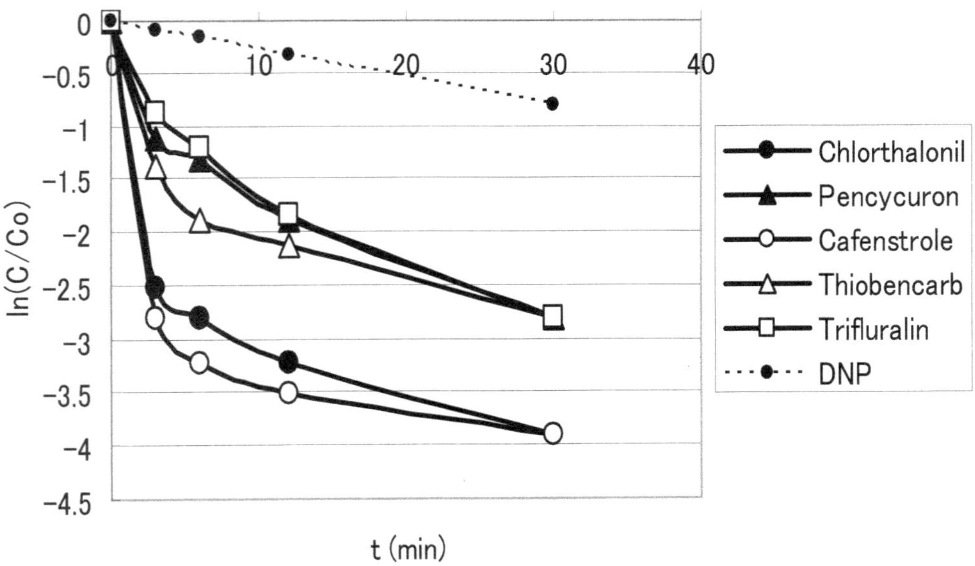

Fig. 11. Relationship between $\ln(C/C_0)$ and t photocatalyst.

Figure 11 shows the relationship between $\ln(C/C_0)$ and t. As these pesticides were decomposed immediately, the linear relationship between $\ln(C/C_0)$ and t could not be obtained.

Every pesticide has a biologically or chemically changeable structure in molecule. For example, a N-CO-N bond (Pencycuron and Cafenstrole) and a N-CO-S bond (Thiobencarb) are easily hydrolyzed. Cyano group (Chlorthalonil) is easily oxidized. Carbon-Cl bond, benzene ring and alkyl group are biologically hydroxylized (Haque & Freed, 1975; Hutson & Roberts, 1981; Matsumura & Murti, 1982; Uesugi et al., 1997; Yamamoto & Fukami, 1979). On the other hand, the scissions of a C-Cl bond, a C-F bond, a C-NO_2 bond, a C-NR_2 bond and a N-N bond, especially a C-Cl bond and a N-N bond, are easily occurred photochemically (Ishikawa et al., 1989, 1992; Ishikawa & Suetomi, 1993; Ishikawa, 1996; Tanizaki et al., 2005). Moreover, the photochemical scission of a C-Cl bond is faster than the photohydrolyses of organic phosphate esters (Ishikawa et al., 1992). The differences of the photodecomposition rates in these parts would have caused the differences of the photodecomposition rate or the removal efficiency after 30 min UV irradiation of each pesticide. As the photodecomposition products could not be detected by the GC/MS analysis, it was considered that these pesticides converted into the high polar compounds.

4. References

Ando, M. (2004). Standards for Drinking Water Quality. *J. Food Hyg. Soc. Japan,* Vol.45, No.2, J-151-J-155

Coleman, W. E., Melton, R. G., Kopfler, F. C., Barone, K. A., Aurand, T. A. & Jellison, M. G. (1980). Identification of Organic Compounds in a Mutagenic of a Surface Drinking Water by a Computerized Gas Chromatography Mass Spectrometry System (GC/MS/COM). *Environ. Sci. Technol.,* Vol.14, 576-588

Haque, R. & Freed, V.H. (1975). Environmental Dynamics of Pesticides. Plenum Press, New York and London, USA and UK

Hutson, D.H. & Roberts, T.R. (1981). Progress in Pesticide Biochemistry Volume 1. A Wiley-Interscience Publication, Chichester

Ishikawa, S. & Suetomi, R. (1993). Utilization of Photochemical Reaction in Environmental Chemistry. *J. Environ. Chem.,* Vol.3, No.2, 295-304

Ishikawa, S. (1996). Utilization of Photochemical Reaction in Environmental Chemistry. *KITAKYUSHU ENVIRONMENTOPIA,* Vol.11, No.1, 2-5

Ishikawa, S., Baba, K., Hanada, Y., Uchimura, Y. & Kido, K. (1989). Photodecomposition of *o*-Chloroaniline in Aqueous Solution with Low Pressure Mercury Lamp. *Bull. Environ. Contam. Toxicol.,* Vol.42, 65-70

Ishikawa, S., Eguchi, Y. & Ito, S. (2008). Decomposition Behavior of DMSO and Methanol. Report on Development of Photo-supersonic Wave Complex Technique with TNT (I), FS Study. The University of Kitakyushu, Japan

Ishikawa, S., Naetoko, E., Kawamura, S., Yamaguchi, R., Higuchi, M., Kojima, T., Yamato, Y. & Takahashi, M. (2004). Investigation of Pesticide Residues in Foods Distributed in Kitakyushu City. *J. Food Hyg. Soc. Japan,* Vol.45, No.2, 87-94

Ishikawa, S., Uchimura, Y., Baba, K., Eguchi, Y. & Kido, K. (1992). Photochemical Behavior of Organic Phosphate Esters in Aqueous Solutions Irradiated with a Mercury Lamp. *Bull. Environ. Contam. Toxicol.,* Vol.49, 368-374

Ishikawa, S., Ueda, N., Okumura, U., Iida, Y., Higuchi, M., Naetoko, E., Tokunaga, Y. & Baba, K. (2006). Removal Efficiency for Pesticides on Coagulation and Sedimentation Using Coagulant Recovered from Water Supply Sludge. *J. Japan Soc. Water Environ.,* Vol.29, No.10, 653-658

Japan Industrial Standards Committee (1995). JIS Handbook: Measurement of Environment-1995. Japanese Standards Association, Tokyo, Japan, pp. 973-1188

Matsuda, A., Taketomi, M., Tanizaki, T., Hashimoto, A., Eguchi, M., Suzuki, M., Ishibashi, M., Ichida, K., Nakano, K. & Takagi, S. (2002). Application of Solar Photodecomposition to Water Purification Treatment Using Titanium Dioxides, *Proceedings of JWWA 52th Annual National Conference on Water Works,* pp. 204-205

Matsumura, F. & Murti, C.R.K. (1982). Biodegradation of Pesticides. Plenum Press, NewYork, USA

Nakano, K., Obuchi, E., Takagi, S., Yamamoto, R., Tanizaki, T., Taketomi, M., Eguchi, M., Ichida, K., Suzuki, M. & Hashimoto, A. (2004). Photocatalytic Treatment of Water Containing Dinitrophenol and City Water over TiO_2/SiO_2. *Separation and Purification Tech.,* Vol.34, 67-72

Nakasugi, O. (1993). Pollution and Control of Groundwater for Volatile Organic Chlorine Compounds. Kyushu branch seminar-1992. Japan Society on Water Environment, pp. 6-15

Okumura, Y., Iida, Y., Ichida, K., Kadokami, K., Suzuki, M., Okayama, Y., Haraguchi, K. & Hashimoto, A. (2004). Photodecomposition of Musty Odor Cause Substances Using Titanium Dioxides, *Proceedings of JWWA 55th Annual National Conference on Water Works,* pp. 594-595

Okumura, Y., Taketomi, M., Ichida, K., Kadokami, K., Suzuki, M., Okayama, Y. & Haraguchi, K. (2003). Application of Solar Photodecomposition to Water Purification Treatment Using Titanium Dioxides, *Proceedings of JWWA 54th Annual National Conference on Water Works,* pp. 270-271

Tanada, K., Haraguchi, K. & Takeuchi, Y. (1995). Behaviors of Pesticides in Water Purification Process, *Proceedings of JWWA 46th Annual National Conference on Water Works,* pp. 478-479

Tanizaki, T., Hashimoto, A., Matsuoka, Y., Ishikawa, S., Hanada, Y., Kadokami, K. & Shinohara, R. (2005). Photodegradation of Organic Compounds in Tap Water Using High Reactive Titanium Dioxide. *J. Environ. Chem.,* Vol.15, No.4, 847-853

Tanizaki, T., Kadokami, K., Sugishima, S., Suzuki, M. & Shinohara, R. (1997). Study on Degradation of Trihalomethanes in Water with Photocatalysts Coated with Titanium Dioxide. *J. Environ. Lab. Assoc.,* Vol.22, 104-108

U. S. EPA (1993). Innovative Technology. Annual Status Report EPA 542-R-93-003

Uesugi, Y., Ueji, M. & Koshioka, M. (1997). Pesticide Data Book 3rd Edition. Soft Science Publications, Tokyo, Japan

Yamaguchi, S., Eto, S., Eguchi, M., Kido, K., Hisamatsu, Y. & Nakazawa, H. (1997). Screening Analysis for Multiple Pesticide Residues in Agricultural Products with the GC/MS-SCAN Method. *Bunseki Kagaku,* Vol.46, 905-914

Yamamoto, I. & Fukami, J. (1979). Pesticide Design, Strategy and Tactics. Soft Science Publications, Tokyo, Japan

4

Emulsifying Properties of Hydrolized Sunflower Lecithins by Phospholipases A_2 of Different Sources

D. M. Cabezas[1], R. Madoery[2],
B. W. K. Diehl[3] and M. C. Tomás[1]
[1]*Centro de Investigación y Desarrollo en Criotecnología de Alimentos (CIDCA) – CCT La Plata – CONICET - Facultad de Ciencias Exactas (FCE), Universidad Nacional de La Plata (FCE - UNLP), La Plata*
[2]*Cátedra de Química Orgánica – Facultad de Ciencias Agrarias, Universidad Nacional de Córdoba (FCA - UNC), Córdoba*
[3]*Spectral Service GmbH Laboratorium für Auftragsanalytik, Cologne*
[1,2]*Argentina*
[3]*Germany*

1. Introduction

Lecithins are a mixture of acetone insoluble phospholipids, containing mainly phosphatidylcholine (PC), phosphatidylethanolamine (PE), phosphatidylinositol (PI), minor compounds such as phosphatidic acid (PA), and other minor substances such as carbohydrates and triglycerides (Schneider, 1989). The production of sunflower oil, in Argentina, is of utmost importance from an economic point of view (Franco, 2008). In this country, sunflower lecithin might represent an alternative to soybean lecithin because it is considered a non-GMO product, which is in accordance with the preference of some consumers.

The introduction of changes in the original concentration of these phospholipids, by chemical or enzymatic modification of their structure can lead to obtain lecithins with different physicochemical and functional properties, with respect to native lecithin (van Nieuwenhuyzen & Tomás, 2008). The modification processes usually applied on native lecithins are the fractionation with ethanol (Sosada, 1993; Wu & Wang, 2004; Cabezas et al., 2009a, 2009b) and the enzymatic hydrolysis (Schmitt & Heirman, 2007; Cabezas et al., 2011a). Native and modified lecithins are used in a wide range of industrial applications: nutritional, pharmaceutical applications, food, cosmetics, etc. (Prosise, 1985; Wendel, 2000). In the food industry, lecithin represents a multifunctional additive in the manufacture of chocolate, bakery and instant products, margarines, and mayonnaise, due to the characteristics of its phospholipids (van Nieuwenhuyzen, 1981).

In particular, enzymatic hydrolyzed lecithin may present technological and commercial advantages over native lecithins: (1) enhanced O/W emulsifying property; (2) increased emulsion stability under acid conditions and in the coexistence with salts; (3) improved capability to bind proteins and starch; (4) excellent mold- or pan-releasing property. Consequently, the demand for lysolecithins was increasing in recent years (Hirai et al., 1998; Erickson, 2008).

The main application of lecithin at the food industry is associated with its rol as emulsifier agent for dispersions or emulsions (Hernández & Quezada, 2008). Emulsions are thermodynamically unstable systems from a physicochemical point of view. In virtue of that, it is important to characterize and know their behaviour against different destabilization processes (flocculation, coalescence, creaming, etc.) (McClements, 1999).

Enzymatic hydrolysis is carried out mainly by two groups of enzymes: phospholipases and lipases (Mustranta et al., 1995). Phospholipases A_2 catalyze the hydrolysis of the ester bond in the sn-2 position of glycerophospholipids, producing free fatty acids and the corresponding lysophospholipid. Advances in biotechnology and certain requirements of consumers (*kosher* or *halal* foods) have influenced the development of the production of microbial enzymes (bacteria, fungi, yeasts) which could be substitute of the traditionally obtained from porcine pancreas (Minchiotti, 2006; Cabezas et al., 2011b).

The aim of this work was analyze the emulsifying activity of sunflower lysolecithins obtained by phospholipases A_2 from diverse sources: bacterial (LysoMax PLA_2, Danisco) and porcine pancreas (Lecitase 10L, Novo Nordisk) in O/W systems. This study seeks to contribute to the oil industry with useful information for rescaling of the mentioned hydrolysis process, with the aim of increasing the aggregated value of sunflower lecithins.

2. Materials and methods

2.1 Materials

Native sunflower lecithin was used as starting material, and was provided by a local oil industry (Vicentin S.A.I.C.). Enzymatic hydrolysis processes were carried out using a porcine pancreatic PLA_2 (Lecitase 10L, Novo Nordisk) and a microbial PLA_2 (*Streptomyces violaceoruber*, LysoMax PLA_2, Danisco). All solvents used were of analytical grade.

The sunflower lecithin used as starting material presented the following composition: 43.1% phospholipids (16.5% PI, 16.2% PC, 5.3% PE, and 5.1% minor phospholipids), 33.4% oil, and 23.5% of other compounds (glycolipids, complex carbohydrates).

2.2 Enzymatic hydrolysis process

Enzymatic hydrolysis was carried out in a thermostated reactor at laboratory scale, using 27 g of native sunflower lecithin and 18 ml of 0.4 M $CaCl_2$. Initial pH was adjusted to 7 or 9 by adding 4 N NaOH solution. Then, the resulting mixture was set to the optimal temperature of each phospholipase, i.e. 60 °C for porcine pancreatic PLA_2 and 50 °C for microbial PLA_2, which were incorporated in a concentration of 2.0% ml lipase per 100 g lecithin. Next, continuous agitation (50 rpm) was applied during 5 h. Evolution of hydrolysis process was followed by measuring pH, using a pH meter for solid samples (840049 Puncture Tip, Saen S.R.L.). Products of enzymatic hydrolysis were subjected to a sudden decrease in temperature to stop the process of hydrolysis and then deoiled using acetone, according to AOCS Official Method Ja 4-46, procedures 1-5 (Cabezas et al., 2011a). After that, samples were stored at 0 °C. The hydrolysis process was carried out in duplicate.

Also, native sunflower lecithin was deoiled with acetone obtaining the deoiled sunflower lecithin (DSL). DSL was used as a sample control. Deoiling procedure was performed by duplicate.

2.3 Phospholipid composition

2.3.1 Sample preparation

100 mg of each hydrolyzed sample were diluted in 1 ml of deuterated chloroform, 1 ml of methanol and 1 ml of Cs-EDTA (pH 8). The organic layer was separated after 15 min shaking, and analyzed by ^{31}P NMR (Cabezas et al., 2009a).

2.3.2 Quantitative ^{31}P NMR analysis

Quantitative ^{31}P NMR analysis was carried out in a Bruker Avance 600 MHz automatic spectrometer using triphenyl phosphate as internal standard (Spectral Service GmbH, Köln, Germany) (Diehl, 1997; 2001; 2008). Phospholipid content of samples obtained under different conditions of enzymatic hydrolysis, was determined by this spectroscopic technique.

2.4 Oil-in-water (O/W) emulsions preparation

Commercial sunflower oil was used to prepare oil-in-water (O/W) emulsions with a formulation of 30:70 (wt/wt) according to Pan et al., 2004. Emulsions were prepared at room temperature in an Ultra-Turrax T25 homogenizer using S 25 N-10 G dispersing tool (7.5 mm rotor diameter) at 10,000 rpm for 1 min, with the addition of the different modified sunflower lecithins in a range of 0.1-2.0% (wt/wt). This process was carried out in triplicate for each case.

2.5 Optical characterization of emulsions

The backscattering of light was measured using a QuickScan Vertical Scan Analyzer (Coulter Corp., Miami, FL). The backscattering of monochromatic light (λ = 850 nm) of the emulsions was determined as a function of the height of the sample tube (ca. 65 mm) in order to quantify the rate of the different destabilization processes during the first 90 min. This methodology allowed to discriminate between particle migration (sedimentation, creaming) and particle size variation (flocculation, coalescence) processes (Pan et al., 2002). The basis of the vertical scan analyzer profiles has been exhaustively studied by Mengual (Mengual et al., 1999).

2.6 Particle size measurements

Particle size distribution, and De Brouckere (D[4,3]) and Sauter (D[3,2]) mean diameters of particles of the emulsions were determined with a particle size analyzer (Malvern Mastersizer 2000E, Malvern Instruments Ltd., Worcestershire,U.K.). Samples were diluted in the water bath of the dispersion system (Hydro 2000MU), which is a laser diffraction based particle size analyzer (Márquez & Wagner, 2010). This determination was carried out in triplicate for each case.

2.7 Statistical analysis

Data were evaluated by analysis of variance (ANOVA) using the software Systat® 12.0 (Systat, 2007). For this purpose, differences were considered significant at $p < 0.05$.

	DSL	SHLM9	SHLM7	SHLP9	SHLP7
PC	36.7	7.4	6.7	< 0.1	< 0.1
1-LPC	< 0.1	< 0.1	< 0.1	< 0.1	< 0.1
2-LPC	1.5	39.2	39.4	46.1	45.5
PI	35.3	11.2	12.3	< 0.1	< 0.1
LPI	< 0.1	18.6	15.5	26.5	27.9
PE	15.1	3.7	6.4	1.0	0.7
LPE	< 0.1	8.8	9.0	10.7	13.0
APE	1.8	< 0.1	< 0.1	< 0.1	< 0.1
PA	5.2	3.1	3.4	2.5	2.0
LPA	< 0.1	1.0	1.0	5.1	4.1
Other	4.5	7.1	6.4	8.0	6.7
PL / 100g lecithin	62.8	40.6	45.9	38.7	44.4

Table 1. Phospholipid (PL) composition of modified sunflower lecithins by ^{31}PNMR[a]

3. Results and discussion

3.1 Compositional analysis of modified lecithins

The phospholipid composition of different modified sunflower lecithins obtained in this chapter is shown in Table 1. The sunflower hydrolyzed lecithins with a porcine pancreatic (SHLP) and microbial (SHLM) phospholipase A_2, at different levels of pH (7, 9), recorded a marked difference regarding the phospholipid composition in relation to the native sunflower lecithin (DSL). The hydrolyzed lecithins presented a high concentration of major lysophospholipids (> 64.9 mol LPL/ mol of total PL) compared to the native sunflower lecithin (≈ 1.5%), showing the efficiency of the enzymatic hydrolysis processes. In particular, the pancreatic PLA_2 (SHLP7, SHLP9) produced a higher hydrolysis degree of the main phospholipids in comparison with the microbial phospholipase (SHLM7, SHLM9).

PC presented a high degree of hydrolysis in all performed conditions on the hydrolysis processes. These results can be correlated with the ones described by Penci, 2010. In that work, it was reported that phosphatidylcholine is the phospholipid with higher tendency to be hydrolyzed when using a porcine pancreatic PLA_2 with a very low amount of sunflower lecithins (1 mg). In this way, the residual PC and PI concentration in the hydrolyzed lecithin was lower than the detection limit of the ^{31}P NMR equipment when the pancreatic porcine PLA_2 was used.

3.2 Optical characterization of O/W emulsions

Stability of the different O/W emulsions (30:70 wt/wt) was studied recording the backscattering (BS) profiles as a function of the cell length and time, by a vertical scan analyzer (QuickScan). For instance, Figure 1 shows two typical profiles obtained for emulsions with the addition of 0.1% of DSL and SHLP9.

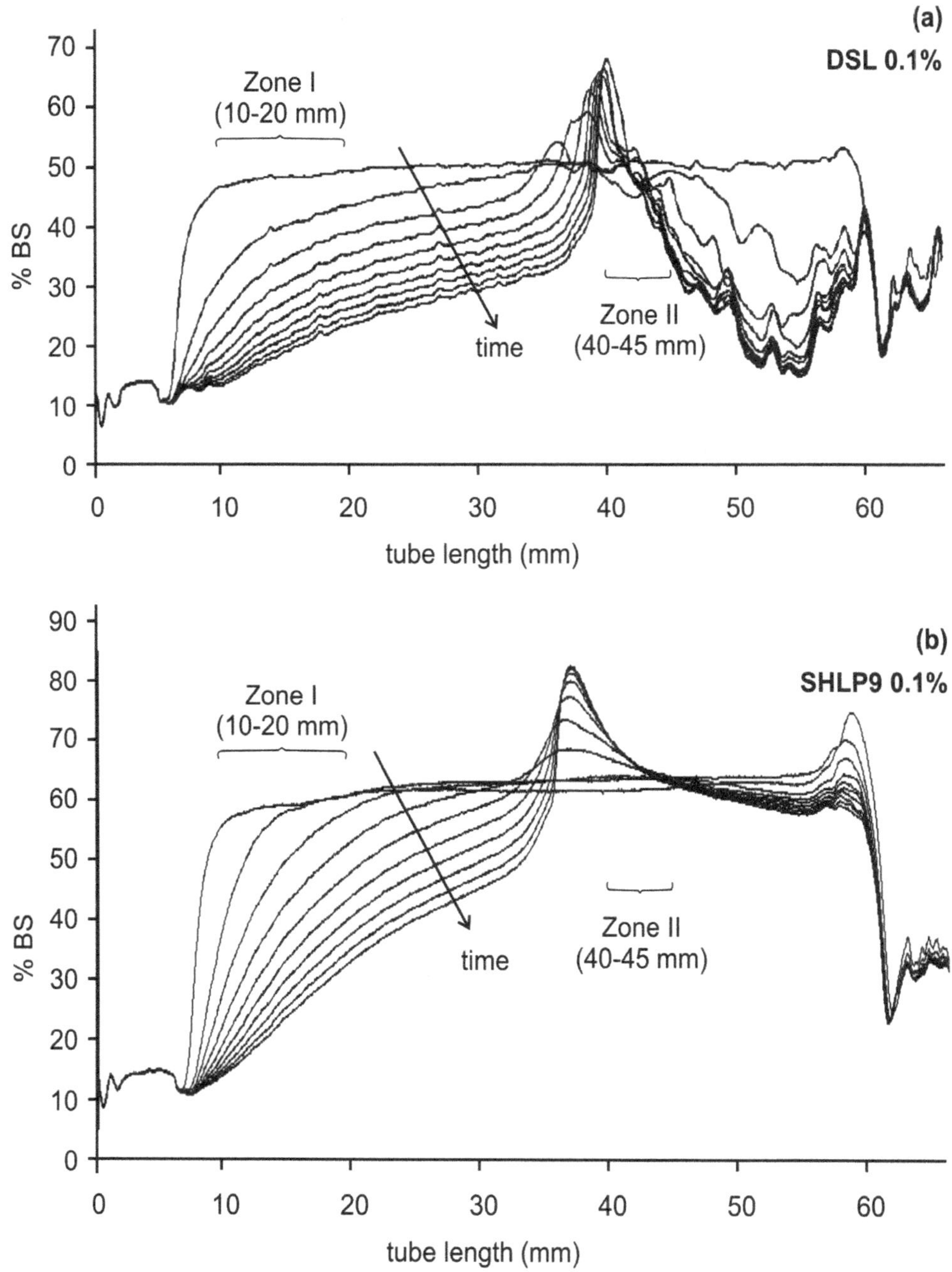

[a] Mean values are shown (n = 3). The coefficient of variation was lower than 5%

Fig. 1. Backscattering (%BS) profiles of O/W emulsions (30:70 wt/wt) with the addition of: (a) DSL, 0.1%; (b) SHLP, 0.1%

The creaming destabilization process (i.e. migration of oil particles to the upper portion of the tube) is evidenced by a decrease of %BS values at the bottom of the tube. The QuickScan profiles corresponding to the zone I (10-20 mm) showed an increase of the emulsion stability against the creaming process, as a function of increasing concentration of different modified lecithins (Fig. 2). In particular, the hydrolyzed lecithins (SHLP and SHLM) generated a high stability in O/W emulsions than DSL, over the studied range of concentration. Moreover, O/W emulsions with 0.1-0.5% of DSL showed a sharp decrease of %BS in the Zone I.
The tube zone between 40-45 mm (Zone II) is characterized by the accumulation of oil droplets after the creaming process (cream phase); Figure 3 shows the %BS values vs. time in Zone II. Emulsions formulated with hydrolyzed lecithins presented higher %BS values than those obtained using DSL, for all concentrations studied. The higher levels of %BS and the greater stability of these emulsions would be associated with the formation of dense cream phases with a lower proportion of continuous phase inside (Palazolo, 2006). However, emulsions with 0.1 -0.5% of DSL did not allow the formation of the cream phase. These results are related to the rapid decrease of %BS and the formation of an oil layer in the upper part of the tube (Fig. 1a) suggesting the occurrence of a cream phase destabilization by coalescence (Pan et al., 2002).

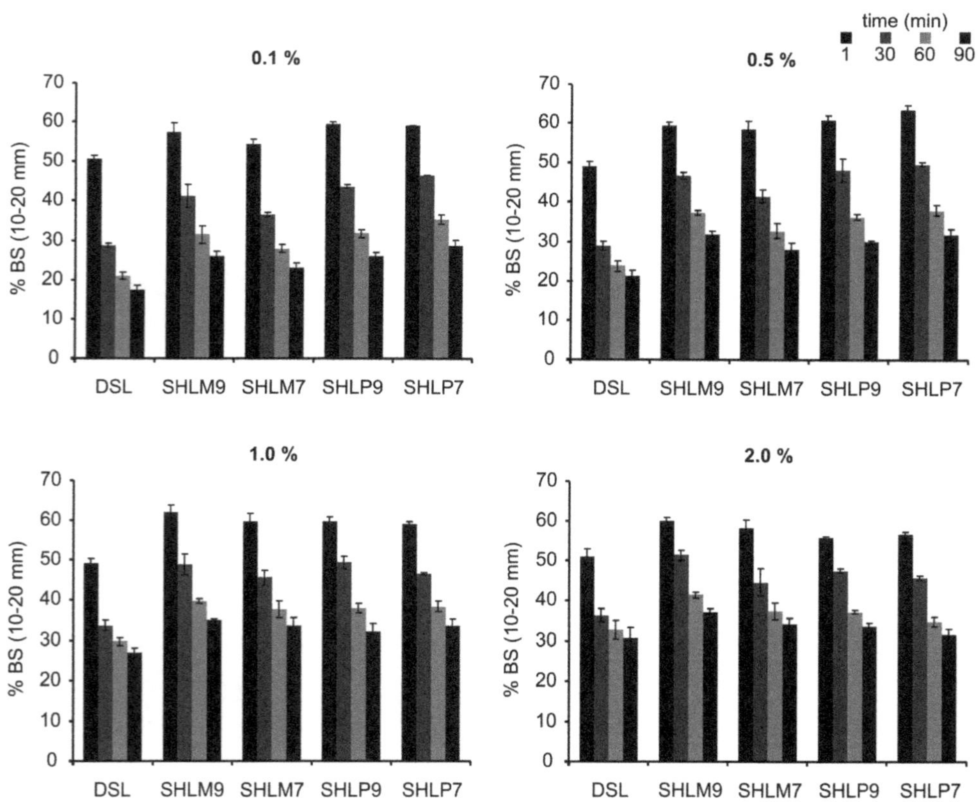

Fig. 2. Backscattering (%BS) values of O/W emulsions (30:70 wt/wt) with the addition of different modified sunflower lecithins in Zone I (10-20 mm). Mean values (n = 3) ± sd

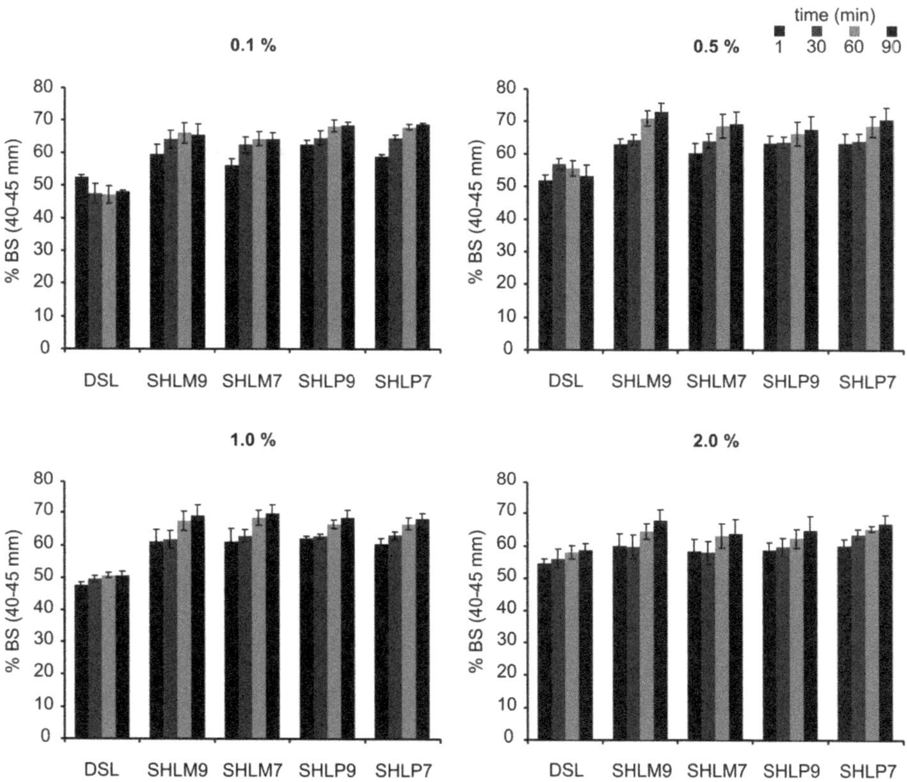

Fig. 3. Backscattering (%BS) values of O/W emulsions (30:70 wt/wt) with the addition of different modified sunflower lecithins in Zone II (40-45 mm). Mean values (n = 3) ± sd

3.3 Particle size distribution

Particle size distribution in volume and surface of O/W emulsions obtained with different modified lecithins was measured just after emulsification (t=0); corresponding results can be seen in Figures 4 and 5, respectively. These distributions presented a bimodal or trimodal character depending on the concentration of aggregated lecithin, and the following particle size populations: (I) particle size < 4 μm; (II) particle size between 4 and 30 μm; (III) particle size > 30 μm. In this sense, only SHLP7 showed a trimodal character for all concentrations assayed. It should be noted that the SHLM7 presented a bimodal character for concentration in the range 0.1-0.5%, but with a high percentage of particles of the population II in comparison with DSL.

In order to complete the analysis of particle size distribution, Figure 6 depicts the evolution of De Brouckere (D [4,3]) and Sauter (D [3,2]) mean diameters as a function of the concentration of the different emulsifiers. Hydrolyzed lecithins generated values of D [4,3] and D [3,2] significantly lower than those corresponding to DSL. These results are correlated with the high stability of the O/W emulsions recorded when using hydrolyzed lecithins, considering the main destabilization processes determined by the corresponding QuickScan profiles (creaming or coalescence). It is worth to note that a high concentration of small particles produces a slow creaming process, according to the Stokes' law (McClements, 1999; Palazolo, 2006).

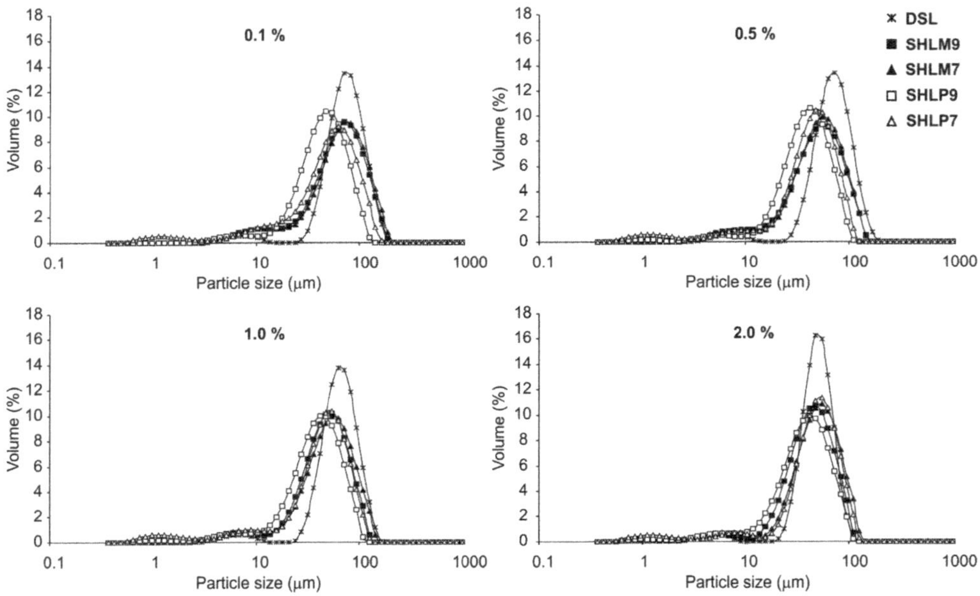

Fig. 4. Volume particle size distribution for O/W emulsions with the addition of different modified sunflower lecithins. Mean values (n = 3)

The hydrophilic-lipophilic balance value (HLB) is often used in connection with the performance of emulsifiers (McClements, 1999). The high concentration of hydrophilic phospholipids (lysophospholipids) presented in the hydrolyzed lecithins (SHLP and SHLM) increase this empirical value. In this sense, according to Carlsson (Carlsson, 2008), these modified lecithins with higher HLB values presented best properties as O/W emulsifying agents.

Also, the phase structure at the interface of the different phospholipids influences the emulsion formation and stability (van Nieuwenhuyzen & Tomás, 2008). LPC and LPE form hexagonal wide spread clusters. These structures have a great importance for the stabilisation of O/W emulsions. This behaviour is in relation to the low mean diameters and the high concentration of small particle populations recorded in emulsions using sunflower hydrolyzed lecithins (Figs. 4-6). However, PE gives reversed hexagonal phase, which are more difficult to arrange at the interface (van Nieuwenhuyzen, 1998). The presence of PE could explain the minor characteristics as emulsifying agent of DSL, and the high mean diameters when was using SHLM in contrast to when using SHLP (Fig. 6).

Taking into account the results presented in Figures 2 to 6, the addition of a concentration between 0.5 and 1.0% of hydrolyzed lecithin (SHLP, SHLM) is enough for covering all droplets surface. High concentrations of this modified lecithin do not show significant differences in the % BS values, nor in the mean particle sizes. On the other hand, concentration levels higher than 0.1% of hydrolyzed lecithins from different phospholipases at different initial pH levels showed similar characteristics in terms of their emulsifying activity. However, DSL presented an improved in the stability of O/W emulsions as a function of increasing concentration.

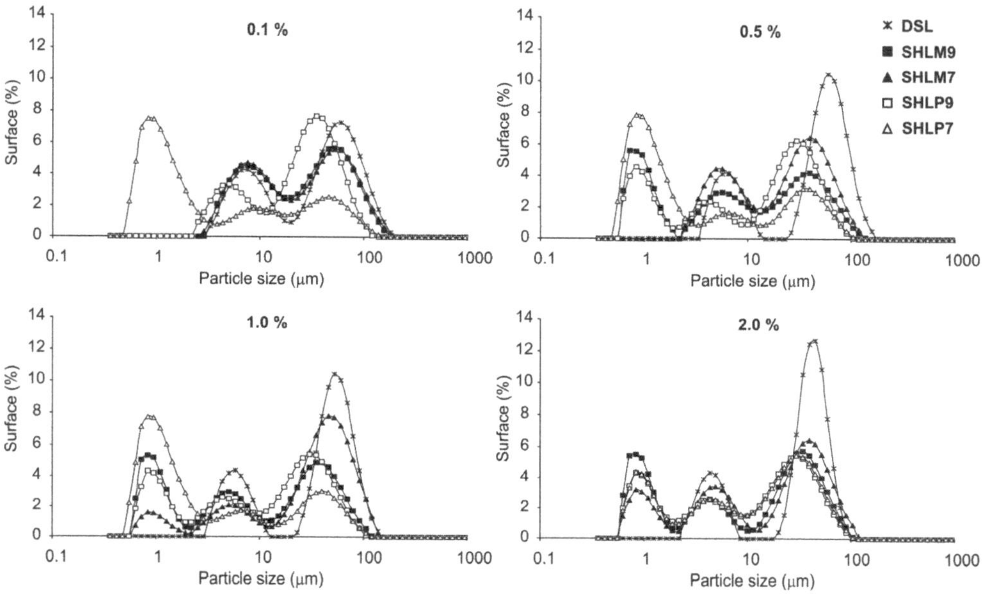

Fig. 5. Surface particle size distribution for O/W emulsions with the addition of different modified sunflower lecithins. Mean values (n = 3)

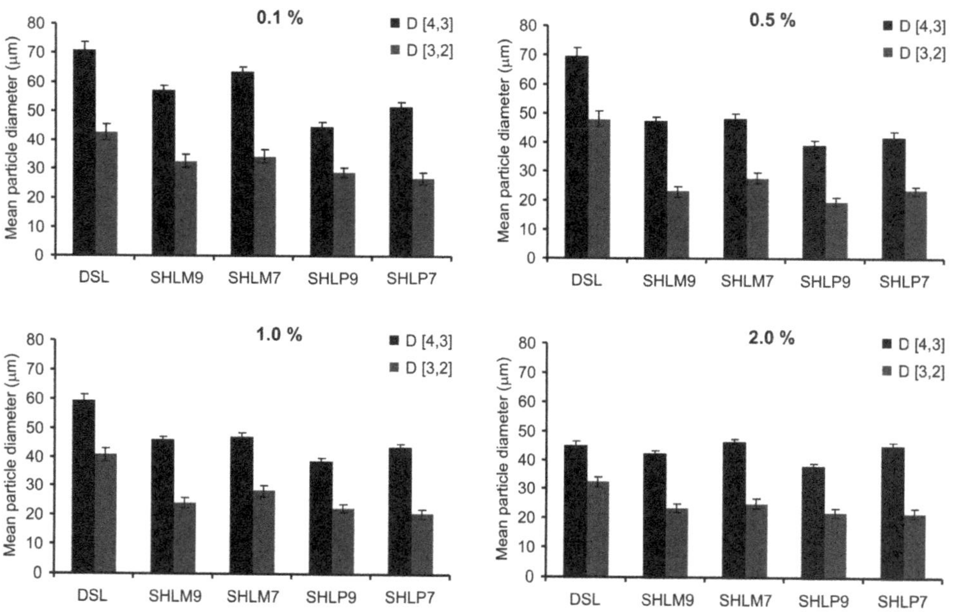

Fig. 6. De Brouckere (D[4,3]) and Sauter (D[3,2]) mean diameters for O/W emulsions with the addition of different modified sunflower lecithins. Mean values (n = 3) ± sd

4. Conclusions

The best emulsifying properties exhibited by hydrolyzed sunflower lecithins (LHM7, LHP7, LHM9, LHP9) in comparison with the native lecithin could indicate that their application by the local oil industry could revalue this by-product of the refining process of crude sunflower oil.

The use of a microbial phospholipase A_2 gives the possibility to obtain diverse sunflower lysolecithins, which functionality could be applied to the development of foods with *kosher* and *halal* certification.

5. Acknowledgments

This work was supported by grants from Agencia Nacional de Promoción Científica y Tecnológica (ANPCyT), Argentina, PICT 1085 (ANPCyT); Consejo Nacional de Investigaciones Científicas y Técnicas (CONICET), Argentina, PIP 1735 (CONICET); and Universidad Nacional de La Plata (UNLP), Argentina, 11/X502 (UNLP).

M. C. Tomás is a member of the Career of Scientific and Technological Researcher of Consejo Nacional de Investigaciones Científicas y Técnicas (CONICET), Argentina. D. M. Cabezas is a recipient of a postdoctoral fellowship from Consejo Nacional de Investigaciones Científicas y Técnicas (CONICET). R. Madoery is a Scientific and Technological Researcher, and Professor at Universidad Nacional de Córdoba (UNC). B. W. K. Diehl is Director of Spectral Service GmbH, Cologne, Germany.

Sunflower lecithin and microbial PLA_2 were provided by Vicentin S.A.I.C. and Danisco, respectively. Thorsten Buchen is acknowledged for his technical assistance.

6. References

Cabezas, D.M., Diehl, B. W. K. & Tomás M.C. (2009a). Effect of processing parameters on sunflower PC enriched fractions extracted with aqueous-ethanol. *European Journal of Lipid Science and Technology,* No 111, pp. 993-1002

Cabezas, D.M., Diehl, B. W. K. & Tomás M.C. (2009b). Sunflower lecithin: Application of a fractionation process with absolute ethanol. *Journal of the American Oil Chemists' Society,* No 86, pp. 189–196.

Cabezas, D.M., Madoery, R., Diehl, B.W.K. & Tomás, M.C. (2011a). Application of enzymatic hydrolysis on sunflower lecithin using a pancreatic PLA_2. *Journal of the American Oil Chemists' Society,* No 88, pp. 443-446

Cabezas, D.M., Madoery, R., Diehl, B.W.K. & Tomás, M.C. (2011b). Enzymatic hydrolysis of sunflower lecithins using a microbial PLA_2. In: Hughes VC (ed) Sunflowers: cultivation, nutrition, and biodiesel uses. Nova Science Publishers, New York (USA), Chapter 9, pp. 303-316

Carlsson, A. (2008). Physical properties of phospholipids. In: Gunstone FD (ed) Phospholipid technology and applications. The Oily Press, Bridgwater (UK), Chapter 5, pp. 104

Diehl, B. W. K. (1997). Multinuclear high-resolution magnetic resonance spectroscopy. In: Lipid Analysis in Oils and Fats. Ed. R. J. Hamilton, Blackie Academic & Professional, London (UK), pp. 87–135

Diehl, B. W. K. (2001). High resolution NMR spectroscopy. *European Journal of Lipid Science and Technology*, No 103, pp. 830–834

Diehl, B. W. K. (2008). NMR spectroscopy of natural substances. In: Holzgrabe U, Wawer I, Diehl B (eds). NMR spectroscopy in pharmaceutical analysis, Chapter 5. Elsevier, Jordan Hill (UK), pp. 194–196

Erickson, M.C. (2008). Chemistry and function of phospholipids. In: Akoh CC, Min DB (eds) Food lipids: chemistry, nutrition and biotechnology, Chapter 2, 3rd edn. CRC Press, Boca Raton, pp. 39–62

Franco, D. (2008). Informe de Cadenas Alimentarias: Sector Oleaginoso. Ministerio de Economía y Producción Argentina. http://www.alimentosargentinos.gov.ar/0-3/revistas/r_41/cadenas/Oleaginosos_Oleaginosas.htm (online)

Hernández, E. & Quezada, N. (2008). Uses of phospholipids as functional ingredients. In: F.D. Gunstone (Ed.), Phospholipid Technology and Applications, Chapter 4, Bridgwater, The Oily Press

Hirai, H., Sono, R. & Koh, H. (1998). Process for manufacturing vegetable lysolecithins. European Patent EP0870840

Márquez, A.L. & Wagner, J.R. (2010). Rheology of double (W/O/W) emulsions prepared with soybean milk and fortified with calcium. *Journal of Texture Studies*, No 41, pp. 651-671

McClements, D.J. (1999). Food emulsions: principles, practice and techniques. CRC Press, Florida (USA)

Mengual, O., Meunier, G., Cayré, I., Puech, K. & Snabre, P. (1999). Turbiscan MA 2000: multiple light scattering measurement for concentrated emulsion and suspension instability analysis. *Talanta*, No 50, pp. 445-456

Minchiotti, M.C. (2006). Purificación, caracterización y estudio de propiedades de fosfolipasa A2 presente en soja (Glycine max). Doctoral Thesis. Facultad de Ciencias Agropecuarias, Universidad Nacional de Córdoba (UNC), Argentina

Mustranta, A., Forssell, P. & Poutanen, K. (1995). Comparison of lipases and phospholipases in the hydrolysis of phospholipids. *Process Biochemistry*, No 30, pp. 393–401

Palazolo, G.G. (2006). Formación y estabilidad de emulsiones O/W preparadas con proteínas nativas y desnaturalizadas de soja. Doctoral Thesis, Universidad Nacional de La Plata, Argentina

Pan, L.G., Tomás, M.C. & Añón, M.C. (2002). Effect of sunflower lecithins on the stability of water in Oil (W/O) and Oil in Water (O/W) emulsions. *Journal of Surfactants and Detergents*, No 5, pp. 135 - 143

Pan, L.G., Tomás, M.C. & Añón, M.C. (2004). Oil-in-Water emulsions formulated with sunflower lecithins: vesicle formation and stability. *Journal of the American Oil Chemists' Society*, No 81, pp. 241-244

Penci, M.C., Constenla, D.T. & Carelli, A.A. (2010). Free-fatty acid profile obtained by enzymatic solvent-free hydrolysis of sunflower and soybean lecithins. *Food Chemistry*, No 120, pp. 332-338

Prosise, W.E (1985). Commercial lecithin products: Food use of soybean lecithin. In: Lecithins. Eds. B. F. Szuhaj, G. List, AOCS Press, Champaign (USA), pp. 163–181

Schmitt, H. & Heirman, M. (2007). Enzymatic modification of lecithin. US Patent US7189544B2

Schneider, M. (1989). Fractionation and Purification of Lecithin. In Lecithins: Sources, Manufacture, & Uses. Ed. B.F. Szuhaj, AOCS Press, Champaign (USA), pp. 109-130

Sosada, M. (1993). Optimal conditions for fractionation of rapeseed lecithin with alcohols. *Journal of the American Oil Chemists' Society*, No. 70, pp. 405–410

Systat, Version 12 (2007). Systat Software Inc.. San José, California (USA).

van Nieuwenhuyzen W. and Tomás, M. C. (2008). Update on vegetable lecithin and phospholipid technologies. *European Journal of Lipid Science and Technology,* No 110, pp. 472-486

van Nieuwenhuyzen, W. (1981). The industrial uses of special lecithins. *Journal of the American Oil Chemists' Society,* No 58, pp. 886–888

van Nieuwenhuyzen, W. (1998). Effects of lecithins and proteins on the stability of emulsions. *Fett/Lipid,* No 100, pp. 282-291

Wendel, A. (2000). Lecithin: The first 150 years. Part I: From discovery to early commercialization. *Inform,* No 11, pp. 885–892

Wu Y. & Wang T. (2004). Fractionation of crude soybean lecithin with aqueous ethanol. *Journal of the American Oil Chemists' Society,* No 81, pp. 697–704

5

Acid-Induced Aggregation and Gelation of Bovine Sodium Caseinate-Carboxymethylcellulose Mixtures

María Eugenia Hidalgo[1], Bibiana D. Riquelme[1,2], Estela M. Alvarez[1], Jorge R. Wagner[3] and Patricia H. Risso[1,2,4]
[1]*Facultad de Ciencias Bioquímicas y Farmacéuticas, Universidad Nacional de Rosario*
[2]*Instituto de Física Rosario (IFIR), CONICET-UNR, (2000), Rosario,*
[3]*Departamento de Ciencia y Tecnología,*
[4]*Facultad de Ciencias Veterinarias,Universidad Nacional de Rosario, Rosario*
Universidad Nacional de Quilmes, Buenos Aires,
Argentina

1. Introduction

The main protein fraction in bovine and ovine milk is represented by caseins (76-83% of total proteins). Caseins (CN) occur in milk as stable colloidal aggregates known as casein micelles, mainly composed by α_{S1}-, α_{S2}-, β- and κ-CN (Walstra et al., 1984). Among different types of CN, there are some important characteristics that make the difference between them, based on their charge distribution and their sensitivity to be precipitated by Ca^{2+}. κ-CN fraction, insensitive to Ca^{2+}, acts as protection that attempts to prevent other CN from Ca^{2+}-induced precipitation (Qi et al., 2001). From a nutritional point of view, caseins have all the essential aminoacids and play an important role in calcium and phosphate transport, representing an easily digestible source of nutrients, contributing to a carefully balanced diet (Linde, 1982). CN and their derived salts, the caseinates, are extensively used in food industry because of their physicochemical, nutritional and functional properties that make them valuable ingredients in complex food preparations.

Caseinates (CAS) are prepared by acid precipitation of milk casein at its isoelectric point (pH 4.6) and resolubilized by increasing the pH. If the increase in the pH is carried out by the addition of NaOH, it is possible to end up obtaining sodium caseinate (NaCAS), a more soluble form of CN. In these conditions, the micellar structure is destroyed and the NaCAS form aggregates or sub-micelles due to the high proportion of hydrophobic amino acid side chains that self-associate in aqueous solutions (Farrell et al., 1990). Further association of submicelles to form the large casein micelles present in milk is prevented by the removal of most of the calcium (Oakenfull et al., 1999). NaCAS is commonly employed as additive in a great variety of food products because of its high emulsifying, water-binding and gelation capabilities, its heat stability and its contribution to the food texture and juiciness. Water-holding capacity and gelling properties are used to improve rheological properties, texture, stability, and appearance of many food products such as processed meats, surimi, cheese, yogurt and confectionary products (Corzo-Martínez et al., 2010). Some of these properties make caseinates useful and desirable ingredients in the preparation of bakery and confectionery products, where they can be used as milk substitutes (Gaucheron, 1997).

Dissociation and a further aggregation step of casein fractions due to caseinate acidification results in the formation of a gel structure. A possible explanation to this effect is that as the pH is adjusted toward the isoelectric point it causes a decrease of the repulsive interactions, resulting in a destabilization of the colloidal aggregates as the pH drops slightly below 5 at a given temperature (Braga et al., 2006; Ruis et al., 2007). Nowadays, a process that has gained the attention of food industry is direct acidification by the addition of a lactone, such as glucono-δ-lactone (GDL), which slowly hydrolyzes to gluconic acid with a resulting reduction in pH. GDL allows us to overcome some of the difficulties associated with the traditional process of using bacteria. In fact, the final pH of the system is a function of the amount of GDL added, whereas starter bacteria produce acid until they inhibit their own growth as the pH becomes lower (de Kruif, 1997; Lucey et al., 1998). Gels made with the two types of acidifying precursors, bacterial cultures (lactic fermentation) or the addition of GDL, differ in their rheological properties partly as a function of the velocity of acidification. In GDL gels, the isoelectric point (pH 4.6) can be reached faster and remains stable, thus allowing longer aging near this point. This phenomenon contributes to the continuous fusion and rearrangement of casein particles (Ribeiro et al., 2004). Acid gel formation of NaCAS dispersions has been examined leading to quantitative structural information for testing ideas about the fractal properties of casein gels (Bremer et al., 1993).

The effect of different processing parameters (heat treatment, temperature and pH conditions), the presence of other ingredients or the GDL concentration on the microstructure of acid gels has been investigated (Belyakova et al., 2003; Braga et al., 2006; Lucey et al., 2001; Nespolo et al., 2010; Perrechil et al., 2009). Particularly, protein/polysaccharide/water mixtures are frequently used in the food industry as thickening agents for low or zero fat products (Semenova et al., 2009). However, the interaction of proteins with polysaccharides in solution could influence in a positive or negative way, depending on the colloidal system in question, the functionality of the protein and, therefore, the food properties, due to the balance of the protein–protein and protein–solvent interactions. The rheological and structural properties of protein–polysaccharide gels depend on biopolymer interactions that can be influenced by the concentration and molecular structure of biopolymers. Three different systems can result from the mixture of proteins and polysaccharides in aqueous solution: a) stable homogeneous solutions; b) associative phase separation or coacervation, in which both components are concentrated in the same phase due to the formation of a complex; c) segregative phase separation, where the two components are in different phases due to the limited thermodynamic compatibility (Tolstoguzov, 1991).

In the case of coacervation, phase more concentrated in colloid component is the coacervate and the other phase is the equilibrium solution. Associative phase separation of two polymers in water occurs if there is an electrostatic attraction. Complex coacervation is caused by the interaction of two oppositely charged colloids (de Kruif et al., 2004).

When both polymers have the same charge, repulsive interactions lead to incompatibility between proteins and polysaccharides as a result of differences in their molecular properties, such as shape, size or charge and may cause phase separation. In the case of gelation of the proteins and polysaccharides, the balance between phase separation and gelation process determines the micro-structure and the mechanical properties of gels (de

Jong et al., 2009). The relative concentration of a biopolymer mixture is critical for the gelling process. Increasing the macromolecule concentration can improve the gelation process, since the macromolecules become closer to each other, facilitating aggregate formation and contributing to the strengthening of the structure (Picone & da Cunha, 2010; Yamamoto & Cunha, 2007). However, above a critical value, thermodynamic incompatibility takes place and phase separation is observed (de Jong & van de Velde, 2007).

Carboxymethylcellulose (CMC) is an anionic linear polysaccharide that comes from cellulose, and has been widely used as a stabilizer in food products, for example in acidified milk drinks (Du et al., 2009). CMC is generally used in aqueous solutions, where useful characteristics such as high viscosity at low concentrations, defoaming, surfactant, and bulking abilities are applicable (de Britto & Assis, 2009). Delhen and Stefancich reported the formation of soluble complexes at low pH values in β-casein-CMC systems. The backbone of CMC seems to be too rigid for interacting appreciably with proteins. However, lowering the pH, which reduces the free charges on the polymer backbone and hence the stiffness of the macromolecule, enables CMC to interact with the protein (Delben & Stefancich, 1997).

Yu et al. informed that the addition of CMC to calcium caseinate, enhance its aggregation and seem to prevent protein precipitation during storage (Yu et al., 2004). Du et al., investigated the interaction between CMC and casein micelles and the influence on the stability of acidified milk drinks (Du et al., 2007; Du et al., 2009). They found that at pH 6.7, there was no interaction between caseins and CMC due to charge repulsion and mixtures of casein and CMC were stable at low CMC concentrations. Above a certain CMC concentration, depletion flocculation occurred leading to phase separation. Electrosorption of CMC onto casein micelles took place below pH 5.2 and the adsorbed CMC layer on the surface of casein could prevent flocculation of casein micelles by steric repulsion. In addition, the non-adsorbed CMC increased the viscosity of serum and slowed down the sedimentation of casein particles. In the case of low CMC concentrations, CMC/casein micelles mixture was phase separated via bridging flocculation. With increasing CMC concentrations, the casein micelles were effectively coated and consequently electrostatic and sterically stabilized.

In previous work, we found that the compactness and average size of the aggregates formed at the end of the acidification process of ovine caseinate depend on the kinetics of the aggregation phenomena. As the aggregation process becomes slower, the more easily a polypeptide chain could acquire different orientations, leading to the formation of a more compact aggregates and gels with more elasticity and hardness (Nespolo et al., 2010). Therefore, given that CMC affects the stability of colloidal particles in solution such as NaCAS particles, this polysaccharide can affect the kinetics of acid aggregation and gelation processes, and thus control the microstructure of the aggregates and gels formed. The aim of this work was to investigate conformational, aggregating and gelling behaviours of NaCAS aqueous solutions in the presence of different concentrations of CMC.

2. Materials and methods

2.1 Materials

Bovine sodium caseinate powder, CMC low viscosity (50-200 cP, 4 % in H_2O, 25 °C), GDL, tris(hydroxymethyl)aminomethane (Tris), 8-anilino-1-naphthalenesulfonate (ANS) as an ammonium salt, and sodium azide were purchased from Sigma-Aldrich Co. (Steinheim, Germany). HCl, and NaOH were provided by Cicarelli SRL (San Lorenzo, Argentina). The CMC stock solutions were also prepared in water and stored at 4 °C.

NaCAS suspensions were prepared from dissolution of commercial drug in distilled water (isoionic pH) at room temperature. After concentration measurements, 0.15 $g.L^{-1}$ sodium azide was added as a bacteriostatic agent, and the solutions were stored at 4 °C. Protein concentration was determined by the Kuaye's method which is based on the ability of strong alkaline solutions to shift the spectrum of the amino acid tyrosine to higher wavelength values in the UV region (Kuaye, 1994).
Stock solutions nearly 6 mM for ANS were prepared in distilled water and stored in the dark at 4 °C; the concentration was determined by absorbance measurements, using a molar extinction coefficient (ε) of 4,950 M^{-1} cm^{-1} at 350 nm.

2.2 NaCAS–CMC phase diagram

Phase diagram was established at pH 6.8 (buffer Tris-HCl 10 mM) and 35 °C, and it was constructed by determining the transition from single to two-phase systems. The thermodynamic compatibility study was carried out using the method proposed by Spyropoulos et al. They propose to carefully prepare a series of polysaccharide/protein aqueous solutions to give binary systems and incubate them a certain time at a given temperature, and then evaluate a single or two-phase systems formation (Spyropoulos et al., 2010).
These binary systems were prepared with the same CMC concentration but with NaCAS concentrations ranging from 0 to 4 wt% in one case, and with the same protein concentration but with polysaccharide concentrations ranging from 0 to 4.5 wt% in the other. A total of three samples were taken from each of these binary solutions and kept in sealed tubes at 35 °C for at least 24 h, after which the occurrence of phase separation (or not) was verified by visual inspection.

2.3 Thermal stability of NACAS in the presence or absence of CMC

The effects of thermal treatments on NaCAS, CMC and their mixtures were monitored through spectrophotometry with the aim of evaluating the biopolymer aggregation by heating at different temperatures. Measurements at increasing temperature were made at 650 nm from 10 to 100 °C with a heating rate of 0.5 °C per minute. The equipment used was a Jasco V-550 doublebean spectrophotometer equipped with a cell holder heated by Peltier effect and controlled by a programmable unit. The cell was filled with a 0.02 wt% NaCAS solution in buffer Tris-HCl 10mM pH 6.8 up to a final volume of 2.5 mL and sealed with a teflon stopper to avoid evaporation during each experiment. The spectrophotometer compartment was continuously purged with nitrogen to prevent the condensation of water vapor on the cell walls. This method was also performed in the presence of CMC at NaCAS:CMC proportion of 8:1, 4:1, 2:1, 1:1 and 1:1.5.

2.4 Spectrofluorimetric determinations

Fluorescence excitation and emission spectra of the NaCAS (0.1 wt%) were obtained using a spectrofluorometer Aminco Bowman Series 2. Measurements were carried out in the presence and absence of CMC, in order to detect any spectral shifts and/or changes in the relative intensity of fluorescence (FI). Previously, the excitation wavelength (λ_{ex}) and the range of concentration with a non significant internal filter effect were determined. The samples (3 mL) used for the spectral analysis and FI measures were transferred into a fluorescence cell with a light path length of 1 cm and placed into a cell holder keeping the temperature constant at the fixed values desired. Values of FI were registered within the range of 300-400 nm at 35°C using a λ_{ex} of 286 nm.

The surface hydrophobicity (S_0) was estimated according to the method of Kato and Nakai (Haskard & Li-Chan, 1998; Kato & Nakai, 1980), using ANS as an hydrophobic fluorescent marker. The measurements were carried out in an Aminco Bowman Series 2 spectrofluorometer, using an λ_{ex} of 396 nm and an emission wavelength (λ_{em}) of 489 nm, previously determined from excitation and emission spectra of protein-ANS complex, at a constant temperature of 35°C. The FI was measured in samples containing ANS 0.04 mM and consecutive aggregates of 0.1 wt% NaCAS with or without CMC (FI_b). The FI was also determined in samples containing only protein (i.e. without the addition of the fluorescent probe) in the presence or absence of CMC at the same concentrations (FI_p). The difference between FI_b and FI_p (ΔF) was calculated and S_0 was determined as the initial slope in the ΔF vs. NaCAS concentration (wt%) curve.

2.5 Acid aggregation

Kinetics of NaCAS aggregation induced by the acidification with GDL, in the presence or absence of CMC, was analyzed by measuring turbidity (τ) in the range of 450 to 650 nm, in a Spekol 1200 spectrophotometer with a diode arrangement and a thermostatized cell. The amount of GDL added was calculated using the following relation:

$$R = \frac{\text{wt\% GDL}}{\text{wt\% NaCAS}} \qquad (1)$$

R used for this experiment was 0.5, at a temperature of 35 °C. Acidification was initiated by the addition of solid GDL to 10 g of NaCAS suspension (0.5 wt%).

2.6 Changes in size and compaction of particles

Changes in the protein average size were followed by the dependence of τ on wavelength (λ) of the suspensions, determined according to:

$$\beta = 4.2 + \frac{d(\log\tau)}{d(\log\lambda)} \qquad (2)$$

β is a parameter that has a direct relationship with the average size of the particles, can be used to easily detect and follow rapid size changes, and was obtained from the slope of log τ versus log λ plots, in the 450 to 650 nm range, where the absorption due to the protein chromophores is negligible allowing the estimation of τ as absorbance (Camerini-Otero & Day, 1978; Risso et al., 2007). Absorption spectra and absorbance at 650 nm (A_{650}) were registered as a function of time until a maximum and constant value of A_{650} was reached; simultaneously the pH decrease was measured. The measurements of pH were carried out on digital pH meter Orion SA 720, equipped with proton-selective glass membrane electrode combined with saturated calomel reference electrode. On the other hand, it has been shown that β, for a system of aggregating particles of the characteristics of caseinates, tends, upon aggregation, toward an asymptotic value that can be considered as a fractal dimension (D_f) of the aggregates (Horne, 1987; Risso et al., 2007).

To verify if β was actually related to the average size of the particles, the size distribution functions and the hydrodynamic diameters of NaCAS particles were determined by dynamic light scattering (DLS) using a Brookhaven 9863 Model equipment with a He-Ne laser (λ_0 = 632.8 nm) with a maximal power of 15 mV, and using 90° as the measuring angle. Hydrodynamic diameters were calculated using the BI9000AT 6,5 Version software processing. To carry out this determination an amount of solid GDL was added to 8 mL of a 0.5 wt% NaCAS solution in order to obtain a GDL/protein relation of 0.5. Measurements at different times were performed until the maximum of τ allowed by the instrument was reached while pH was simultaneously monitored.

2.6.1 Effect of CMC on the viscosity of media

The aggregation process is limited by diffusion, which depends on the medium viscosity (η). Therefore, is important to determine the effect that the presence of the polysaccharide exerts on that property. The η was measured in triplicate, using a rotational viscosimeter Brookfield LV Master (LVDV-III) with cone/plate geometry and thermostatically controlled at a temperature of 35.00 ± 0.05°C. The relative viscosity (η_r) was calculated as:

$$\eta_r = \eta_{sol}/\eta_0 \tag{3}$$

where η_{sol} is the solution viscosity and η_0 is the water ones.

2.7 Acid gelation

Above a certain protein concentration, the loss of electrostatic stability by acidification results in the formation of a three-dimensional gel network. Effect of CMC concentration on the kinetic of gelation, rheological properties and microstructure of gels were investigated.

2.7.1 Rheological properties of acid gels

Rheological properties of NaCAS samples (3 wt%), in the absence or presence of CMC, were determined in a stress and strain controlled rheometer AR G2 model using a cone geometry (diameter: 40 mm, cone angle: 2°, cone truncation: 55 mm) and a system of temperature control with a recirculating bath (Julabo model ACW 100) connected to a Peltier plate. An amount of solid GDL according to a certain R was added to initiate the acid gelation.

Measurements were performed each 20 sec with a constant oscillation stress of 0.1 Pa and a frequency of 0.1 Hz. The Lissajous figures at various times were plotted to ensure that the measurements of storage or elastic modulus (G') and loss or viscous modulus (G'') were always obtained within the linear viscoelastic region.

The G'-G'' crossover times (t_g) of acidified caseinate systems were considered here as the gel times, since most studies of milk/caseinate gelation have adopted this criterion (Braga et al., 2006; Curcio et al., 2001). pH at t_g was also determined considering the pH value at the G'-G'' crossover (pH_g).

2.7.2 Conventional inverted microscopy

The degree of compactness of gels was evaluated through digital image analysis. For this, bottom surface image of gel were obtained by conventional inverted microscopy. To obtain the microscopic images, 90 µL of each sample were placed in compartments of the LAB-TEK II cells. The samples were obtained by duplicate under a constant temperature set at 35°C. Transmission images of gels were obtained using a conventional inverted microscopy (Union Optical) with an objective 100x and a digital camera (Canon PowershotA640) with a zoom 7.1x and microscope adapter of 52 mm.

The average pore diameters of gels were determined using the program Image J. To do this, straight lines were drawn on the digital images and values of pore size were measured in pixels. These values were averaged (n=5) and obtained the average pore size. To determine the pixel width in μm, linear calibration was carried out using a micrometer rule. The final system resolution for the protein gels images was:

$$1\text{pixel width} = (0.0645 \pm 0.0005)\,\mu m \Rightarrow Resolution = 15.5\,pixel/\mu m \quad (4)$$

2.8 Statistical analysis

The data are reported as the average values ± their standard deviations. Statistical analysis was performed with Sigma Plot 10.0 and Image J softwares. Relationship between variables was statistically analyzed by correlation analysis using Pearson correlation coefficient (r). The differences were considered statistically significant at $p < 0.05$ values.

3. Results and discussions

3.1 Thermodynamic compatibility of NaCAS:CMC mixtures

The results obtained for mixtures of NACAS and CMC are shown in Fig. 1. The polysaccharide and protein concentrations, in each of the prepared binary solutions, correspond to a single point on the phase diagram. This approach provides a "map" of the transition from the single-phase to the two-phase region of the phase diagram.

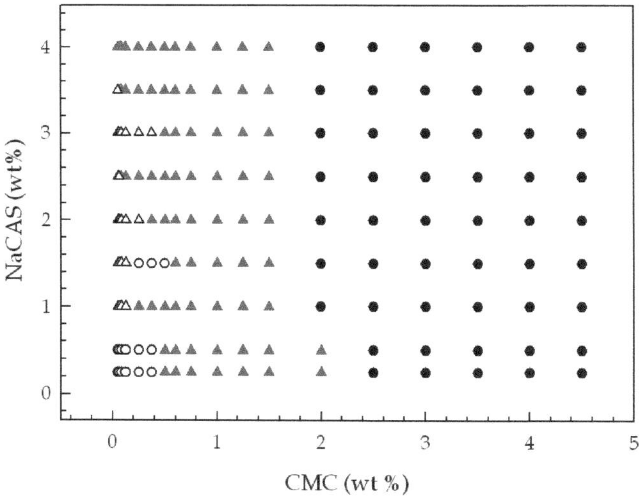

Fig. 1. Approach used for the determination of the phase diagrams for NaCAS:CMC systems after 24 h at 35 °C. Key: (○) one-phase clear solution, (Δ) one-phase turbid solution, (▲) two-phase samples, (•) two-phase gel-like systems

3.2 Thermal stability of NACAS: CMC mixtures

Both the CMC as all mixtures NaCAS:CMC in all relations tested were not affected by rising temperature within the temperature range studied (10-100 °C). This would indicate that the polysaccharide is thermally stable in this range, and that the addition of CMC to NACAS increases its thermal stability, since the NaCAS starts aggregating at about 60 °C in the absence of the polysaccharide.

3.3 Analysis of conformational changes and surface hydrophobicity of NaCAS

Emission spectra of intrinsic fluorescence of NaCAS and mixtures at different NaCAS:CMC ratios were analyzed. In the presence of CMC, a slightly decrease in the fluorescence intensity without changes in emission peaks was observed (data no shown). This would indicate no significant changes in the environment of the intrinsic protein fluorophores when the protein is in the presence of the polysaccharide.

S_0 of the NaCAS was determined in the presence of different CMC concentrations, and it is listed in Table 1.

CMC (wt%)	S_0 (wt%$^{-1}$) ± 0.2
0	64.3
0.0625	59.1
0.1250	54.2
0.2500	45.8
0.5000	26.6
0.7500	16.1

Table 1. S_0 values of NaCAS in the presence of different concentrations of CMC, at 35°C.

S_0 decreased as CMC concentration increased, which would indicate a higher exposure of hydrophilic groups in the protein surface that protrude towards the aqueous environment. These results point to the adsorption of CMC on the surface of the protein.

3.4 Effect of CMC on the viscosity of media

Due to the fact that aggregation is limited by particles diffusion, it was determined the effect on the viscosity caused by the addition of CMC. An increment of η_r with the concentration of the polysaccharide is shown in Fig. 2, especially at CMC concentrations higher than 1.5 wt%.

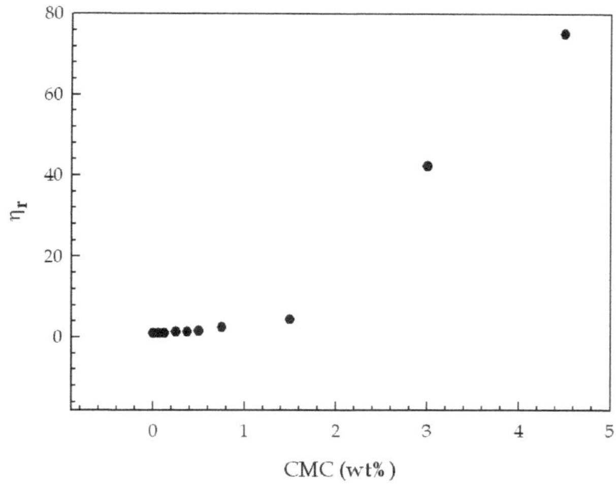

Fig. 2. Relative viscosity (η_r) variations of the medium in the presence of different concentrations of CMC, T 35°C.

3.5 Effect of CMC on the NaCAS acid aggregation

After addition of GDL, NaCAS solutions start a number of changes that lead to protein aggregation. The influence of CMC on this acid aggregation at 35°C, in conditions that no significantly changes on the rate at which pH becomes lower, is shown in Fig. 3.

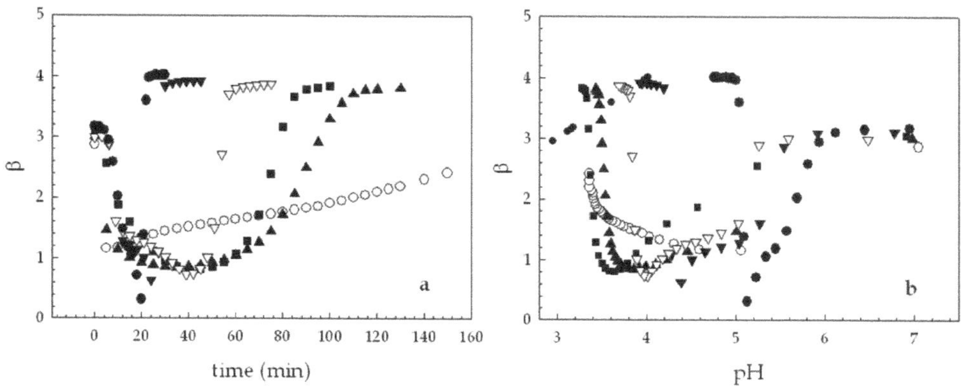

Fig. 3. Variations of parameter β as function of time (a) and pH (b) after GDL addition, at 35°C. NaCAS concentration: 0.5 wt%; (•) NaCAS R 0.7; (▼) NaCAS:CMC 8:1 R 1; (∇) NaCAS:CMC 4:1 R 1; (■) NaCAS:CMC 2:1 R 3; (▲) NaCAS:CMC 1:1 R 6; (○) NaCAS:CMC 1:1.5 R 10.

The acid aggregation, induced by addition of GDL, showed two well-defined steps. At the beginning, a slow phase with a decrease of average size of protein particles is observed. The second step presents a sharp increase in the average size of particles due to formation of colloidal aggregates (aggregation time, t_{ag}) that grow until they reach a limit value, i.e., a fractal dimension of aggregates.

It is known that bovine sodium caseinate in aqueous solution has a considerable level of self-association, like sub-micelles or micelles (Farrell HM, 1996; Fox PF, 1983). Other authors have suggested that bovine sodium caseinate associates into small well-defined aggregates with an aggregation number that depends on the environmental conditions such as temperature, pH, or ionic strength. Probably star-like aggregates are formed with a hydrophobic centre and a hydrophilic (charged) corona (Pitkowski et al., 2008). The profiles in Fig. 3 suggest a slow dissociation of original caseinate aggregates or sub-micelles to form a large number of small particles, which finally aggregate to form bigger particles.

These results show that t_{ag} increases as CMC proportion rises, partially due to a decrease in aggregation pH (pH_{ag}). Because the colloidal particles of NaCAS in suspension have a negative net charge, the addition of CMC would increase its electrostatic stability hindering their aggregation by a consequent increment of the net charge of the soluble particles. On the other hand, this effect can be related to an increase of the viscosity in the medium and a decrease of S_0 in the presence of the polysaccharide. Since the rate of aggregation is limited by the diffusion of particles, an increment of η generates a slower movement giving rise to an increase of t_{ag}. A decrease of S_0 diminishes the participation of hydrophobic interactions during the formation of aggregates.

On the other hand, the degree of compactness of acid aggregates, estimated by D_f, slightly diminishes as CMC:NaCAS ratio increases.

Fig. 4 shows, as an example, the average hydrodynamic diameters of NaCAS particles measured by DLS and β variations during the acid aggregation at 35°C. These profiles confirm the existence of the two stages mentioned above. In addition, the average hydrodynamic diameters determined by DLS showed a good linear correlation (r=0.9082; p<0.0018) with the β values, allowing us to corroborate that the parameter β can be used to estimate the average size of the particles. Therefore the use of simple spectrophotometric techniques could produce reliable results in studying processes of aggregation or gelling of proteins as caseinates.

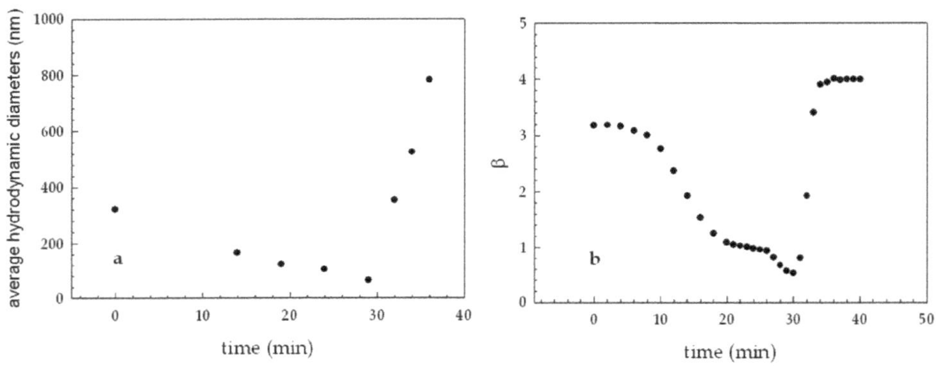

Fig. 4. Average hydrodynamic diameters (A) and parameter β (B) of NaCAS particles during the acid aggregation of 0.5 wt% NaCAS, R 0.5, at 35°C.

3.6 Rheological properties of acid gels

Table 2 shows t_g, pH_g and the maximum G′ (G'_{max}) reached during protein gel formation after addition of GDL at 35°C. Gel times increase and the pH_g decrease as CMC percentage becomes higher revealing a stabilizing effect of CMC.

System	t_g (min ± 0.01)	G'_{max} (± 0.1)	pH_g (± 0.01)
NaCAS 3%	7.54	31.6	4.72
NaCAS 3%-CMC 0.375% (8:1)	17.73	74.4	4.49
NaCAS 3%-CMC 0.50% (6:1)	21.50	51.4	3.90
NaCAS 3%-CMC 0.75% (4:1)	24.99	26.3	3.77
NaCAS 3%-CMC 1.5% (2:1)	81.36	4.8	3.79

Table 2. Values of t_g, G'_{max} and pH_g of gels obtained from NaCAS:CMC mixtures at 35°C and R 1.

As mentioned, it have been reported that an adsorbed CMC layer on the surface of casein micelles gives rise to a repulsive interaction between the casein micelles at low pH in the same way as κ-casein at neutral pH. These can be the reason of the increase on the stability of NaCAS:CMC mixtures against acid aggregation and gelation. The degree of compactness and the elasticity of NaCAS aggregates and gels respectively were higher at low CMC proportion but underwent a sharp decrease when the polysaccharide amount rises. On the other hand, the degree of thermodynamic compatibility affected the final elasticity of mixed gels. At 0.375 wt% of CMC, the two biopolymers are in the same phase, but at higher proportions of CMC there is a thermodynamic incompatibility and phase separation occur. This incompatibility appears to induce the formation of weaker gels.

3.7 Digital images of gels

The microstructure of protein gels can be characterized through optical analysis (Lucey, 2002).

Fig. 5 shows the transmission images of gels obtained for mixed gels at constant NaCAS concentration (3 wt%) and different CMC proportions. From the digital images of gels, it was possible to observe differences in the internal microstructure of gels.

Fig. 5. Images of gels obtained using a conventional inverted microscopy with an objective 100x and a digital camera with a zoom 7.1x, at NaCAS 3 wt%, 35 °C and R 1, for different NaCAS:CMC ratios: a) without CMC, b) NaCAS:CMC 8:1, c) NaCAS:CMC 6:1, and d) NaCAS:CMC 4:1.

Performing a qualitative analysis of these images, it is possible to observe different degrees of structure of the gels formed at different ratios of CMC. Table 3 shows the values of mean pore size of NaCAS gels in the absence and presence of CMC obtained from digital images. In the presence of lower concentration of CMC, the slower rate of gelation (higher t_g) produced gels more structured, more compact and with smaller pores. This is due to, if the process is performed slowly, the gel mesh can be restructured by breaking of some interactions and formation of new ones, forming a tighter mesh and, therefore, progressively smaller pores. Other authors have also reported that processing speed can affect the hardness and elasticity of the gel formed (Cavallieri & da Cunha, 2008). But with increasing CMC concentration, there was an increase in the average pore diameter. Mixtures NaCAS:CMC 2:1 failed to gel consistency.

System	Average pore sizes (μm)
NaCAS 3%	3.1 ± 0.4
NaCAS 3%-CMC 0.375% (8:1)	2.7 ± 0.2
NaCAS 3%-CMC 0.50% (6:1)	3.0 ± 0.2
NaCAS 3%-CMC 0.75% (4:1)	3.4 ± 0.3

Table 3. Average pore sizes of NaCAS gels in the absence and presence of different concentrations of CMC, at 35°C and R 1.

These results are consistent with the values of G'_{max} (Table 2) obtained for the different mixtures. Gels with larger pores will be less elastic.

4. Conclusion

As CMC proportion rises, the aggregation and gel times of NaCAS:CMC mixtures increased and the pH at which these processes begin decreased, revealing a stabilizing effect of CMC. The degree of compactness diminished when the CMC proportion increased. This effect can be linked to protein conformational changes in the presence of CMC that lead to a decrease of surface hydrophobicity, which difficult the establishment of hydrophobic interactions. The gels also showed lower elasticity at CMC:NaCAS high ratios.

Therefore, it is possible to obtain acid gels with different textures varying the protein:polysaccharide proportions due to surface hydrophobicity and electrostatic stability modification of NaCAS particles and due to changes on the kinetic of aggregation and gelation processes.

5. Acknowledgment

This work was supported by grants from the Universidad Nacional de Rosario. Thanks to MSc. Luis Martínez of the National University of Quilmes for their advice in the rheological measurements. Thanks to Prof. Mirta Armendariz of the National University of Rosario for their advice in the statistical analysis. The authors would like to thank Lic. Romina Ingrassia for the English revision.

6. References

Belyakova, L.E.; Antipova, A.S.; Semenova, M.G.; Dickinson, E.; Matia Merino, L. & Tsapkina, E.N. (2003). Effect of sucrose on molecular and interaction parameters of sodium caseinate in aqueous solution: relationship to protein gelation. *Colloids and Surfaces B: Biointerfaces*, Vol. 31, No. 1-4, (November 2002), pp. 31-46, 0927-7765

Braga, A.L.M.; Menossi, M. & Cunha, R.L. (2006). The effect of the glucono-[delta]-lactone/caseinate ratio on sodium caseinate gelation. *International Dairy Journal*, Vol. 16, No. 5, (June 2005), pp. 389-398, ISSN 0958-6946

Bremer, L.G.B.; Bijsterbosch, B.H.; Walstra, P. & van Vliet, T. (1993). Formation, properties and fractal structure of particle gels. *Advances in Colloid and Interface Science*, Vol. 46, No. (December 1993), pp. 117-128, ISSN 0001-8686

Camerini-Otero, R.D. & Day, L.A. (1978). The wavelength dependence of the turbidity of solutions of macromolecules. *Biopolymers*, Vol. 17, No. 9, (September 1978), pp. 2241-2249, ISSN 0006-3525

Cavallieri, A.L.F. & da Cunha, R.L. (2008). The effects of acidification rate, pH and ageing time on the acidic cold set gelation of whey proteins. *Food Hydrocolloids*, Vol. 22, No. 3, (January 2007), pp. 439-448, 0268-005X

Corzo-Martínez, M.; Moreno, F.J.; Villamiel, M. & Harte, F.M. (2010). Characterization and improvement of rheological properties of sodium caseinate glycated with galactose, lactose and dextran. *Food Hydrocolloids*, Vol. 24, No. 1, (August 2009), pp. 88-97, ISSN 0268-005X

Curcio, S.; Gabriele, D.; Giordano, V.; Calabrò, V.; de Cindio, B. & Iorio, G. (2001). A rheological approach to the study of concentrated milk clotting. *Rheologica Acta*, Vol. 40, No. 2, (March 2001), pp. 154-161, ISSN 0035-4511

de Britto, D. & Assis, O.B.G. (2009). Thermal degradation of carboxymethylcellulose in different salty forms. *Thermochimica Acta*, Vol. 494, No. 1-2, pp. 115-122,

de Jong, S.; Klok, H.J. & van de Velde, F. (2009). The mechanism behind microstructure formation in mixed whey protein-polysaccharide cold-set gels. *Food Hydrocolloids*, Vol. 23, No. 3, (March 2008), pp. 755-764, 0268-005X

de Jong, S. & van de Velde, F. (2007). Charge density of polysaccharide controls microstructure and large deformation properties of mixed gels. *Food Hydrocolloids*, Vol. 21, No. 7, (September 2006), pp. 1172-1187, ISSN 0268-005X

de Kruif, C.G. (1997). Skim Milk Acidification. *Journal of Colloid and Interface Science*, Vol. 185, No. 1, (August, 1996), pp. 19-25, ISSN 0021-9797

de Kruif, C.G.; Weinbreck, F. & de Vries, R. (2004). Complex coacervation of proteins and anionic polysaccharides. *Current Opinion in Colloid & Interface Science*, Vol. 9, No. 5, (September 2004), pp. 340-349, ISSN 1359-0294

Delben, F. & Stefancich, S. (1997). Interaction of food proteins with polysaccharides, I. Properties upon mixing. *Journal of Food Engineering*, Vol. 31, No. 3, (September 1996), pp. 325-346, ISSN 0260-8774

Du, B.; Li, J.; Zhang, H.; Chen, P.; Huang, L. & Zhou, J. (2007). The stabilization mechanism of acidified milk drinks induced by carboxymethylcellulose. *Lait*, Vol. 87, No. 4-5, (October 2007), pp. 287-301, ISSN 1958-5586

Du, B.; Li, J.; Zhang, H.; Huang, L.; Chen, P. & Zhou, J. (2009). Influence of molecular weight and degree of substitution of carboxymethylcellulose on the stability of acidified milk drinks. *Food Hydrocolloids*, Vol. 23, No. 5, (October 2008), pp. 1420-1426, ISSN 0268-005X

Farrell HM, C.P., King G, Hoagland PD, Groves ML, Kumosinski TF, Chu B. (1996). Particle sizes of casein submicelles and purified κ-casein, In: *Macromolecular Interactions in Food Technology*, K. A. Parris N, Creamer LK, Pearce J Eds., pp. 61, American Chemical Society, Washington

Farrell, H.M.; Pessen, H.; Brown, E.M. & Kumosinski, T.F. (1990). Structural Insights into the bovine casein micelle: Small angle X-ray scattering studies and correlations with spectroscopy. *Journal of Dairy Science*, Vol. 73, No. 12, (December 1990), pp. 3592-3601, ISSN 0022-0302

Fox PF, M.D. (1983). Physico-chemical aspects of dehydrated protein-rich milk products, *Proceedings of International Dairy Federations Symposium*, 188, Hillerod, Denmark, 17-19 May, 1983

Gaucheron, F.L.G., Y.; Boyaval, E.; Piot, M. (1997). Binding of cations to casein molecules. Importance of physicochemical conditions. *Milchwissenschaft*, Vol. 52, No. pp. 322-327, ISSN 0026-3788

Haskard, C.A. & Li-Chan, E.C.Y. (1998). Hydrophobicity of Bovine Serum Albumin and Ovalbumin Determined Using Uncharged (PRODAN) and Anionic (ANS-) Fluorescent Probes. *Journal of Agricultural and Food Chemistry*, Vol. 46, No. 7, (April 1998), pp. 2671-2677, ISSN 1570-9639

Horne, D.S. (1987). Determination of the fractal dimension using turbidimetric techniques. Application to aggregating protein systems. *Faraday Discussions of the Chemical Society*, Vol. 83, No. 0, pp. 259–270, ISSN 1364-5498

Kato, A. & Nakai, S. (1980). Hydrophobicity determined by a fluorescence probe method and its correlation with surface properties of proteins. *Biochimica et Biophysica Acta (BBA) - Protein Structure*, Vol. 624, No. 1, (July 1980), pp. 13-20, ISSN 1570-9639

Kuaye, A.Y. (1994). An ultraviolet spectrophotometric method to determine milk protein content in alkaline medium. *Food Chemistry*, Vol. 49, No. 2, (June 1993), pp. 207-211, ISSN 0308-8146

Linde, A. (1982). Calcium metabolism in dentinogenesis, In: *The Role of Calcium in Biological Systems*, L. J. T.-A. Anghileri, A. M., (Eds.), pp. 55-71, CRC Press, ISBN 978-084-9362-74-3, Boca Raton

Lucey, J.A. (2002). Formation and Physical Properties of Milk Protein Gels. *Journal of Dairy Science*, Vol. 85, No. 2, (August 2001.), pp. 281-294, 0022-0302

Lucey, J.A.; Tamehana, M.; Singh, H. & Munro, P.A. (1998). A comparison of the formation, rheological properties and microstructure of acid skim milk gels made with a bacterial culture or glucono-[delta]-lactone. *Food Research International*, Vol. 31, No. 2, (March 1998), pp. 147-155, ISSN 0963-9969

Lucey, J.A.; Tamehana, M.; Singh, H. & Munro, P.A. (2001). Effect of heat treatment on the physical properties of milk gels made with both rennet and acid. *International Dairy Journal*, Vol. 11, No. 4-7, (July 2001), pp. 559-565, ISSN 0958-6946

Nespolo, C.R.; Reggiardo, A.D.; Mancilla Canales, M.A.; Wagner, J.R.; Alvarez, E.M.; Brandelli, A. & Risso, P.H. (2010). Effect of calcium on ovine caseinate functional properties. *Journal of Chemical & Engineering Data*, Vol. 55, No. 11, (November 2010), pp. 4624-4631, ISSN 0021-9568

Oakenfull, D.; Miyoshi, E.; Nishinari, K. & Scott, A. (1999). Rheological and thermal properties of milk gels formed with [kappa]-carrageenan. I. Sodium caseinate. *Food Hydrocolloids*, Vol. 13, No. 6, (August 1999), pp. 525-533, ISSN 0268-005X

Perrechil, F.A.; Braga, A.L.M. & Cunha, R.L. (2009). Interactions between sodium caseinate and LBG in acidified systems: Rheology and phase behavior. *Food Hydrocolloids*, Vol. 23, No. 8, (May 2009), pp. 2085-2093, ISSN 0268-005X

Picone, C.S.F. & da Cunha, R.L. (2010). Interactions between milk proteins and gellan gum in acidified gels. *Food Hydrocolloids*, Vol. 24, No. 5, (December 2009), pp. 502-511, ISSN 0268-005X

Pitkowski, A.; Durand, D. & Nicolai, T. (2008). Structure and dynamical mechanical properties of suspensions of sodium caseinate. *Journal of Colloid and Interface Science*, Vol. 326, No. 1, (July 2008), pp. 96-102, ISSN 0021-9797

Qi, P.X.; Brown, E.M. & Farrell, H.M. (2001). New views' on structure-function relationships in milk proteins. *Trends in Food Science & Technology*, Vol. 12, No. 9, (September 2000), pp. 339-346, ISSN 0924–2244

Ribeiro, K.O.; Rodrigues, M.I.; Sabadini, E. & Cunha, R.L. (2004). Mechanical properties of acid sodium caseinate-[kappa]-carrageenan gels: effect of co-solute addition. *Food Hydrocolloids*, Vol. 18, No. 1, (January 2003), pp. 71-79, ISSN 0268-005X

Risso, P.; Relling, V.; Armesto, M.; Pires, M. & Gatti, C. (2007). Effect of size, proteic composition, and heat treatment on the colloidal stability of proteolyzed bovine casein micelles. *Colloid & Polymer Science*, Vol. 285, No. 7, (November 2006), pp. 809-817, ISSN 0303-402X

Ruis, H.G.M.; Venema, P. & van der Linden, E. (2007). Relation between pH-induced stickiness and gelation behaviour of sodium caseinate aggregates as determined by light scattering and rheology. *Food Hydrocolloids*, Vol. 21, No. 4, (June 2006), pp. 545-554, ISSN 0268-005X

Semenova, M.G.; Belyakova, L.E.; Polikarpov, Y.N.; Antipova, A.S. & Dickinson, E. (2009). Light scattering study of sodium caseinate + dextran sulfate in aqueous solution: Relationship to emulsion stability. *Food Hydrocolloids*, Vol. 23, No. 3, (June 2008), pp. 629-639, ISSN 0268-005X

Spyropoulos, F.; Portsch, A. & Norton, I.T. (2010). Effect of sucrose on the phase and flow behaviour of polysaccharide/protein aqueous two-phase systems. *Food Hydrocolloids*, Vol. 24, No. 2-3, (September 2009), pp. 217–226, ISSN 0268-005X

Tolstoguzov, V.B. (1991). Functional properties of food proteins and role of protein-polysaccharide interaction. *Food Hydrocolloids*, Vol. 4, No. 6, (March 1991), pp. 429-468, ISSN 0268-005X

Walstra, P.; Jenness, R. & Badings, H.T. (1984). *Dairy chemistry and physics* John Wiley & Sons, ISBN 978-047-1097-79-2, New York, USA

Yamamoto, F. & Cunha, R.L. (2007). Acid gelation of gellan: Effect of final pH and heat treatment conditions. *Carbohydrate Polymers*, Vol. 68, No. 3, (November 2006), pp. 517-527, ISSN 0144-8617

Yu, H.; Sabato, S.F.; D'Aprano, G. & Lacroix, M. (2004). Effect of the addition of CMC on the aggregation behaviour of proteins. *Radiation Physics and Chemistry*, Vol. 71, No. 1-2, (October 2004), pp. 131-135, ISSN 0969-806X

Electrochemical Behaviour of AISI 304 Stainless Steel Immersed in Mixtures Consisting by Biocide and Fungal Suspensions

Maricica Stoica[1], Petru Alexe[1], Rodica Dinica[2] and Geta Cârâc[2]
[1]*Dunarea de Jos University of Galati/ Department of Biochemistry and Technologies*
[2]*Dunarea de Jos University of Galati/ Department of Chemistry Physics and Environment*
Romania

1. Introduction

Chemical disinfection of industrial facilities in food bioprocessing is a major consumer of biocides and represents an essential technological issue. Due to this requirement, the metallic surfaces of the equipments used in bioprocessing inevitably interact with electrolyte environments that are exposed to washing and disinfecting solutions with or without microorganisms through on electrochemical mechanism (Landoulsi et al., 2008; Osarolube et al., 2008). Recently the electrochemical behaviour of stainless steel surfaces have become interested to the many researchers (Hiromoto & Hanawa, 2006; Stoica et al. 2010a). However, only a few research studies were devoted to the electrochemical behaviour of stainless steels used in bioprocesses having a synergic effect on biocides and microorganism (Stoica et al., 2009; Stoica et al., 2010b). The electroanalytical techniques used in previous studies through discharge of an electric field can generate some chemical and physical processes, reversible or irreversible because of the fungi present in the environments (Shen et al., 2008 ; Yang et al., 2008) and on metallic surfaces. These processes are strongly influenced by many factors, such as: biological factors (microorganisms type; cell wall; size and shape of the cell; cells density, arrangement and cell position, fluid medium properties in medium conductivity, electric field waveforms and the number of electric pulses (Yang et al., 2008). The complex phenomena occurring in the electrochemical biocide-fungi-metallic surface system are studied as electrochemical interface processes occurring at the limit between molecules of aqueous solution (biocide solution) coming into contact with the metallic electrode 'live' (fungal cell membranes) and the metallic surface (AISI 304 Stainless Steel). The electron transfer in these electrochemical systems respects the general laws of charge transfer, but also presents the specific properties based on the dynamic environment in which the electron transfer occurs, at the processes of adsorption/desorption and at the surface reactions. These can occur between molecules of biocide and biological surfaces as well as between biocide penetration through these surfaces endowed with distinct architecture, composition and characteristics. The study of these complex processes requires a multidisciplinary approach regarding the metallic surfaces, fungi, biocides and electrochemical processes interface. The aim of this chapter is to systematically present the relevant aspects about the interface electrochemical processes on metallic surfaces with fungi and biocides. A relevant study on the electrochemical behaviour of the AISI 304

Stainless Steel in mixtures consisting by biocide and fungal suspensions is presented. The results show that there is a synergic effect between the active substances from the disinfectant and fungal suspensions and the applied electric potential during tests, thus this effect can be taken into consideration in the food bioprocessing safety.

2. Theoretical aspects regarding the stainless steel, some factors that influence the hygiene of food contact surfaces and metallic surface corrosion

A variety of materials are used in the construction and fabrication of different food equipments. Various metals as well as non metals (*e.g.* plastics, rubber) are used as materials depending on the applications and because of the high demands imposed on the corrosion resistance and the hygiene the food processing industry. These materials vary in their properties regarding their workability, compatibility with type of food product, processing conditions and sanitary design features, depending the applications. The stainless steel is the obvious choice of better material for equipments used on food contact surfaces (Holah & Thorpe, 1990; Saikhwan et al., 2006). This choice is due to the corrosion resistance of stainless steels coupled with their strength and durability, their ability to be readily cleaned and sterilized without deterioration using a wide range of cleaning/sterilizing systems, and the fact that they impart neither color nor flavor to foodstuffs and beverages.

2.1 Stainless steel categories

There are more than 70 standard types of stainless steel and many special alloys, fabricated by multiple processing steps which modify their properties. These steels are produced in the rough form (AISI - American Iron and Steel Institute - types). Generally, stainless steels are mainly iron based with 12% to 30% chromium, up to 22% nickel and minor amounts of carbon, copper, molybdenum, selenium and titanium. The AISI designation for these materials is well known with the number series 300 referring to austenitic stainless steels and the 400 series covering the ferritic and martensitic stainless steels. These stainless steels can be classified into distinct categories including ferritic, martensitic, austenitic and duplex stainless steels (SS) (Figure 1).

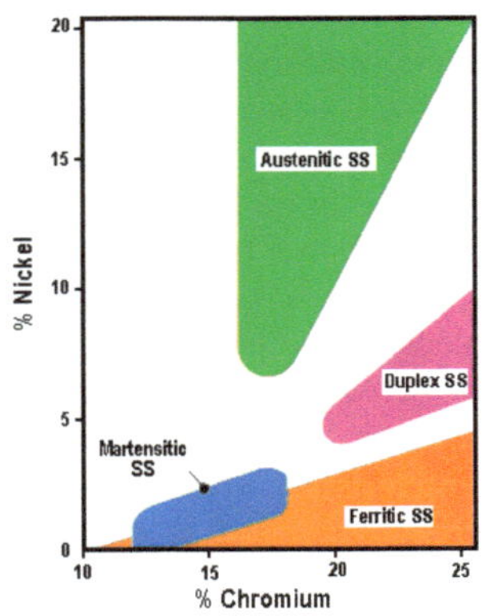

Fig. 1. The stainless steels categories

2.1.1 Ferritic stainless steels

The ferritic Stainless Steel is a magnetic type possessing a microstructure which is primarily ferritic. There are more economical from a cost point of view, but they have limited corrosion resistance compared to the more used austenitics. Similarly the ferritics have limited toughness, formability and weldability in comparison to the austenitics. An example of ferritic grades is AISI 430 used widely for applications in cutlery, kitchen sinks and catering/gastronomy industry and enclosured equipments, where the corrosion resistance requirements are not so demanding due to material economic advantages (Carvalho et al., 2006).

2.1.2 Martensitic stainless steels

Martensitic stainless steels are similar in structure to the ferritics, but by the addition of more carbon, they can be hardened and strengthened by heat treatment in the same way to carbon steels. They are classified as a "hard" ferro-magnetic group. Their corrosion resistance is inferior to that of austenitic stainless steels, therefore they are generally used in mildly corrosive environments. The martensitic grades commonly used are AISI 403, AISI 410 and AISI 420 and they are widely used for cutting and grinding applications, especially for knives.

2.1.3 Austenitic stainless steels

Austenitic stainless steel is non-magnetic (*i.e.*, has a low "permeability") and has excellent ductility, formability and toughness, even at cryogenic temperatures. Depending on the nickel content the austenitics respond to cold working by increases in strength, which can surprisingly be useful in severe forming operations, avoiding premature tearing and cracking. The most representative austenitic grades are AISI 304 and AISI 316. Most stainless steel containers, pipeworks and food contact equipments are manufactured from the most representative austenitic grades either 304 or 316 type austenitic stainless steels. They are widely used in food processing, beverage industry (Mai et al., 2006) and others: bulk storage and transportation and many other applications. For example the sugar, starch and wine industry requires equipments with good corrosion resistance and thus have adopted the AISI 304 and AISI 316 stainless steel. Beer is produce using raw materials like water, barley, hops, malt by fermentation, filtration, canning and sterilization process. Beer, wort and mashed grain is generally not corrosive to stainless steel such as 304, even though the process vessels and pipe systems during brewing operate from low temperatures up to the boiling point. In sections with temperatures above 60°C, there is a risk of chloride-induced stress corrosion cracking, often from the outside, in case the insulation material gets wet.

2.1.4 Duplex stainless steels

The duplex stainless steels have a balanced or mixed structure of austenite and ferrite and as a result have characteristics of both "basic" types. Just like the ferritics, they are ferro-magnetic with a good formability and weldability as the austenitics. In adition, the duplex stainless steels have the general corrosion resistance similar to or better than that of AISI 304 and 316 (Tavares et al., 2010). Examples of duplex grades are AISI 2304, AISI 2205, AISI 2507.

2.2 Factors that influence the hygiene of food contact surfaces

Food contact surfaces include any item that comes in direct contact or could potentially be in contact with the exposed food. The surfaces must be cleaned and disinfected before beginning each day's work, after each use and before changing to a different type of food, such as raw meats to vegetables. The cleaning and disinfection in food industries is an issue of utmost importance since high hygienic standards assure the safety and quality of end products and therefore the consumer's health.` This subchapter will shortly describe the passive film, the finishing, roughness and cleanability on the surfaces.

2.2.1 Passive film

The stainless steel corrosion resistance arises from a protective layer formed on metallic surfaces. The protective layer, sometimes only a few nm in thickness is called a passive film and it is formed instantaneously in an oxidizing atmosphere such as air, water or other fluids that contain oxygen (Olsson et Landolt, 2003). Once the layer is formed, the metallic surface becomes *passivated* and the oxidation or *rusting* rate will slow down to less than 0.05 mm/year. The passive film has normally considerable practical importance also in food production quality since the thin film protects the underlying metal from corrosion. The protective film is strongly adherent on the metallic surfaces and presents longer stability in time having some impact on food safety.

2.2.2 Surfaces finishing and cleanability

The surfaces finishing grade renders the quality of the surfaces described by roughness. The surfaces roughness is characterized in two main directions: one perpendicular to the surface, described as height deviation and the second one in the plane of the surface, described by spatial parameters and identified as texture (Round et al., 2001). The surfaces status was defined by the following characteristic parameters: R_a known as the arithmetic average height parameter, R_q known as the root mean square and R_{max} is the difference in height between the highest and lowest points on the surface relative to the mean plane. The R_a is the most universally roughness parameter used (Buchalla et al., 2000; Turssi et al., 2001) to with a general quality surface control (Whitehead & Verran, 2006). The variation of the average roughness can be useful to express the corrosion process on the surfaces (Sánchez et al., 2008). At the same time the roughness is important for surfaces cleanability (Leclerq-Perlat & Lalande,1994; Boulange-Petermann et al., 2004). A lower roughness allows for easy cleaning intended to eliminate bacteria, which considerably reduces service cost of materials. Surfaces in contact with the food product should have an standard roughness (R_a) value. For surfaces coming in contact with a product having a large area, the R_a should be less than 0.8 µm (Holah and Thorpe, 1990). Under special conditions a roughness higher than 0.8 µm can be accepted in case test results prove that the required cleaning capacity is reached.

2.3 The metallic surfaces corrosion

The corrosion is the destruction or the deterioration of metallic materials by direct chemical, electrochemical or biochemical reactions with different environments (Landoulsi et al., 2008; Osarolube et al., 2008). The corrosion processes are a very broad and complex phenomenon which can be uniform (*e.g.* general corrosion), located (*e.g.* crevice and pitting corrosion) and or in any other way.

2.3.1 Electrochemical corrosion

In food industry due to specific requirements (moisture, aqueous solutions, significant concentrations of organic/inorganic acids, sulfiting agents used to treat foods, high in salinity content, highly concentrated disinfecting solutions, *etc.*) the stainless steel surface is corroded through exposure in different environments. Most corrosion processes are electrochemical (corrosion of metals involves electrons transfer) (see equations: 1 - 4).

$$Me \rightarrow Me^{n+} + ne^{-} \text{ (anodic reaction; metal dissolution)} \quad (1)$$

$$e.g.\ Fe \rightarrow Fe^{+2} + 2e^{-} \quad (2)$$

$$O_2 + H_2O + 4e^{-} \rightarrow 4\,OH^{-} \text{ (catodic reaction; consume the electrons)} \quad (3)$$

$$Me + nH_2O \rightarrow Me(OH)n + nH^{+} + ne^{-} \quad \text{(overall reaction)} \quad (4)$$

The excess of electrons resulting from the reaction loads the metal negatively and the anodic process can not continue within measurable intensity, wheareas on the metallic surface there is a cathodic process of depolarizing agent reduction (*e.g.* hydrogen ions). The reduction of hydrogen ions is the most common reaction that accompanies aqueous corrosion of metals. During corrosion processes the hydrogen ions can be absorbed on the metallic surface and diffuses inside, while the rest of hydrogen forms gas molecules and escapes. The hydrogen dissolved in metals significantly affects their mechanical properties, composition and structure of passive films formed on the surface. The electrochemical corrosion initiation at sub-microscopic level involves several phenomena starting from breakdown of passive film in a stochastic and sporadic way to localized dissolution of oxide covered metal and mass transport of atoms across the surface to support the continuing dissolution process (Marcus et al., 2008).

2.3.2 Corrosion severity

The surface corrosion is an issue in many industries and it is even a greater challenge on the food processing industry, where it can cause direct (equipments failure) or indirect damage through the loss of production time for maintenance of equipments, the risk of food products contamination by corrosion products or endanger workers' safety and operational security (Holah & Thorpe, 1990; Ofoegbu et al., 2011). When the equipments begin the long walk down the dark road of corrosion, small amounts of metallic elements in the alloy may migrate into foodstuffs from equipment leading to human ingestion and can cause adverse health effects. The severity of the corrosion can be estimated easier through corrosion rate (V_{corr}). A corrosion rate for food equipments surfaces which exceeds 0.02 mm/y reduces the equipments lifetime (Fontana, 1987). The corrosion on metallic surfaces can lead to the formation and expansion of cavities and grooves. This in turn provides breeding sites for microorganisms, thereby compromising the efficacy of cleaning and disinfection procedures and encourages more biofilm adhesion and biofilm resistance to detachment. The biofilms probably do not participate directly in the corrosion process, but they can lead to some changes of the interfacial environment by the increase of cells concentration that facilitates the corrosion process on the surface (Yuan & Pehkonen, 2007).

3. The fungi relevance on food industry

The living world is classified in three domains such as the *Bacteria*, the *Archaea* and the *Eukaryota* (Woese, 1990). Some authors argued that these species represent separate lines of descent that diverged early from an ancestral colony of organisms (Woese & Gupta, 1981; Woese, 1998). The Eukaryota domain include three major multicellular groups: animals, plants and fungi. Fungi including yeasts and molds are unicellular or multicellular eukaryotic microrganisms (Tournas et al., 2006; Pommerville, 2010). The fungal cells are larger than bacteria and structurally more complex than other microorganisms (Hugo et al., 2004). Some strains are important in food biotechnology as starter cultures with the ability to modify food characteristics and in industrial biotechnology to produce antibiotics and other beneficial by-products, such as enzymes, vitamins, organic acids (Deacon, 1997; Maier et al., 2009; Gautam et al., 2011) and others. *Saccharomyces cerevisiae* plays an important role in various fermentation processes to produce enzymes, antioxidants and vitamins. *Candida mycoderma* and *Aspergillus niger* have a major role to forming the citric acid (Tisnadjaja et al., 1996; Papagianni, 2007). Moreover, *Aspergillus niger* is important as a producer of proteins and a great variety of enzymes such as catalase, celullase, endoglucanase, glucosoxidase, invertase and pectinase (Lu et al., 2010). On the other hand, the food spoilage by fungi raises an economic issues and it is estimated that annually between 5% and 10% of the world's food production is lost due to fungal biodeterioration (Pitt & Hocking, 1997). The risk of health problems can appear due to mycotoxins produced as secondary metabolites by fungi during the stationary phase of growth in some specific physico-chemical conditions. The control of fungal spoilage of food products is therefore an essential and decisive matter to prevent different biological safety risks.

4. Biocides

Biocides represent a very diversified group of chemical substances with a crucial role in pharmaceutical and food industry and this role is becoming increasingly significant (Zani et al., 1997; Bridier et al., 2011). In food industry biocides are commonly used to disinfect the processing areas, equipments, containers, surfaces or pipes associated with the production, transport and storage of food or drink, including drinking water.

4.1 Biocides effectiveness

Biocides act in different ways (sometimes several biocides are combined in a product to increase the overall effectiveness) to destroy, render harmless, prevent the action, or otherwise exert a controlling effect on any harmful organism by chemical or biological means (Table 1). The biocides effectiveness is limited and much dependent on application conditions (Bessems 1998). The efficacy of biocides greatly depend on contact time, concentration temperature, *p*H and microorganisms type (Russell and McDonnell, 2000; Russel, 2003; Kitis 2004).

All biocides show varying degrees of activity against bacteria, bacterial spores, fungi, viruses and protozoa and at least some have algicidal activity (Russell, 2003). The biocides efficacy depends on a large of intrinsic and extrinsic factors. The intrinsic factors are characterized by the biocide type and its concentration, contact time and application (Russell & McDonnell, 2000). Furthermore, the relationship between contact time and concentration biocides determines the microbial reduction. The stability of the active biocide compounds in the environment also has an influence on the microbial action. The temperature is important, as most substances have a lower efficiency at low temperature. The contact mode also influences the biocides efficacy, as well as the contact time (mechanical effects) and the *p*H which plays an important role. Extrinsic factors consist in the presence of organic matters, the concentration and the age of the microorganisms, the microorganisms ability to transform the biocide into inactive forms (McDonnell & Russell, 1999). The biocides have a broad spectrum of antimicrobial activity and generally act to several targets sites in microbial cells (Figure 2) and damage them as a results of the biocidal effect (Russell, 2003).

Biocides	Action mode	References
Alcohols	Membrane damage, cell lysis	Larson & Morton, 1991
Chlorine compounds and halogens	Oxidising agents	McDonnell & Russell, 1999
Peroxydes		
Quaternary Ammonium Compounds	Membrane destabiliser, cell lysis	
Biguanides	Chlorhexidine specifically inhibits membrane - bound ATPase	Breslin & Tharp, 2001

Table 1. Biocides types in food processing and action mode

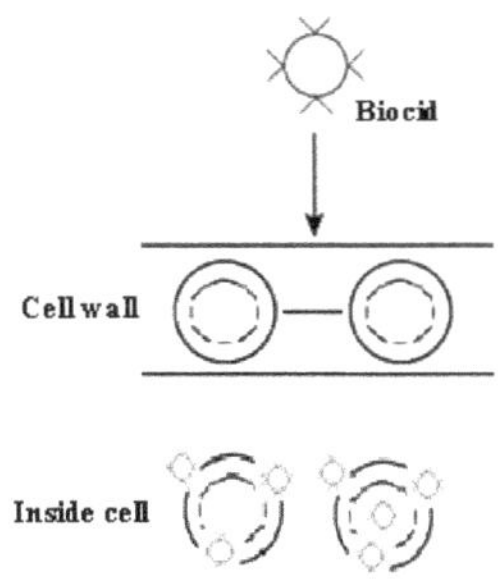

Fig. 2. Action mode of biocide entry into a microorganism type (adapted from Russell, 2003).

4.2 Legislative framework of using biocide

During the past decades, the food production has become more complex; the production volume is larger, the operations are more mechanical, the food is more processed and the time and distance between production and consumption are longer. The new trends in food production and consumption lead to an increased need for efficient safety practices in the food processing industry. Currently, the biocides industry as a whole is governed by end-industry growth, technological developments, regulatory changes and the growing use of biocides as an aid in the hygiene processes. Science 2009 the world biocide demand is estimated to grow to 6,880 million dollars (5.4% annual growth from 2004) and by 2014 to 9,050 million dollars (5.6% annual growth from 2009). Accordingly, emerging EU (European Union) regulations such as Biocidal Products Directive 98/8/EC (BPD) (from May 14th 2000) and Registration, Evaluation and Authorisation and Restriction of Chemicals (REACH), strive to increase the safety and the eco-efficiency of chemical products and production processes (BPD, 1998; REACH, 2006). This directive has as objectives: (i) to harmonise the

European market for biocidal products and their active substances such that product authorisation_content in one Member State can be recognized in other Member States; (ii) to provide a high level of protection for people, animals and the environment (from the use of biocidal products) through risk assessment. These objectives requires the submission and evaluation of data relating to substances' chemistry, toxicity to humans, and toxicity and fate in the environment and (iii) to ensure products are sufficiently effective against the target species. The Registration REACH - a new chemical regulation was implemented by the EU on June 1st 2007 and is being implemented in stages to be completed by 1 st June 2018. The main objective of REACH is a high level of protection for human health and the environment, while maintaining the competitiveness and innovation of EU chemicals industry. REACH provides a single regulatory framework for the control of chemicals, replacing the previous patchwork of controls, and ensures that information on the properties of chemicals is transmitted down the supply chain, thus enabling them to be safely handled.

5. Study of electrochemical behaviour of AISI 304 stainless steel immersed in mixtures consisting of *Neoseptal* and fungal suspensions

In this study the corrosion behaviour of AISI 304 Stainless Steel (*SS*) treated by biocide *Neoseptal* solution (noted as *Neo*) and in mixtures consisting of *Neo* with fungal suspension (*Neo - Aspergillus niger, Neo - Candida mycoderma* and *Neo - Saccharomyces cerevisiae*) was investigated.

5.1 Materials and methods

5.1.1 AISI 304 stainless steel

Tests were performed using AISI 304 Stainless Steel in rectangular samples and pretreated by mechanically polishing with abrasive paper of increasingly finer grit between 800 and 2000 μm and finally chemical cleaning. The samples as working electrode were mounted in a tetrafluoroethylene-perfluoroalkylvinylether copolymer (PFA), sleeve sample holder except for an exposed test area of 12 cm^2 (Stoica et al., 2010b).

5.1.2 Biocide solution

The biocide (*Neoseptal*) is a commercial disinfectant, effective against all types of microorganisms in food industry manufactured by Dr.Weigert (Germany) based on H_2O_2 (dihydrogen dioxide - 30% wt.). Fresh solutions of *Neoseptal* were prepared by dilution of commercially *Neoseptal* biocide. The concentration of *Neoseptal* was 0.2% performed at 20±2°C and acted for at least 30 min.

5.1.3 Fungi strains

Three fungal types were tested: a) *Aspergillus niger* ATCC 16404 (provided by the "Ion Cantacuzino" Institute (Bucharest-Romania), with the spores concentration of 1.41×10^7spores/mL; b) *Candida mycoderma* isolated from spoiled wine, of suspension containing 1.38×10^7 cells/mL; c) *Saccharomyces cerevisiae* (Pakmaya, Rompak), with a cell number in the suspensions of 1.40×10^7 cells/mL. The cells concentrations were measured using a Thoma cytometer. An aliquot volume of fungal suspension (5 mL) was used in the electrochemical experiments (50 mL as total volume).

5.1.4 Electrochemical measurements

Tests were performed on the corrosion behaviour of samples of *SS* in *Neo* with and without fungal suspensions. All electrochemical measurements were carried out in a glass electrochemical-cell (Metrohm, Switzerland) equipped with three electrodes, at room temperature (22±1°C).The working electrode was AISI 304 Stainless Steel type, the counter electrode (CE) was a Platinum foil and as the reference electrode (RE) was a saturated calomel electrode (SCE). The entire three-electrode assembly was placed in a Faraday cage to limit the noise disturbance and then connected to potentiostat-galvanostat Bio-Logic SP-150 (France). The investigations are carried out using EC-Lab® Express v 9.46 software. The electrochemical measurements were carried out using the Linear Polarization Technique (LP). The polarization measurements were initiated after 60 seconds immersions to access an equilibrium open potential on the sample surface. In the preliminary study the potential was tested for different ranges, as ±2V; ±1.0 V (SCE) a variation of scan rate between 100÷10 mV/s. There were no large modifications of the results in the polarization curves tested. The following measurements recorded for a potential of ±1.5V (SCE) at the 20 mVs^{-1} were take into consideration because of larger scan rate appeared in the distorsion of the curves also for higher potential range. At the same time all solutions were examined for conductivity measurements, using the Metrohm 712 Conductometer.

5.2 Results and discussions

5.2.1 Electroactivity of working solutions

The electrochemical behaviour of *SS* samples in *Neo* with and without fungal suspensions was investigated at room temperature. Figure 3 shows the polarization curves obtained for *SS* immersed in *Neo* and also in different fungal suspensions.

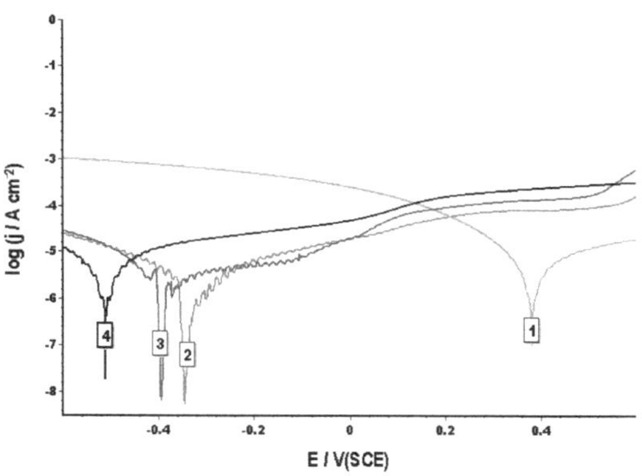

Fig. 3. Polarization curves of SS in *Neo* (1) and different fungal suspensions: *Aspergillus niger* (2), *Candida mycoderma* (3) and *Saccharomyces cerevisiae* (4).

A larger range passivation of the processes could be observed in both electrochemical reactions (anodic and cathodic). From the polarization curves (Fig. 3, curves 2-4), that the potential E_{corr} values show a more negative range for all fungal suspensions tested, whereas for *Neo* E_{corr} the value is situated at a positive potential range (Fig. 3, curve 1). Electrochemical parameters for *Neo* (Fig. 3, curve 1) are E_{corr} +390 mV and corrosion current density, i_{corr} 24 μA/cm². The potential E_{corr} and i_{corr} values are -316 mV respectively 30.27 μA/cm² for *Aspergillus niger* suspension and -338 mV, respectively 37 μA/cm² for *Candida mycoderma* suspension. In case of *Saccharomyces cerevisiae* suspension a - 511 mV value for E_{corr} respectively 70.40 μA/cm² for i_{corr} were obtained. These data confirm on the one hand the electroactivity behaviour of the tested biocide solutions and on the other hand a high electrons density, in the electrochemical system which contains only fungal suspensions. Just like electrons in wires, these ions contribute to the transport charge in the electric field and thus to the current flow. In this situation the system responds through a current predominant anodic and it can speed up the metal dissolution. A more current density is produced by *Saccharomyces cerevisiae* suspension (Fig. 3, curve 4). This fact could explain a more intense sensibility of these cells for the applied potential, due to their size. These cells appear bigger in comparison to *Aspergillus niger* spores and *Candida mycoderma* cells (Bui et al., 2008). Some authors consider that the larger cells are more sensitive to lower electric potential field, thus stronger than the smaller cells (Teissie et al., 1999; Kotnik et al., 2001). The applied potential for the acceleration of corrosion tests and biocide can produce a permeabilization of the plasma membrane and appear an electrochemical contamination in mixtures, due to the intracellular ions transit out of cells.

5.2.2 The corrosion effect of mixtures solutions

The electrochemical behaviour of the *SS* in the *Neo* with fungal suspension (without nitrogen) was evaluated at room temperature at different contact times after adding the fungal suspensions. The polarization curves are presented in Figures 4-6. Using Tafel fit. and R_p fit. analysis tool, made it possible to obtain the electrochemical parameters like corrosion potential (E_{corr}), current density (j_{corr}), polarization resistance (R_p), corrosion rate (V_{corr}). The results are presented in Tables 2-4. Figure 4 presents the variation of E_{corr} potential values during polarization curves of *SS* immersed in *Neo* and in *Neo - Aspergillus niger* at different contact time from the immersion of samples.

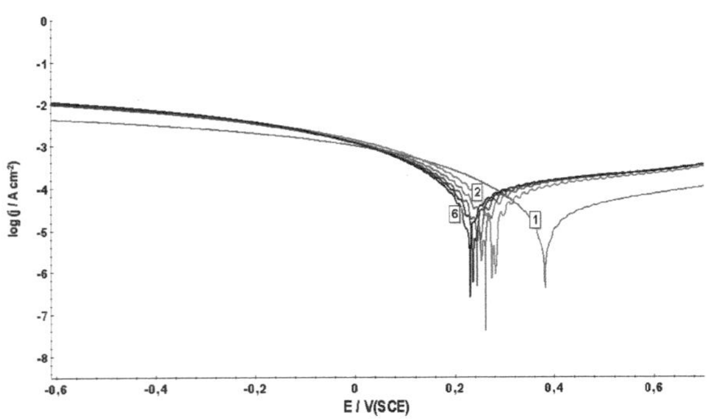

Fig. 4. Polarization curves of *SS* immersed in *Neo* (1) and *Neo - Aspergillus niger* at 3 min. (2); 6 min.(3); 9 min.(4); 12 min. (5) and 30 min. (6).

The potential E_{corr} values are in a negative range for the *Neo - Aspergillus niger* (Fig. 4, curves 2-6), whereas E_{corr} value is situated in the positive range for *Neo* (Fig. 4, curve 1). The absence of the parallelism of cathodic and anodic branches between the set of curves 2-6

corresponding to the *Neo - Aspergillus niger* and curve 1 for the only *Neo* biocide suggests that the processes that occur on *SS* surface are not similar.
Table 2 shows the electrochemical parameters of *SS* immersed in *Neo* and in *Neo - Aspergillus niger.*

Solution	Time (min.)	E_{corr} (mV)	j_{corr} (μA/cm²)	βc (mV/dec.)	βa (mV/dec.)	R_p (Ω·cm²)	V_{corr} (mm/y)
Neo	3	390	24	182	405	1722	0.0207
Neo - Aspergillus niger	3	296	70	203	478	595	0.0608
	6	263	80	198	536	580	0.0691
	9	257	77	201	497	562	0.0668
	12	247	77	198	509	553	0.0666
	30	239	76	189	519	573	0.0657

Table 2. Tafel parameters of *SS* immersed in *Neo* and *Neo - Aspergillus niger.*

As it can be observed from the Table 2 the potential E_{corr} value of *SS* immersed in *Neo* is +390 mV(SCE). By adding the *Aspergillus niger* suspension in *Neo*, the spores are exposed initially to a chemical attack. The spores wall due to its affinity to the active molecules oxidizing – H_2O_2, may restrict acces to the plasma cells membrane and to the cytoplasm. The oxidizing molecules have a damaging action to the spores wall, more or less crossed, then act at a cytoplasmic level where it causes the first serious damage. The spores wall penetration is physically a charge transfer between the wall (regarded as electrode) and the spores cytoplasm (regarded as electrolyte) and has as effect the charge modification of the double layer near the spores surface and also shift in potential. Under these circumstances, the wall can be crossed by a net current which raises with increasing the concentration of ionic species participating in the process. The application of electrical potential exposes the *Aspergillus niger* spores (chemically attacked) to the aggression of the electric field. The potential effects manifest through dielectric breakdown and migration of the cytoplasma, followed by electrolytic contamination of the *Neo* biocide. At the same time, the electrical applied potential leads to the occurrence of some oxidations and reductions followed by a potential E_{corr} shift in the negative range with an amplitude of about 94 mV (*i.e.* from +390 mV (SCE) at +296 mV) (SCE). During the experiment the E_{corr} values increases slightly, through a shift in positive range with an amplitude of about 8 mV to the end of the exposure. The shift of potential E_{corr}, suggests that the *Aspergillus niger* suspension added in *Neo* biocide is a depolarization agent for *SS* surfaces.
The electrochemical system *Neo - Aspergillus niger* indicate a current flow which generating the orderly movement between the surface and biocide solution whithout an interference. Morever ther is a current resulting after oxidation of some biomolecules in solution (not attacked intracytoplasmic) and a current after oxidation of intracytoplasmic electrolysis products, but go into solution because of the dielectric breakdown. This fact shown that in *Neo - Aspergillus niger* the current which flows through working electrode has a higher intensity than the current which flows through working electrode in *Neo* only. The addition of *Aspergillus niger* suspension is characterized by an increase of current density with 46 μA/cm², *i.e.* from 24 μA/cm² to 70 μA/cm². During the experiment, a slight breakthrough

corrosion current density was observed with an amplitude of about 10 μA/cm^2 at 6 minutes from exposure and after that it begins to decrease slightly up to the end of the exposure (Table 2). The decrease of current density suggests that the absorption of OH^- ions occurs and a possible surface passivation may appear. However, the current density does not return to the initial value and it maintains at levels three times higher than the initial value.
An increasing content of hydrogen and oxygen from oxidation of biomolecules by biocide substance could appear in the electrochemical system which contains *Neo - Aspergillus niger*. A large flow of electrons have as effect the cell current increase an electrode polarization. This polarization is confirmed by the shift in the cathodic (β_c) and anodic (β_a) Tafel slopes (Stoica et al., 2010b). As it can be observed from the Table 2, the *Neo - Aspergillus niger* move the cathodic (β_c) slope from 182 mV/dec. to 203 mV/dec. and the anodic (β_a) slope from 405 mV/dec. to 478 mV/dec. The shift of the Tafel slopes observed throughout the experiment (Table 2) reveals that the *Aspergillus niger* fungal suspension induces a corrosion mechanism on metallic surface (Stoica et al., 2010b). These results suggest that the *Neo - Aspergillus niger* controlling predominantly the anodic reactions.
There is a shift of the potential E_{corr} and the increase of j_{corr} and the decrease of polarization resistance, during experiment from the Figure's 4 curves and from the Table's 2 data. These results indicate a synergic effect between H_2O_2, *Aspergillus niger* spores and applied electric potential. This fact can substantially accelerate the corrosion process of metallic surfaces immersed in the mixture consisting of biocide and fungal suspension. The predominant anodic parameters could be an answer of the degradation metallic surfaces and this phenomenon is in good agreement with data previously reported in literature (Stoica et al, 2010b).
Figure 5 presents polarization curves of *SS* immersed in *Neo* and *Neo - Candida mycoderma* at different contact time.

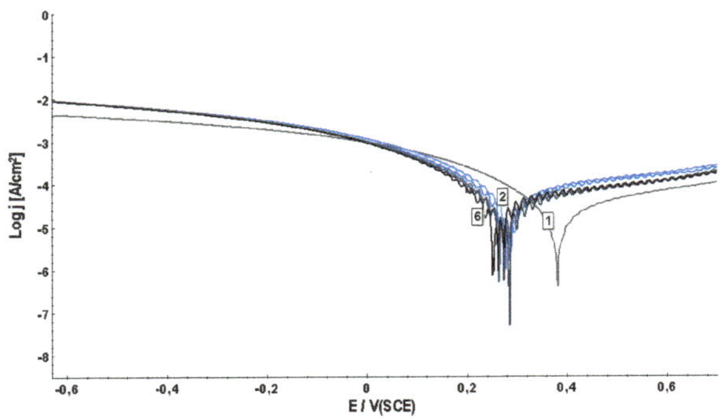

Fig. 5. Polarization curves of *SS* immersed in *Neo* (1) and *Neo - Candida mycoderma* at 3 min. (2); 6 min.(3); 9 min.(4); 12 min. (5) and 30 min. (6).

In Figure 5, it can be observed that the potential E_{corr} values are in negative range for the *Neo - Candida mycoderma* (curves 2-6), whereas E_{corr} value is situated in the positive range for *Neo* (Fig. 5, curve 1). Absence the parallelism of cathodic and anodic branches between the set of curves 2-6 which corresponding the *Neo - Candida mycoderma* and curve 1 which corresponding the *Neo* suggests, that the processes that occur on *SS* immersed in *Neo -*

Candida mycoderma are not similar with the processes that occur on *SS* immersed in *Neo* in good agreement to *Neo - Aspergillus niger.*

Table 3 presents the electrochemical parameters of *SS* immersed in *Neo* and *Neo - Candida mycoderma.*

Solution	Time (min.)	E_{corr} (mV)	j_{corr} (μA/cm²)	βc (mV/dec.)	βa (mV/dec.)	R_p (Ω·cm²)	V_{corr} (mm/y)
Neo	3	390	24	182	405	1722	0.0207
Neo - Candida mycoderma	3	294	58	196	488	746	0.0503
	6	289	50	195	484	827	0.0437
	9	296	41	194	566	961	0.0354
	12	276	33	161	488	1222	0.0285
	30	281	88	185	546	949	0.0334

Table 3. Tafel parameters of *SS* immersed in *Neo* and *Neo - Candida mycoderma.*

Similar to the electrochemical system (which contains *Neo - Aspergillus niger)* the *Neo - Candida mycoderma* system random currents circulate, whose intensity is determined by different source. The applied electric potential exposes the *Candida mycoderma* cells to the aggression of the electric field. The experimental potential effects manifest through dielectric breakdown and migration of the cytoplasmic content, followed by an electrolytic contamination of the *Neo* biocide. At the same time, the potential leads to the occurrence of some oxidations and reductions, followed by a potential E_{corr} shift in negative range with an amplitude of about 96 mV *i.e.* from +390 mV (SCE) at +294 mV (SCE). During exposure was observed that the corrosion potential fluctuated suggesting alternative passivation and activation of the surface and conducting its the spontaneous reversible oxidations. The shift of E_{corr} suggests that the *Candida mycoderma* suspension added in *Neo* biocide is a depolarization agent for *SS* surface. The addition of *Candida mycoderma* fungal suspension is characterized by an increase of current density with 34 μA/cm², *i.e.* from 24 μA/cm² to 58 μA/cm². During the experiment, the current density showed a reduction value after at 6 minutes (Table 3). This suggests the initiation of a partial surface passivation process, but which is insignificantly in an amplitude of 47 μA/cm² at exposure end (Table 3). In electrochemical system of *Neo - Candida mycoderma,* the polarization is confirmed by the shift in the cathodic (β_c) and anodic (β_a) Tafel slopes (Stoica et al., 2010b). As it can be observed from the Table 3, *Neo - Candida mycoderma* moves the cathodic (β_c) slope from 182 mV/dec. to 196 mV/dec. and the anodic (β_a) slope from 405 mV/dec. to 488 mV/dec. The shift of the Tafel slopes observed throughout the experiment (Table 3) reveals that the *Candida mycoderma* fungal suspension induces a corrosion mechanism on metallic surface (Stoica et al., 2010b). These results suggest that the *Neo - Candida mycoderma* controlling predominantly the anodic reactions.

There is a shift of the potential E_{corr}, the increase of j_{corr} and the decrease of polarization rezistance, during experiment (Figure 5 curves and Table 3). The results indicate a synergic effect between H_2O_2 *Candida mycoderma* cells and apllied electric potential, which manifests through a distorted current predominantly anodic. The predominant anodic mechanism observed in this case, is in good agreement with experimental results previously reported in case of H_2O_2 – *Aspergillus niger* – applied electric potential.

Figure 6 presents the polarization curves of *SS* immersed in *Neo* and *Neo - Saccharomyces cerevisiae* at different contact time.

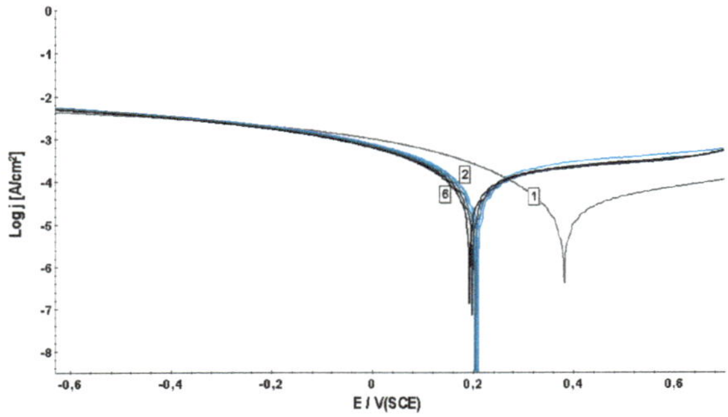

Fig. 6. Polarization curves of *SS* immersed in *Neo* (1) and *Neo - Saccharomyces cerevisiae* at 3 min. (2); 6 min.(3); 9 min.(4); 12 min. (5) and 30 min. (6).

The potential E_{corr} values are in negative range for the *Neo - Saccharomyces cerevisiae* (Fig. 6, curves 2-6), whereas E_{corr} value is situated in the positive range for *Neo* (Fig. 5, curve 1). The absence of the parallelism of cathodic and anodic branches between the curves 2-6 corresponding the *Neo - Saccharomyces cerevisiae* and curve 1 corresponding the *Neo* suggests that the processes are not similar with the processes that occur on *SS* immersed in *Neo - Aspergillus niger* and *Neo - Candida mycoderma* system.
Table 4 reveals the electrochemical parameters of *SS* immersed in *Neo* and *Neo - Saccharomyces cerevisiae.*

Solution	Time (min.)	E_{corr} (mV)	j_{corr} (μA/cm²)	βc (mV/dec.)	βa (mV/dec.)	R_p (Ω·cm²)	V_{corr} (mm/y)
Neo	3	390	24	182	405	1722	0.0207
Neo-Saccharomyces cerevisiae	3	213	114	247	449	418	0.0993
	6	216	98	238	492	475	0.0852
	9	210	95	236	482	507	0.0824
	12	205	93	236	514	507	0.0802
	30	198	94	236	495	499	0.0816

Table 4. Tafel parameters of *SS* immersed in *Neo* and *Neo-Saccharomyces cerevisiae.*

Similar to electrochemical systems (*Neo - Aspergillus niger* and *Neo - Candida mycoderma*), through electrochemical system *Neo - Saccharomyces cerevisiae*, the random currents circulate, by their source of origin. The applicated potential exposes the *Saccharomyces cerevisiae* cells to the aggression of the electric field. The experimental potential effects manifest through dielectric breakdown and migration of the cytoplasmic content, followed by an electrolytic contamination of the *Neo* biocide. At the same time, the potential lead to the occurrence of some oxidations and reductions, followed by a potential E_{corr} shiftting in negative range with an amplitude of about 180 mV *i.e.* from +390 mV (SCE) at +213 mV (SCE). During the exposure the corrosion potential fluctuated suggesting alternative passivation and activation of the surface and conducting its spontaneous reversible oxidations. The shift of

E_{corr} suggests that the *Saccharomyces cerevisiae* suspension added in *Neo* biocide is a depolarization agent for *SS* surface.

The addition of *Saccharomyces cerevisiae* fungal suspension is characterized by an increase of current density with 90 μA/cm², *i.e.* from 24 μA/cm² to 114 μA/cm². During the experiment, the current density showed a value reduction after 6 minutes. This suggests of initiation of partial surface passivation process with insignificantly the amplitude of about 1 μA/cm² at exposure end (Table 4). In electrochemical system *Neo-Saccharomyces cerevisiae*, the polarization is confirmed by the shift in the cathodic (β_c) and anodic (β_a) Tafel slopes (Stoica et al., 2010b). The cathodic (β_c) slope from 182 mV/dec. to 247 mV/dec. and the anodic (β_a) slope from 405 mV/dec. to 449 mV/dec., for the first 3 minutes from exposure moves (Table 4). The shift of the Tafel slopes observed throughout the experiment reveals that the *Saccharomyces cerevisiae* fungal suspension induces a corrosion mechanism on metallic surfaces (Stoica et al., 2010b) and suggests that *Neo-Saccharomyces cerevisiae* system controls the cathodic and anodic reactions, being predominantly anodic.

There is a shift of the potential E_{corr}, the increase of j_{corr} and the decrease of polarization rezistance, during the experiment (Fig. 6 and Table 4). The results indicate a synergistic effect between H_2O_2 *Saccharomyces cerevisiae* cells and apllied potential. The predominant anodic mechanism observed in this case, is in good agreement with experimental results previously reported for the systems: H_2O_2 - *Aspergillus niger* - electric potential and H_2O_2 - *Candida mycoderma* - applied potential.

5.2.3 Conductivity variations of solutions

Conductivity is simply the ability of a liquid to conduct electricity, which is related directly to the concentration of ions in the liquid. Figure 7 gives the variation of solutions conductivity of *Neo* compared with *Neo* with fungal suspensions tested.

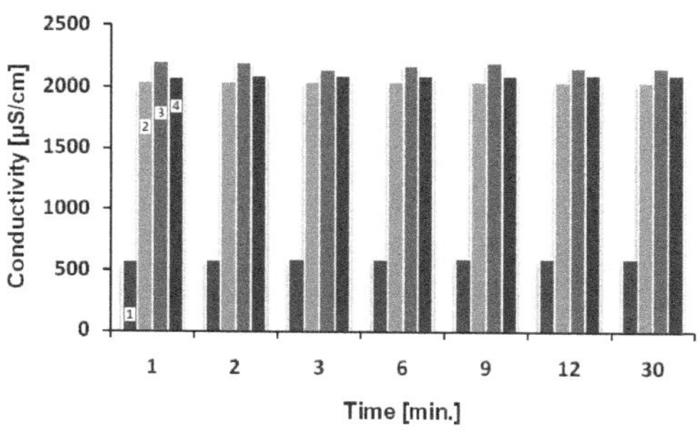

Fig. 7. Conductivity variation in *Neo* (series 1) and *Neo* with different fungal suspension: *Aspergillus niger* (series 2), *Candida mycoderma* (series 3) and *Saccharomyces cerevisiae* (series 4).

The presence of the fungal suspensions can be easily observed in the *Neo* biocide (Fig. 7, series 2-4) with a strong effect on the conductivity values in comparison with those observed for *Neo* without fungal cells (Fig. 7, series 1). This fact could suggest that in *Neo* with fungal cells an electrolytic contamination is produced, which can be explained through synergistic action between H_2O_2 and the applied potential inside the fungal cells, although the ions efflux remains difficult to elucidate (Sukhorukov et al., 2007). As a result the mixtures consisting of *Neo* biocide with all fungal suspensions are very conductive systems and they can accelerate the corrosion of AISI 304 stainless steel immersed in them.

5.2.4 Synergic of working parameters

The corrosion behaviour on *SS* was tested in the biocide solution with and without fungal suspensions. The working parameters taken into account in accelerated corrosion tests (LP) were: nature and size of cells, cellular density of suspensions and conductivity of solutions, applied potential and pretreatment of working electrode of *SS*. Eletctrochemical parameters can justify the shift of E_{corr} potential in the anodic direction and increase of i_{corr} for mixtures consisting by *Neo* biocide with fungal suspensions. The corrosion behaviour of *SS* surface is specific to each system and it is made evident through R_p (polarization resistance) and V_{corr} (corrosion rate) variations from Tables 2 - 4.
Figure 8 shows the R_p values of *SS* immersed in *Neo* and *Neo* with fungal suspension at different contact time after immersion.

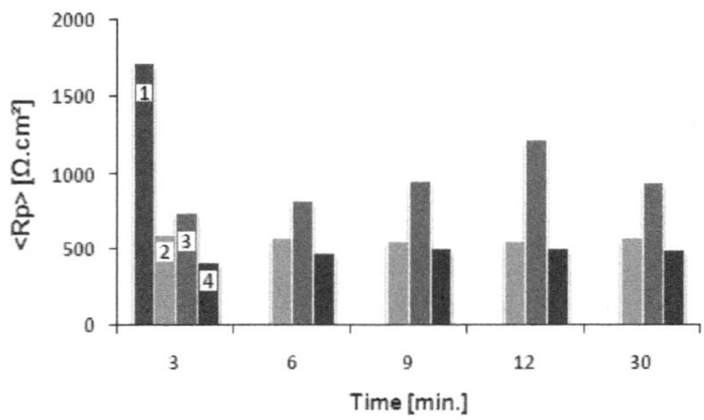

Fig. 8. R_p values of *SS* immersed in *Neo* and *Neo* with fungal suspension: *Aspergillus niger* (series 2), *Candida mycoderma* (series 3) and *Saccharomyces cerevisiae* (series 4).

The R_p values are lower in *Neo* with fungal suspension than in *Neo* only. The order of magnitude of R_p is 10^4 Ω.cm^2 for *Neo*. In case of *Neo - Aspergillus niger* is observed a greatly decrease of R_p with the increase in time. In *Neo - Candida mycoderma* system the R_p values decrease up to minute 3 and afterwards a significant increase was obtained up to minute 12 followed by a lower decrease at final contact time (Fig. 8, series 3). In *Neo - Saccharomyces cerevisiae* a significant decrease in the R_p values was observed (Fig. 8, series 4). The order of magnitude of R_p is 10^3 Ω.cm^2 in *Neo* with fungal suspensions.Thus, the *SS* surface is more susceptible in electrochemical system containing H_2O_2 from *Neo* biocide with *Aspergillus niger* suspension, respectively *Saccharomyces cerevisiae* suspension.
Figure 9 presents the V_{corr} values of *SS* immersed in *Neo* and *Neo* with fungal suspension at different contact time after immersion.
In case of *Neo - Aspergillus niger* the V_{corr} values increase more in time up to minutes 6 and afterwards a smaller decrease was obtained up to minutes 30 (Fig. 9, series 2). The V_{corr} values decrease more in *Neo - Candida mycoderma* system up to 12 minutes followed by an increase at final contact time (Fig. 9, series 3). The highest increase of V_{corr} values was obtained when *Saccharomyces cerevisiae* was added in *Neo*, at the beginning of the measurements and then there was a slight decrease up to minute 12 from immersion contact time (Fig. 9, series 4). Thus, the *SS* surface is more corrosive in electrochemical system

containing H_2O_2 from *Neo* biocide with *Aspergillus niger* respectively *Saccharomyces cerevisiae.* This fact suggests that during food line disinfection the corrosion rate of AISI 304 Stainless Steel surfaces exceeds 0.02 mm/y (Fontana, 1987) and reduces the equipments lifetime. It is possible that the killing action showed in these experiments is due to other chemical reactions occurring while the electrical currents pass through the liquid medium. The results obtained could suggest that an electrochemical disinfection is quite attractive as a promising alternative technology. The conductivity and potential sensors constantly monitor the concentration of the biocide solutions with cells contamination are benefic in cleaning the food processing. The *SS* surface behaviour is complex and requires further investigation for its understanding under the action of different biocides and fungal suspensions.

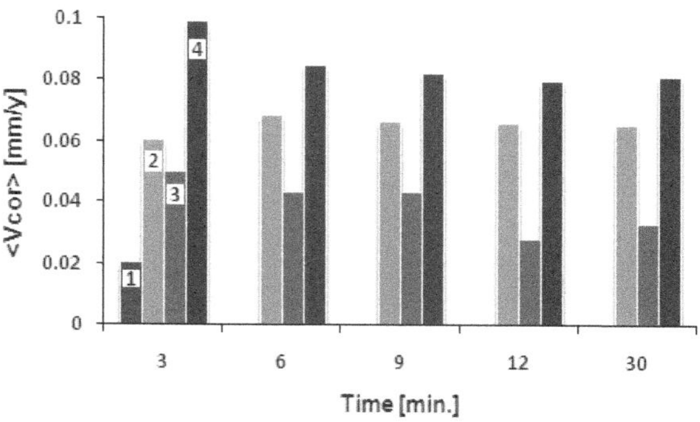

Fig. 9. V_{corr} values of *SS* immersed in *Neo* (1) and trend V_{corr} values of *SS* immersed in *Neo* with fungal suspensions (10% vol.): *Aspergillus niger* (series 2), *Candida mycoderma* (series 3) and *Saccharomyces cerevisiae* (series 4).

5.3 Remarks

The work deals with the corrosion behaviour of AISI 304 Stainless Steel into biocide solution (*Neoseptal* with hydrogen peroxide as active substance) through artificial contamination with three (10% vol.) fungal suspensions as: *Aspergillus niger, Candida mycoderma* and *Saccharomyces cerevisiae.* At the applied electrical potential the biocide can work better within the fungal cells, and thus disturbing the present microorganisms homeostasis in several ways such as increasing the environmental conductivity of solutions and the corrosion rate of metallic support. A synergic effect achieved through the mixture of biocide, fungal suspension and applied electric potential, is more destructive than each parameters by its self. The E_{corr} values of AISI 304 Stainless Steel in the mixtures decreased during the contact time after artificial contamination. The fungal suspension has a significant influence on the synergic effect of the AISI 304 Stainless Steel corrosion in the following order: *Saccharomyces cerevisiae*>*Aspergillus niger*>*Candida mycoderma.* A more influence on the synergic effect of the surfaces immersed at the mixture consisting on biocide *Neoseptal* solution with *Saccharomyces cerevisiae* could be explained through the less resistance at the chemical attack from biocide. The synergic effect between the active substance of the disinfectant, fungal suspensions and the applied electric potential should taken into account for the hygienic and safety food bioprocessing industry.

6. Conclusions

Experimental data conclude that the mixtures consisting of biocide with fungal suspensions has an major effect to increase the anode current density, leading to further degradation by corrosion of the AISI 304 Stainless Steel surfaces. The results showed that there is a synergic effect between the active substances from the disinfectant, fungal suspensions and applied electric potential. This chapter could be considered a pioneer research and brinks possible ideas for further researches and raises more issues that need to be explored on the food processing.

7. Acknowledgment

This work has benefited from financial support through the 2010 POSDRU/89/1.5/S/52432 project, ORGANIZING THE NATIONAL INTEREST POSTDOCTORAL SCHOOL OF APPLIED BIOTECHNOLOGIES WITH IMPACT ON ROMANIAN BIOECONOMY, project co-financed by the European Social Fund through the Sectoral Operational Programme Human Resources Development 2007-2013.

8. References

Bessems, E. (1998). The effect of practical conditions on the efficacy of disinfectants. *International Biodeterioration and Biodegradation*, Vol.41, No. 3-4, (April 1998), pp. 177-183, ISSN 0964-8305

Boulange-Petermann, L., Jullien, C., Dubois, P.E., Benezech, T. & Faille, C. (2004). Influence of Surface Chemistry on the Hygienic Status of Industrial Stainless Steel. *Biofuling*, Vol.20, No.1, (February 2004), pp. 25–33, ISSN 0892-7014

BPD (1998). Biocidal Products Directive 98/8/EC, Date of access 22 August 2011; Available from: < ec.europa.eu/environment/biocides/>

Breslin, P.A.S. & Tharp, C.D. (2001). Reduction of Saltiness and Bitterness After a Chlorhexidine Rinse. *Chemical Senses*, Vol.26, No.2, (February 2001), pp. 105–116, ISSN 0379-864X

Bridier, A., Briandet, R., Thomas, V. & Dubois-Brissonnet, F. (2011). Comparative biocidal activity of peracetic acid, benzalkonium chloride and ortho-phthalaldehyde on 77 bacterial strains. *Journal of Hospital Infection*, Vol.78, No.3, (July 2011), pp. 208-213, ISSN 0195-6701

Buchalla, W., Attin, T. & Hellwig, E. (2000). Brushing Abrasion of Luting Cements under Neutral and Acidic Conditions. *Operative Dentistry*, Vol.25, No.6, (November/December 2000), pp. 482–487, ISSN 0361-7734

Bui, V.C., Kim, Y.U. & Choi, S. (2008). Physical Characteristics of Saccharomyces cerevisiae. *Surface and Interface Analysis*, Vol.40, No.10, (October 2008), pp. 1323–1327, ISSN 0142-2421

Carvalho, C.E.R. de, Costa, G.M. da, Cota, A.B. & Rossi, E.H. (2006). High temperature oxidation behavior of AISI 304 and AISI 430 stainless steels. *Materials Researh*, Vol.9, No.4, (Octomber/December 2006), pp. 393-397, ISSN 1516-1439, Available from

<http://www.scielo.br/scielo.php?script=sci_arttext&pid=S1516 14392006000400009&lng=en&nrm=iso>. http://dx.doi.org/10.1590/S1516-14392006000400009>

Deacon, J.W. (1997). *Modern Mycology* (3rd Edition), Blackwell Science Ltd, London , UK

Fontana, M. G. (1987). *Corrosion Engineering* (3rd Edition), McGraw-Hill Book Company, New York, USA

Gautam, A.K., Sharma, S., Avasthi, S. & Bhadauria, R. (2011). Diversity, pathogenicity and toxicology of *A. niger*: An important spoilage fungi. *Research Journal of Microbiology*, Vol.6, No.3, (March 2011), pp. 270-280, ISSN 1816-4935

Hiromoto, S. & Hanawa T. (2006). Electrochemical properties of 316L stainless steel with culturing L929 fibroblasts. *Journal of the Royal Society Interface*, Vol.3, No.9, (August 2006), pp. 495–505, ISSN 1742-5689

Holah, J.T., & Thorpe, R.H. (1990). Cleanability in Relation to Bacterial Retention on Unused and Abraded Domestic Sink Materials. *Journal of Applied Bacteriology*, Vol.69, No.4, (October 1990), pp. 599–608, ISSN 0021-8847

Hugo, W.B., Denyer, S.P., Hodges, N.A. & Gorman, S.P. (2004). *Hugo and Russell's pharmaceutical microbiology* (7th Edition), Wiley-Blackwell Publishing, Oxford, UK

Kitis, M. (2004). Disinfection of wastewater with peracetic acid: a review. *Environment International*, Vol.30, No.1, (March 2004), pp. 47-55, ISSN 0160-4120

Kotnik, T., Mir, L.M., Flisar, K., Puc, M. & Miklavčič, D. (2001). Cell Membrane Electropermeabilization by Symmetrical Bipolar Rectangular Pulses: Part I. Increased Efficiency of Permeabilization. *Bioelectrochemistry*, Vol.54, No.1, (August 2001), pp. 83–90, ISSN 1567-5394

Landoulsi, J., El-Kirat, K., Richard, C., Feron, D. & Pulvin, S. (2008). Enzymatic Approach in Microbial Influenced Corrosion: A Review Based on Stainless Steels in Natural Waters. *Environmental Science and Technology*, Vol.42, No.7, (April 2008), pp. 2233–2242, ISSN 0013-936X

Larson, E.L. & Morton, H.E. (1991). Alcohols, In: *Disinfection, Sterilization and Preservation*, S. S. Block, (Ed.), 191–203, Philadelphia: Lea & Febiger, Philadelphia, USA

Leclercq-Perlat, M.N. & Lalande, M. (1994). Cleanability in Relation to Surface Chemical Composition and Surface Finishing of some Materials Commonly Used in Food Industries. *Journal of Food Engineering*, Vol.23, No.4, pp. 501–517, ISSN 0260-8774

Lu, X., Sun, J., Nimtz, M., Wissing, J., Zeng A-P. & Rinas, U. (2010). The intra- and extracellular proteome of *Aspergillus niger* growing on defined medium with xylose or maltose as carbon substrate. *Microbial Cell Factories*, Vol.9, (April 2010), pp. 23, ISSN 1475-2859

Mai, T.L., Sofyan, N.I., Fergus, J.W., Gale, W.F. & Conner, D.E. (2006). Attachment of Listeria monocytogenes to an austenitic stainless steel after welding and accelerated corrosion treatments. *Journal of Food Protection*, Vol.69, No.7, (July 2006), pp. 1527-1532, ISSN 0362-028X

Maier, R.M., Pepper, I.L. & Gerba, C.P. (2009). *Environmental Microbiology* (2nd Edition). Academic Press, London, UK

Marcus, P., Maurice, V. & Strehblow, H.-H. (2008). Localized corrosion (pitting): A model of passivity breakdown including the role of the oxide layer nanostructure. *Corrosion Science*, Vol.50, No.9, (September 2008), pp. 2698–2704, ISSN 0010-938X

McDonnell, G. & Russell, A.D. (1999). Antiseptics and Disinfectants: Activity, Action, and Resistance. *Clinical Microbiology Reviews*, Vol.12, No.1, (January 1999), pp. 147–179, ISSN 0893-8512

Ofoegbu, S.U., Ofoegbu, P.U., Neife, S.I. & Okorie, B.A. (2011). Corrosion Behaviour of Steels in Nigerian Food Processing Environments, *Journal of Applied Sciences and Environmental Management*, Vol.15, No.1, (March 2011), pp. 135 - 139, ISSN 1119-8362

Olsson, C.O.A. & Landolt, D. (2003). Passive Films on Stainless Steels--chemistry, Structure and Growth. *Electrochimica Acta: Microscopic and Nanoscopic Aspects of Corrosion and Corrosion Protection*, Vol.48, No.9, (April 2003), pp. 1093–1104, ISSN 0013-4686

Osarolube, E., Owate, I.O. & Oforka, N.C. (2008). Corrosion Behaviour of Mild and High Carbon Steels in Various Acidic Media. *Scientific Research and Essay*, Vol.3, No.6, (June 2008), pp. 224–228, ISSN 1992-2248

Papagianni, M. (2007). Advances in citric acid fermentation by *Aspergillus niger*: Biochemical aspects, membrane transport and modeling. *Biotechnology Advances*, Vol.25, No.3, (May 2007), pp. 244–263, ISSN 0734-9750

Pitt, J.I. & Hocking, A.D. (1997). *Fungi and Food Spoilage* (2nd Edition), Blackie Academic & Professional, London, UK

Pommerville, J.C. (2010). *Alcamo's Fundamentals of Microbiology*, (9th Edition), Jones and Bartlett Publishers, Sudbury, USA

REACH (2006). Date of access 22 August 2011; Available from: <ec.europa.eu/.../reach/reach_intro.htm>

Round, A.N., Rigby, N.M., Ring, S.G. & Morris, V.J. (2001) Investigating the nature of branching in pectins by atomic force microscopy and carbohydrate analysis. *Carbohydrate Research*, Vol. 331, No.3, (April 2001), pp. 337-342, ISSN 0008-6215

Russell, A. D. (2003). Similarities and Differences in the Responses of Microorganisms to Biocides. *Journal of Antimicrobial Chemotherapy*, Vol.52, No.5, (September 2003), pp. 750–763 ISSN 0305-7453

Russell, A.D. & McDonnell, G. (2000). Concentration: a major factor in studying biocidal action. *Journal of Hospital Infection*, Vol.44, No.1, (January 2000), pp. 1-3, ISSN 0195-6701

Saikhwan, P., Geddert, T., Augustin, W., Scholl, S., Paterson, W.R. & Wilson, D.I. (2006). Effect of Surface Treatment on Cleaning of a Model Food Soil. *Surface and Coatings Technology*, Vol.201, No.3-4, (October 2006), pp. 943–951, ISSN 0257-8972

Shen, Y.J., Lei, Le C. & Zhang, X.W. (2008). Evaluation of energy transfer and utilization efficiency of azo dye removal by different pulsed electrical discharge modes. *Chinese Science Bulletin, Vol.*53, No.12, (June 2008), pp. 1824-1834, ISSN 1001-6538

Stoica, M., Brumă, M. & Cârâc, G. (2010b). Electrochemical study of AISI 304 SS at disinfectants with fungi. *Materials and Corrosion*, Vol.61, No.12, (December 2010), pp. 1017-1025, ISSN 0947-5117

Stoica, M., Cârâc, G., Tofan, C., Constantin, O.E. & Enache, G. (2009). Effect of fungal suspensions in NaDCC disinfectant on the corrosion behaviour AISI 304 stainless steel. *Journal of Agroalimentary Processes and Technologies*, Vol.15, No.4, (December 2009), pp. 543-546, ISSN 1453-1399

Sukhorukov, V.L., Endter, J.M., Zimmermann, D., Shirakashi, R., Fehrmann, S., Kiesel, M., Reuss, R., Becker, D., Hedrich, R., Bamberg, E., Roitsch, Th. & Zimmermann, U. (2007). Mechanisms of Electrically Mediated Cytosolic Ca^{2+} Transients in Aequorin-Transformed Tobacco Cells. *Biophysical Journal*, Vol. 93, No.9, (August 2007), pp. 3324-3337, ISSN 0006-3495

Tavares, S.S.M., Pardal, J.M., Guerreiro, J.L., Gomes, A.M. & da Silva M.R. (2010). Magnetic detection of sigma phase in duplex stainless steel UNS S31803. *Journal of Magnetism and Magnetic Materials*, Vol.322, No.17, (March 2010), pp. L29-L33, ISSN 0304-8853

Teissie, J., Eynard, N., Gabriel, B. & Rols, M.P. (1999). Electropermeabilization of Cell Membranes. *Advanced Drug Delivery Reviews*, Vol.35, No.1, (January 1999), pp. 3–19, ISSN 0169-409X

Tisnadjaja, D., Gutierrez, N.A. & Maddox, I.S. (1996). Citric Acid Production in a Bubble-column Reactor using Cells of the Yeast Candida guilliermondii immobilized by Adsorption onto Sawdust. *Enzyme and Microbial Technology*, Vol.19, No.5, (October 1996), pp. 343–347 ISSN 0141-0229

Tournas, V.H., Heeres, J. & Burgess, L. (2006). Moulds and yeasts in fruit salads and fruit juices. *Food Microbiology*, Vol.23, No.7, (October 2006), pp. 684-688, ISSN 0740-0020

Turssi, C.P., Magalhães, C.S., Serra, M.C. & Rodrigues, A.L. (2001). Surface Roughness Assessment of Resin-based Materials during Brushing Preceded by pH Cycling Simulations. *Operative Dentistry*, Vol.26, No.6, (November/December 2001), pp. 576–584, ISSN 0361-7734

Whitehead, K.A. & Verran, J. (2006). The effect of substratum surface topography on the retention of microorganisms and organic material on defined surfaces, *Fouling, Cleaning and Disinfection in Food Processing 2006: Proceedings of a Conference held at Jesus College*, ISBN 9780954248314, Cambridge, UK, March 20-22, 2006

Woese, C. (1998). The universal ancestor. *Proceedings of the National Academy of Sciences*, Vol.95, No.12, (June 1998), pp. 6854–6859, ISSN 0027-8424

Woese, C.R. & Gupta, R. (1981). Are archaebacteria merely derived 'prokaryotes'?". *Nature*, Vol. 289, No. 5793, (January 1981), pp. 95–96, ISSN 0028-0836

Woese, C., Kandler, O. & Wheelis, M. (1990). Towards a natural system of organisms: proposal for the domains Archaea, Bacteria, and Eucarya. *Proceedings of the National Academy of Sciences*, Vol.87, No.12, (June 1990). pp. 4576–4579, ISSN 0027-8424

Yang, L., Banada, P.P., Bhunia, A.K. & Bashir, R. (2008). Effects of Dielectrophoresis on Growth, Viability and Immuno-reactivity of Listeria Monocytogenes. *Journal of Biological Engineering*, Vol.2, No.1, (April 2008), pp. 2–6, ISSN 1754-1611

Yuan, S.J. & Pehkonen, S.O. (2007). Microbiologically influenced corrosion of 304 stainless steel by aerobic Pseudomonas NCIMB 2021 bacteria: AFM and XPS study. *Colloids and Surfaces B: Biointerfaces*, Vol.59, No.1, (September 2007), pp. 87-99, ISSN 0927-7765

Zani, F., Minutello, A., Maggi, L., Santi, P. & Mazza, P. (1997). Evaluation of preservative effectiveness in pharmaceutical products: the use of a wild strain of Pseudomonas cepacia. *Journal of Applied Microbiology,* Vol.83, No.3, (August 1997), pp. 322-326, ISSN 1364-5072

7

Review: Potential Antioxidants from Tropical Plants

Lai Teng Ling[1] and U.D. Palanisamy[2]
[1]*University of Malaya*
[2]*Monash University Sunway Campus*
Malaysia

1. Introduction

1.1 Background

Higher plants have been utilised as important sources of medicinal drugs and health products since ancient days. Our ancestors practice the use of plants in their daily life as medicines. Therefore, the inherited knowledge about the traditional medicine is a key factor that could promote the development of modern drugs. Advances in modernisation and progress are expected to maximise the benefits of traditional herbal medicines for public health care (Chan, 2003). Investigations about plants have yielded amazing discoveries and development in modern medicine. Scientists carry out chemical investigation and purification of plants to get purified compounds which contribute to its medicinal properties.

The Malaysian rainforest is rich in diverse species of flora and fauna. Our forests store plenty of plant species which are important sources of traditional medicine. From about 10,000 species of higher plants and 2000 species of lower plants available in Peninsular Malaysia, approximately 16% are identified to be useful in medicines (Lattif et al., 1984). There are still a great number of unexplored plants that have high potential to be developed into medicines.

To date, a huge number of plants have been studied for their potential sources of antioxidants. Plants contain a wide variety of free radical scavenging molecules, such as polyphenols, dietary glutathione, vitamins and endogenous metabolites. These natural products make good antioxidants. Plant-derived antioxidants have been shown to function as singlet and triplet oxygen quenchers, peroxide decomposers, enzyme inhibitors and synergists (Larson, 1988). Studies have proven the correlation between the intake of fruits and vegetables and the morbidity and mortality from degenerative diseases (Rimm et al., 1996). It is not known which specific dietary constituents are responsible for this association, but antioxidants are assumed to be the major compounds that play an important role (Gey et al., 1991). Epidemiological studies that analyse the health implications of dietary components rely on the intake estimates in sample populations found in databases that list the component's content in commonly consumed foods (Pellegrini et al., 2003).

Besides dietary sources, antioxidants can be obtained in bulk from food processing industries and agricultural-by-products. The respective by-products are seeds, peels, bark, mill wastes and trimming wastes. In the citrus industries, industrial-by-products may account for up to 50% of the total fruit weight (Bocco et al., 1998). The utilisation of the by-products in those industries is beneficial to both the economy and the environment. The industrial-by-products, like peel and seed, are proven to have high antioxidant level which is even higher than the flesh and other parts of the fruit with the presence of high polyphenol content. This is true for *Viburnum opulus* seed (Cam et al., 2007), peach peel (Chang et al., 2000), apple peel (He & Liu, 2007), mangosteen peel (Moongkarndi et al., 2004) and grape seed and skin (Rockenbach et al., 2011).

Numerous studies have been carried out on other potential agricultural-by-products such as trees. They serve as one of the cheapest available source of antioxidants. Among different parts of the plants, leaves receive special attention, e.g. Etlingera genus (Chan et al., 2011), Olea europaea (Silva et al., 2006), *Ligustrum vulgare* (Agati et al., 2009) and *Stevia rebaudiana* (Tadhani et al., 2007) ; bark from *Casuarina equisetifolia* (Zhang et al., 2010), *Acacia confusa* (Chang et al., 2001), *Populus tremuloides Michx* (Diouf et al., 2009); root of *Medicago sativa* (Dalton et al., 1998) and *Carissa spinarum* (Hegde & Joshi, 2010) were also reported to contain antioxidants.

Despite the utilisation of agricultural-by-products for human consumption, the safeties of the products have raised much attention. Dietary exposure to heavy metals, especially cadmium (Cd), lead (Pb), zinc (Zn) and copper (Cu), has been identified as a risk to human health. Heavy metals may be present in trace amounts occurring naturally in plants grown in the soil (Boruvka et al., 1997). Heavy metals have also been found in herbal medicines from Malaysia, e.g. *Eurycoma longifolia* products (Ang et al., 2003). Studies in Malaysia showed that only 92% of the products complied with the quality requirement for traditional medicines in the country, however, they cannot be assumed to be safe from lead contamination because of batch-to-batch inconsistency (Ang et al., 2003). Cadmium is reported to accumulate in the kidney. There is overwhelming evidence that the cadmium induced tubular damage which is irreversible (Järup et al., 1998). Therefore, it is important to ensure that plants that are consumed by humans do not contain heavy metals higher than the permissible levels. This has to be monitored carefully to ensure the safety of plant parts used as nutraceuticals.

Generally, antioxidants extracted from plants show prooxidant activity at low concentration and antioxidant activity at higher concentrations (Yen et al., 1997). However, the opposite effect was observed in the case of ascorbic acid in the presence of transition irons (Halliwell, 1996). These findings remind us the importance of quantifying the prooxidant capacity of an extract that exerts high antioxidant activity and to interpret net antioxidant potential.

To date, there is no information concerning the profile of Malaysian plants, its antioxidant/prooxidant activity, cytotoxicity, heavy metal contamination and method of standardisation. The purpose of this study is to bridge this gap of knowledge. Direct beneficiaries of this research would be the general public, the herbal industries and the natural product researchers in Malaysia and elsewhere.

2. Assessment of antioxidant capacity and cytotoxicity of selected malaysian plants

Malaysia is rich in its biodiversity with over 12,000 flowering plant species, many of which are currently being used in traditional medicine. To date, a huge number of plants have been studied for its potential source of antioxidants. However, there exist no reports on the antioxidant activity, heavy metal and elemental analysis and cytotoxicity of Malaysian plants which are potential sources of new antioxidants. A study was carried out in our laboratory to evaluate selected Malaysian plants for its free radical scavenging, inhibition in lipid peroxidation, phenolic and content. This was followed with the heavy metal and elements analysis and its cytotoxicity against several cell lines. The plants were chosen based on its use in traditional application within the region. The investigation is essential to establish the phenolic content of selected Malaysian plants, its capability as potential antioxidant and ensure its safety (Ling et al., 2010a). Figure 1 shows the cytotoxicity activity of the selected Malaysian plants.

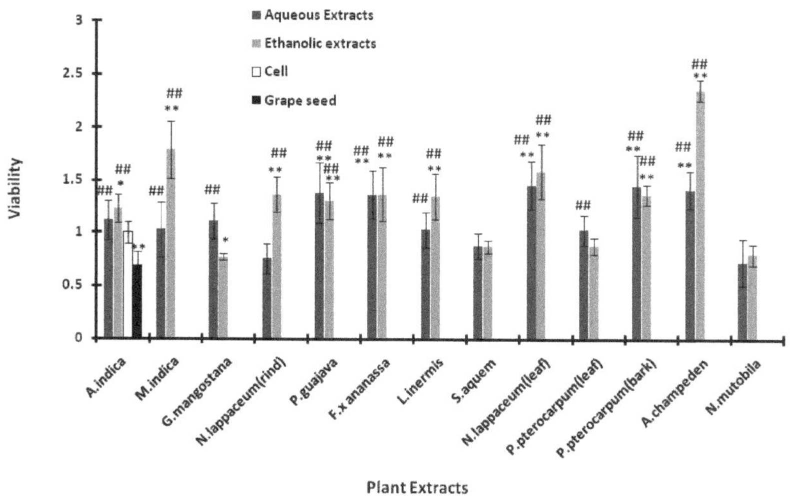

Fig. 1. Cytotoxicity activity of selected Malaysian plants on 3T3 mouse fibroblast cell at the concentration of 100µg/mL. * designates significant difference from cell alone ($P< 0.05$), ** designates a significant difference from cell alone, ($P< 0.01$), ## designates a significant difference from grape seed ($P<0.01$).

In our study, ethanolic and aqueous extracts of more than seventy Malaysian plants were screened using the DPPH assay. After the initial screen, thirteen plant extracts were selected with IC_{50} values lower than 5mg/mL and further tested with other free radical scavenging assays and for its ability to inhibit lipid peroxidation.. The extracts were also evaluated for its total phenolic content, heavy metal content and cytotoxicity. In general, the ethanolic extracts were observed to be far better free radical scavengers than the aqueous extracts. Some of the extracts were more potent (IC_{50} values) than the commercial grape seed antioxidant preparation (Agricultural Research Institute Speyer, Germany) and vitamin C. Generally, ethanolic extracts showed better activity in free radical scavenging assays and inhibition in lipid peroxidation compared to their aqueous extracts as well as higher than the commercial grape seed preparation. Most of the selected plant extracts showed hardly any heavy metal contamination in the powderised plants, some extracts even showed the presence of essential trace mineral. Majority of the plant extracts did not exhibit anti-proliferative effects on cultured mouse fibroblast and breast cancer cell indicating that most of the plant extracts are not cytotoxic to the cell been studied. We also observed a positive correlation between the ethanolic extract, phenolic content and antioxidant activity. We can conclude that although a broaden use of these plants are in aqueous form, its commercial preparation can be achieved using ethanol since a high total phenolic content and antioxidant activity was seen in this preparation. It is desirable that these extracts be further purified to gain a better understanding of the active compounds contributing to its antioxidant activity.

Similar results were observed in the lipid peroxidation inhibition studies (Palanisamy et al., 2008). There was a strong correlation between antioxidant activity and the total phenolic content of the extracts. The high antioxidant extracts had below the permissible value of heavy metal content for nutraceutical application. Most of the extracts were also not cytotoxic to 3T3 and 4T1 cells at concentration as high as 100µg/mL (Ling et al., 2010a).

3. Prooxidant/antioxidant ratio (proantidex) as a better index of net free radical scavenging potential

Antioxidants are substances that protect other chemicals of the body from damaging oxidation reactions by reacting with free radicals and other reactive oxygen species within the body, hence hindering the process of oxidation (Halliwell & Gutteridge, 1995). Plants contain active components namely phenolics and polyphenolics that are known to act as antioxidants (Cai et al., 2003).

Every antioxidant is in fact a redox agent and might become a pro-oxidant to accelerate lipopolysaccarides and induce DNA damage under special conditions and concentrations. Studies have revealed pro-oxidant effects of antioxidant vitamins and several classes of plant-derived polyphenols such as flavonoids (Rahman et al., 1990) and tannins (Singh et al., 2001). As reported earlier, resveratol (Lastra & Villegas, 2007), phloroglucinols from *Garcinia subelliptica* (Wu et al., 2008) and curcumin (Ahsan & Hadi, 1998) can exhibit pro-oxidant properties, leading to oxidative breakage of cellular DNA in the presence of transition metal ions such as copper. Therefore, it is essential to discover natural compounds that good antioxidant activity but low pro-oxidant capabilities.

Pro-oxidant and antioxidant effect of plant extracts are due to the balance of two activities, free radical-scavenging activity and reducing power on iron ions, which may drive the Fenton reaction via reduction of iron ions. In a Fenton reaction, Fe^{2+} reacts with H_2O_2, resulting in the production of hydroxyl radical, which is considered to be the most harmful radical to biomolecules. Fe^{2+} is oxidized to Fe^{3+} in the Fenton reaction initially. By the action of many reductants, such as ascorbic acid, the oxidized forms of iron ion can be reduced to reduced forms (Fe^{2+}) later, which can enhance the generation of hydroxyl radicals. A predominant reducing power (on iron ions) over the free radical-scavenging activity in a mixture of compounds results in the pro-oxidant effect (Tian & Hua, 2005). In this study, the pro-oxidant capacity of the extracts were compared to the IC_{50} (mg/mL) of the antioxidant scavenging activity of DPPH radical. This ratio of pro-oxidant/antioxidant activity enabled us to evaluate the net antioxidant capacity of the extracts as this index will include not only the effective free radical-scavenging ability, taking into account pro-oxidant effect of the extracts as shown in equation (1).

$$\text{ProAntidex}=\frac{\text{Prooxidant capacity at the absorbance set at arbitrary 1.0 } (\text{mg / mL})}{IC_{50}(\text{mg / mL}) \text{ from DPPH scavenging assay}} \quad (1)$$

Ethanolic Extract	Plant Part	DPPH (IC_{50}, mg/ml)	Pro-oxidant (mg/ml)	Pro-Antidex
Azadirachta indica	leaf	0.74±0.46	0.50±0.02	0.91±0.58
Mangifera indica	leaf	0.17±0.02	0.22±0.03	1.32±0.29
Garcinia mangostana	peel	0.11±0.02	0.17±0.04	1.62±0.40
Nephelium lappaceum	**peel**	**0.12±0.05**	**0.05±0.04**	**0.48±0.37**
Psidium guajava	leaf	0.18±0.08	0.20±0.03	1.27±0.60
Fragaria x ananassa	**leaf**	**1.87±0.80**	**0.99±0.34**	**0.64±0.45**
Lawsonia inermis	**leaf**	**1.3±0.18**	**0.54±0.07**	**0.43±0.11**
Syzygium aqueum	**leaf**	**0.22±0.02**	**0.13±0.03**	**0.62±0.13**
Nephelium lappaceum	leaf	0.33±0.03	0.76±0.42	2.37±1.40
Peltophorum pterocarpum	leaf	0.17±0.12	0.11±0.03	0.83±0.44
Peltophorum pterocarpum	bark	0.10±0.04	0.08±0.03	0.94±0.66
Artocarpus champeden	leaf	0.30±0.21	0.75±0.51	3.82±4.24
Nephelium mutobile	leaf	0.24±0.03	0.18±0.06	0.76±0.28
Vitis vinifera	**seed**	**0.15±0.10**	**0.07±0.02**	**0.59±0.33**

Table 1. DPPH scavenging activity, Pro-oxidant activity and ProAntidex in ethanolic extracts of selected Malaysian plants and standard. ProAntidex was devised using the ratio of pro-oxidant activities to the IC_{50} DPPH scavenging activity. All values represent means ± SD, n=3.

Aqueous Extract	Plant Part	DPPH (IC_{50}, mg/ml)	Pro-oxidant (mg/ml)	Pro-Antidex
Azadirachta indica	leaf	0.96±0.14	1.13±0.01	1.20±0.18
Mangifera indica	leaf	0.49±0.39	1.03±0.88	2.03±1.38
Garcinia mangostana	peel	1.66±2.4	3.04±0.08	7.26±6.12**
Nephelium lappaceum	peel	0.54±0.15	0.55±0.37	1.08±0.65
Psidium guajava	leaf	0.22±0.01	0.42±0.33	1.89±1.43
Fragaria x ananassa	leaf	0.37±0.07	0.58±0.003	1.60±0.29
Lawsonia inermis	**leaf**	**3.71±0.34**	**1.41±0.21**	**0.38±0.06**
Syzygium aqueum	leaf	0.33±0.07	0.26±0.09	0.88±0.51
Nephelium lappaceum	leaf	0.67±0.02	>2	NA
Peltophorum pterocarpum	leaf	0.16±0.05	0.17±0.01	1.14±0.45
Peltophorum pterocarpum	bark	0.20±0.12	0.23±0.01	1.48±0.88
Artocarpus champeden	leaf	0.22±0.01	0.20±0.001	0.93±0.05
Nephelium mutobile	**leaf**	**3.76±0.27**	**0.18±0.01**	**0.05±0.01**
Vitis vinifera	**seed**	**0.46±0.18**	**0.24±0.11**	**0.59±0.35**
Green tea	NA	0.28±0.04	0.23±0.07	0.82±0.28
Emblica™	NA	0.31±0.07	0.27±0.12	0.69±0.18
Vitamin C	NA	0.01±0.00	0.03±0.35	4.10±3.36

Table 2. DPPH scavenging activity, pro-oxidant and ProAntidex in aqueous extracts of selected Malaysian plants and standards. ProAntidex was devised using the ratio of pro-oxidant activities to the IC_{50} DPPH scavenging activity. All values represent means ± SD, n=3. **Designates a significance difference from Emblica™, $p<0.01$.

The plant extracts having a high antioxidant activity were simultaneously analyzed for its pro-oxidant capability. Interestingly, in our laboratory, we established a Pro-oxidant/Antioxidant ratio (ProAntidex) which represents an index of the net free radical scavenging ability of whole plant extracts. The ethanolic extracts, *Nephelium lappaceum* peel, *Fragaria x ananassa* leaf, *Lawsonia inermis* leaf, *Syzygium aqueum* leaf and grape seed had lower Pro-Antidex than the commercially available Emblica™ extract which is a commercially available extract from *Phyllantus emblica* claims to have high antioxidant but low pro-oxidant activity (Table 1) (Ling et al., 2010b).

Among the aqueous extracts on the other hand; *Lawsonia inermis* leaf, *Nephelium mutobile* leaf and grape seed had low pro-oxidant activity (Table 2). In this study, Emblica™, green tea, vitamin C and grape seed were used as the positive controls in comparison to other plant extracts as shown in Figure 2. Among these extracts, the aqueous extract of *Nephelium mutobile* leaf had a very low ProAntidex of 0.05 compared to 0.69 for Emblica™. Most of the extracts had a far lower ProAntidex value compared to vitamin C. This index enables us to identify extracts with high net free radical scavenging activity potential. The ProAntidex is therefore beneficial as a screening parameter that can be used in food and healthcare industries.

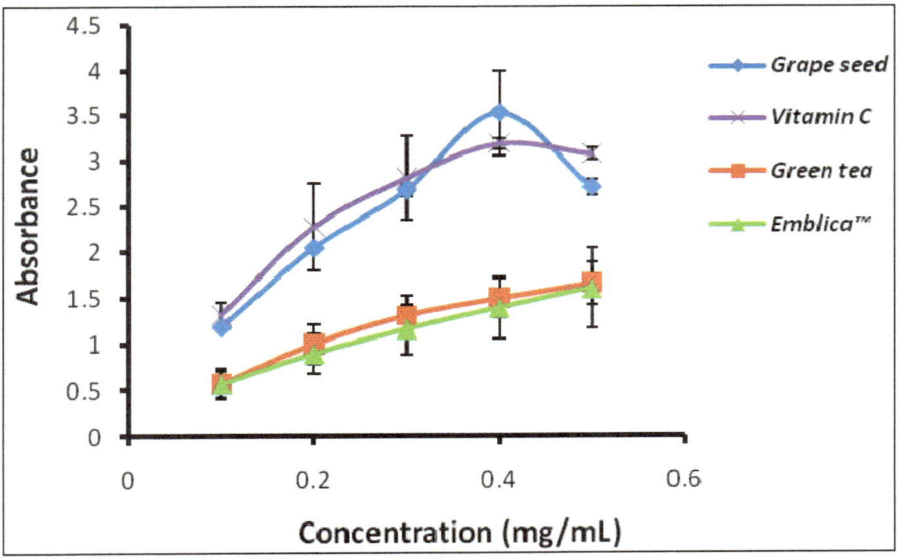

Fig. 2. Prooxidant activity of the standards used in the study. The prooxidant assay was carried out by measuring reducing power on Fe3+ in the Fenton reaction. Grape seed, green tea, Emblica ™ and vitamin C were used as positive controls. All the values represent means ± SD, *n*=3.

Pro-Antidex is a useful indicator in free radical research. The ratio of pro-oxidant to the antioxidant activity capacity gives a better picture of the real antioxidant capacity of the plant extracts. The pro-oxidant assay will enable the nutritionists and chemists to formulate antioxidant mixtures that balance between the two activities, which is higher antioxidant activity with lower pro-oxidant capacity. In other words, the net ProAntidex should be low to reflect that the particular plant extract has good overriding antioxidant property.

4. Standardised *Mangifera indica* leaf extract as an ideal antioxidant

Mangifera indica, commonly known as the mango plant has been the focus of many researchers for the next source of potent anti-oxidants. Previously, a standardised aqueous extract from the bark of *Mangifera indica* was reported to contain anti-inflammatory activity immunomodulatory and antioxidant activities (Garrido et al., 2004). The extract, is composed of a variety of phenolic acids, phenolic esters, flavanols and the xanthone mangiferin (Janet et al., 2006). When fed orally to mice that have been induced to have ear oedema by arachidonic acid and phorbol myristate acetate injection, the ear edema was observed to reduce markedly. *In vitro* studies showed that the extract also inhibited the induction of prostaglandin E (PG_E) and leukotriene-B_4 (LTB_4) release by macrophages (Garrido et al., 2004).

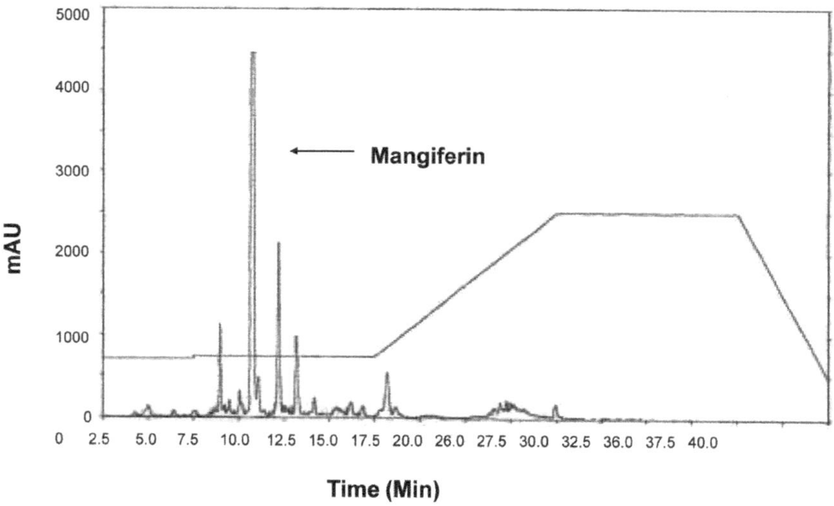

Fig. 3. HPLC chromatogram of *M. indica* extracts and standard, mangiferin. The mobile phase consisted of solvent A: 3% acetic acid in water and solvent B : acetonitrile; starting from 90%B for 5 minutes, 80%B for 15minutes and finally 100% B for 10minutes for washing and recondition the column (a) Mangiferin (b) ethanolic *M indica* extract and (c) aqueous *M indica* extract.

In our laboratory, the standardised ethanolic and aqueous extracts of Mangifera indica leaf was analyzed for its free radical scavenging activity using a variety of other assays. Its IC50 values using the DPPH assay was 0.17mg/mL ± 0.02 and 0.49mg/mL ± 0.4 respectively. Mangiferin, the main active compound in M.indica plant has been established to contribute to its biological activities (Ling et al., 2009). Standardised ethanolic extracts of the Mangifera indica leaf was found to have a mangiferin concentration of 71mg/g extract, free radical scavenging activity (IC50) of 0.17mg/mL ± 0.02 and total phenolic content of 590mg/g ± 48.08 of extract. The protection seen by Mangifera indica extracts against lipid peroxidation was observed to be far better than butlylated hydroxytoluene (BHT; a commercial anti oxidant used to prevent rancidity of oils) and commercial grape seed extract. The Mangifera indica extracts at higher concentrations did not exhibit pro-oxidant activities when compared to Vitamin C is yet another interesting feature of this extract. We also found that the aqueous and ethanolic extracts of Mangifera indica leaf protects the mouse fibroblasts cells, NIH/3T3, from oxidant-induced cell death by about 84% and it is also non-toxic to cultured splenocytes .

5. Rind of rambutan, *Nephelium lappaceum*, a potential source of natural antioxidant

Nephelium lappaceum L. belongs to the same family (Sapindaceae) as the sub-tropical fruits lychee and longan and it is native to Southeast Asia. This fruit is an important commercial crop in Asia, where it is taken freshly or processed. In Southeast Asia, the dried fruit rind has been employed in traditional medicine for centuries. Additionally, the rind is used in cooking and the manufacture of soap. The roots, bark, and leaves have various uses in medicine and in the production of dyes. Previous studies have shown *N.lappaceum* rind extract to exhibit high antioxidant activity (Palanisamy et al., 2008), antibacterial activity (Thitilertdecha et al., 2010) and anti-Herpes Simplex virus type 1 (Nawawi A, 1999). Recently in our laboratory, *N.lappaceum* rind was also shown to have anti hyperglycemic potential (Palanisamy, Uma et al., 2011). The utilisation of *Nephelium lappaceum* rind to manage hyperglycemia is seen as an important finding not only in traditional medicine but also in aspects of valorisation of food waste.

The rind of *Nephelium lappaceum* (rambutan) was selected as the rind contains extremely high antioxidant activity when assessed using several free radical scavenging methods. Although having a yield of only 18%, the ethanolic rambutan rind extract has a total phenolic content of 762 ± 10 mg GAE/g extract, which is comparable to the commercial grape seed extract. The rambutan rind had lower pro-oxidant activities compared to vitamin C, α-tocopherol, grape seed and green tea in a dose response experiment. In addition, the rind extract at 100 μg/ml reduced oxidant-induced cell death (DPPH at 50 μM) by apoptosis to an extent similar to that of grape seed extract. The rind extracts were not cytotoxic to normal mouse fibroblast cells or splenocytes. Powderised rind had low heavy metal content far below the permissible levels for nutraceuticals. This study is the first to show a unique combination of high phenolic content, low pro-oxidant capacity and strong antioxidant activity of the rind extract of *Nephelium lappaceum*.

Whole extracts of the rind of *N.lappaceum* was standardized using a reverse phase column on analytical HPLC. Bioassay-guided fractionation of the extract was attempted to establish the most effective method to extract fractions with high antioxidant activity. Two fractionated extracts of the rind having DPPH activity of 0.01±0.001 and 0.01±0.003mg/ml and total phenolic content of 6662±240 and 1761±239 mg/g GAE respectively were established using preparative HPLC.

Bioassay guided fractionation was found to be time and labour consuming; therefore we investigated a rapid purification method to isolate and purify the bioactive compound from *N.lappaceum* rind extract. It was pertinent that we isolate and identify the active compound(s) in this extract that contribute(s) to the said biological activities. Structural characterization of purified compounds can lead to the formulation of the new therapeutic products. Geraniin was found to be the major phenolic compound in the *N.lappaceum* rind extract. A composition of 13% geraniin contributed to the high free radical scavenging activity in the extract. The compound exhibited radical scavenging activity of IC_{50}; 3.8 μg/mL (DPPH radical test), 1.7 μg/mL(ABTS radical test) and 1.7 μg/mL (Galvinoxyl radical test). The compound also displayed very low pro-oxidant capabilities.

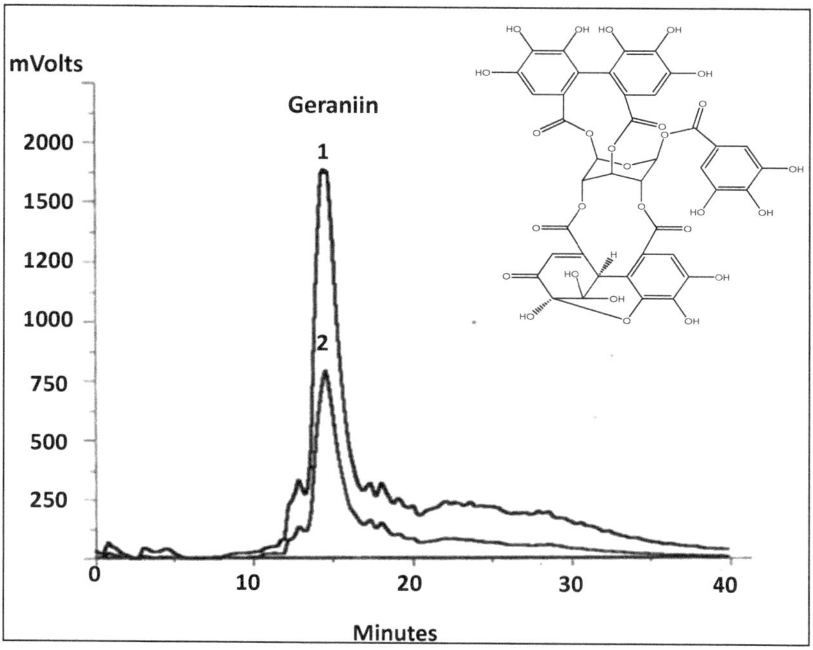

Fig. 4. Purification on Prep-HPLC showing geraniin as the major compound in the ethanolic *Nephelium lappaceum* rind extract. The solvent gradient consisted of 0-10% acetonitrile for 3 minutes, 10-40% for 12 minutes and finally 100% acetonitrile for 5 minutes to recondition the column. at a flow rate of 18mL/min. Geraniin was obtained at the retention time of 13minutes. 1 was detected at 210nm and 2 at 275nm. Insert shows the chemical structure of geraniin (Palanisamy, Uma D. et al., 2011).

Sample/Fraction	Extraction Method	Yield (%)	Geraniin (%) in sample
N.lappaceum rind	Ethanol extraction	30.58	3.79
Ethanolic extract	LiChroprep RP-18	60.00	12.68
F1	Preparative HPLC	21.15	21.13

Table 3. Quantification of geraniin in the rapid purification method

The rind of *N.lappaceum* extract was standardized to 13% of geraniin, the active hydrolysable ellagitannins responsible for over 50% of the antioxidant potential of the ethanolic extract of *Nephelium lappaceum* rind. In a single dose acute toxicity studies, oral LD_{50} of the rind extract in ICR mice was found to be greater than 5 g/kg body weight. In a subsequent study, Sprague Dawley rats were given via oral gavage 0 (control), 1000 mg/kg body weight/day of the extract for 28 days to evaluate the subacute toxicity of the extract to animals. Animals in a satellite group scheduled for follow-up observations were kept for 14 days without treatment to detect for any delayed effects. At the end of the experiment, kidney, liver, brain and testis were collected and followed by histopathological studies. No behavioural or organs to body weight changes were found in all the groups. Furthermore, no obvious abnormal changes were observed histologically in all the groups (unpublished data).

6. Conclusion

Malaysian plant extracts are a potential source of natural antioxidants. The chapter focuses on the identification of selected Malaysian plants that exhibit high antioxidant capability. We provide information concerning the complete profile of selected Malaysian plants on their antioxidant/pro-oxidant activity, cytotoxicity, heavy metal content and method of standardisation. In conclusion, it was established that *Nephelium lappaceum* rind and *Mangifera indica* leaf extract have great potential to be developed into an antioxidant nutraceutical. In future study, studies of membrane interaction and the regulation of antioxidant gene expression in the presence of extracts and their pure compounds in the cells will provide better understanding of the mode of actions of the antioxidant activity exhibited. *In vivo* subacute and chronic toxicity studies will need to be carried out to determine the effect of long term intake of *Nephelium lappaceum* rind extract and geraniin in the animals.

7. References

Chan, K. (2003). Some Aspects of Toxic Contaminants in Herbal Medicines. *Chemosphere,* Vol. 52, No. 9, pp.1361-1371.

Lattif, A. G.; Omar, I. M.; I.M., S.. & Kadri, A. (1984). A Multi-Variate Approach to the Study of Previous Term Medicinal Plants Next Term in Malaysia. *Journal of Singapore Natural Academy of Science* Vol. 13, No., pp.101-105.

Larson, R. A. (1988). The Antioxidants of Higher Plants. *Phytochemistry,* Vol. 27, No. 4, pp.969-978.

Rimm, E. B.; Ascherio, A.; Giovannucci, E.; Spiegelman, D.; Stampfer, M. J.. & Willett, W. C. (1996). Vegetable, Fruit, and Cereal Fiber Intake and Risk of Coronary Heart Disease among Men. *The Journal of the American Medical Association,* Vol. 275, No. 6, pp.447-451.

Gey, K. F.; Puska, P.; Jordan, P.. & Moser, U. K. (1991). Inverse Correlation between Plasma Vitamin E and Mortality from Ischemic Heart Disease in Cross-Cultural Epidemiology. *The American Journal of Clinical Nutrition,* Vol. 53, No. 1, pp.326S-334S.

Pellegrini, N.; Serafini, M.; Colombi, B.; Del Rio, D.; Salvatore, S.; Bianchi, M.. & Brighenti, F. (2003). Total Antioxidant Capacity of Plant Foods, Beverages and Oils Consumed in Italy Assessed by Three Different *in Vitro* Assays. *The Journal of Nutrition,* Vol. 133, No. 9, pp.2812-2819.

Bocco, A.; Cuvelier, M.-E.; Richard, H.. & Berset, C. (1998). Antioxidant Activity and Phenolic Composition of Citrus Peel and Seed Extracts. *Journal of Agricultural and Food Chemistry,* Vol. 46, No. 6, pp.2123-2129.

Cam, M.; Hisil, Y.. & Kuscu, A. (2007). Organic Acid, Phenolic Content, and Antioxidant Capacity of Fruit Flesh and Seed of *Viburnum Opulus. Chemistry of Natural Compounds,* Vol. 43, No. 4, pp.460-461.

Chang, S.; Tan, C.; Frankel, E. N.. & Barrett, D. M. (2000). Low-Density Lipoprotein Antioxidant Activity of Phenolic Compounds and Polyphenol Oxidase Activity in Selected Clingstone Peach Cultivars. *Journal of Agricultural and Food Chemistry,* Vol. 48, No. 2, pp.147-151.

He, X.. & Liu, R. H. (2007). Triterpenoids Isolated from Apple Peels Have Potent Antiproliferative Activity and May Be Partially Responsible for Apple's Anticancer Activity. *Journal of Agricultural and Food Chemistry,* Vol. 55, No. 11, pp.4366-4370.

Moongkarndi, P.; Kosem, N.; Kaslungka, S.; Luanratana, O.; Pongpan, N.. & Neungton, N. (2004). Antiproliferation, Antioxidation and Induction of Apoptosis by *Garcinia Mangostana* (Mangosteen) on Skbr3 Human Breast Cancer Cell Line. *Journal of Ethnopharmacology,* Vol. 90, No. 1, pp.161-166.

Rockenbach, I. I.; Gonzaga, L. V.; Rizelio, V. M.; Goncalves, A. E. d. S. S.; Genovese, M. I.. & Fett, R. (2011). Phenolic Compounds and Antioxidant Activity of Seed and Skin Extracts of Red Grape (*Vitis Vinifera* and *Vitis Labrusca*) Pomace from Brazilian Winemaking. *Food Research International,* Vol. 44, No. 4, pp.897-901

Chan, E.; Lim, Y.. & Wong, S. (2011). Antioxidant Properties of Ginger Leaves: An Overview. *Free Radicals and Antioxidants,* Vol. 1, No. 1, pp.6-16.

Silva, S.; Gomes, L.; Leitao, F.; Coelho, A. V.. & Boas, L. V. (2006). Phenolic Compounds and Antioxidant Activity of *Olea Europaea* L. Fruits and Leaves. *Food Science and Technology International,* Vol. 12, No. 5, pp.385-395.

Agati, G.; Stefano, G.; Biricolti, S.. & Tattini, M. (2009). Mesophyll Distribution of 'Antioxidant' Flavonoid Glycosides in *Ligustrum Vulgare* Leaves under Contrasting Sunlight Irradiance. *Annals of Botany,* Vol. 104, No. 5, pp.853-861.

Tadhani, M. B.; Patel, V. H.. & Subhash, R. (2007). *In Vitro* Antioxidant Activities of *Stevia Rebaudiana* Leaves and Callus. *Journal of Food Composition and Analysis,* Vol. 20, No. 3-4, pp.323-329.

Zhang, S.-J.; Lin, Y.-M.; Zhou, H.-C.; Wei, S.-D.; Lin, G.-H.. & Ye, G.-F. (2010). Antioxidant Tannins from Stem Bark and Fine Root of *Casuarina Equisetifolia*. *Molecules,* Vol. 15, No. 8, pp.5658-5670.

Chang, S. T.; Wu, J. H.; Wang, S. Y.; Kang, P. L.; Yang, N. S.. & Shyur, L. F. (2001). Antioxidant Activity of Extracts from *Acacia Confusa* Bark and Heartwood. *Journal of Agricultural and Food Chemistry,* Vol. 49, No. 7, pp.3420-3424.

Diouf, P.; Stevanovic, T.. & Cloutier, A. (2009). Antioxidant Properties and Polyphenol Contents of Trembling Aspen Bark Extracts. *Wood Science and Technology,* Vol. 43, No. 5, pp.457-470.

Dalton, D. A.; Joyner, S. L.; Becana, M.; Iturbe-Ormaetxe, I.. & Chatfield, J. M. (1998). Antioxidant Defenses in the Peripheral Cell Layers of Legume Root Nodules. *Plant Physiology,* Vol. 116, No. 1, pp.37-43.

Hegde, K.. & Joshi, A. B. (2010). Hepatoprotective and Antioxidant Effect of *Carissa Spinarum* Root Extract against Ccl4 and Paracetamol-Induced Hepatic Damage in Rats. *Bangladesh Journal of Pharmacology* Vol. 5, No. 1, pp.73-76.

Boruvka, L.; Kozák, J.. & Krištoufková, S. (1997). Heavy Metal Accumulation in Plants Grown in Heavily Polluted Soils. *Folia Microbiologica,* Vol. 42, No. 5, pp.524-526.

Ang, H.; Lee, E.. & Matsumoto, K. (2003). Analysis of Lead Content in Herbal Preparations in Malaysia. *Human & Experimental Toxicology,* Vol. 22, No. 8, pp.445-51.

Järup, L.; Berglund, M.; Elinder, C.; Nordberg, G.. & Vahter, M. (1998). Health Effects of Cadmium Exposure--a Review of the Literature and a Risk Estimate. *Scandinavian Journal of Work, Environment & Health,* Vol. 24, No. 1, pp.1-51.

Yen, G.-C.; Chen, H.-Y.. & Peng, H.-H. (1997). Antioxidant and Pro-Oxidant Effects of Various Tea Extracts. *Journal of Agricultural and Food Chemistry,* Vol. 45, No. 1, pp.30-34.

Halliwell, B. (1996). Commentary: Vitamin C: Antioxidant or Pro-Oxidant *in Vivo*? *Free Radical Research,* Vol. 25, No. 5, pp.439-454.

Ling, L. T.; Radhakrishnan, A. K.; Subramaniam, T.; Cheng, H. M.. & Palanisamy, U. D. (2010a). Assessment of Antioxidant Capacity and Cytotoxicity of Selected Malaysian Plants. *Molecules,* Vol. 15, No. 4, pp.2139-2151.

Palanisamy, U.; Cheng, H. M.; Masilamani, T.; Subramaniam, T.; Ling, L. T.. & Radhakrishnan, A. K. (2008). Rind of the Rambutan, *Nephelium Lappaceum*, a Potential Source of Natural Antioxidants. *Food Chemistry*, Vol. 109, No. 1, pp.54-63.

Halliwell, B.. & Gutteridge, J. M. C. (1995). The Definition and Measurement of Antioxidants in Biological Systems. *Free Radical Biology and Medicine*, Vol. 18, No. 1, pp.125-126.

Cai, Y.; Sun, M.. & Corke, H. (2003). Antioxidant Activity of Betalains from Plants of the Amaranthaceae. *Journal of Agricultural and Food Chemistry*, Vol. 51, No. 8, pp.2288-2294.

Rahman, A.; Shahabuddin; Hadi, S.. & Parish, J. (1990). Complexes Involving Quercetin, DNA and Cu(Ii). *Carcinogenesis*, Vol. 11, No. 11, pp.2001-2003.

Singh, S.; Farhan Asad, S.; Ahmad, A.; Khan, N. U.. & Hadi, S. M. (2001). Oxidative DNA Damage by Capsaicin and Dihydrocapsaicin in the Presence of Cu(Ii). *Cancer Letters*, Vol. 169, No. 2, pp.139-146.

Lastra, C. A. d. l.. & Villegas, I. (2007). Resveratrol as an Antioxidant and Pro-Oxidant Agent: Mechanisms and Clinical Implications. *Biochemical Society Transactions*, Vol. 35, No., pp.1156-1160.

Wu, C. C.; Lu, Y. H.; Wei, B. L.; Yang, S. C.; Won, S. J.. & Lin, C. N. (2008). Phloroglucinols with Prooxidant Activity from Garcinia Subelliptica. *Journal of Natural Products*, Vol. 71, No. 2, pp.246-250.

Ahsan, H.. & Hadi, S. M. (1998). Strand Scission in DNA Induced by Curcumin in the Presence of Cu(Ii). *Cancer Letters*, Vol. 124, No. 1, pp.23-30.

Tian, B.. & Hua, Y. (2005). Concentration-Dependence of Prooxidant and Antioxidant Effects of Aloin and Aloe-Emodin on DNA. *Food Chemistry*, Vol. 91, No. 3, pp.413-418.

Ling, L. T.; Palanisamy, U. D.. & Cheng, H. M. (2010b). Prooxidant/Antioxidant Ratio (Proantidex) as a Better Index of Net Free Radical Scavenging Potential. *Molecules*, Vol. 15, No. 11, pp.7884-7892.

Garrido, G.; Gonzalez , D.; Lemus, Y.; Garcia, D.; Lodeiro, L.; Quintero, G.; Delporte, C.; Nunez-Selles, A. J.. & Delgado, R. (2004). *In Vivo* and *in Vitro* Anti-Inflammatory Activities of *Mangifera Indica* L. Extract (Vimang). *Pharmacological Research*, Vol. 50, No. 2, pp.143-149.

Janet, R.; Donato, D. P.; Magda, G.; Susanna, M.; René, D.; Massimiliano, C.. & Stefano, M. (2006). Effects of a Natural Extract from *Mangifera Indica* L, and Its Active Compound, Mangiferin, on Energy State and Lipid Peroxidation of Red Blood Cells. *Biochimica et Biophysics Acta*, Vol. 1760, No. 9, pp.1333–1342.

Ling, L. T.; Yap, S.-A.; Radhakrishnan, A. K.; Subramaniam, T.; Cheng, H. M.. & Palanisamy, U. D. (2009). Standardised *Mangifera Indica* Extract Is an Ideal Antioxidant. *Food Chemistry*, Vol. 113, No. 4, pp.1154-1159.

Thitilertdecha, N.; Teerawutgulrag, A.; Kilburn, J. D.. & Rakariyatham, N. (2010). Identification of Major Phenolic Compounds from Nephelium Lappaceum L. And Their Antioxidant Activities. *Molecules*, Vol. 15, No. 3, pp.1453-1465.

Nawawi A, N. N., Hattori M, Kurokawa M, Shiraki K. (1999). Inhibitory Effects of Indonesian Medicinal Plants on the Infection of Herpes Simplex Virus Type 1. *Phytotherapy Research*, Vol. 13, No. 1, pp.37-41.

Palanisamy, U.; Manaharan, T.; Teng, L. L.; Radhakrishnan, A. K. C.; Subramaniam, T.. & Masilamani, T. (2011). Rambutan Rind in the Management of Hyperglycemia. *Food Research International*, Vol. 44, No. 7, pp.2278-2282.

Palanisamy, U. D.; Ling, L. T.; Manaharan, T.. & Appleton, D. (2011). Rapid Isolation of Geraniin from *Nephelium Lappaceum* Rind Waste and Its Anti-Hyperglycemic Activity. *Food Chemistry*, Vol. 127, No. 1, pp.21-27.

Part 2

Biotechnological Aspects

Genetically Engineered Lactobacilli for Technological and Functional Food Applications

María J. Yebra, Vicente Monedero,
Gaspar Pérez-Martínez and Jesús Rodríguez-Díaz
Departamento de Biotecnología de los Alimentos
IATA- CSIC
Spain

1. Introduction

Lactic acid bacteria (LAB) are Gram-positive microorganisms that produce lactic acid as a major product of their metabolism. Among them the genus *Lactobacillus* comprises a large heterogeneous group of low G+C DNA content, anaerobic and nonsporulating bacteria, that includes species widely used in the food industry. They play key roles in fermented dairy, meats and vegetables products. Due to their claimed health-promoting properties certain lactobacilli species are used as probiotics and they are commonly applied to dairy and functional foods products. Lactobacilli have a relatively simple fermentative metabolism focused to rapidly convert carbohydrates into lactic acid, and exhibited a limited biosynthetic capacity. In addition, several tools and strategies to manipulate them genetically are available. All those characteristics make lactobacilli specially suited for genetic engineering aimed to increase existing compounds or to produce novel metabolites of interest for the food industry. Regarding probiotic lactobacilli, through genetic manipulation, the health attributes of probiotic strains can be enhanced or new probiotic activities can be developed and additionally, an understanding of the underlying molecular mechanisms can be obtained. Here, we review metabolic engineering strategies in lactobacilli that have successfully been used to efficiently reroute sugar metabolism to compounds such as L-lactic acid, aroma compounds (acetoin, diacetyl), low-calorie sugars (mannitol, sorbitol) and exopolysaccharides. We will also describe strains of probiotic lactobacilli that have been developed to exploit their adherence and immunomodulatory properties, and to delivery proteins at the intestinal mucosa.

1.1 Metabolic potential of lactobacilli

Because of their fermentative metabolism and global utilization in food fermentations, LAB are specially suited for rerouting sugar metabolism to produce industrially important food compounds. During fermentation, monosaccharides are catabolized through glycolysis (Embden-Meyerhoff pathway) and related pathways (Figure 1). Glycolytic catabolism of sugars involves phosphorylation reactions that drive hexoses to fructose-1,6-bisP, which is then hydrolysed to glyceraldehyde-3P (GADH-3P) and dihydroxyacetone-P. Then, GADH-

3P undertakes dephosphorylation and oxidation, which yields 2 pyruvate, 2 ATP and the reduction of 2 NAD^+ to 2 NADH per glucose. Under normal metabolic conditions NADH is used mainly to reduce pyruvate. Some LAB exclusively produce lactate from pyruvate (homofermentative), while other species (heterofermentative) produce lactate, acetate, ethanol and CO_2. There are other differences in heterofermentative species, since they can shift sugar catabolism towards the so called pentose phosphate or phosphoketolase pathway that renders GADH-3P and acetyl-P. Then, GADH-3P enters the lower part of the glycolysis. In this pathway two additional NADH molecules are oxidised by means of alcohol dehydrogenase that produces ethanol from acetyl-CoA. Therefore, the global balance of the heterofermentation of one mol of glucose is one mol of lactate, one mol of ethanol and one mol of CO_2, with a net energy yield of one mol of ATP.

An important strategy frequently used during metabolic engineering consists in blocking the formation of natural proton sinks, such as lactate or ethanol in the final steps of glycolysis. However, dissipation of the H^+ pool has such a great relevance that LAB normally has several isoenzymes of L-lactate dehydrogenase (L-LDH) as showed by the analysis of different genomes such as *Lactobacillus plantarum* and *Lactobacillus casei* (Kleerebezem *et al.*, 2003; Rico *et al.*, 2008), and alternative dehydrogenases yielding D-lactate have been found in most LAB genomes. An additional difficulty is imposed by the fact that glycolysis is subject to a strict allosteric regulation by its own intermediate and final metabolites, as well as by Pi, fructose-1,6bisP, phospho*enol*pyruvate (PEP), ADP, ATP and NADH/NAD^+ ratio. Its robustness and flexibility would assure an efficient bacterial growth, so that its activity rate would always respond to the cell's energy demand. It has also been observed that there are three enzyme activities especially sensitive to allosteric modulation, which are most relevant in the pathway's regulation and they are: phosphofructokinase (PFK), GADH-3P dehydrogenase and pyruvate kinase. PFK is strongly inhibited by PEP and pyruvate kinase activity is inhibited by Pi in *Lactococcus lactis* and this enzyme is stimulated by ADP and fructose-1,6bisP in *Lactobacillus bulgaricus* (Branny *et al.*, 1998). Furthermore, in *L. lactis*, GADH-3P dehydrogenase has a remarkable role in the modulation of the carbon flux, which is regulated by the NADH/NAD^+ ratio, as it happens with LDH (Garrigues *et al.*, 1997).

LAB suitability as starter cultures in dairy fermentations highly depends on their ability to produce small concentrations of volatile compounds derived from the alternative metabolism of pyruvate. The production of diacetyl and acetoin is quite common in LAB. These compounds are produced through decarboxylation of α-acetolactate obtained from pyruvate by the enzyme α-acetolactate syntase (Figure 2). Additionally, under substrate limitation and anaerobiosis, the enzyme pyruvate-formate lyase produces acetyl-CoA and formate from pyruvate and CoA. Acetyl-CoA is an important metabolite, as it can be used as electron acceptor to oxidise NADH or as energy compound to obtain ATP. The enzymatic complex of pyruvate dehydrogenase also produces acetyl-CoA, CO_2 and NADH from pyruvate, coenzyme A and NAD^+. Under aerobic conditions, this is an anabolic enzyme producing acetyl-CoA used for lipid synthesis, but under aerobic conditions, it also has a catabolic function where NADH oxidases can regenerate the excess of NADH produced. Pyruvate oxidase mediates onversion of pyruvate to acetyl-P, CO_2 and H_2O_2. This activity allows to obtain ATP when carbon sources are limiting, by substrate level phosphorylation of acetyl-P.

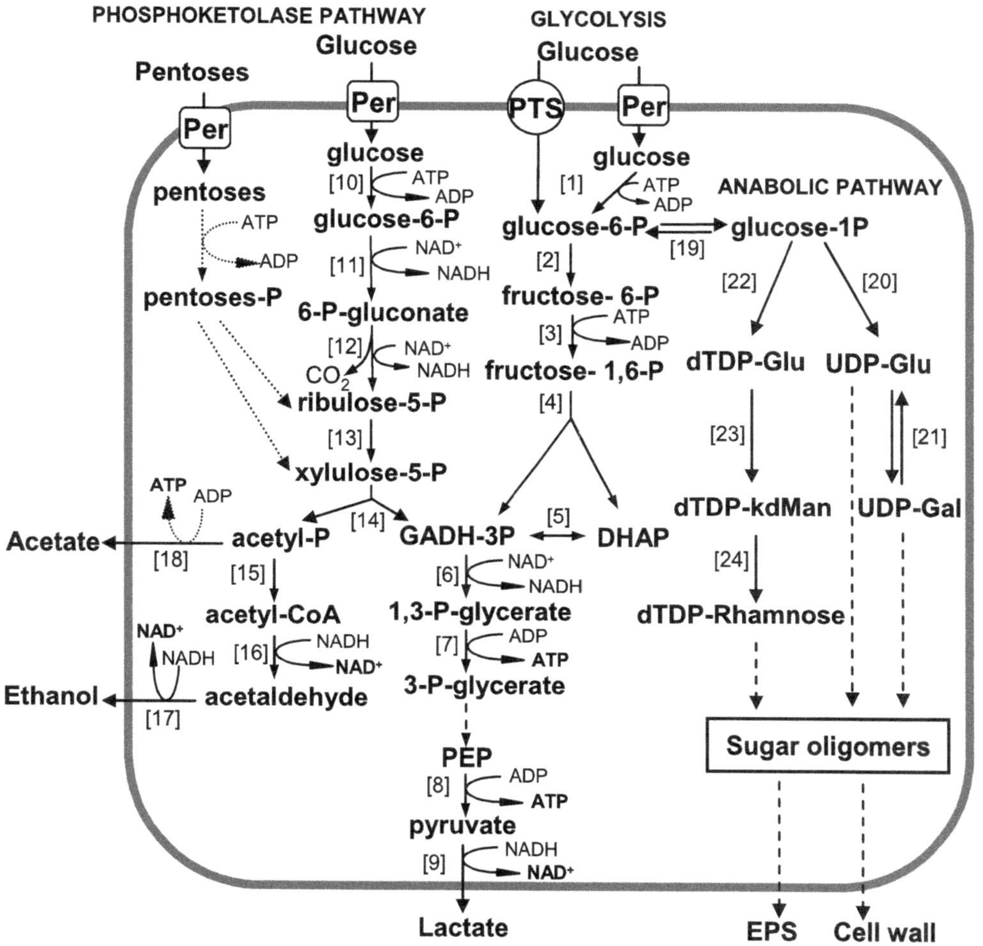

Fig. 1. Scheme of glycolysis, phosphoketolase pathway and anabolic pathway of UDP-sugars in LAB. [1] glucokinase, [2] phosphoglucose isomerase, [3] phosphofructokinase, [4] fructose 1,6-bisP aldolase, [5] triose-phosphate isomerase, [6] glyceraldehyde-3P dehydrogenase, [7] phosphoglycerate kinase, [8] pyruvate kinase, [9] lactate dehydrogenase, [10] hexokinase, [11] glucose-6P dehydrogenase, [12] 6-phosphogluconate dehydrogenase, [13] ribulose-5P 3-epimerase, [14] xylulose-5P phosphoketolase, [15] phosphotransacetylase, [16] acetaldehyde dehydrogenase, [17] alcohol dehydrogenase, [18] acetate kinase, [19] α-phosphoglucomutase, [20] UDP-glucose pyrophosphorylase, [21] UDP-galactose 4-epimerase, [22] deoxyTDP-glucose pyrophosphorylase, [23] deoxyTDP-glucose 4,6-dehydratase, [24] deoxyTDP-rhamnose synthetic enzyme system. PTS, PEP phosphotransferase system; Per: permease; CoA, coenzyme A; DHAP, dihydroxiacetone phosphate; GADH-3P, glyceraldehyde-3-phosphate; UDP-Glu, UDP-glucose; UDP-Gal, UDP-galactose; dTDP-Glu, deoxyTDP-glucose; dTDP-kdMan, deoxyTDP-4-keto-6-deoxymannose; EPS, exopolysaccharide; PEP, phospho*enol*pyruvate.

1.2 Functional properties of lactobacilli

Bacterial populations in the gut of vertebrates have evolved through millions of years to render interdependent functions. They have been studied for long time using classical culturing techniques and recently through molecular approaches and it soon became evident that in humans, the gut's microbioma is formed by numerous bacterial species whose proportions change between individuals (Eckburg *et al.*, 2005). The most abundant genera are *Bacteroides, Faecalibacterium* or *Bifidobacterium*, however, although lactobacilli are not as abundant, they have been proved to play a remarkable role sustaining the global population balance and interact at different levels with the intestinal mucosa. In this environment some strains exerted beneficial health effects and they are considered probiotics. These are defined as live microorganisms that, when administered in adequate amounts, confer a beneficial effect on the health of the host (FAO/WHO, 2001). In addition to probiotics, functional food ingredients also include prebiotics, which are define as selectively fermented ingredients that allow specific changes in the composition and/or activity of the gastrointestinal microbiota that confer benefits upon host wellbeing and health (Roberfroid, 2007). Several beneficial effects of lactobacilli on human host have been reported. Strains of *Lactobacillus rhamnosus, Lactobacillus acidophilus* and *L. bulgaricus* alone or in combination are effectives in reduce the risk of acute infectious diarrhoea and prevent antibiotic-associated diarrhoea (Sazawal *et al.*, 2006). A mixture of probiotics including lactobacilli seems effective in the maintenance of remission of intestinal bowel diseases such as chronic pouchitis and ulcerative colitis, and to decrease symptoms in patient with irritable bowel syndrome (Haller *et al.*, 2010). A synbiotic food composed of the prebiotic oligofructose-enriched inulin, *L. rhamnosus* GG and *Bifidobacterium lactic* Bb12 was able to alter favourably several colorectal cancer markers in patients with cancer of colon (Rafter *et al.*, 2007). Besides to gastrointestinal disorders, lactobacilli have also showed positive effects in other pathologies, such as in the treatment and prevention of bacterial vaginosis (Falagas *et al.*, 2007), the prevention of atopic eczema (Tang *et al.*, 2010) and prevention of dental caries (Stamatova & Meurman, 2009). The health promoting effects of probiotic bacteria are mediated mainly by three mechanisms, (i) microbe-microbe interactions; (ii) beneficial interactions with gut epithelium and (iii) immunomodulatory interactions (Lebeer *et al.*, 2008). Regarding the first mechanism, probiotics can have a beneficial effect on the host by modifying the microbiota trough competition and cooperation for nutrients, production of antimicrobial compounds (lactic acid, bacteriocins, H_2O_2), competition with pathogens for attachment sites to the host mucosal surface and by bacterial cell-host cell communication. With respect to the beneficial interactions of probiotics with gut epithelium, this constitutes the main target tissue of probiotic action, and *Lactobacillus* molecules can modify it by affecting the metabolic and barrier functions of the epithelial cells. The preservation of the epithelial barrier by probiotic lactobacilli has been attributed to induction of mucin secretion, enhancement of tight-junction functioning, upregulation of cytoprotective heat-shock proteins and prevention of apoptosis of ephitelial cells.

The gut mucosal surface is continuously exposed to pathogens, beneficial mutualistic and commensal bacteria, and it is armoured with the largest part of the immune system in the organism, with lymphocytes scattered in the lamina propria or in organized gut-associated lymphoid tissues (GALT) such as the Peyer's patches of the small intestine and mesenteric lymph nodes (MLNs). Those immune cells can discriminate pathogens from harmless antigens, preventing an inappropriate immune response to harmless bacteria, thorough regulatory mechanisms known as "oral tolerance", which is an still incompletely understood *active nonresponse* to dietary and commensal enteric bacteria or food derived antigens administered orally, also related to the maintenance of homeostasis in the gut (Murphy *et al.*, 2007). The subepithelial dendritic cells (DCs), B cells and T cells, in the lamina propria and GALT express a wide range of pattern-recognition receptors (PRRs), surface Toll like receptors (TLRs) and intracellular nucleotide-binding oligomerization domain (NOD)-like receptors (NLRs), to acquire antigens from the intestinal lumen. Then, secreted cytokines and chemokines from DCs will determine *tolerance* and *active immune responses* against a particular antigen, and whether lymphocyte differentiation, innate, adaptive or allergy immune responses will be displayed (Hart *et al.*, 2004). Also intestinal epithelial cells (IECs) at the mucosal surface express PRRs and can also secrete cytokines and regulatory molecules, therefore, they participate actively in the discrimination of both pathogenic and commensal bacteria (Artis, 2008). In the gut, certain *Lactobacillus* strains have been proved to play a remarkable role sustaining the global population balance through their ability to synthesize antagonistic compounds that restrain the proliferation of a number of pathogens. However, their mutualistic behaviour with the host involves different levels of interaction with the intestinal mucosa, resulting in an anti-inflammatory effect and restoration of the mucosal homeostasis (Haller *et al.*, 2010). The proinflammatory cytokine profiles occasionally induced by some lactobacilli in model systems (Dong *et al.*, 2010) be related to the moderate degree of inflammation (*physiological inflammation*) elicited by some commensals and moderate pathogens, and it has been conceived as a beneficial feature that creates a state of awareness for a rapid immune defence response against possible infective aggressions, while preserving homeostasis (Sansonetti & Medzhitov, 2009).

1.3 Tools and strategies for genetic manipulation of lactobacilli

1.3.1 Gene cloning vectors, genetic markers and promoters

Several constitutive or inducible gene expression systems have been developed for lactobacilli (Fang & O'Toole, 2009). The vectors have different parameters such as copy-number and host-range, they are derivatives of the rolling-circle plasmids pWVO1 or pSH71 from *L. lactis* or the theta-type plasmids pAMβ1 from *Enterococcus faecalis* (Perez-Arellano *et al.*, 2001) and pRV500 from *Lactobacillus sakei* (Crutz-Le Coq & Zagorec, 2008). Vectors for controlled expression are mainly based on genes and promoters involved in bacteriocin production, sugar utilization genes and stress resistance. In addition, following the production of proteins by LAB using a specific expression vector they should be properly folded, targeted and sometimes recovered. Several vectors included secretion cassettes, such as those based on the secretion signal of the lactococcal Usp45 protein (Schotte *et al.*, 2000), the expression and secretion signals of S-layer proteins (Savijoki *et al.*, 1997), the PrtP signal sequence (Kajikawa *et al.*, 2010) or the M6 carboxy-terminal domain to anchor proteins to the cell wall (Reveneau *et al.*, 2002).

The nisin-controlled expression (NICE) system, based on the autoregulation mechanism of the bacteriocin nisin, is a very effective expression vector for production of heterologous proteins in LAB (Mierau & Kleerebezem, 2005). The NICE system contains the promoter *nisA* conducting gene expression under the control of the transcriptional regulator NisR, which is modulated by phosphorylation due to the histidine-protein kinase NisK immersed in the cytoplasmic membrane. The expression of the genes placed behind the P_{nisA} is induced by the addition of subinhibitory concentrations of nisin into the culture medium, in such a way that increasing amounts of nisin resulted in a linear dose-response curve. Optimization of the NICE system includes the incorporation in the vectors of the nisin immunity gene *nisI*, which resulted in better tolerance of the cells to high amounts of the inducer nisin (Oddone *et al.*, 2009). The NICE system was created for expression of genes in *L. lactis* but it has been adapted to other low-GC Gram-positive bacteria including *Lactobacillus helveticus* (Kleerebezem *et al.*, 1997), *L. plantarum* (Pavan *et al.*, 2000), *Lactobacillus brevis* (Avall-Jaaskelainen *et al.*, 2002), *L. casei* (Hazebrouck *et al.*, 2007), *L. salivarius* (Sheehan *et al.*, 2006) and *L. reuteri* (Wu *et al.*, 2006). In these species different strategies have been used to express the *nisRK* genes: on a different plasmid in relation to the *nisA* promoter with the target gene, both on the same plasmid or with the *nisRK* genes inserted into the chromosome. Similar to the NICE system, in *L. plantarum* (Mathiesen *et al.*, 2004) and *L. sakei* (Axelsson *et al.*, 2003) vectors have been developed using a pheromone-regulated bacteriocin promoter and the regulatory system of sakacin A production, respectively. The pSIP vector series, based on the genes and promoters involved in sakacin A and P, used erythromycin as selection marker (Sorvig *et al.*, 2005). In order to developed a potential food-grade expression system the erythromycin gene in the pSIP vectors has been replaced by the *alr* gene, which encodes the alanine racemase enzyme that is essential for cell wall biosynthesis (Nguyen *et al.*, 2011). In *L. casei* an integrative vector, pIlac, has been constructed that allowed stable gene insertion in the chromosomal lactose operon. The vector is based on the nonreplicative plasmid pRV300 and it contains the 3' end of *lacG* and the complete *lacF* gene (Gosalbes *et al.*, 2000). Both vectors, pSIP and pIlac, are based on the complementation host/marker system, a gene in the host is mutated or deleted, and a wild copy is inserted into the vector. Other potential food-grade vectors are based on a selection marker that confers a new ability to the host strain. In this sense a vector has been recently developed that contains a bile salt hydrolase gene from *L. plantarum* and which allows the host to grow in media containing bile salts (Yin *et al.*, 2011). Bioluminiscence markers have also been used in lactobacilli and they are based on genes encoding enzymes that produce light as *lux*, which encodes bacterial luciferase, and *gfp* that encodes green fluorescence protein (Chang *et al.*, 2003; Perez-Arellano & Perez-Martinez, 2003).

1.3.2 DNA mutagenesis systems: integration and insertion systems, and random mutagenesis systems

There are two principal methods to generate mutations in lactobacilli: (i) integration, which is a rec-dependent recombination of cloned DNA with an homologous locus; (ii) rec-independent, which involves transposons and insertion elements (Fang & O'Toole, 2009). The integration procedures mostly used in LAB are based on vectors able to integrate by homologous recombination with known chromosomal genes, causing their disruption by inserting foreign genes. The integrative vectors developed for lactobacilli are either based in temperature-sensitive replicons such as pG+host, pIP501, pTNI and pGID or non-replicative plasmid such as pUC18/19 and pBlueScript SK$^-$. As well, a two plasmids system have also been used to direct integration into *Lactobacillus* chomosomes via homologous recombination (Russell & Klaenhammer, 2001). This system utilizes pOWV01-derived vectors from which the *repA* gene has been removed. The *repA* is supplied in trans in a temperature-sensitive helper vector. A subsequent temperature shift selects for loss of the helper plasmid and integration of the pOWV01-derived vector. In addition, there are other mutagenesis systems as that of the Cre-*lox*-based system used in *L. plantarum* (Lambert *et al.*,

2007) and site-specific integrative vectors based on prophage fragments (Martin *et al.*, 2000). Other important genetic tool used to study chromosomal genes and their regulation in lactobacilli is random transposon mutagenesis. The insertional sequence ISS1, combined with the thermosensitive pG+ replicon, was used to inactive genes involved in the regulation of phenolic acid metabolism in *L. plantarum* (Gury *et al.*, 2004) and several genes in *L. salivarius* (Mason *et al.*, 2005). Tn5 transposome system was also efficiently used to generate a library of transposon insertion mutants in *L. casei* (Ito *et al.*, 2010). As well, factors affecting the reduction of serum cholesterol by *L. acidophillus* were identified by random transposon mutagenesis (Lee *et al.*, 2010).

2. Lactic acid production

Lactic acid produced by many LAB is a racemic mixture of L-lactate and D-lactate isomers. D-lactate is not metabolized by humans, then L-lactate is the most important isomer for food biotechnological applications, and also for pharmaceutical and biopolymers industries. Many efforts have been made to construct LAB strains affected in one or several of the identified *ldh* genes, as they can be used in the production through fermentation of non-racemic, optically active lactic acid. In *L. casei* BL23, a strain that has been widely used for genetic, physiological and biochemical studies, five genes encoding proteins with LDH activity have been described (Rico *et al.*, 2008). Mutant strains for those genes demonstrated the involvement of each *ldh* gene in L- and D-lactate formation in this bacterium. Gene *ldh1* codes for an L-LDH responsible for the main synthesis of L-lactate, whilst *hicD* encodes a D-hydroxyisocaproate dehydrogenase which renders D-lactate. However, an *L. casei* BL23 *ldh1* mutant still produced substantial amounts of L-lactate and an increase in the production of D-lactate was observed (Viana *et al.*, 2005). D-lactate was probably synthesized via the activity of HicD, since it was abolished in a *Δldh1 hicD* double mutant. *ldh2, ldh3* or *ldh4* single mutations or combined with an *ldh1* deletion (*Δldh1 ldh2, Δldh1 ldh3, Δldh1 ldh4*) had a low impact on L-lactate synthesis showing that *ldh2, ldh3* and *ldh4* genes play a minor role in lactate synthesis (Rico *et al.*, 2008). Comparable behaviour has been reported for many LAB where *ldhs* have been deleted. In this sense, mutation of the genes encoding L- and D-LDHs from *L. plantarum,* an organism which produces a mixture of 50% D- and 50% L-lactate, never resulted in a complete lack of lactate production (Ferain *et al.*, 1996). An *ldhL* mutation in *L. sakei,* a lactic acid bacterium which lacks D-lactate dehydrogenase activity, rendered a strain with strongly reduced L- and D- lactate production (the D isomer was a consequence of the presence of a racemase activity able to transform L- into D-lactate), but small amounts of lactate were still produced (Malleret *et al.*, 1998). Recombinant strategies have also been used in *Lactobacillus* strains to produce lactic acid from sugars others than glucose and from biomass such as starch and cellulose. In an *L. plantarum ΔldhL1* strain, that only produced D-lactate from glucose, the phosphoketolase gene was substituted by a transketolase gene from *L. lactis,* and the resulting *L. plantarum ΔldhL1-xpk1::tkt* strain produced 38.6 g/l of D-lactate from 50 g/l of arabinose (Okano *et al.*, 2009). The production of D-lactate from xylose was also achieved in *L. plantarum* by disrupting a phosphoketolase 2 gene in the *L. plantarum ΔldhL1-xpk1::tkt* strain and transforming it with a plasmid that contains the genes *xylAB*. The *L. plantarum ΔldhL1* strain was transformed with plasmids expressing amylolytic or cellulolytic enzymes, and the resulted strains were able to produce D-lactate from raw corn starch or cellulosic compounds, respectively (Okano *et al.*, 2010).

In addition to rational methods of metabolic engineering, lactic acid production has also been enhanced by a combination of classical strain improvement methods (nitrosoguanidine and ultraviolet mutagenesis) with whole-genome shuffling by recursive protoplast fusion. In this way, shuffled strains derived from an industrial strain of *Lactobacillus* have been isolated, and they produce threefold more lactic acid than the wild type at pH 4.0 (Patnaik *et al.*, 2002). Shuffled *L. rhamnosus* strains with improved tolerance to glucose and enhanced L-lactate production has also been obtained (Yasuda *et al.*, 2008). In the same way, a fusant derived from *Lactobacillus delbueckii* able of growing at low pH and utilizing starch from cassava bagasse was obtained and it produced large amounts of L-lactic (John *et al.*, 2008).

3. Diacetyl and acetoin production

Diacetyl and acetoin are important compounds of buttery flavor in fermented foods and are used as additives in the food industry. Both compounds are derived from pyruvate, which is converted to α-acetolactate by the action of α-acetolactate synthase or acetohydroxyacid synthase. Then, acetoin is formed by the activity of α-acetolactate decarboxylase on α-acetolactate and diacetyl results from a non-enzymatic oxidative decarboxilation of α-acetolactate (Figure 2). Most metabolic engineering approaches to produce diacetyl/acetoin by fermentation have been developed in the model LAB *L. lactis*, in which strains that divert an important part of pyruvate flux towards the production of α-acetolactate have been constructed (Hugenholtz *et al.*, 2000; Lopez de Felipe *et al.*, 1998). *ilvBN* genes, encoding acetohydroxyacid synthase from *L. lactis*, have been expressed from the lactose operon in *L. casei*, an organism which shows marginal production of diacetyl/acetoin, resulting in increased diacetyl formation (Gosalbes *et al.*, 2000). In addition, to enhance diacetyl/acetoin production, the amount of pyruvate available for IlvBN was increased by blocking pyruvate alternative pathways in *L. casei*. Thus, the *L. casei* strain that expresses the *ilvBN* genes was mutated in the *ldh* gene and in *pdhC*, encoding the E2 subunit of the pyruvate dehydrogenase enzyme. The introduction of these mutations resulted in an increased capacity to synthesize diacetyl/acetoin from lactose fermentation in whey permeate (1400 mg/l at pH 5.5) (Nadal *et al.*, 2009).

4. Mannitol and sorbitol production

Sugar alcohols are hydrogenated carbohydrates widely used in the food industry as sugar replacers. Mannitol and sorbitol are used as food additives due to their sweetening effect (about half as sweet as sucrose) and low calorie content. They are also used in the food and pharmaceutical industries due to their technological properties, such as texturing agents, humectants, softeners and color stabilizers. In nature, mannitol is found in some plants, algae and mushrooms, and sorbitol is found in many fruits and vegetables. Those polyols are also produced by fungi, yeast and bacteria, where they play several roles in carbon storage and protection during osmotic and oxidative stresses. Industrial production of most sugar alcohols is performed by catalytic reduction of sugars with hydrogen gas and nickel at high temperature and pressure, for which highly pure sugar substrates and costlychromatographic purification steps are required. Regardless the limitations of this chemical method, it is until now the only process able to assume the high market demand of sorbitol and mannitol, estimated to be thousands of tons per year. However, processes using bacteria and yeasts have demonstrated that biotechnological production may represent an efficient and cost-effective alternative to the chemical production.

The production of polyols by using genetically engineered LAB has been recently reviewed (Monedero *et al.*, 2010). Mannitol is a natural fermentation product in heterofermentative LAB, in which the NADH generated during sugar metabolism is regenerated by the

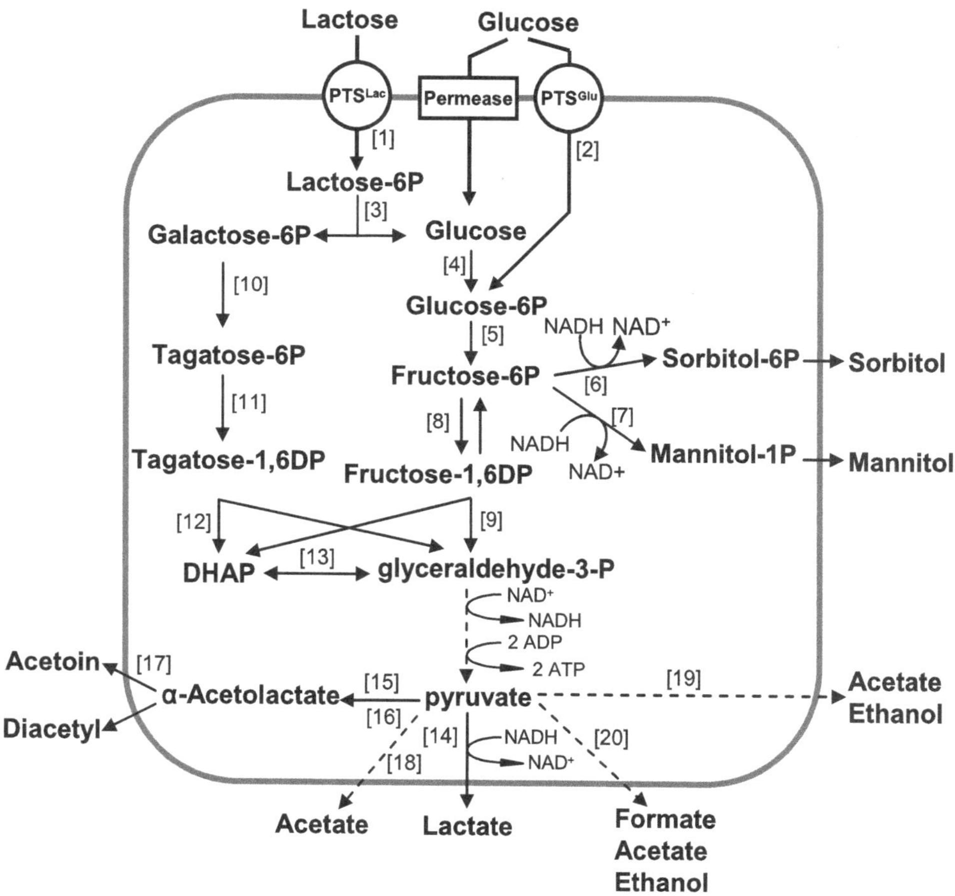

Fig. 2. Proposed pathways for sorbitol, mannitol, acetoin and diacetyl production by engineered lactic acid bacteria. [1] phosphoenolpyruvate: lactose phosphotransferase system, [2] phosphoenolpyruvate: glucose phosphotransferase system, [3] β-phosphogalactosidase, [4] glucokinase, [5] phosphoglucose isomerase, [6] sorbitol-6P dehydrogenase, [7] mannitol-1P dehydrogenase, [8] phosphofructokinase, [9] fructose 1,6-bisP aldolase, [10] galactose-6P isomerase, [11] tagatose-6P kinase, [12] tagatose-1,6DP aldolase, [13] triose-phosphate isomerase, [14] lactate dehydrogenase, [15] α-acetolactate synthase, [16] acetohydroxyacid synthase , [17] α-acetolactate decarboxylase, [18] pyruvate oxidase, [19] pyruvate dehydrogenase complex, [20] pyruvate-formate lyase. DHAP, dihydroxiacetone phosphate.

production of lactate and ethanol. However, in the presence of fructose, a mannitol dehydrogenase activity (MDH) can account for NADH recycling with the concomitant production of mannitol. Homofermentative LAB, which use the glycolytic pathway for sugars fermentation and lack MDH, are also able to produce mannitol under special circumstances (Figure 2). Mutants of *L. plantarum* and *L. casei* impaired in NADH regeneration by the lactate dehydrogenase were able to produce small amounts of mannitol from glucose due to a mannitol-1-P dehydrogenase (M1PDH) activity on fructose-6P (Ferain *et al.*, 1996; Viana *et al.*, 2005). M1PDH can recycle NADH rendering mannitol-1P that can be dephosphorylated to mannitol and excreted from the bacterial cell. M1PDH activity is generally low, because its gene (*mtlD*) is only induced by the presence of mannitol. Furthermore, mannitol is also a common carbon and energy source that can be fermented. Therefore, subsequent re-uptake and metabolism of the produced mannitol should be avoided. In bacteria, mannitol is usually taken up by a mannitol-specific phospho*enol*pyruvate: sugar phosphotransferase system (PTS^{Mtl}) which catalizes the simultaneous mannitol uptake and phosphorylation to mannitol-1P. *L. lactis ldh* mutants have been constructed which were deleted in *mtlA* and *mtlF*, encoding the $EIICB^{Mtl}$ and $EIIA^{Mtl}$ components of the PTS^{Mtl}, respectively, involved in mannitol uptake (Gaspar *et al.*, 2004). This resulted in strains unable to utilize mannitol which converted 33% of the fermented glucose into mannitol. In another approach, the M1PDH encoding gene from *L. plantarum* and a gene encoding a mannitol-1P phosphatase from the protozoan parasite *Eimeria tenella* were overexpressed by using the NICE system in an *L. lactis ldh* mutant (Wisselink *et al.*, 2005). This strategy avoided the main bottleneck in mannitol production: most mannitol was accumulated inside the cell as mannitol-1P, which could reach concentrations up to 76 mM in high density non-growing cells of an *L. lactis ldh* mutant (Neves *et al.*, 2000). In this new strain 50% of the glucose was converted to mannitol (maximum theoretical yield of 67%). Other alternatives comprise the expression of MDH genes from heterofermenters. The *mdh* gene from *L. brevis* was expressed in a *L. plantarum* strain deficient in both *ldhL* and *ldhD* genes, and resulted in an increase in mannitol synthesis from glucose (Liu *et al.*, 2005).

The sorbitol (*gut*) operon of *L. casei* contained the genes *gutCBA*, encoding the EII component of the sorbitol-specific PTS involved in sorbitol transport and phosphorylation, two regulatory genes, *gutR* and *gutM*, and the gene *gutF*, encoding a sorbitol-6P dehydrogenase (S6PDH) (Alcantara *et al.*, 2008). A recombinant strain of *L. casei* with the *gutF* gene integrated in the chromosome at the lactose operon produces sorbitol from fructose-6P by reversing the sorbitol catabolic pathway (Nissen *et al.*, 2005) (Figure 2). Resting cells of this strain synthesized small amounts of sorbitol from glucose, with a conversion rate of 2.4 %. Subsequent inactivation of *ldh1* gene, encoding the main LDH (Rico *et al.*, 2008) promoted an increment in the conversion rate (4.3 %), suggesting that the engineered route provides an alternative pathway for NAD^+ regeneration. Once glucose was depleted, reutilization of the produced sorbitol by *L. casei* recombinant strains was avoided by deleting *gutB* gene that encodes the IIBC component of the PTS^{Gut} (De Boeck *et al.*, 2010). *L. casei* recombinant strains produced mannitol in addition to sorbitol and this polyol mixture was avoided by inactivation of the *mtlD* gene that encodes a M1PDH. The engineered *L. casei* strain (*lac::gutF* Δ*ldh1* Δ*gutB mtlD*) produced sorbitol from lactose, the milk sugar, in non-growing cells or in growing cells under pH control. Fed-batch fermentations using whey permeate, a waste product from the dairy industry with high concentration of lactose, resulted in a conversion rate of 9.4% of lactose into sorbitol (De Boeck *et al.*, 2010). *L. plantarum* has also been metabolically engineered to produce sorbitol by constitutive overexpression of either *srlD1* or *srlD2* genes that encode S6PDH activities in a mutant strain deficient in LDH activity (Ladero *et al.*, 2007). Using non-growing or growing cells under pH control resulted in a very efficiency conversion rate of about 65% and 25%, respectively, of sugar into sorbitol. The different efficiencies were suggested to be the result of a higher ATP demand for biomass production in growing cells.

5. Exopolysaccharides (EPS) production

Some LAB produced EPS, which are extracellular polysaccharides, with important characteristics for the dairy industry. They are used to improve the rheological and textural properties of fermented foods. EPS have also potential as food additives and functional food ingredients. In this sense they are claimed to act as prebiotics in the intestine (Bello *et al.*, 2001) and to stimulate the immune system (Vinderola *et al.*, 2006). The synthesis of EPS in LAB starts at the glycolytic intermediate glucose-6P, which connects the anabolic pathways of biosynthesis of sugar nucleotides, the precursors of the EPS, and the catabolic pathways for obtaining energy through the glycolysis (Figure 1). Glucose-6P is converted to glucose-1P by the α-phospoglucomutase (α-Pgm) activity, and this sugar phosphate is further metabolized to UDP-glucose and UDP-galactose by the consecutively action of enzymes UDP-glucose pyrophosphorylase (GalU) and UDP-galactose 4-epimerase (GalE) (Boels *et al.*, 2001). Glucose-1P is also substrate for dTDP- glucose pyrophosphorylase to produce dTDP-glucose, which will be further metabolized to dTDP-rhanmnose. Glucose, galactose and rhamnose are the principal sugars found in the EPS produced by LAB. The subsequent steps in the synthesis of EPS is the assembly of the monosaccharide repeating unit by specific glycosyltransferases, the polymerization of the repeating units and the secretion from the cell (Welman *et al.*, 2006). The enzymes that participate in these stages are encoded by genes that form part of *eps* gene clusters. Genetic engineering strategies could be applied to one or more of those stages involved in the EPS biosynthesis in order to increase the EPS production or to modify its composition, however, until now only strains of *L. lactis* and *Streptococcus thermophilus* have been genetically modified to enhance EPS biosynthesis. In *S. thermophilus* the modification of the levels of the GalU, PgmA and the Leloir route enzymes resulted in increased levels of EPS (Levander *et al.*, 2002). Homologous overexpression of a complete *eps* operon in *L. lactis* resulted in about fourfold increase in EPS production (Boels *et al.*, 2003). In *Lactobacillus* species there are no examples of metabolic engineering strategies aimed to produce EPS. In this species the synthesis of EPS has been improved by modifying the culture conditions, such as carbon source and pH. As well, chemically induced mutants of *Lactobacillus* species that produce higher amounts of EPS than the parental strain have been isolated. The synthesis of EPS by *L. casei* strain CRL 87 was improved by using galactose as carbon source at a controlled pH of 5.0, and the high EPS production was correlated with high activity level of the enzymes involve in the synthesis of UDP-sugars (Mozzi *et al.*, 2003). Similar approaches were applied for *L. helveticus* strain ATCC 15807, which produces a higher amount of EPS from lactose at pH 4.5 than at pH 6.2, which was correlated with higher levels of α-Pgm activity (Torino *et al.*, 2005). A *L. delbrueckii* subesp. *bulgaricus* mutant with improved EPS production has been isolated, and it showed higher amounts of GalU activity, glucose-6P and ATP than the parent strain. These characteristics suggest that GalU and α-Pgm enzymes play important roles in the synthesis of high EPS production. The elevated concentration of ATP in the mutant indicated that the glycolysis influence the anabolic route of EPS biosynthesis (Welman *et al.*, 2006). A metabolic engineering strategy aimed to direct the carbon flux towards UDP-glucose and UDP-galactose biosynthesis was successfully applied in *L. casei*. The *galU* gene coding for GalU enzyme in *L. casei* strain BL23 was cloned under control of the inducible *nisA* promoter, and the resulting strain showed about an 80-fold increase in GalU activity, a 9-fold increase of UDP-glucose and a 4-fold increase of UDP-galactose (Rodriguez-Diaz & Yebra, 2011). *L. casei* strain BL23 does not produce EPS, hence it would be an adequate host for the production of heterologous EPS.

6. Improvement of probiotic properties

6.1 Adhesion to intestinal epithelial cells

Adhesion of probiotic bacteria has been employed as a criterion for strain selection and, although it is not indispensable for some probiotic traits, it has positive effects on strain persistence at the gastrointestinal tract and in pathogen inhibition by displacement and competition for adhesion sites. Also, it has been suggested that the capacity to adhere to mucosal surfaces influences the cross-talk established between probiotic bacteria and host cells (Sanchez *et al.*, 2008; Velez *et al.*, 2007).

Probiotic strains have shown the ability to bind to intestinal epithelial cultured cells (e.g. Caco-2, HT-29), to mucus components and to proteins of the extracellular matrix (ECM) such as collagen, fibronectin or laminin. Although theses late molecules are not commonly found at the mucosal surface, they may be shed into the mucus or may be exposed in case of trauma or inflammation. They are common targets for pathogen adhesion during the process of infection and adhesion to them by probiotic bacteria can compete with pathogen binding. In contrast to the knowledge about adhesive factors in bacteria causing infectious diseases in humans and animals, the knowledge about adhesion mechanisms in probiotics is very limited. Some molecules from probiotics have been identified as responsible for adhesion, including lipoteichoic acid and exopolysaccharides. However, surface proteins are the major responsible for adherence. Typical surface adhesins from pathogenic bacteria with binding capacity to cultured cells and ECM components are not found in probiotics although it is hypothesized that they may share similar mechanisms for attachment. Similar to some pathogens, probiotic lactobacilli display on their surface *moonlighting* proteins which are in most cases of cytoplasmic location and are exported and retained on bacterial surfaces by yet unknown mechanisms. These include glycolytic enzymes such as enolase and glyceraldehyde 3-phosphate dehydrogenase (GAPDH), molecular chaperones (DnaK, GroEL) and translational elongation factors (EF-Tu, EF-G) (Sanchez *et al.*, 2008; Velez *et al.*, 2007). These proteins have demonstrated binding ability to ECM proteins and epithelial cells and in some cases interfere with host pathways (plasminogen activation mediated by enolase from *Lactobacillus crispatus* and *Lactobacillus johnsonii* (Antikainen *et al.*, 2007) or immunoregulation by GroEL from *L. johnsonii* (Bergonzelli *et al.*, 2006). Other surface proteins with identified roles in adhesion are surface layer (S-layer) proteins or Slp. S-layers from lactobacilli such as *L. acidophilus* or *L. brevis* are formed by small basic proteins which form a crystalline matrix on the bacterial surface. The real function of the S-layer is uncertain but is has been demonstrated that some Slp proteins display binding capacity to ECM proteins and possess immunoregulatory capabilities. In general the adhesive properties of the above extracellular proteins did not show a strict specificity for substrate binding and it is postulated that they may possess lectin-like characteristics that allow their binding to highly glycosylated ECM proteins and mucosal surfaces (Velez *et al.*, 2007).

6.1.1 Specific proteins implicated in mucus binding

Species of lactobacilli from intestinal origin (*L. plantarum, L. reuteri, Lactobacillus gasseri, L. acidophilus, L. johnsonii*) express surface proteins covalently anchored to the cell wall by a sortase-dependent mechanism with mucus binding ability (Msa and Mub proteins) (Boekhorst *et al.*, 2006). They are large multidomain proteins containing up to fifteen tandem copies of a mucin-binding domain (MucBP) and act as mannose-dependent adhesins with the capacity to aggregate *Saccharomyces cerevisiae* cells. The presence of these proteins in certain *L. reuteri* isolates correlates with the binding ability to mucus. Similarly, the presence in *L. plantarum* of *msa* genes is the sole requisite for mucus binding in this species, but domain composition and subtle amino acids changes in each specific Msa protein account for the diverse adhesion properties reported in different strains. In *L. rhamnosus* GG (LGG), a well characterized probiotic with established mucus adhesion properties, the product of the LGG_02337 gene is the only protein encoded in the genome which contains four MucBP domains about 50 amino acids shorter in length compared to the large mucus binding proteins Msa and Mub. This protein is anchored at the bacterial surface and possesses *in vitro* binding activity to mucus (von Ossowski *et al.*, 2011).

A genome analysis of several *L. rhamnosus* strains identified the *spaCBA* cluster as another trait responsible for mucus adherence in LGG. This cluster codes for the three components of pili structures similar to pili described in Gram-positive pathogens that can be identified by immune electron microscopy at the LGG surface (Kankainen *et al.*, 2009). SpaA is the major pilin protein that forms the pilus shaft, while SpaC and SpaB are ancillary pilus proteins which are present at the pilus tip or along the pilus structure and possess adhesive properties. Adhesion experiments with purified proteins, specific antibodies, and mutant construction have demonstrated that SpaC and SpaB are responsible for the mucus binding activity displayed by LGG. This is the first example of the presence of pili adhesive structures in a probiotic strain and exemplifies the adaptation of these bacteria to persist in host tissues.

6.1.2 Engineered lactic acid bacteria with enhanced adhesion

Some of the adhesion factors characterized in probiotic bacteria may be targets for strain engineering aimed to enhance bacterial adhesion. In addition, heterologous expression of well characterized adhesins from different sources can be envisaged. This can be useful to increase residence times at the gastrointestinal tract, enhance interactions with the mucosal immune system and promote competitive exclusion of pathogens by probiotics. Some probiotic strains like *L. casei* Shirota have been engineered to express a fibronectin binding domain from the Sfb protein of *Streptococcus pyogenes*, allowing this strain, which barely binds fibronectin, to bind this ECM substrate, fibrinogen and human fibroblasts (Kushiro *et al.*, 2001). However, to date most genetic engineering strategies aimed to increase lactic acid bacteria adhesion have been carried out in the model *L. lactis* species. This bacterium is not a normal inhabitant of the gastrointestinal tract but it has been used as a food grade delivery vehicle for presenting bioactive molecules to mucosal surfaces, including antigens, cytokines or enzymes. Expression of a protein containing a chitin-binding domain from *L. plantarum* on the surface of *L. lactis* resulted in enhanced capacity to attach to natural compounds carrying polymers of *N*-acetylglucosamine, such as human mucins (Sanchez *et al.*, 2011). The recombinant strain also showed increased attachment to epithelial Caco-2 cells. In another approach, the receptor binding domain of FedD adhesin from *E. coli* F18 fimbriae was expressed and anchored to the bacterial surface by creating a fusion with the surface anchoring domain of the *L. lactis* autolysin AcmA. This fusion protein promoted the binding of *L. lactis* to porcine intestinal epithelial cells (Lindholm *et al.*, 2004). Finally, expression in *L. lactis* of either a fibronectin binding protein from *Staphylococcus aureus* or Internalin A from *Listeria monocytogenes* promoted its binding to human epithelial cells and bacterial internalization, providing a tool for DNA delivery into eukaryotic cells (Innocentin *et al.*, 2009).

6.2 Inmunomodulation of colonic mucosa

In the case of functional properties of lactobacilli, due to legal restrictions and public opinion attitudes against the use of genetically modified microorganisms, the most general strategy has been the selection of naturally competent probiotic strains, nevertheless, some examples of mutants and genetically engineered strains can be found with specific and improved properties (see below). The molecular mechanisms underlying this process are still unknown, however, intervention studies using probiotics in controlled placebo double blind clinical assays are very abundant and different meta-analysis confirmed that several specific beneficial effects of probiotics pass very stringent examination criteria (Williams, 2010), however, they are costly in time and resources and cannot be used to test a great number of strains. Hence, this review will initially describe the general features that characterise the recognition of bacteria by the mucosa and, then, it will focus on the characterisation of the mechanisms of action and the understanding of the effect of probiotics on model systems, as a mean efficiently select funtional strains.

As described in the introduction, the mucosal surface is continuously exposed to both potential pathogens and beneficial commensal microorganisms. This creates a requirement for a homeostatic balance between tolerance and immunity that represents a unique regulatory challenge to the mucosal immune system. Dendritic cells (DCs) in the lamina propria efficiently recognise microbial components from the intestinal lumen through PRRs, TLRs and NLRs. Then, DC migrate to draining lymph nodes, where they have the unique ability to activate and influence functional differentiation of naïve T cells. Signals from DC can determine whether tolerance or active immune responses occur to a particular antigen and furthermore influence whether a T helper (T_h) cells of the type T_h1 (innate immune response), T_h2 (adaptive immune response and allergy), T_h17 or T_{reg} (lymphocyte differentiation) predominates. The DC subtype, whether CD11c+ (myeloid) or CD11c- (plasmacytoid), maturation status, and cytokine production contribute to the type of T cell response. For example, when DCs upregulate the coestimulatory molecules CD80 and CD86, produce IL-12 which contributes a Th1 response, but if DCs produce IL-10 and IL-4, they will promote the generation of a Th2 or regulatory T cells (Hart *et al.*, 2004).

Furthermore, intestinal epithelial cells (IECs) are not just a simple physical barrier. They express TLRs as well as intracellular NLRs and they can secrete cytokines and regulatory molecules (TSLP, TGFβ, IL-10, etc) that regulate cytokine secretion by DCs and macrophages. Therefore, EICs actively participate in the discrimination of both pathogenic and commensal bacteria, they are crucial in triggering lymphocyte differentiation, maintaining intestinal immune homeostasis and mechanisms of innate defense (Artis, 2008). As consequence, commensal bacteria and pathogens are detected at different levels, in IECs, DC and macrophages. Different receptors recognise different bacterial ligands, so that the mucosa would integrate the information to recognise the microorganisms approaching the mucosa. PRRs in IECs and DC binding to bacterial molecular patterns (PAMPs) are expressed at the cell surface (TLR2, TLR4, CD14, TLR5) or in specialised endosomes (TLR3, TLR7, TLR8, TLR9). They can recognise single bacterial ligands or act synergistically to recognise others. As a quick overview, TLR3 recognises double stranded viral RNA, TLR9 hypomethylated CpG bacterial DNA, TLR7 and TLR8 were reported to recognize small imidazoquinoline compounds and TLR4, with the aid of CD14, recognises lipopolysaccharides (LPS) and lipoarabinomannans. Peptidoglycan (PGN) of different grampositive bacteria have been shown to interact with TLR2 (Iwaki *et al.*, 2002), however, TLR2 recognises other PAMPs (PGN, lipoteicoic acid, mycobacterial cell walls, protozoan parasite GPI anchors, lipoproteins, glycoproteins, glycolipids, LPS, etc). Furthermore, the complex TLR2/TLR6 recognises dipalmitoylated mycoplasma lipoprotein (MALP2), phenol soluble modulin from *Staphylococcus epidermidis* and fungal zymosan, and also, TLR2 associated to TLR1 recognises triacylated lipoproteins such as *Borrelia burgdorferi* OspA (for review see (Qureshi & Medzhitov, 2003)). In addition, cell wall components in lactobacilli and firmicutes are recognized by intracellular pattern-recognition molecules members of the nucleotide-binding oligomerization domain (NOD) family. The NLR family (also called Nod-leucine-rich repeats (LRRs)) are responsible for the signalling response induced by bacterial PGN and bacterial surface components, for instance, Nod1/CARD4 receptor in macrophages recognises Meso-diaminopimelic acid (meso-DAP), Nod2/CARD15 recognises muramyl dipeptide (MDP), Nod2 acting as a general sensor for bacteria, Ipaf/CLAN/CARD12 recognises intracellular flagellin (independently of TLR5, which senses extracellular flagellin) and cryopyrin/PYPAF1/NALP3 recognises bacterial RNA (and endogenous danger signals) and among others (for review, see (Franchi *et al.*, 2006). In particular, Nod2 recognizes a PGN motif present on both Gram-positive and Gram-negative bacteria.

6.2.1 Mechanisms of immunomodulation and probiotic factors involved

Nuclear factor κB (NF-κB) is a transcriptional regulator, or rather a regulator family, that controls the expression of hundreds of genes related to different cellular processes, including innate and adaptive immune responses. NF-κB signalling is the major pro-inflammatory pathway controlling the expression of cytokines, chemokines, enzymes that produce secondary inflammatory mediators and inhibitors of apoptosis. It is activated by numerous proteins, among them pathogen-associated molecular patterns (PAMPs). Various lactobacilli have been described to inhibit NF-κB activity (Kim *et al.*, 2008; Petrof *et al.*, 2004). Although the initial steps of the process are not known, it was determined that *L. casei* DN-114.001 can maintain intestinal homeostasis after an inflammatory stimulus through a process that controls the ubiquitin/proteasome pathway upstream of I-kBα resulting in the stabilization of it, therefore blocking NF-κB for nuclear translocation (Tien *et al.*, 2006). Such effects possibly occur through targeting of multiple effectors and, in some cases, through complementary pathways such as NF-κB and p38 MAPK signaling pathways as shown in *L. casei* and *L. reuteri* where they could play important roles in the augmentation of innate

immunity (Iyer *et al.*, 2008; Kim *et al.*, 2006). Two proteins of *L. rhamnosus* GG (p40 and p75) (Yan *et al.*, 2007) have been suggested to preserve the tight junction (TJ) integrity, an effect mediated by the PKC and MAPK pathways. They display antiapoptotic activity, prevent epithelial barrier damage caused by several agents and show *in vivo* effect by decreasing the susceptibility to dextran sulphate sodium (DSS)-induced colon epithelial injury (Yan *et al.*, 2007). Interestingly, recent studies (Bäuerl *et al.*, 2010) showed that these proteins are active cell wall hydrolases present in the seven genome sequences available for strains in the *L. casei-paracasei/rhamnosus* group. PGNs are cell wall components that interact with the intracellular receptor Nod2. Their *in vivo* effect changes with their composition composition, which also varies between strains. In the case of *Lactobacillus salivarius* Ls33 the muropeptide, M-tri-Lys, acts as ligand that protected mice from colitis in a NOD2-dependent but MyD88-independent manner (Macho Fernandez *et al.*, 2011).
Two strains of *L. reuteri* and *L. casei*, but not *L. plantarum*, had the ability to prime DC to drive the development of Treg cells which produced increased levels of the anti-inflammatory cytokine IL-10 (Smits *et al.*, 2005). This ability was mediated by binding to the C-type lectin DC-specific intercellular adhesion molecule 3-grabbing non-integrin (DC-SIGN). Targeting of DC-SIGN by certain probiotic bacteria could explain their beneficial effect in the treatment of several immune diseases, including atopic dermatitis and Crohn's disease. However, a DC-SIGN ligand capacity has been found in an extracellular protein (LP_2145) in *L. plantarum* WCFS1(patent application WO/2009/035330) and in the surface-layer protein A (SlpA) from *L. acidophilus* NCFM, inducing IL-10, IL-4 and decreasing IL12p70 synthesis (proinflammatory) (Konstantinov *et al.*, 2008).
There are also experiments that reported the *in vivo* effect on gene transcription in human volunteers where biopsies were analysed by microarrays. Strains of *L. rhamosus* GG (Di Caro *et al.*, 2005) and *L. plantarum* WCFS1 (van Baarlen *et al.*, 2011) were used administrated and a differential expression of analysed, noticing that expression affected mainly to genes involved in the immune and inflammatory response, as well as to genes related to apoptosis, growth and cell differentiation, signalling, adhesion, etc. Remarkably, striking differences were found in the modulation of NF-κB related pathways.

6.2.2 Engineered lactic acid bacteria

An efficient expression of IL-10 was achieved in *L. lactis* , that showed a striking effect in the remediation of DSS (dextran sodium sulphate) induced colitis in IL-10$^{-/-}$ mice (Steidler *et al.*, 2000). Then, the genetic manipulation strategy was improved and the human IL-10 encoding gene used to replace the thymidylate synthase gene (*thy*A) in the bacterial chromosome, which achieved a stable genetic construction and a self contained recombinant strain requiring thymidine to proliferate. This strain was used in a double blind assay on Crohn's disease patients (phase I), proving that the intake of this strain significantly decreased the disease activity (Braat *et al.*, 2006). Other anti-inflammatory and epithelium repairing peptide, such as the trefoil factor, was similarly successful in mice (Vandenbroucke *et al.*, 2004).
Mutants obtained by site directed mutagenesis, holding genetic changes precisely introduced, have been tested for improved health features. Teichoic acids (TA) activate NF-κB through TLR-2 binding and they are one of the main immunostimulatory components of pathogenic grampositive bacteria. This effect was also observed in *L. plantarum* NCIMB8826, however, a mutant (*dlt*) deficient in D-alanylation of TA was much more anti-inflammatory than the parental strain on peripheral blood mononuclear cells and mice (Grangette *et al.*, 2005).

In other cases mutants have been very useful to demonstrate the functional effect of certain cell components. In *L. casei* Shirota (LcS), a mutant strain was very useful to prove that inhibition of the pro-inflammatory cytokine IL-6, through Nod2/ NF-κB, was due to a cell wall-derived polysaccharide- PGN complex (PSPG) (Matsumoto *et al.*, 2009).

6.3 Engineered probiotics to delivery bioactive proteins

The ability of lactobacilli to survive on the mucosal surfaces of humans and animals has been utilized for the delivery and presentation of bioactive molecules at these surfaces. These bacteria have several advantages, including their recognised GRAS/QPS status, their capacity to interact with the host at several levels and their public acceptance. Studies of lactobacilli as delivery vehicles have mainly focused on the development of mucosal vaccines. In addition, interleukins have been also co-expressed with antigens in lactobacilli to enhance the immune response. Other applications of *Lactobacillus* species as delivery systems include anti-infectives, therapies for allergic diseases and therapies for gastrointestinal diseases. The ability of lactobacilli and other LAB to express these antigens/bioactive molecules at mucosal surfaces have been widely reviewed (Monedero & Pérez-Martínez, 2008; Wells & Mercenier, 2008).

Some recent examples include the use of *Lactobacillus jensenii* strains isolated from human vaginal mucosa for the delivery vehicle of a surface-anchored two-domain CD4 (2D CD4) molecule for the mitigation of heterosexual transmission of HIV (Liu *et al.*, 2008). *L. casei* was the host to express the viral proteins from porcine rotavirus and porcine parvovirus fused the heat-labile toxin B subunit from *Escherichia coli*. The results showed that mice responded producing increased levels of anti-viral antibodies (Liu *et al.*, 2011). *L. casei* Zhang was engineered to stably express the p23 immunodominant surface protein of *Cryptosporidium parvum* sporozoites. Recombinant *L. casei* Zhang-p23 was able to activate the mucosal immune system and to elicit specific serum immunoglobulin G (IgG) and mucosal IgA in mice. The expression of cytokines such as IL4, IL6, and IFN-gamma was detected in splenocytes of mice by real-time PCR after oral immunization with this strain (Geriletu *et al.*, 2011). A recombinant *L. casei* strain secreting biologically active murine interleukin-1β has been constructed. This strain was able to induce IL8 secretion in Caco2 cells and IL6 in vivo using a ligated-intestinal-loop assay in mice after oral administration. The increased adjuvant properties of this strain were confirmed after intragastric immunization with heat-killed *Salmonella enterica* serovar *Enteritidis* (Kajikawa *et al.*, 2010).

Another application of lactobacilli has been their use to express molecules to deliver passive immunity against pathogens, such as single-chain antibodies (lactobodies). Martin and co-workers have been able to construct a series of expression cassettes stably integrated in the chromosome to mediate the secretion or surface display of antibody fragments in *Lactobacillus paracasei*. These new constructed strains, producing surface-anchored variable domain of llama heavy chain (VHH) (ARP1) directed against rotavirus, showed efficient binding to rotavirus and protection in the mouse model of rotavirus infection (Martin *et al.*, 2011).

Lactobacilli can also be engineered to produce an increased immune response against cancer cells. *L. rhamnosus* GG (LGG) has been used to successfully induce tumor regression in an orthotopic model of bladder cancer. The potential of LGG to induce a directed anti-tumor response was evaluated with modified LGG secreting the prostate specific antigen (PSA) or IL15 and PSA (IL-15-PSA). Recombinant LGG activated neutrophils, induced dendritic cells maturation, T cell proliferation and PSA specific cytotoxic T lymphocytes activity leading to the tumor regression (Kandasamy *et al.*, 2011).

In spite of the dozens of reports describing the capacity of LAB to deliver bioactive molecules to the host mucosa, important issues, such as potential side-effects should be always addressed when working with recombinant LAB. The expression of the *Salmonella* OmpC in a recombinant *L. casei* was fortuitously found to reduce the secretion of tumor necrosis factor alpha (TNF-α) from murine macrophages (Kajikawa & Igimi, 2009). Another non desired effect of a recombinant *Lactobacillus* was found when *L. acidophilus* was engineered to express the interferon-β to study the local delivery of this molecule in a colitis mouse model. Surprisingly the administration of the recombinant bacteria secreting IFN-β has an immunological effect that resulted in the exacerbation of colitis in this model (McFarland *et al.*, 2011).

7. References

Alcantara, C., Sarmiento-Rubiano, L. A., Monedero, V., Deutscher, J., Perez-Martinez, G. & Yebra, M. J. (2008). Regulation of *Lactobacillus casei* sorbitol utilization genes requires DNA-binding transcriptional activator GutR and the conserved protein GutM. *Appl Environ Microbiol* 74, 5731-5740.

Antikainen, J., Kuparinen, V., Lahteenmaki, K. & Korhonen, T. K. (2007). Enolases from Gram-positive bacterial pathogens and commensal lactobacilli share functional similarity in virulence-associated traits. *FEMS Immunol Med Microbiol* 51, 526-534.

Artis, D. (2008). Epithelial-cell recognition of commensal bacteria and maintenance of immune homeostasis in the gut. *Nat Rev Immunol* 8, 411-420.

Avall-Jaaskelainen, S., Kyla-Nikkila, K., Kahala, M., Miikkulainen-Lahti, T. & Palva, A. (2002). Surface display of foreign epitopes on the *Lactobacillus brevis* S-layer. *Appl Environ Microbiol* 68, 5943-5951.

Axelsson, L., Lindstad, G. & Naterstad, K. (2003). Development of an inducible gene expression system for *Lactobacillus sakei*. *Lett Appl Microbiol* 37, 115-120.

Bäuerl, C., Pérez-Martínez, G., Yan, F., Polk, D. B. & Monedero, V. (2010). Functional Analysis of the p40 and p75 Proteins from *Lactobacillus casei* BL23. *Journal of Mol Microbiol and Biotechnol* 19, 231-241.

Bello, F. D., Walter, J., Hertel, C. & Hammes, W. P. (2001). In vitro study of prebiotic properties of levan-type exopolysaccharides from Lactobacilli and non-digestible carbohydrates using denaturing gradient gel electrophoresis. *Syst Appl Microbiol* 24, 232-237.

Bergonzelli, G. E., Granato, D., Pridmore, R. D., Marvin-Guy, L. F., Donnicola, D. & Corthesy-Theulaz, I. E. (2006). GroEL of *Lactobacillus johnsonii* La1 (NCC 533) is cell surface associated: potential role in interactions with the host and the gastric pathogen Helicobacter pylori. *Infect Immun* 74, 425-434.

Boekhorst, J., Helmer, Q., Kleerebezem, M. & Siezen, R. J. (2006). Comparative analysis of proteins with a mucus-binding domain found exclusively in lactic acid bacteria. *Microbiology* 152, 273-280.

Boels, I. C., Ramos, A., Kleerebezem, M. & de Vos, W. M. (2001). Functional analysis of the *Lactococcus lactis galU* and *galE* genes and their impact on sugar nucleotide and exopolysaccharide biosynthesis. *Appl Environ Microbiol* 67, 3033-3040.

Boels, I. C., Van Kranenburg, R., Kanning, M. W., Chong, B. F., De Vos, W. M. & Kleerebezem, M. (2003). Increased exopolysaccharide production in *Lactococcus lactis* due to increased levels of expression of the NIZO B40 eps gene cluster. *Appl Environ Microbiol* 69, 5029-5031.

Braat, H., Rottiers, P., Hommes, D. W. & other authors (2006). A Phase I Trial With Transgenic Bacteria Expressing Interleukin-10 in Crohn's Disease. *Clinical Gastroenterology and Hepatology* 4, 754-759.

Branny, P., de la Torre, F. & Garel, J. R. (1998). An operon encoding three glycolytic enzymes in *Lactobacillus delbrueckii* subsp. *bulgaricus*: glyceraldehyde-3-phosphate dehydrogenase, phosphoglycerate kinase and triosephosphate isomerase. *Microbiology* 144 (Pt 4), 905-914.

Crutz-Le Coq, A. M. & Zagorec, M. (2008). Vectors for Lactobacilli and other Gram-positive bacteria based on the minimal replicon of pRV500 from *Lactobacillus sakei*. *Plasmid* 60, 212-220.

Chang, T. L., Chang, C. H., Simpson, D. A. & other authors (2003). Inhibition of HIV infectivity by a natural human isolate of *Lactobacillus jensenii* engineered to express functional two-domain CD4. *Proc Natl Acad Sci U S A* 100, 11672-11677.

De Boeck, R., Sarmiento-Rubiano, L. A., Nadal, I., Monedero, V., Perez-Martinez, G. & Yebra, M. J. (2010). Sorbitol production from lactose by engineered *Lactobacillus casei* deficient in sorbitol transport system and mannitol-1-phosphate dehydrogenase. *Appl Microbiol Biotechnol* 85, 1915-1922.

Di Caro, S., Tao, H., Grillo, A., Elia, C., Gasbarrini, G., Sepulveda, A. R. & Gasbarrini, A. (2005). Effects of *Lactobacillus* GG on genes expression pattern in small bowel mucosa. *Dig Liver Dis* 37, 320-329.

Dong, H., Rowland, I., Tuohy, K. M., Thomas, L. V. & Yaqoob, P. (2010). Selective effects of *Lactobacillus casei* Shirota on T cell activation, natural killer cell activity and cytokine production. *Clin Exp Immunol* 161, 378-388.

Eckburg, P. B., Bik, E. M., Bernstein, C. N., Purdom, E., Dethlefsen, L., Sargent, M., Gill, S. R., Nelson, K. E. & Relman, D. A. (2005). Diversity of the Human Intestinal Microbial Flora. *Science* 308, 1635-1638.

Falagas, M. E., Betsi, G. I. & Athanasiou, S. (2007). Probiotics for the treatment of women with bacterial vaginosis. *Clin Microbiol Infect* 13, 657-664.

Fang, F. & O'Toole, P. W. (2009). Genetic tools for investigating the biology of commensal lactobacilli. *Front Biosci* 14, 3111-3127.

FAO/WHO (2001). Evaluation of Health and Nutritional Properties of Powder Milk and Live Lactic Acid Bacteria *Food and Agriculture Organization of the United Nations and World Health Organization Expert Consultant Report*, 1–34.

Ferain, T., Schanck, A. N. & Delcour, J. (1996). 13C nuclear magnetic resonance analysis of glucose and citrate end products in an ldhL-ldhD double-knockout strain of *Lactobacillus plantarum*. *J Bacteriol* 178, 7311-7315.

Franchi, L., McDonald, C., Kanneganti, T. D., Amer, A. & Nunez, G. (2006). Nucleotide-Binding Oligomerization Domain-Like Receptors: Intracellular Pattern Recognition Molecules for Pathogen Detection and Host Defense. *The Journal of Immunology* 177, 3507-3513.

Garrigues, C., Loubiere, P., Lindley, N. D. & Cocaign-Bousquet, M. (1997). Control of the shift from homolactic acid to mixed-acid fermentation in *Lactococcus lactis*: predominant role of the NADH/NAD+ ratio. *J Bacteriol* 179, 5282-5287.

Gaspar, P., Neves, A. R., Ramos, A., Gasson, M. J., Shearman, C. A. & Santos, H. (2004). Engineering *Lactococcus lactis* for production of mannitol: high yields from food-grade strains deficient in lactate dehydrogenase and the mannitol transport system. *Appl Environ Microbiol* 70, 1466-1474.

Geriletu, Xu, R., Jia, H., Terkawi, M. A., Xuan, X. & Zhang, H. (2011). Immunogenicity of orally administrated recombinant *Lactobacillus casei* Zhang expressing Cryptosporidium parvum surface adhesion protein P23 in mice. *Curr Microbiol* 62, 1573-1580.

Gosalbes, M. J., Esteban, C. D., Galan, J. L. & Perez-Martinez, G. (2000). Integrative food-grade expression system based on the lactose regulon of *Lactobacillus casei. Appl Environ Microbiol* 66, 4822-4828.

Grangette, C., Nutten, S., Palumbo, E. & other authors (2005). Enhanced antiinflammatory capacity of a *Lactobacillus plantarum* mutant synthesizing modified teichoic acids. *Proc Natl Acad Sci U S A* 102, 10321-10326.

Gury, J., Barthelmebs, L. & Cavin, J. F. (2004). Random transposon mutagenesis of *Lactobacillus plantarum* by using the pGh9:IS S1 vector to clone genes involved in the regulation of phenolic acid metabolism. *Arch Microbiol* 182, 337-345.

Haller, D., Antoine, J. M., Bengmark, S., Enck, P., Rijkers, G. T. & Lenoir-Wijnkoop, I. (2010). Guidance for substantiating the evidence for beneficial effects of probiotics: probiotics in chronic inflammatory bowel disease and the functional disorder irritable bowel syndrome. *J Nutr* 140, 690S-697S.

Hart, A. L., Lammers, K., Brigidi, P. & other authors (2004). Modulation of human dendritic cell phenotype and function by probiotic bacteria. *Gut* 53, 1602-1609.

Hazebrouck, S., Pothelune, L., Azevedo, V., Corthier, G., Wal, J. M. & Langella, P. (2007). Efficient production and secretion of bovine beta-lactoglobulin by *Lactobacillus casei. Microb Cell Fact* 6, 12.

Hugenholtz, J., Kleerebezem, M., Starrenburg, M., Delcour, J., de Vos, W. & Hols, P. (2000). *Lactococcus lactis* as a cell factory for high-level diacetyl production. *Appl Environ Microbiol* 66, 4112-4114.

Innocentin, S., Guimaraes, V., Miyoshi, A., Azevedo, V., Langella, P., Chatel, J. M. & Lefevre, F. (2009). *Lactococcus lactis* expressing either Staphylococcus aureus fibronectin-binding protein A or *Listeria monocytogenes* internalin A can efficiently internalize and deliver DNA in human epithelial cells. *Appl Environ Microbiol* 75, 4870-4878.

Ito, M., Kim, Y. G., Tsuji, H., Kiwaki, M., Nomoto, K., Tanaka, R., Okada, N. & Danbara, H. (2010). A practical random mutagenesis system for probiotic *Lactobacillus casei* using Tn5 transposition complexes. *J Appl Microbiol* 109, 657-666.

Iwaki, D., Mitsuzawa, H., Murakami, S., Sano, H., Konishi, M., Akino, T. & Kuroki, Y. (2002). The Extracellular Toll-like Receptor 2 Domain Directly Binds Peptidoglycan Derived from *Staphylococcus aureus. Journal of Biological Chemistry* 277, 24315-24320.

Iyer, C., Kosters, A., Sethi, G., Kunnumakkara, A. B., Aggarwal, B. B. & Versalovic, J. (2008). Probiotic *Lactobacillus reuteri* promotes TNF-induced apoptosis in human myeloid leukemia-derived cells by modulation of NF-+ÝB and MAPK signalling. *Cellular Microbiology* 10, 1442-1452.

John, R. P., Gangadharan, D. & Madhavan Nampoothiri, K. (2008). Genome shuffling of *Lactobacillus delbrueckii* mutant and *Bacillus amyloliquefaciens* through protoplasmic fusion for L-lactic acid production from starchy wastes. *Bioresour Technol* 99, 8008-8015.

Kajikawa, A. & Igimi, S. (2009). Reduction of tumor necrosis factor alpha-inducing capacity of recombinant *Lactobacillus casei* via expression of *Salmonella* OmpC. *Appl Environ Microbiol* 75, 2727-2734.

Kajikawa, A., Ichikawa, E. & Igimi, S. (2010). Development of a highly efficient protein-secreting system in recombinant *Lactobacillus casei. J Microbiol Biotechnol* 20, 375-382.

Kandasamy, M., Bay, B. H., Lee, Y. K. & Mahendran, R. (2011). Lactobacilli secreting a tumor antigen and IL15 activates neutrophils and dendritic cells and generates cytotoxic T lymphocytes against cancer cells. *Cell Immunol*. doi:10.1016/j.cellimm.2011.06.004

Kankainen, M., Paulin, L., Tynkkynen, S. & other authors (2009). Comparative genomic analysis of *Lactobacillus rhamnosus* GG reveals pili containing a human- mucus binding protein. *Proc Natl Acad Sci U S A* 106, 17193-17198.

Kim, J. M., Kim, J. S., Kim, Y. J., Oh, Y. K., Kim, I. Y., Chee, Y. J., Han, J. S. & Jung, H. C. (2008). Conjugated linoleic acids produced by *Lactobacillus* dissociates IKK-[gamma] and Hsp90 complex in *Helicobacter pylori*-infected gastric epithelial cells. *Lab Invest* 88, 541-552.

Kim, Y. G., Ohta, T., Takahashi, T., Kushiro, A., Nomoto, K., Yokokura, T., Okada, N. & Danbara, H. (2006). Probiotic *Lactobacillus casei* activates innate immunity via NF-[kappa]B and p38 MAP kinase signaling pathways. *Microbes and Infection* 8, 994-1005.

Kleerebezem, M., Beerthuyzen, M. M., Vaughan, E. E., de Vos, W. M. & Kuipers, O. P. (1997). Controlled gene expression systems for lactic acid bacteria: transferable nisin-inducible expression cassettes for *Lactococcus*, *Leuconostoc*, and *Lactobacillus* spp. *Appl Environ Microbiol* 63, 4581-4584.

Kleerebezem, M., Boekhorst, J., van Kranenburg, R. & other authors (2003). Complete genome sequence of *Lactobacillus plantarum* WCFS1. *Proc Natl Acad Sci U S A* 100, 1990-1995.

Konstantinov, S. R., Smidt, H., de Vos, W. M. & other authors (2008). S layer protein A of *Lactobacillus acidophilus* NCFM regulates immature dendritic cell and T cell functions. *Proc Natl Acad Sci U S A* 105, 19474-19479.

Kushiro, A., Takahashi, T., Asahara, T., Tsuji, H., Nomoto, K. & Morotomi, M. (2001). Lactobacillus casei acquires the binding activity to fibronectin by the expression of the fibronectin binding domain of *Streptococcus pyogenes* on the cell surface. *J Mol Microbiol Biotechnol* 3, 563-571.

Ladero, V., Ramos, A., Wiersma, A., Goffin, P., Schanck, A., Kleerebezem, M., Hugenholtz, J., Smid, E. J. & Hols, P. (2007). High-level production of the low-calorie sugar sorbitol by *Lactobacillus plantarum* through metabolic engineering. *Appl Environ Microbiol* 73, 1864-1872.

Lambert, J. M., Bongers, R. S. & Kleerebezem, M. (2007). Cre-lox-based system for multiple gene deletions and selectable-marker removal in *Lactobacillus plantarum. Appl Environ Microbiol* 73, 1126-1135.

Lebeer, S., Vanderleyden, J. & De Keersmaecker, S. C. (2008). Genes and molecules of lactobacilli supporting probiotic action. *Microbiol Mol Biol Rev* 72, 728-764.

Lee, J., Kim, Y., Yun, H. S., Kim, J. G., Oh, S. & Kim, S. H. (2010). Genetic and proteomic analysis of factors affecting serum cholesterol reduction by *Lactobacillus acidophilus* A4. *Appl Environ Microbiol* 76, 4829-4835.

Levander, F., Svensson, M. & Radstrom, P. (2002). Enhanced exopolysaccharide production by metabolic engineering of *Streptococcus thermophilus*. *Appl Environ Microbiol* 68, 784-790.

Lindholm, A., Smeds, A. & Palva, A. (2004). Receptor binding domain of *Escherichia coli* F18 fimbrial adhesin FedF can be both efficiently secreted and surface displayed in a functional form in *Lactococcus lactis*. *Appl Environ Microbiol* 70, 2061-2071.

Liu, S., Saha, B. & Cotta, M. (2005). Cloning, expression, purification, and analysis of mannitol dehydrogenase gene mtlK from *Lactobacillus brevis*. *Appl Biochem Biotechnol* 121-124, 391-401.

Liu, X., Lagenaur, L. A., Lee, P. P. & Xu, Q. (2008). Engineering of a human vaginal *Lactobacillus* strain for surface expression of two-domain CD4 molecules. *Appl Environ Microbiol* 74, 4626-4635.

Liu, D., Wang, X., Ge, J., Liu, S. & Li, Y. (2011) Comparison of the immune responses induced by oral immunization of mice with *Lactobacillus casei*-expressing porcine parvovirus VP2 and VP2 fused to Escherichia coli heat-labile enterotoxin B subunit protein. *Comp Immunol Microbiol Infect Dis* 34, 73-81.

Lopez de Felipe, F., Kleerebezem, M., de Vos, W. M. & Hugenholtz, J. (1998). Cofactor engineering: a novel approach to metabolic engineering in *Lactococcus lactis* by controlled expression of NADH oxidase. *J Bacteriol* 180, 3804-3808.

Macho Fernandez, E., Valenti, V. , Rockel, C., Hermann, C., Pot, B., Boneca, I. G. & Grangette, C. (2011). Anti-inflammatory capacity of selected lactobacilli in experimental colitis is driven by NOD2-mediated recognition of a specific peptidoglycan-derived muropeptide. *Gut*. 60, 1050-1059.

Malleret, C., Lauret, R., Ehrlich, S. D., Morel-Deville, F. & Zagorec, M. (1998). Disruption of the sole ldhL gene in *Lactobacillus sakei* prevents the production of both L- and D-lactate. *Microbiology* 144 (Pt 12), 3327-3333.

Martin, M. C., Alonso, J. C., Suarez, J. E. & Alvarez, M. A. (2000). Generation of food-grade recombinant lactic acid bacterium strains by site-specific recombination. *Appl Environ Microbiol* 66, 2599-2604.

Martin, M. C., Pant, N., Ladero, V. & other authors (2011). Integrative expression system for delivery of antibody fragments by lactobacilli. *Appl Environ Microbiol* 77, 2174-2179.

Mason, C. K., Collins, M. A. & Thompson, K. (2005). Modified electroporation protocol for Lactobacilli isolated from the chicken crop facilitates transformation and the use of a genetic tool. *J Microbiol Methods* 60, 353-363.

Mathiesen, G., Sorvig, E., Blatny, J., Naterstad, K., Axelsson, L. & Eijsink, V. G. (2004). High-level gene expression in *Lactobacillus plantarum* using a pheromone-regulated bacteriocin promoter. *Lett Appl Microbiol* 39, 137-143.

Matsumoto, S., Hara, T., Nagaoka, M., Mike, A., Mitsuyama, K., Sako, T., Yamamoto, M., Kado, S. & Takada, T. (2009). A component of polysaccharide peptidoglycan complex on *Lactobacillus* induced an improvement of murine model of inflammatory bowel disease and colitis-associated cancer. *Immunology* 128, e170-e180.

Schotte, L., Steidler, L., Vandekerckhove, J. & Remaut, E. (2000). Secretion of biologically active murine interleukin-10 by *Lactococcus lactis*. *Enzyme Microb Technol* 27, 761-765.

Sheehan, V. M., Sleator, R. D., Fitzgerald, G. F. & Hill, C. (2006). Heterologous expression of BetL, a betaine uptake system, enhances the stress tolerance of *Lactobacillus salivarius* UCC118. *Appl Environ Microbiol* 72, 2170-2177.

Smits, H. H., Engering, A., van der Kleij, D. & other authors (2005). Selective probiotic bacteria induce IL-10-producing regulatory T cells in vitro by modulating dendritic cell function through dendritic cell-specific intercellular adhesion molecule 3-grabbing nonintegrin. *Allergy Clin Immunol.* 115, 1260-1267.

Sorvig, E., Mathiesen, G., Naterstad, K., Eijsink, V. G. & Axelsson, L. (2005). High-level, inducible gene expression in *Lactobacillus sakei* and *Lactobacillus plantarum* using versatile expression vectors. *Microbiology* 151, 2439-2449.

Stamatova, I. & Meurman, J. H. (2009). Probiotics: health benefits in the mouth. *Am J Dent* 22, 329-338.

Steidler, L., Hans, W., Schotte, L., Neirynck, S., Obermeier, F., Falk, W., Fiers, W. & Remaut, E. (2000). Treatment of Murine Colitis by *Lactococcus lactis* Secreting Interleukin-10. *Science* 289, 1352-1355.

Tang, M. L., Lahtinen, S. J. & Boyle, R. J. (2010). Probiotics and prebiotics: clinical effects in allergic disease. *Curr Opin Pediatr* 22, 626-634.

Tien, M. T., Girardin, S. E., Regnault, B., Le Bourhis, L., Dillies, M. A., Coppee, J. Y., Bourdet-Sicard, R., Sansonetti, P. J. & Pedron, T. (2006). Anti-Inflammatory Effect of *Lactobacillus casei* on Shigella-Infected Human Intestinal Epithelial Cells. *The Journal of Immunology* 176, 1228-1237.

Torino, M. I., Mozzi, F. & Font de Valdez, G. (2005). Exopolysaccharide biosynthesis by *Lactobacillus helveticus* ATCC 15807. *Appl Microbiol Biotechnol* 68, 259-265.

van Baarlen, P., Troost, F., van der Meer, C., Hooiveld, G., Boekschoten, M., Brummer, R. J. M. & Kleerebezem, M. (2011). Human mucosal in vivo transcriptome responses to three lactobacilli indicate how probiotics may modulate human cellular pathways. *Proc Natl Acad Sci U S A.* 108 Suppl 1, 4562-4569.

Vandenbroucke, K., Hans, W., Van Huysse, J., Neirynck, S., Demetter, P., Remaut, E., Rottiers, P. & Steidler, L. (2004). Active delivery of trefoil factors by genetically modified *Lactococcus lactis* prevents and heals acute colitis in mice. *Gastroenterology* 127, 502-513.

Velez, M. P., De Keersmaecker, S. C. & Vanderleyden, J. (2007). Adherence factors of *Lactobacillus* in the human gastrointestinal tract. *FEMS Microbiol Lett* 276, 140-148.

Viana, R., Yebra, M. J., Galan, J. L., Monedero, V. & Perez-Martinez, G. (2005). Pleiotropic effects of lactate dehydrogenase inactivation in *Lactobacillus casei*. *Res Microbiol* 156, 641-649.

Vinderola, G., Perdigon, G., Duarte, J., Farnworth, E. & Matar, C. (2006). Effects of the oral administration of the exopolysaccharide produced by *Lactobacillus kefiranofaciens* on the gut mucosal immunity. *Cytokine* 36, 254-260.

von Ossowski, I., Satokari, R., Reunanen, J., Lebeer, S., De Keersmaecker, S. C., Vanderleyden, J., de Vos, W. M. & Palva, A. (2011). Functional Characterization of a Mucus-Specific LPXTG Surface Adhesin from Probiotic *Lactobacillus rhamnosus* GG. *Appl Environ Microbiol* 77, 4465-4472.

Welman, A. D., Maddox, I. S. & Archer, R. H. (2006). Metabolism associated with raised metabolic flux to sugar nucleotide precursors of exopolysaccharides in *Lactobacillus delbrueckii subsp. bulgaricus*. *J Ind Microbiol Biotechnol* 33, 391-400.

Wells, J. M. & Mercenier, A. (2008). Mucosal delivery of therapeutic and prophylactic molecules using lactic acid bacteria. *Nat Rev Microbiol* 6, 349-362.

Williams, N. T. (2010). Probiotics. *American J. of Health-System Pharmacy* 67, 449-458.

Wisselink, H. W., Moers, A. P., Mars, A. E., Hoefnagel, M. H., de Vos, W. M. & Hugenholtz, J. (2005). Overproduction of heterologous mannitol 1-phosphatase: a key factor for engineering mannitol production by *Lactococcus lactis*. *Appl Environ Microbiol* 71, 1507-1514.

Wu, C. M., Lin, C. F., Chang, Y. C. & Chung, T. C. (2006). Construction and characterization of nisin-controlled expression vectors for use in *Lactobacillus reuteri*. *Biosci Biotechnol Biochem* 70, 757-767.

Yan, F., Cao, H., Cover, T. L., Whitehead, R., Washington, M. K. & Polk, D. B. (2007). Soluble proteins produced by probiotic bacteria regulate intestinal epithelial cell survival and growth. *Gastroenterology* 132, 562-5675.

Yasuda, E., Serata, M. & Sako, T. (2008). Suppressive effect on activation of macrophages by *Lactobacillus casei* strain Shirota genes determining the synthesis of cell wall-associated polysaccharides. *Appl Environ Microbiol* 74, 4746-4755.

Yin, S., Zhai, Z., Wang, G., An, H., Luo, Y. & Hao, Y. (2011). A novel vector for lactic acid bacteria that uses a bile salt hydrolase gene as a potential food-grade selection marker. *J Biotechnol* 152, 49-53.

Biotechnological Utilisation of Fusel Oil for Biolubricant Production

Nándor Nemestóthy, Tamás Bányai,
Katalin Bélafi-Bakó, László Bartha[1] and László Gubicza
*University of Pannonia, Research Institute on Bioengineering,
Membrane Technology and Energetics,
[1]Department of MOL Hydrocarbon and Coal Processing
Hungary*

1. Introduction

Complete utilisation of by-products from agricultural and food industrial production is an important task both from economical and environmental aspects. There are numerous possibilities for manufacture of bioproducts by biotechnological processes, among them biofuels attracted the greatest attention (Watanabe et al., 2000). Industrial by-products can be, however, processed by several biotechnological methods, for example their utilisation as a food additive, which requires intensive research and development work (Filipini & Hogg, 1997, Demirbas, 2000).

Fusel oil is a by-product of distilleries, its average composition is 10% ethanol, 13% n-propanol, 15% i-butanol, 51% i-amyl-alcohol, 11% miscellaneous alcohols and water. Nowadays fusel oil is usually burned to cover the energy demand of the distilleries. Researches have been carried out to utilise it as an additive to improve octane number in gasoline or for production of natural flavours and lubricants (Özgülsün & Karaosmanoglu, 1999).

Esterification of fusel oil with oleic acid using sulphuric acid catalyst was studied by Turkish researchers and bio-lubricant—according to ASTM (American Society Testing and Materials) standard—was manufactured for industrial purposes. Pure, natural lubricants manufactured by environmental-safe processes, however, have gained more and more attention recently, since they do not contain toxic compounds and are biological degradable. The demands against a bio-lubricant are that it should provide maximal protection during its usage, do not pollute the environment and do not accumulate (Özgülsün et al., 2000).

Unfortunately the used lubricants are usually deposited in the environment, endangering our planet. To solve the problem lubricants should be manufactured from plant oil derivatives. There are several industrial application possibilities for fatty acid esters, as natural compounds. Oleic acid (cis-9-octadecenoic acid) is one of the most important fatty acids in nature, it can be obtained from plant oils (Bélafi-Bakó et al., 1994), its esters produced by enzyme catalysis can be applied as lubricant (Linko et al., 1998).

Modern enzymology has achieved improvements in the development and application of lipase as catalyst. New immobilisation techniques make possible to use enzymes in industrial processes in a similar way to the classical catalysis for heterogeneous reactions.

For example, esters produced from long-chain fatty acids (12–20 carbon atoms) and short-chain alcohols (3–8 carbon atoms) have been used increasingly in the food, detergent, cosmetic and pharmaceutical industries. Esters prepared from the reaction of long-chain acids with long-chain alcohols (12–20 carbon atoms) also have important applications as plasticizers and lubricants (Zaidi et al., 2002, Dossat et al., 2002). The direct effect of the ester group on the physical properties of a lubricant is to lower the volatility and raise the flash point.

Compared with conventional chemical synthesis from alcohols and carboxylic acids using mineral acids as a catalyst, the use of enzymes such as lipases to produce these high value-added fatty acid esters in solvent-free media may offer many significant advantages (Yadav & Lathi, 2003). These include the use of any hydrophobic substrate, higher selectivity, milder processing conditions and the ease of product isolation and enzyme reuse. The ecological properties of oleochemical esters have been intensively studied within the last couple of years. In general, their aquatic toxicity is very low or almost negligible. For the aquatic compartment the fish, daphnia, algae and bacteria are the most relevant test organisms and standardized test methods, such as laid down in the OECD methods 201–210 (Willing, 1999).

Esterification reactions by lipase in non-conventional media have been studied in our laboratory for long (Gubicza et al., 2003). Enzymatic esterification of fatty acids and ingredients of fusel oil was studied by (Gulati et al., 2003) using lipase from *Aspergillus tereus*. They found that in n-hexane solvent the alcohols were able to react with the fatty acids (miristic acid, palmitic acid, stearic acid), except oleic acid. Using other lipase preparations (*Candida antarctica, Candida rugosa, Rhizomucor miehei, Porcine pancreas*), however, made it possible the successful oleic acid esterification with similar low molecular weight alcohols. Description of the correct kinetics on the particular esterification reaction is even more difficult due to the various possible inhibition effects.

In our earlier work natural aroma esters were produced by enzymatic esterification in organic solvents and in solvent-free media (Gubiza, 2000). In this work the purpose was to find a utilisation of fusel oil, where bio-lubricants can be manufactured. The alcohol compounds of fusel oil were esterified with oleic acid using enzyme catalysis in non-conventional, solvent-free media (section 2) and in ionic liquid (section 4), moreover the kinetics of the reaction was described (section 3).

2. Esterification in solvent-free system

After determining the optimal working parameters in the solvent-free system (molar ratio, temperature, enzyme concentration, initial water content) the reactions were completed in integrated system, where pervaporation - and effective and mild membrane separation process (Garcia, 1999) - was connected to the bioreactor.

2.1 Experimental

The catalyst used was Novozyme 435, a commercial *Candida antarctica* lipase (E.C. 3.1.1.3. Triacylglycerol acylhydrolase) immobilized on a macroporous acrylic resin with a water content of 1-2 % w/w. The enzyme was provided as a gift by Novo Nordisk A/S (Denmark). The nominal activity of the catalytic preparation was approximately 7000 Propyl Laurate Units (PLU/g). One propyl laurate unit (PLU) is defined as the number of μmol of n-propyl laurate obtained in the standard test corresponding to the esterification of lauric acid with n-propyl alcohol, after 15 min at atmospheric pressure.

The fusel oil was provided as a gift by Distillery Győr (Hungary). All the other chemicals used in analysis were of analytical grade and purchased from Reanal Ltd. (Hungary) and Sigma Chemical Co. (USA).

Two different procedures were used for ester production. Firstly, synthesis of esters was carried out in shaking flasks (150 rpm) containing 25 ml solution of several alcohols and oleic acid mixture with different molar ratios, different temperatures and various amounts of enzyme by using New Brunswick Scientific (USA) shaking incubator to study the esterification kinetics. The starting time of the reaction was the addition of the enzyme.

In the other procedure a 200 ml round flask reactor was thermostated and connected with a pervaporation unit using hydrophilic membrane for continuous removal of water produced. The reaction mixture was circulated through the pervaporation unit by a peristaltic pump. The vacuum pump, manometer and the cooled traps were connected to the pervaporation unit.

The laboratory scale pervaporation unit was purchased from Carbone Lorraine (Germany) and it was jacketed later. The membrane surface area was $2.0*10^{-2}$ m^2. The membranes used for the pervaporation experiments (PERVAP 2201, PERVAP 2202, CMC-VP-43) were provided by GFT (Germany) and Celfa (Switzerland).

Aliquots of the reaction mixture were withdrawn periodically and the residual acid content was assayed by titrating against potassium hydroxide (0.1 M) using phenolphthalein as an indicator and ethanol as a quenching agent. The percentage esterification was calculated from the values obtained for the blank and the test samples. The fusel oil esters were confirmed by chromatographic analyses of the samples using a Hewlett Packard Model 5890 Series II GC equipped with a flame ionisation detector and a 30 m HP-FFAP capillary column. The percentage esterification calculated by both GC analysis (which showed product formation) and titrimetry (which showed acid consumption) were found to be in good agreement. The water content of the reaction mixture was measured by Mettler DL 35 automatic Karl-Fischer titrator

2.2 Results

The esterification reaction of oleic acid with the fusel oil fraction occurs as follows:

oleic acid + fraction of fusel oil = oleate esters + water

In this reversible reaction, the molar ratio of reactants, temperature, enzyme and removal of water from the reaction mixture are the variables affecting the conversion and the reaction rate.

2.2.1 Water content

Water level is critical for enzymes and affects enzyme action in various ways: by influencing enzyme structure via noncovalent bonding and disruption of hydrogen bonds; by facilitating reagent diffusion; and by influencing the reaction equilibrium. Too low water content usually reduces enzyme activity. High water content can also decrease reaction rates by aggregating enzyme particles and causing diffusional limitations. The optimal amount of water is often within a narrow range.

Optimal water content is not only important to preserve the catalytic activity of an enzyme, but also to achieve high reaction rates and yields, and stability of the enzyme. Water requirements for enzymes in organic media vary greatly; therefore each enzyme must be examined at various levels of hydration.

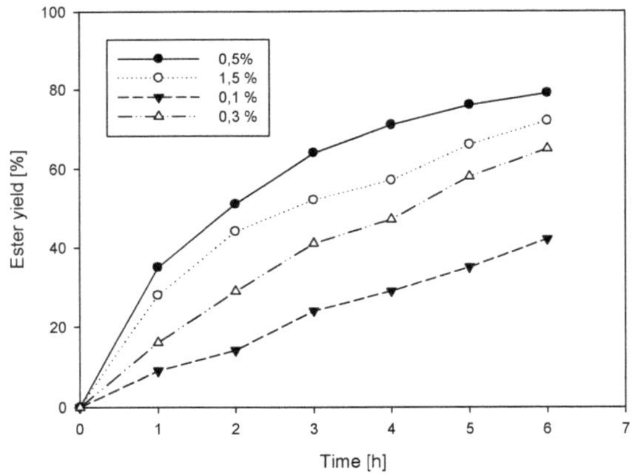

Fig. 1. Influence of water content on the synthesis of isoamyl oleate. Oleic acid, 1.84 mol/l; i-amyl-alcohol, 3.78 mol/l; temperature, 40 °C; Novozym 435, 0.5 %; speed, 150 rpm

The role of the water content in the reaction mixture was studied in the range of 0.1 - 1.5 %. As it can be seen in Figure 1 the ester yield as a function of the initial water content has a minimum. In very low water content (0.1 %) the amount of ester produced is small: in this case water present in the reaction mixture is not enough for the enzyme's optimal work. Increasing the water content the yield is growing, and a maximal value is reached at about 0.5 % initial water content. Then at higher and higher water contents the ester yields obtained are gradually decreased, which can be explained by the fact that at high water content the opposite reaction, the hydrolysis is favoured. Based on the results of these measurements, 0.5 % initial water content was used in the further experiments.

2.2.2 Effect of temperature

The effect of temperature was studied in a series of experiment in the range of 30 - 60 °C under the same conditions (initial water content, molar ratio, enzyme concentration). No thermal deactivation was observed up to 60 °C, and - as it is shown in Figure 2 - the concentration of the oleates produces increased in higher temperatures. This curve is typical of enzyme with high thermostability and which thermal denaturation, during the time of the assay, is negligible. After 12 hours reaction time (not shown) 92 % conversion was obtained in each case.

The activation energy of the enzymatic reaction was determined based on the well-known Arrhenius-equation. The logarithm of the reaction rate data were plotted as a function of reciprocal temperature and the activation energy was calculated from the slope of the regression line. It's value was 16.2 kJ/mol, which is similar to the value reported by (Garcia, et al., 1999) (11.7 kJ/mol), determined for the esterification of isopropyl alcohol and palmitic acid in solvent-free system by the same Novozyme 435 lipase preparation.

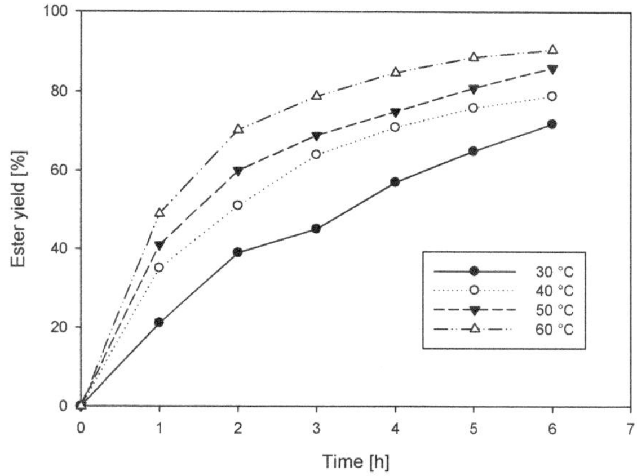

Fig. 2. Influence of reaction temperature on the synthesis of isoamyl-oleate. Oleic acid, 1.84 mol/l; i-amyl-alcohol, 3.78 mol/l; Novozym 435, 0.5 %; speed, 150 rpm.

2.2.3 Effect of the molar ratio

It is well-known, that acid/alcohol molar ratio is one of the most important parameters in enzymatic esterifications. Since the reaction is reversible, an increase in the amount of one of the reactants will result higher ester yields and as expected, this will shifts the chemical equilibrium towards the product side. One way of shifting the reaction toward the synthesis is to increase the alcohol concentration. However, high alcohol concentration may slow down the reaction rates due to inhibition. Therefore, it is necessary to optimize the actual excess nucleophile concentration to be employed in a given reaction. In order to determine the effect of molar ratio, oleic acid was esterified at molar ratios of 1:1, 1:2, 1:5 and 2:1 oleic acid/iso-amyl alcohol under the following conditions: 40 °C temperature applying 0.5 % enzyme concentration, 0.5 % initial water content.

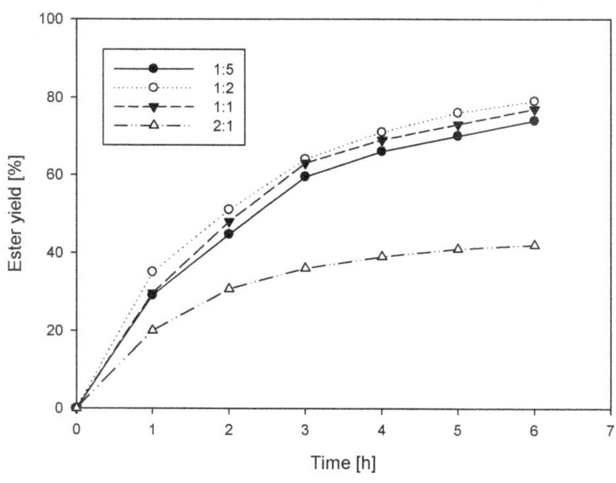

Fig. 3. Influence of the acid/alcohol molar ratio on the synthesis of isoamyl-oleate. Novozym 435, 0.5 %; temperature, 40 °C; speed, 150 rpm.

As expected, a higher ester conversion was obtained in a shorter period of time for 1:2 and 1:5 molar ratios compared to 1:1. The highest acid conversion was achieved in case of 1:2 molar ratio, as it can be seen from Figure 3, thus alcohol excess was used in the further experiments.

2.2.4 The effect of chain length of alcohols in fusel oil on the esterification reaction

The main alcohol compounds of fusel oil, two with linear chains (ethanol, propanol) and three with a branch chains (isopropyl alcohol, isobutyl alcohol and isoamyl alcohol) were applied in the next serial of experiments to determine the effect of chain length. In the measurements the conditions were as follows: 1:2 acid/alcohol molar ratio, 0.5 % enzyme and 60 °C temperature.

The results shown in Figure 4, that the esterification rates of oleic acid with ethanol are smaller than with isoamyl alcohol, indicating some effect of the length and structure of the alcohol molecule. The difference between the isoamyl oleate and ethyl oleate was 33 %. Based on this result, it was assumed that esterification of oleic acid with model solution of fusel oil will be similar to the ester synthesis with i-amyl alcohol, since it is the main compound of fusel oil.

2.2.5 Removal of excess water produced by pervaporation

The optimal parameters of the reaction was determined in the shaken flasks experiments: 0.5 % water concentration, 1:2 oleic acid-isoamyl alcohol molar ratio, 60 °C temperature. According to the analysis data without water removal, the reaction reaches equilibrium after 12 hours reaction time.

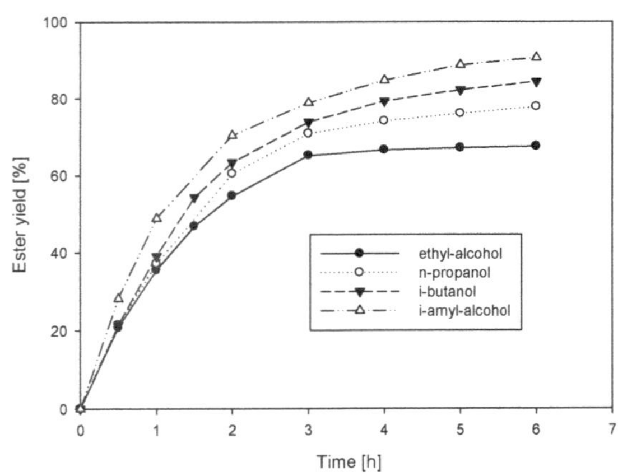

Fig. 4. Influence of chain length of alcohols. Acid/alcohol molar ratio, 1:2; Novozym 435, 0.5 %; temperature, 60 °C; speed, 150 rpm.

The esterification of oleic acid resulted in 92 % yield, while water produced as excess during the process and it accumulated as a second phase in the bottom of the flask. Since water excess has a strong inhibition effect on the enzyme, stops the further conversion of substrate, therefore it should be removed. Pervaporation seemed a suitable membrane separation technique for continuous water removal, helping to convert all the substrate. Thus an

integrated system was designed for the enzymatic reaction, where a thermostated bioreactor is combined with a pervaporation unit The membrane unit consists of two stainless steel plate, the membrane is located between them. The permeate passed through the membrane was condensed in traps cooled by dry ice – acetone mixture. The permeate side was kept under 0.8 kPa vacuum pressure. During the procedure the feed stream was recirculated in liquid phase on the primary side of the membrane module, while the permeate was obtained in vapour phase. The reaction mixture was recirculated through a glass filter to keep the immobilised enzyme in the reactor. In the integrated system the bioreactor was thermostated at 60 °C temperature, and the experiments were carried out under the optimal conditions determined earlier. Samples were taken from the reactor and the analysis has shown that the water content was managed to keep at 0.5 %. In the integrated system 99.8 % conversion of the oleic acid was achieved.
Firstly hydrophilic pervaporation membranes were tested. It was turned out that the CMC-VP-43 type membrane has the highest flux (0.05 kg/m^{2}h) and selectivity (58.94), so this membrane was used for the further experiments.

2.2.6 Tribological properties of the biolubricant

The tribological properties of the biolubricant manufactured were tested according to the IP and ASTM standards. The chemical, physical parameters of the product were given in Table 1. The thermal and oxidation stabilities were measured by the modofied IP 8.6 method using an oxidative and thermal treating. The results obtained were compared to a commercial, synthetic, DB-32 dicarboxylic acid ester used as a reference lubricating oil (Table 2).

Properties		Biolubricant	DB-32
Density	g/cm^3	0.864	0.925
ASTM colour	–	0	0
Flash point (Cleveland)	°C	211	204
Pour point	°C	−27	n.d.
Viscosity at 100 °C ($KV_{100\,°C}$)	mm^2/s	2.28	2.7
Viscosity at 40 °C ($KV_{40\,°C}$)	mm^2/s	6.39	9.03
Viscosity index (VI_E)	–	207	148
Acid value	mg KOH/g	0.01	0.1
Iodine–bromine value	g I/100 g	68.4	n.d.

n.d.: no data.

Table 1. Properties of the biolubricant compared to the reference oil

As it can be seen from Table 1, the bioester produced is considered as a low flashpoint lubricant. Due to its low pour point, it can be applied also at low temperatures. Its viscosity

index is quite high, thus the viscosity of the ester is not highly influenced by temperature. The good conversion of oleic acid resulted in the low acid number, while the iodin-bromine number refers to the quite low unsaturation.
The data in Table 2 show that the stability of the measured ester based on thermal and oxidative treatment is higher than the reference oil, i.e. the increase in viscosity and acid number is lower, the change of its colour is acceptable. According to these data the biolubricant is suggested to use mainly at the high speed and low load regime of the tribological circumstances. In the mechanical industry it can be applied e.g. as a cooling lubricant compound for metalworking processes, moreover in particular lubrication processes where lubricant loss may occur, e.g. mist lubrication, chain lubrication, launch engine lubrication.

Lubricant	Thermal treatment (200 °C, 12 h)					Oxidative treatment (200 °C, 12 h, airflow 20 dm³/h)				
	$VK_{100\,°C}$ (mm²/s)	$VK_{40\,°C}$ (mm²/s)	VI_E	Colour	Acid value (mg KOH/g)	$VK_{100\,°C}$ (mm²/s)	$VK_{40\,°C}$ (mm²/s)	VI_E	Colour	Acid value (mg KOH/g)
DB-32	2.75	9.45	140	7	5.5	5.79	29.47	143	8	50.8
Biolubricant	2.29	6.41	209	0	0.1	5.57	25.91	162	8	10.6

Table 2. The effect of thermal and oxidative treatments

2.2.7 Study on toxicity of biolubricant by zebrafishes

To study the effect of the particular biolubricant on the water environment, its acute toxicity was assessed with Acute Fish Toxicity Test on Zebrafish (*Brachydanio rerio*), over an exposure period of 96 hours in a static system. A limit test was conducted according to the OECD Guidelines for the Testing of Chemicals, Procedure No. 203 (1992). In this kind of test the limit in LC_{50} number is 100 mg/l. In the experiment two 20 l aquaria were used, in the control there was no biolubricant, while in the other aquarium its nominal concentration was 100 mg/l. Zebrafishes (7 fishes in both aquaria) were used in the test, which are extremely sensitive to the waste compounds in living waters. Analytical methods were applied to measure the parameters of the water (chemical properties, temperature, oxigen saturation, pH...etc.). The observation of fishes were carried out at regular intervals (3, 6, 24, 48, 72 and 96 hours) and mortality was determined.
As a result of the experiment no signs of reaction or mortality were detected. Thus the 24 h, 48 h, 72 h and 96 h LC_{50} number of the particular biolubricant is > 100 mg/l. So it can be concluded that the biolubricant is not toxic for the living water.

3. Kinetics approach

Before developing a complete method for enzymatic manufacture of this biolubricant, a detailed kinetic analysis should be carried out on the reaction mechanism. Enzymatic esterification of fatty acids and ingredients of fusel oil was studied by (Gulati et al., 2003) using lipase from *Aspergillus tereus*. They found that in n-hexane solvent the alcohols were able to react with the fatty acids (miristic acid, palmitic acid, stearic acid), except oleic acid. Using other lipase preparations (*Candida antarctica, Candida rugosa, Rhizomucor miehei, porcine pancreas*), however, made it possible the successful oleic acid esterification with similar low molecular weight alcohols. In Table 3 results published on the kinetics of enzymatic esterification of oleic acid with short chain alcohols in organic solvents are summarized.

Garcia et al. (1996) studied the kinetics of i-propyl-oleate formation by *Candida antarctica* lipase. The model used was an ordered bi-bi type containing 13 kinetic parameters. Thus the model seems too complicated having high uncertainty. Esterification of butyl alcohol by *Candida rugosa* lipase was studied by (Zaidi et al., 2002), where ping-pong bi-bi mechanism was assumed in the kinetical model with 5 parameters. However, the range of substrate concentration measured was quite narrow (0.1 - 1 mol/L), and the error of the modelling was found very high (28 %).

Alcohol	Enzyme	Model	Inhibition	Reference	Note
i-propyl alcohol	Novozym 435 *Candida antarctica*	ordered bi-bi 13 parameters	both S and P competitive	(Garcia et al., 1996)	too many parameters fitted to one measurement; high error of fitting
butyl alcohol	immobilised *Candida rugosa*	ping-pong bi-bi 5 parameters	both acid and alcohol	(Zaidi et al., 2002)	error is 28 %; narrow range of substrate concentration (0.1-1 mol/L)
ethyl alcohol	immobilised *Rhizomucor miehei*	random bi-bi 4 parameters	alcohol	(Oliveira et al., 2001)	Difference between the parameters is 15 order of magnitude; quasi one-substrate kinetics
ethyl alcohol	immobilised *Rhizomucor mihei*	Michaelis-Menten 2+1 parameters	alcohol	(Goddard et al.,2000)	Different kinetics for each alcohol concentration; pseudo one-substrate kinetics
ethyl alcohol	soluble, from porcine pancrease	ping-pong bi-bi 4 parameters	alcohol	(Hazirka et al., 2002)	Narrow range of substrate concentration (0.3-0.8 mol/L)

Table 3. Kinetic studies on esterification of oleic acid

Immobilized *Rhizomucor miehei* lipase was applied for ethyl oleate synthesis by (Oliveira et al., 2001). In order to describe the kinetics, random bi-bio model was used, which contained 4 parameters. However, the difference between the parameter values determined was 15 order of magnitude, implying that the effect of certain parameter is nearly negligible comparing to the others. Esterification of ethyl alcohol and oleic acid by immobilized

Rhizomucor miehei lipase was studied by (Goddard et al., 2000) as well. Michaelis-Menten model was used for the description of the reaction, however, different kinetics was used in each alcohol concentrations, which is considered as a pseudo-one substrate model.
Soluble porcine pancreatic lipase was applied for the ethyl oleate synthesis by (Hazarika et al., 2002). They assumed ping-pong mechanism, as well, containing 4 parameters, however the range of substrate concentration studied was even narrower (0.3 - 0.8 mol/L) than in case of (Zaidi et al., 2002). Description of the correct kinetics on the particular esterification reaction is even more difficult due to the various possible inhibition effects. As it is shown in Table 3, ethyl alcohol as a substrate was in all cases considered as inhibitor, while in the esterifications with other alcohols, both substrates were regarded as inhibitors.
As a summary, it seems from Table 3 that the kinetic models/parameters published so far can not be considered as a proper, detailed kinetic description of the enzymatic process for oleic acid esterification with short chain alcohols, moreover there has been no data found on esterification with i-amyl alcohol. Therefore our aim was to elaborate a proper, sophisticated model for this particular reaction.

3.1 Kinetics model

Kinetics of enzymatic reactions can be described by the well-known Michaelis-Menten model. For reactions having 2 substrates and 2 products (bi-bi reactions), its application is quite complicated since various mechanisms can be considered according to the order and rate both substrates' binding and products' releasing to/of the enzyme active sites (random, ordered, ping-pong...etc.). Since majority of the kinetical studies suggested ping-pong bi-bi mechanism for the enzymatic esterification of oleic acid and short chain alcohols, we also considered it as an initial point for the description. According to Cleland (Cleland, 1979) the ping-pong bi-bi mechanism can be outlined as follows:

$$A + E \underset{k_2}{\overset{k_1}{\rightleftarrows}} \begin{pmatrix} EA \\ \overline{FP} \end{pmatrix} \underset{k_4}{\overset{k_3}{\rightleftarrows}} P + F + B \underset{k_6}{\overset{k_5}{\rightleftarrows}} \begin{pmatrix} FB \\ \overline{EQ} \end{pmatrix} \underset{k_8}{\overset{k_7}{\rightleftarrows}} Q + E \tag{1}$$

In the first step the enzyme reacts with substrate A forming an AE enzyme-substrate complex, which is transformed into FP modified complex - by an internal rearrangement. Product P comes off the complex, then the modified enzyme molecule F is able to react with substrate B, forming a new enzyme-substrate complex FB. It is transformed into an enzyme-product complex EQ, then Q moves to the bulk solution. Finally, enzyme E becomes free and can react with another substrate A molecule. In the equation the reaction rate constants are shown (k_1 - k_8), among them k_1, k_3, k_5 and k_7 belong to the towards direction of the reaction, while the others to the backwards direction (with negative sign). Summarising the steps, the formation rate of product can be written as follows:

$$v = \frac{(k_1 k_2 k_5 k_7 AB - k_2 k_4 k_6 k_8 PQ)E_0}{E + (EA + FP) + F + (FB + EQ)} \tag{2}$$

where E_0 is the initial enzyme concentration and E is the actual enzyme concentration (the other capital letters mean concentration of the particular compound). This model, however, is too complicated to apply it in practice. To simplify the situation, the main parameters influencing the rate of product formation are selected, as follows (Janssen et al., 1996):

the concentration of substrate A (first),
the concentration of substrate B (second),
enzyme concentration,
the amount of products formed altogether.

Initial reaction rate can be described by including only the first three factors, taken into account the fact that product concentration is 0 at the beginning of the reaction:

$$v = \frac{V_m}{1 + \frac{K_A}{[A]} + \frac{K_B}{[B]}} \tag{3}$$

If substrate inhibition is considered, as well, the three-parameter equation should be completed with another constant (Dezbaradica et al., 2006) and the following equation can be used:

$$v = \frac{V_m}{1 + \frac{K_A}{[A]}\left(1 + \frac{[B]}{K_{iB}}\right) + \frac{K_B}{[B]}} \tag{4}$$

This model contains four parameters, which can be further supplemented by the product inhibition factors. In this way two more parameters are added into the equation (Eq. 5). In the equation K'AB parameter is the rate of the apparent product formation, the ping-pong parameter of the reaction.

$$v = \frac{V_m}{1 + \left(\frac{K_A}{[A]} + \frac{K'_{AB} * [P]}{[A] * [B]}\right)\left(1 + \frac{[B]}{K_{iB}}\right) + \frac{K_B}{[B]} * \left(1 + \frac{[P]}{K_{iP}}\right)} \tag{5}$$

where

$$V_m = \frac{k_3 k_7}{k_3 + k_7} \quad K_A = \frac{k_7(k_2 + k_3)}{k_1(k_3 + k_7)} \quad K_B = \frac{k_3(k_6 + k_7)}{k_5(k_3 + k_7)} \quad K_{iP} = \frac{k_3}{k_4}$$

$$K_{iB} = \frac{k_5}{k_6} \quad K'_{AB} = \frac{k_2 k_4 (k_6 + k_7)}{k_1 k_5 (k_3 + k_7)}$$

In our case, water is one of the products in esterification. But it is not only a product, a small amount of water should be present initially in the reaction mixture to keep in an active formation of the enzyme structure. In the beginning of the reaction, however, the initial concentration of water does not change significantly, therefore water content (P) can be considered as constant (its effect is negligible). Thus the equation can be simplified and a 5-parameter equation can be obtained:

$$v = \frac{V_m}{1 + \left(\frac{K_A}{[A]} + \frac{K_{AB}}{[A] * [B]}\right)\left(1 + \frac{[B]}{K_{iB}}\right) + \frac{K_B}{[B]}} \tag{6}$$

It can be seen that the difference between the 4- and 5-parameter equations (Eqs. 4 and 6) is the KAB factor. Its influence is significant only in the case when concentrations of both substrates are very low. However, we do not plan to carry out measurements under these conditions, thus it is assumed that no significant difference will be experienced in the modelling results obtained by using the two systems.

3.2 Reaction rate determination

The operational parameters during the kinetical experiments were 200 rpm and 30 °C. 0.01 g immobilized lipase preparation was added to the incubated homogeneous reaction mixture to initiate the enzymatic reaction. Samples in duplicates were taken from the reaction mixture regularly.

Experiments using substrate concentrations in the range of 0.4 - 6.0 and 0.2 - 2.0 mol/L of i-amyl alcohol and oleic acid concentration, respectively, were carried out at 30 °C temperature, with an optimal initial 0.1 w/w % water content (determined earlier) (Koszorz et al., 2004) in n-heptane solvent. Progress curves: oleic acid consumptions as a function of time were measured.

From the data of progress curves the initial reaction rates were calculated by the Gregory-Newton method (Leskovac, 2003). The initial reaction rate data were modified taken into account the amount of enzyme used. Thus the reaction rate values were obtained as μmol/s.genzyme and summarised in Table 4.

c_{OA} \ c_{iAA}	0.4	1.0	2.0	4.0	6.0
0.2	4.90	5.13	5.45	5.57	3.50
0.5	7.91	7.91	8.26	7.13	7.31
1.0	9.68	12.56	12.80	9.69	10.80
1.5	11.25	15.15	18.45	15.30	13.72
2.0	9.78	14.46	17.65	14.50	14.31

Table 4. Intitial reaction rate data [μmol/s.g_{enzyme}] using various oleic acid (c_{OA}) and i-amyl alcohol concentrations (c_{iAA}) (all the concentrations in mol/L)

3.3 Effect of immobilization on the mass transfer

Since immobilised enzyme preparation was used in the experiments, it was important to decide - before the detailed kinetical analysis - whether the reaction rates measured are the real values of enzymatic reaction or influenced significantly by the diffusion rates of the compounds (from the bulk phase to the solid particle and vice versa).

The reaction and the diffusion take place simultaneously and the rate limiting step is always the one which is slower. Using immobilised lipase preparations, rate of diffusion is usually not the limiting step (Letisse et al., 2003).

In our measurements Yadav-method (Yadav et al., 2003) was applied to determine the rate limiting step, using the Weisz-Prater criteria. This method is based on the calculation and

comparison of the two relevant relaxation times. The ratio of the relaxation time of biocatalysis rate, t_r and that of the diffusion rate, t_d shows which process should be considered as the limiting step.
The relaxation times can be defined as follows :

$$t_r = \frac{C_0}{r(C_0)} \text{ and } t_d = \frac{D}{(k_{sl})^2} \tag{7}$$

Oleic acid - having slower diffusivity - was chosen for the calculations and the highest reaction rate-substrate concentration value-pair was taken from Table 4. Thus t_r was calculate as:

$$t_r = \frac{C_0}{r(C_0)} = 2^5 s \tag{8}$$

Diffusion constant (D) of oleic acid in n-heptane was determined according to Shibel (Perry, 1969):

$$D = k\frac{T}{\eta_b V_s^{1/3}} \tag{9}$$

V_S molar volume density was estimated from its critical volume (V_c):

$$V_S = 0,285 V_C^{1,048} \tag{10}$$

V_c of oleic acid is 1152 cm^3/mol, thus V_s is obtained as 460 cm^3/g. In this way D diffusion coefficient is calculated as 1.61x 10-5 cm^2/s.
The mass transfer coefficient can be calculated - based on the Sherwood number - from the diffusion coefficient and the particle size:

$$k_{SL} = 2D / d_k \tag{11}$$

The average diameter of Novozyme 435 immobilised lipase preparation is 0.06 cm, thus the value of the mass transfer coefficient is 5.3x 10^{-4} cm/s.
The relaxation time for the diffusion is calculated as:

$$t_d = \frac{D}{(k_{SL})^2} = \frac{1,61.10^{-5}\left[cm^2/s\right]}{(5,3*10^{-4}\left[cm/s\right])^2} = 55,9s \tag{12}$$

Comparing the values of t_r and t_d it can be concluded that diffusion rate is three order of magnitude higher than the reaction rate, thus the rates measured in the enzymatic process can be considered as the real reaction rates.

3.4 Kinetical analysis

In Fig. 5a the initial reaction rates are presented as a function of oleic acid (substrate 1) concentration, while in Fig. 5b the same data are shown as a function of i-amyl alcohol (substrate 2) concentration. It can be clearly seen that the i-amyl alcohol has a considerably and oleic acid a slight inhibition effect on the enzymatic reaction.

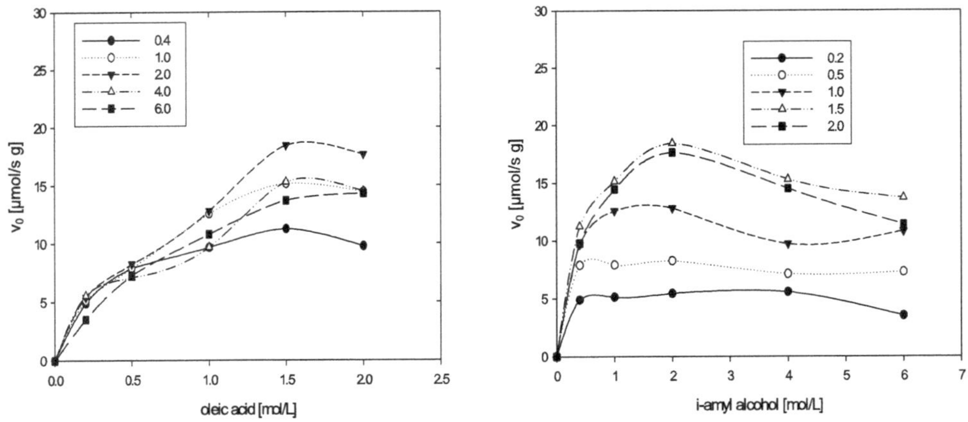

Fig. 5. Initial reaction rates as a (a) function of oleic-acid concentration at different i-amyl alcohol concentrations; (b) a function of i-amyl alcohol concentration at different oleic acid concentrations in mol/L

In the first step of the kinetic analysis the traditional linearization (graphical) methods were applied. Reciprocal data of the initial rates were plotted against the reciprocal data of the initial substrate concentrations (both) (Figs. 6a and 6b). It can be seen that in Fig. 6b in lower substrate concentrations the lines are parallel, implying ping-pong bi-bi mechanism. However, in higher substrate concentrations the lines steeply keep upwards to the ordinate, which means that inhibition (by the alcohol compound) occurred. Thus the kinetic parameters can not be determined graphically and the mechanism of inhibitions can not be doubtlessly decided.

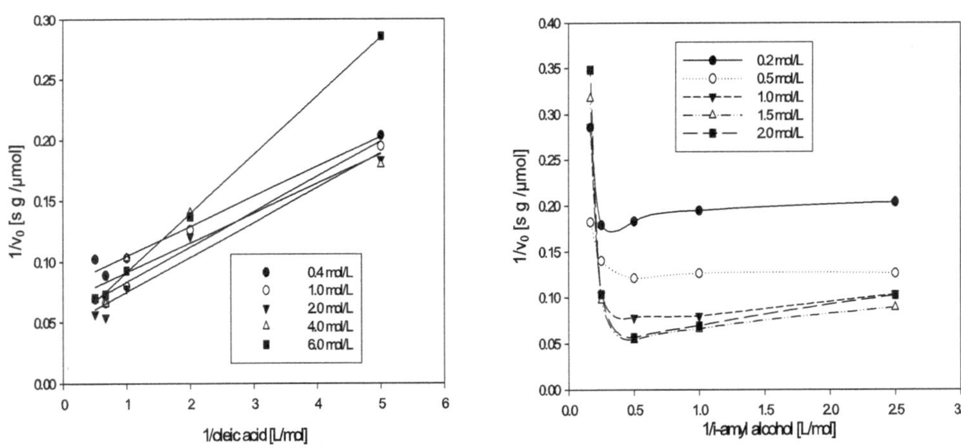

Fig. 6. (a) and (b) Lineweaver-Burk linearization of the reaction rate - substrate concentration data

Ping-ping bi-bi kinetic models having 3 (eq. 3), 4 (eq. 4.) and 5 (eq. 6.) parameters were used for the description of the enzymatic esterification and the parameters were calculated based on the experimental data. In the calculation a variation of simplex method was applied, namely the Nielder-Niemand method (Bailey, 1986), which is more sensitive for the initial values of the parameters and slower than the original simplex method, but it provides more accurate final results.

Since the method is sensitive for the initial values of the parameters, a two-step method was elaborated. In the first step the values of kinetic parameters were estimated by a special programme, applying a simplified model (Luzenc4 programme) with no inhibition, using the experimental data. Thus the instability of the method was eliminated. The estimated kinetic parameters obtained were then used as initial parameters for the extended kinetic model. Having checked the kinetic results by comparing the experimental data, the program either refused the results and started another circle - modifying one or more parameters according to the Nielder-Niemand method - or accepted the results and presented.

During the procedure programme was run on data belonging to only one acid concentration, otherwise divergence occurred. In this way the parameters were obtained for every acid concentration, which were used in a final run of the programme, where errors were minimised.

In Table 3 values of the parameters determined are summarized. It can be seen from the value of r^2, that the four-parameter model (taking into account the inhibitions, as well) described better the kinetics of enzymatic i-amyl oleate synthesis than the three-parameter model.

Parameters	Three-parameter model	Four-parameter model	Five-parameter model
V_m (μmol/s g)	23.5	30.8	29.9
K_A (mol/L)	0.86	0.65	0.55
K_B (mol/L)	0.19	0.58	0.53
K_{iB} (mol/L)	-	3.2	2.7
K_{AB} (mol/L)	-	-	0.055
r^2 (-)	0.952	0.975	0.975
ARE* (%)	12,2	3,1	2,8

Table 5. Parameters of the kinetical models

However, the five-parameter model is not more accurate than the four-parameter one (value of r^2 is the same), therefore it is not reasonable to use the more complicated model This theory evaluated using F-statistics (Bates, 1988) used significance level was P=0.05, as the result shows the 3 parameter model was significance different then the 4 and 5 parameters model while the 4 parameter and 5 parameters model are not significantly different. The results of the modelling were intended to compare with other literature data, however no similar results were found in published materials. Either the substrates of the esterification, or the enzyme applied were different, thus the parameters were not possible to compare.

Kinetic model for description of the enzymatic esterification of oleic acid and i-amyl alcohol was - according to our best knowledge - elaborated and presented in the first time in literature. The model is based on the ping-pong bi-bi mechanism and not only product, but substrate inhibition is taken into account, as well. The model fitted well to the experimental data (checked by r^2 values), though the measurements covered an extremely wide concentration range.

4. Reaction in ionic liquid

Ionic liquids can be suitable media for biocatalytic synthesis because of their enzyme stabilization effect, reusability and negligible vapour pressure (Yang & Pan, 2005). Many enzymes, especially lipases showed higher activity and greater selectivity in ionic liquids than in organic solvents (Jain et al., 2006). The stability of *Candida antarctica* lipase B (CALB) in ionic liquid was found to be higher than in organic solvent (Romero et al., 2005). The main advantages of ionic liquid + enzyme system are milder, lower reaction temperature comparing with conventional chemical synthesis, it needs less energy investment, can be better-controlled, attended with less side-reaction, furthermore the product more readily separable. In this work the purpose was to find a utilization of fusel oil where biolubricants can be manufactured in ionic liquid.
In organic solvents low conversions were achieved, therefore it became necessary to develop a better method, which is energy efficient, able to achieve higher conversion, furthermore environmental friendly, however waste materials may be as natural substrates. Our aim is to optimize the parameters of production, enhance the yield improving the advantages of enzyme+ionic liquid system.

4.1 Ionic liquids

Ionic liquids: 1-butyl-3-methyl-imidazolium hexafluoro phosphate [bmim]PF_6, 1-butyl-3-methyl-imidazolium tetrafluoro-borate [bmim]BF_4, tributyl tetradecyl phosphonium dodecilbenzol sulfonate (Cyphos-201), trihexil tetradecyl phosphonium bis- (2,2,4-trimetilpentil)-phosphate (Cyphos-104), tetradecyl phosphonium trihexil decanoate (Cyphos-105) trihexyl tetradecyl phosphonium hexafluoride, phosphate (Cyphos-110) (IoLiTec GmbH, Germany)
Reaction mixture composition: Under the pre-investigations, the reaction mixture was the follow: 6.36 mmol of oleic acid, 36.95 mmol of isoamyl alcohol, 304.35 mmol n-hexane and 50 mg Novozym 435 lipase. In the course of the main experiment: 1.23 mmol of ionic liquid, 0.16 mmol oleic acid, 1.41 mmol isoamyl-alcohol were used. The reaction mixtures were shaken with an intensity of 150 rpm at 40, 50 and 60 ° C temperatures for 4 hours.

4.2 Effect of ionic liquid

As it well known ionic liquids are not only green solvents, they can have catalytic effect. Our aim was to investigate separately the enzyme stabilization effect, therefore preliminary experiments were necessary. Through these the most important investigated criterion was to verify if the ionic liquid itself had catalytic effect for the reaction without enzyme. Basic criterion was to be work in a two-phase reaction, where separation is easier, since the application of a new solvent can be avoid (Eckstein et al., 2004). Two of the investigated ionic liquids (Cyphos-105 and Cyphos-110) were mixed with the substrates, so they were not investigated further. Henceforth the reactions were carried out at 50 °C, the ester yield was followed by GC and the decreasing oleic acid concentration using titrimetry. The percentage esterification was calculated from the values obtained for the blank and the test samples. The further phosphonium-type ionic liquids, Cyphos-201 and Cyphos-104 without the presence of enzyme greatly catalyze the process as it shown on Fig. 7.

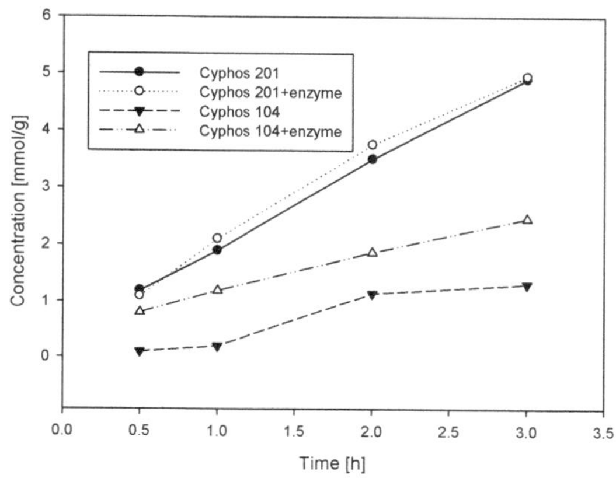

Fig. 7. Oleate ester concentration in case of using different phosphonium-type ionic liquids (reaction time: 180 min)

For imidazolium-cation containing ionic liquids literary data show that these types of ionic liquids are the most suitable for esterification, transesterification reactions (Moniruzzaman at al., 2010). Both investigated imidazolium-type ionic liquids were successfully applied in earlier experiments, where natural aroma esters production was the aim. In case of $[bmim]BF_4$ there were no significant differences in the detected oleate concentration if enzyme was added or not. For $[bmim]PF_6$ - without the presence of the enzyme only negligible product formation was observed, in the presence of enzyme, higher concentrations were achieved than in the experiments where n-hexane was used as solvent.

Therefore for further investigations this ionic liquid was chosen. In case of ionic liquids for water solubility cations are responsible. Comparing the same cation having $[bmim]BF_4$ and $[bmim]PF_6$ shows that while the first not miscible with water, the later has an unlimited solubility in water. Thus, the hydrophilic $[bmim]BF_4$ ionic liquid often distracts the absorbed water layer from the surface of the enzyme which should be necessary for the active conformation. Therefore the enzyme is deactivated (van Rantwijk & Sheldon, 2007). Further advance that in case of using $[bmim]PF_6$ side-reactions were not observed.

4.3 Acid/alcohol molar ratio

The optimal parameters of the batch production were determined using experimental design software application. In doing so Statistica 8.0 program was applied. Based on earlier studies substrate molar ratio, amount of enzyme and ionic liquid were chosen as key factors. Each factor was prepared in two levels: -1 for low level and +1 for high level.

Concrete values were applied in a design matrix. It is evident that increasing reaction temperature enhances the reaction rate, that is way that its affect will be investigated later separately.

During the experiments different, software-defined combinations of the previously selected values of experimental parameters were investigated. Other parameters were fixed: temperature 50 °C, 150 rpm shaking intensity, 5 hours reaction time. Water content of the reaction mixture was also followed as an important parameter of esterification reactions which may shift the equilibrium, but there was not detectable concentration change using Carl-Fischer titration. The aim was to find the optimal parameter values of the isoamyl-oleate production. The results are shown on Fig. 8.

As it shown, increasing amount of ionic liquid results in higher oleate yield, which gives evidence for the advantageous enzyme stabilization effect. Complex investigation of the three chosen key factors shows that the highest ester conversion was obtained in the case of application the highest acid/alcohol molar ratio, amount of enzyme and ionic liquid. Relatively high yield was obtained also around medium values, around the center points.

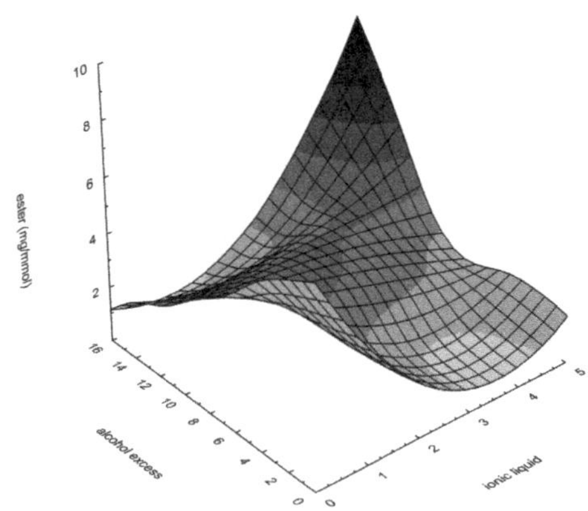

Fig. 8. Influence of the acid/alcohol molar ratio and amount of ionic liquid on the synthesis of isoamyl oleate

There was no inhibition effect observed neither in 1:16 acid/alcohol molar ratio. As none of the point of parameters could be an optimal value, because the highest yield belonging to the highest values, therefore further investigations will necessary to find the optimal values.

4.4 Effect of temperature

In case of using conventional reaction media it is well known that increasing reaction temperature results the same yield in shorter time, but till a limit due to thermal deactivation of enzyme. Beside of the structure stabilization effect of ionic liquids the enzyme can be resist in the active conformation at higher temperatures (van Rantwijk & Sheldon, 2007). For these investigations that mixture was chosen, in which the highest oleate yield was obtained. The results are shown on Fig. 9.

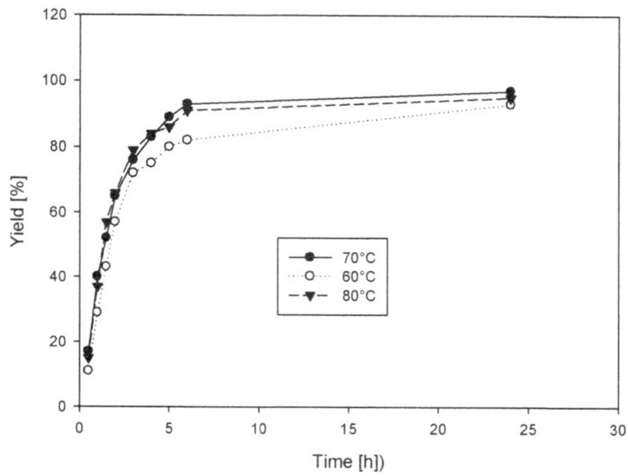

Fig. 9. Influence of reaction temperature on the synthesis of isoamyl-oleate

The yield-time functions in all cases show the saturation curve. Increasing the temperature equilibrium yield was not changed, but shorter time was necessary to achieve it. As it shown neither at 80 °C became the enzyme deactivation significant, but increasing up from 70 °C did not shorten the reaction time till achieving the equilibrium.

5. Conclusions

Our investigations have proven that an ester type biolubricant could be prepared from fusel oils and oleic acid by lipase enzyme in solvent-free system. Compared to the product obtained by acid catalysis, in the biolubricant there was no trace of oleic acid since complete conversion was achieved by continuous water removal by pervaporation. Our product was then tested in an acute toxicological procedure by zebrafishes, which has verified the assumption: the biolubricant is not toxic for the living water, so it is considered as an environmental safe product.

The tribological study has shown that the features of the biolubricant are similar to the DB 32 type synthetic reference lubricating oil, so it can be applied as a low viscosity lubricant, suitable even for special publication processes, where lubricant loss may accur. Although the biological degradation of the product has not been studied, it was manufactured from initial compounds having biological origin (oleic acid from plant oils and alcohols from a by-product in bioalkohol production), thus it is considered as a completely environmental-friend product.

The kinetic model containing the parameters determined can be used in the particular enzymatic esterification reaction for calculations of the optimal conditions of various aspects, like highest yield, lowest acid residue, lowest amount of enzyme, shortest reaction time…etc. Moreover we are planning to apply the model for bioreactor design to realise continuous enzymatic i-amyl oleate synthesis by lipase.

Our work has showenn that an ester type biolubricant could be prepared from isoamyl-alcohol and oleic acid by lipase enzyme in ionic liquid two phase system. Compared to the product obtained in solvent free system, higher conversion in shorter time was achieved. Despite the lack of water removal in the biolubricant there was no trace of oleic acid since complete conversion was achieved. Determining the optimal reaction mixture composition high amount of ionic liquid and enzyme, large excess of alcohol was closest to the optimal. There was no inhibiton effect neither at application of 1:16 acid/alcohol molar ratio. Increasing the temperature to 70 ° C had a positive impact on the process, at 80 °C desactivation of the enzyme was not occurred, although we did not find higher yield.

6. Acknowledgment

The research work was supported by the projects TAMOP-4.2.2/B-10/1-2010-0025 and TÁMOP 4.2.1 / 4.8 B identification number "Mobility and Environment: automotive, energy and environmental research in Central and West Pannon Region".

7. References

Bailey, J.E., Ollis D.F., (1986) Biochemical Engineering Fundamentals, McGraw Hill Co. Singapore

Bates D.M., Watts D.G., (1988) Nonlinear regression analysis and its applications, Wiley Interscience, New York, USA

Bélafi-Bakó, K.,. Dombi, A., Szabó L., Nagy, E., (1994) Triacylglycerol hydrolysis by lipase in a flat membrane bioreactor *Biotechnol. Techn.* 8, pp. 671–674.

Bezbradica, D., Mijin, D., Siler-Marinkovic, S., Knezevic, Z.,(2006) The *Candida rugosa* lipase catalyzed synthesis of amyl isobutyrate in organic solvent and solvent-free system: A kinetic study, *J. Mol. Cat. B: Enzym.* 38 11-16.

Cleland, W.W., (1979) Optimizing Coupled Enzyme Assays, Anal. Biochem. 99 142-145.

Demirbas, A., (2000), Energy from renewable sources *Energ. Convers. Manage.* 42 pp. 1357–1378.

Dossat, A., D. Combes and A. Marty, (2002) J. Biotechnol. 97, pp. 117–124.

Eckstein M., Villela M., Liese A., Kragl U. (2004) Use of an ionic liquid in a two-phase system to improve an alcohol dehydrogenase catalysed reduction. *Chem. Commun.* 9, 1084-1085.

Filipini, M., T. Hogg, (1997), Upgrading of Vegetable Wastes and Applications in the Food Industry. *Med. Fac. Landbouww. Uni. Gent.* 62/a pp. 1329–1331.

Garcia, T., Coteron, A., Martinez, M., (1996) Kinetic Modelling of Esterification Reactions Catalysed by Immobilized Lipases, *J. Aracil, Chem. Eng. Sci.* 51 2841-2846.

Garcia, T., N. Sanchez, M. Martinez, J. Aracil, (1999) Enzymatic synthesis of fatty esters Part I. Kinetic approach *Enzyme Microbial Technol.* 25 584-590.

Goddard, R., Bosley, J., Al-Duri, B., (2000) Esterification of oleic acid and ethanol in plug flow (packed bed) reactor under supercritical conditions: Investigation of kinetics, *J. Supercrit. Fluids.* 18 121-130.

Gubicza L., Kabiri-Badr A., Keoves E., Belafi-Bako K. (2000) Large-scale enzymatic production of natural flavour esters in organic solvent with continuous water removal. *JOURNAL OF BIOTECHNOLOGY* 84:(2) pp. 193-196.

Gubicza, L., Nemestóthy, N., Fráter T., Bélafi-Bakó K., (2003) Enzymatic esterification in ionic liquids integrated with pervaporation for water removal, *Green Chem.* 5 236-239.

Gulati, R., P. Arya, B. Malhotra, A.K. Prasad, R.K. Saxena, J. Kumar, A.C. Watterson, V.S. Parmar, (2003) Novel biocatalytic esterification reactions on fatty acids: synthesis of sorbitol 1(6) - monostearate, ARKIVOC iii 159-170.

Hazarika, S., Goswami, P., Dutta, N.N., Hazarika, A.K., (2002) Ethyl oleate synthesis by Porcine pancreatic lipase in organic solvents, *Chem. Eng. J.* 85 61-68.

Jain N., Kumar A., Chaunhan S., Chaunhan S.M.S. (2005) Chemical and biochemical transformations in ionic liquids. *Tetrahedron* 61, 1015-1060.

Janssen, A.E.M., Vaidya, A.E., Halling, P.J., (1996) Substrate specificity and kinetics of Candida rugosa lipase in organic media, *Enzyme. Microb. Tech.* 18 340-346.

Koszorz, Z., Nemestóthy, N., Ziobrowski, Z., Bélafi-Bakó, K., Krupiczka, R., (2004), Influence of pervaporation process parameters on enzymatic catalyst deactivation, *Desalination*, 162 307-313.

Leskovac V., (2003) Comprehensive Enzyme Kinetics, Kluver Academic, New York, USA

Letisse, F., Lamare, S., Legoy, M.D., Graber, D., (2003) Solid/gas biocatalysis: an appropriate tool to study the influence of organic components on kinetics of lipase-catalized alcoholysis, *BBA-Gen. Sub.* 1652 27-34.

Linko, Y.-Y. M. Lämsä, X. Wu, E. Uosukainen, J. Seppälä and P. Linko, (1998) Biodegradable. Products by Lipase Biocatalysis *J. Biotechnol.* 66, pp. 41–50.

Moniruzzaman M., Kazunori N., Noriho K., Masahiro G. (2010) Recent advances of enzymatic reactions in ionic liquids. *Biochem. Eng. Journal* 48., 295-314

Oliveira, A.C., Rosa, M.F., Aires-Barros, M.R., Cabral, S.M.S., (2001) Enzimatic esterification of ethanol and oleic acid –a kinetic study, *J. Mol. Cat. B Enzym.* 11 999-1005.

Özgülsün A., F. Karaosmanoglu and M. Tüter, (2000), Esterification reaction of oleic acid with fusel oil fraction for production of lubricating oil *JAOCS* 77 pp. 105–110.

Özgülsün A., F. Karaosmanoglu, (1999), Environmentally friendly lubricating oil candidate *Appl. Biochem. Biotechnol.* 77 pp. 83–89.

Perry J.H., (1969) Chemical Engineers' Handbook, McGraw-Hill, New York, USA

Romero M.D., Calvo L., Alba C., Danesfar A., Ghaziaskar H.S. (2005) Enzymatic synthesis of isoamyl acetate with immobilized *Candida antarctica* lipase in n-hexane. *Enzyme Microb. Tech.* 37, 42-48.

Van der Padt, A., J.J.W. Sewalt, K. van't Riet, (1993) On-line water removal during enzymatic triacylglycerol synthesis by means of pervaporation *J. Membr. Sci.* 80 199-208.

Van Rantwijk F., Sheldon R.A. (2007) Biocatalysis in ionic liquids. *Chem. Rev.* 107, 2757-2785.

Watanabe,Y., Shimada, Y., Sugihara A., Noda H., Fukuda H., and Tominaga Y., (2000) Continuous production of biodiesel fuel from vegetable oil using immobilized *Candida antarctica* lipase. *J. Am. Oil. Chem. Soc.* 77, pp. 355-360

Willing, A., (1999) Fett/Lipid 101, pp. 192–198.

Yadav, G.D., Lathi P.S., (2003) Kinetics and mechanism of synthesis of butyl isobutyrate over immobilised lipases, *Biochem. Eng. J.* 16 245-252.

Yang Z., Pan W. (2005) Ionic liquids: Green solvents for nonaqueous biocatalysis. *Enzyme Microb. Tech.* 37, 19-28.

Zaidi, A., J.L. Gainer, G. Carta, A. Mrani, T. Kadiri, Y. Belarbi and A. Mir, (2002) Esterification of fatty acids using nylon-immobilized lipase in n-hexane: kinetic parameters and chain-length effects , *J. Biotechnol.* 93, pp. 209–216.

10

Mycotoxins in Food

Francesca Bosco and Chiara Mollea
Politecnico di Torino
Italy

1. General principles

The term mycotoxin was used for the first time in 1961 in the aftermath of a veterinary crisis in England, during which thousands of animals died. The disease was linked to a peanut meal, incorporated in the diet, contaminated with a toxin produced by the filamentous fungus *Aspergillus flavus* (Bennet & Klich, 2003; Richard, 2007).
In general, mycotoxins are low-molecular-weight compounds that are synthetized during secondary metabolism by filamentous fungi; their chemical structure may range from simple C4 compounds to complex substances (Paterson & Lima, 2010).
Mycotoxins are natural contaminants in raw materials, food and feeds. Mould species that produce mycotoxins are extremely common, and they can grow on a wide range of substrates under a wide range of environmental conditions; they occur in agricultural products all around the world (Bennet & Klich, 2003). Many mycotoxins may be toxic to vertebrates and other animal groups and, in low concentrations, some of them can cause autoimmune illnesses, and have allergenic properties, while others are teratogenic, carcinogenic, and mutagenic (Bennet & Klich, 2003; Council for Agricultural Science and Technology [CAST], 2003).
Apparently, mycotoxins have no biochemical significance on fungal growth; they may have developed to provide a defense system against insects, microorganisms, nematodes, animals and humans (Etzel, 2002).
Exposure to mycotoxins may occur through ingestion, inhalation, and dermal contact, and it is almost always accidental. Most cases of mycotoxicoses (animals and humans) result from eating contaminated food. Human exposure can be direct *via* cereals or indirect *via* animal products (e.g. meat, milk and eggs) (CAST, 2003).
Most mycotoxins are relatively heat-stable within the conventional food processing temperature range (80–121°C), therefore so little or no destruction occurs under normal cooking conditions, such as boiling and frying, or even following pasteurization (Milicevic et al., 2010). The stability of mycotoxins during food processing has been reviewed in the work by Bullerman & Bianchini (2007). In general, the application of a food process reduces mycotoxin concentrations significantly, but does not eliminate them completely. The food processes that have been examined include physical treatments (cleaning and milling) and thermal processing (e.g. cooking, baking, frying, roasting and extrusion). The different treatments have various effects on mycotoxins, and those that utilize the highest temperatures have the greatest effects: roasting or cooking at high temperatures (above 150 °C) appear to reduce mycotoxin concentrations significantly (Bullerman & Bianchini, 2007).

It has been estimated that 25% of the world's crops are affected by fungal growth, and commodities may be, both pre- and post-harvest, contaminated with mycotoxins. The mycotoxins that can be expected in food differ from country to country in relation to the different crops, agronomic practices and climatic conditions (Bryden, 2007). Since climate changes affect the growth of mycotoxigenic fungi, mycotoxin production is also influenced (Magan et al., 2003).

Currently, more than 400 mycotoxins are known. Scientific attention has mainly focused on those that have proven to be carcinogenic and/or toxic in humans and animals. Five classes of mycotoxins are considered the most significant in agriculture and in the food industry: aflatoxins (aflatoxin B1), ochratoxins (ochratoxin A), fumonisins (fumonisin B1), zearalenone, and patulin which are derived from polyketide (PK) metabolism, and trichothecenes (deoxynivalenol), whose biosynthetic pathway is of terpenoid origin. PKs are metabolites that are derived from the repetitive condensation of acetate units or other short carboxylic acids, via an enzymatic mechanism that is similar to that responsible for fatty acid synthesis (Huffman et al., 2010).

Aflatoxin, ochratoxin, fumonisin, trichothecene, zearalenone and patulin are the most widespread mycotoxins in animal feed and human food. The chemical structure, biosynthetic pathway, mycotoxigenic fungi, the influence of environmental factors and toxicology will be briefly described for each class. Zearalenone will not be dealth with in the present work as, because of its hormonal activity, there is considerable knowledge about ZEA and its derivatives which can be found in the literature on growth hormones.

1.1 Toxigenic fungi

Aspergillus, Alternaria, Claviceps, Fusarium, Penicillium and *Stachybotrys* are the recognized genera of mycotoxigenic fungi (Milicevic et al., 2010; Reddy et al., 2010). Many of these genera are ubiquitous and, in some cases, apparently have a strong ecological link with human food supplies. The natural fungal flora associated with food production is dominated by the *Aspergillus, Fusarium* and *Penicillium* genera (Sweeney & Dobson, 1998). *Fusarium* species are pathogens that are found on cereal crops and other commodities, and they produce mycotoxins before, or immediately after, the harvest. Some species of *Aspergillus* and *Penicillium* are also plant pathogens or commensals, but these genera are more commonly associated with commodities and food during drying and storage (Pitt, 2000).

Toxigenic moulds are known to produce one or more of these toxic secondary metabolites. However, not all moulds are toxigenic and not all secondary metabolites from moulds are toxic. Many fungi produce several mycotoxins simultaneously, especially *Fusarium* species.

Moreover, recent studies have demonstrated that the necrotrophic pathogens of wheat, *Stagonospora nodorum, Pyrenophora tritiirepentis* and *Alternaria alternata,* are also capable of synthesizing an array of mycotoxic compounds during disease development (Solomon, 2011).

Nowadays, the identification and quantification of mycotoginenic fungi are carried out by PCR. Diagnostic PCR-based systems are now available for all of the most relevant toxigenic fungi: producers of aflatoxins, trichotecenes, fumonisins and patulin (Niessen, 2007; Paterson, 2006). The primers for mycotoxin pathway sequences have been reviewed in the work by Paterson (2006).

1.2 Influence of environmental factors on mycototoxin production

The production of mycotoxins is highly susceptible to temperature, moisture, water activity (a_w), pH and oxygen concentration, the same environmental factors that affect the growth of toxygenic fungi. Moisture and temperature are two factors that have a crucial effect on fungal proliferation and toxin biosynthesis (Bryden, 2007; Paterson & Lima, 2010). The incidence and level of mycotoxin contamination are closely related to the geographic position and to seasonal factors as well as to the cultivation, harvesting, stocking, and transport conditions (Milicevic et al., 2010).

Mycotoxin contaminations can be divided into the one that occurs in the developing crop (pre-harvest) and the one that develops after maturation (post-harvest). In the pre-harvest period, preventive measures are included in good agronomic practices, such as the careful use of insecticides and fungicides, irrigation to avoid moisture stress, harvesting at maturity and improvement by genetic resistance to fungal attack. During the post-harvest period, the control of the moisture and temperature of the stored commodity will largely determine the degree of fungal activity and consequently the mycotoxin synthesis (Bryden, 2007). Treatments with chemicals, including sodium bisulfite, ozone, and ammonia, acids and bases, represent an opportunity to control fungal growth and mycotoxin biosynthesis in stored grains (Bozoglu, 2009; Magan, 2006; Magan & Aldred, 2007). In recent years, a good control of mycotoxigenic fungi has been achieved using plant products (e.g. extracts and essential oils) as environmental friendly fungicides (Nguefacka et al., 2004; Reddy et al., 2010; Thembo et al., 2010).

Moreover, biological control represents a new opportunity in control strategies: there is evidence that *Bacillus* sp., propionic acid bacteria and lactic acid bacteria (LAB) are able to inhibit fungal growth and mycotoxin production (Bianchini & Bullerman, 2010).

1.3 Toxicology and health

Mycotoxins are toxic to vertebrates and humans at low concentrations. Mycotoxicoses in humans or animals have been characterized as food or feed related, non-contagious, non-transferable, and non-infectious (Zain, 2011).

Mycotoxins have various acute and chronic effects on humans and animals, depending on the species. Within a given species, the impact of mycotoxins on health is influenced by age, sex, weight, diet, exposure to infectious agents, and the presence of other mycotoxins (synergistic effects) and pharmacologically active substances (Milicevic et al., 2010; Zain, 2011).

The majority of mycotoxins currently known are grouped, according to their toxic activity, under chronic conditions as mutagenic, carcinogenic or teratogenic. Grouping according to their site of action results in hemo-, hepato-, nephron-, dermato-, neuro- or immunotoxins (Niessen, 2007).

The most important mycotoxins worldwide are aflatoxins, fumonisins, ochratoxins, deoxinyvalenol and zearalenone. Carcinogenic properties have been recognized with regard to aflatoxin and fumonisins (Mazzoni et al., 2011; Wogan, 1992).

Aflatoxin B1 (AFB1) has been linked to human primary liver cancer, in which it acts synergistically with HBV infection and it has been classified as a carcinogen in humans (Group 1 carcinogen). Fumonisin B1, the most abundant of the numerous fumonisin analogues, was classified as a Group 2B carcinogen (possibly carcinogenic to humans) (Zain, 2011; Wild & Gong, 2010).

The potential role of dietary factors to counteract the toxic effects of mycotoxins has been reviewed by Galvano et al. (2001): the effect of antioxidants, food components and additives on reducing toxicity, by decreasing toxin formation and enhancing the metabolism, has been reported.

A mixture of mycotoxins may occur simultaneously, depending on the environmental conditions and substrate availability (Milicevic et al., 2010). Therefore it can be expected that humans and animals are exposed to a mixture rather than to individual compounds. For example, the interactive (synergistic) cytotoxic effects of Ochratoxin A (OTA), Ochratoxin B (OTB), citrinin, and patulin, which are produced by a number of *Penicillium* and *Aspergillus* species, have recently been evaluated by Heussner et al. (2006).

2. Aflatoxins

Aflatoxins (AFs) are the best known and most widely studied mycotoxins. They were first isolated in the early 1960s when 100,000 turkey poults died after consuming aflatoxin-contaminated peanut meal in the UK (the so-called Turkey X disease); this event was followed by proliferation in research on fungal toxins contaminating food and feeds. AFs were found to be the most potent naturally formed carcinogen, and researchers started their investigating on factors that influence this production (Blount, 1960; CAST, 1989).
AFs are highly toxic, mutagenic, and carcinogenic compounds (Wogan, 1999). They are secondary metabolites that are produced mainly by *Aspergillus parasiticus* and *Aspergillus flavus*; in fact, the name "aflatoxin" is derived from the first letter in Aspergillus, and the first three letters in flavus. These fungi are found in many countries, especially in tropical and subtropical regions, where the temperature and humidity conditions are optimal for the growth of moulds and the production of toxin (Rustom, 1997).

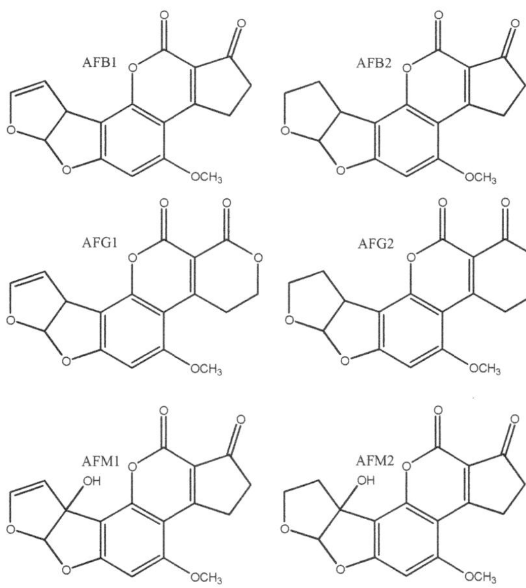

Fig. 1. Principal aflatoxins and metabolites.

AFs are natural contaminants of several agricultural products, such as: corn, peanuts, cottonseed, and other grain crops (Gourama & Bullerman, 1995). Diet is the major way through which humans as well as animals are exposed to these mycotoxins. AFM1 is transformed at the hepatic level by means of cytochrome P450 enzymes and excreted into the milk in the mammary glands of both humans and lactating animals after the animals have ingested feeds contaminated with AFB1 (Oveisi et al., 2007; Prandini et al., 2009).

Structurally, AFs are difurocoumarin derivatives that fluoresce under ultraviolet light. Depending upon colour of the fluorescence, AFs are divided into aflatoxin B1 and B2 (AFB1, AFB2) for blue, and G1 and G2 (AFG1, AFG2) for green (Dalvi, 1986) (Figure 1). Aflatoxin M1 and M2 (AFM1, AFM2), known as milk-AFs, are the metabolites of AFB1 and AFB2, respectively (Carnaghan et al., 1963). Other metabolites of AFB1 are aflatoxin Q1 (AFQ1) and aflatoxicol. Of the known AFs, AFB1 is the most common produced mycotoxin and the most potent; it has been reported to be the most powerful natural carcinogen in mammals (Creppy, 2002).

The biosynthesis of aflatoxins is a complex process, involving multi-enzymatic reactions. Genetic studies on the molecular mechanism of aflatoxin B1 biosynthesis have identified an aflatoxin pathway gene cluster of 70 kilobase pairs in length consisting of at least 24 identified structural genes including a positive regulatory gene as the transcription activator. The structural genes encode cytochrome P450 monooxygenases, dehydrogenases, oxidases, methyltransferases, a polyketide synthase and two unique fatty acid synthases (Yu et al., 2002).

2.1 Fungi

Aflatoxins are closely related to a group of aspergilli: *A. flavus*, *A. parasiticus*, and *A. nomius*; although one report also adds a sclerotium producing strain of *A. tamarii*, which is closely related to *A. flavus*, to the list (Goto et al., 1996). Earlier reports of the production of aflatoxins by other aspergilli, penicillia or even a species of *Rhizopus*, have not been adequately confirmed (Moss, 2002a).

A. flavus and *A. parasiticus*, which are found worldwide in the air and soil, usually infest both living and dead plants and animals, and as a consequence, aflatoxins in agricultural commodities are primarily produced by *A. flavus* and *A. parasiticus*. *A. flavus* produces only B aflatoxins, while *A. nomius* and *A. parasiticus* produce both B and G toxins (Rustom, 1997; Yu et al., 2002).

2.2 Food

Aflatoxin contamination of food and feeds is a serious problem worldwide. Studies focusing on AF contamination in foodstuffs have in fact been reported in many countries, especially those in tropical and subtropical regions, such as Asia and Africa (Bankole et al., 2010; Shundo et al., 2009; Soubra et al., 2009).

Aflatoxin contamination can develop both in the pre- and post-harvest periods, but the highest levels are usually associated with post-harvest spoilage of food commodities, stored under inappropriate high moisture content and high temperature conditions which facilitate the rapid growth of moulds; the level of contamination depends on the plant stress, temperature, water activity, genotype, culture and storage conditions, but appropriate post-harvest treatments, under dry cool settings, should control this source of contamination (Moss, 2002a; Wilson & Payne, 1994).

As far as pre-harvest, is concerned, aflatoxigenic fungi have a complex ecology. The spores of *A. flavus* and *A. parasiticus* can germinate on the stigma surfaces of plants, and the germ tube can penetrate the developing embryo in a manner which mimics pollen germ tubes. The mycelium can establish an endotrophic relationship, which is not harmful to a healthy plant, while if the plant is stressed (e.g. drought), significant levels of aflatoxin may be produced during field growth. Under these circumstances food commodities may already be contaminated at harvesting and, even though the concentrations are never as high as those formed in stored commodities, they can be economically significant and this field contamination is much more difficult to control than post-harvest spoilage (Hill et al., 1983; Moss, 2002a).

Although a wide variety of foods are susceptible to aflatoxin contamination, it has most commonly been associated with peanuts, maize, pistachio, dried fruit, nuts, spices, figs, vegetable oils, cocoa beans, corn, rice and cotton seeds (JECFA, 1998; Reports on Carcinogens [ROC], 2003). Among the agricultural commodities usually infected by aflatoxigenic fungi (Table 1) some are food sources while others are used as animal feeds: the greatest difficulty is that aflatoxin affects the health of the humans and the livestock that consume these commodities and the related products. Speijers & Speijers (2004) reported that AFB1 and OTA are amongst the most frequently observed combinations of mycotoxins in different plant products. According to several other authors, cereals, olives and dried vines are other commodities which could support aflatoxigenic and ochratoxigenic mould growth and OTA and AFB1 production (Molinié et al., 2005; Zinedine et al., 2006).

While aflatoxin B1 is frequently found in contaminated feeds, aflatoxin M1, its hydroxylated metabolite, is normally not present in food, except though carry-over from animal feeds (Fallah, 2010; Kamkar, 2008): following the ingestion of contaminated feedstuffs by lactating dairy cows, AFB1 is biotransformed, by hepatic microsomal cytochrome P450 into AFM1, and is then excreted into the milk (Frobish et al., 1986). Moreover, the AFM1 content in milk is closely correlated to the level of AFB1 in the raw feedstuffs (Bakirci, 2001). AFM1 can be detected in milk 12–24 h after the first ingestion of AFB1; generally, it is deemed that approximately 1–3% of the aflatoxin B1 present in animal feeds appears as AFM1 in milk, depending on the animal, time of milking and many other factors. When the intake of the contaminated source is stopped, the concentration of the toxin in the milk decreases to an undetectable level within 72 h (Gurbay et al., 2006). Additionally, when specific conditions during feed storage are prevalent for the growth of aflatoxigenic species, an additional production and accumulation of AFB1 may occur; this in turn leads to the accumulation of additional AFM1 in the milk. Aflatoxin M1 can survive pasteurization and has even been reported in UHT milk (Unusan, 2006).

AFM1 binds to casein, has a high stability and concentrates in curd during cheese production, in different proportions according to the applied technology (Barbiroli et al., 2007; Brackett & Marth, 1982). In this way, it can also be present in dairy products, manufactured with contaminated milk, at higher concentrations than in the milk (Govaris et al., 2001; Lopez et al., 2001; Oruc et al., 2006). Cheese-making and the ripening period do not result in a reduction in the toxin (Dragacci et al., 1995; Yousef & Marth, 1985). This is why the risk remains, not only in commercially available milk, but also in other derived dairy products. The concentration of AFM1 in cheese varies according to the type of cheese, water content and production technologies (Bakirci, 2001; Lopez et al., 2001). Since the sources of aflatoxin contamination in animal feeds differ because they are location dependent and the incidence and occurrence of AFM1 contamination in animal feeds from different countries varies, there are many reports on AFM1 contamination in cheese and other dairy products from different countries: Slovenia, North Africa, Turkey, Brazil and Portugal (Bakirci, 2001; Elgerbi et al., 2004; Oliveira et al., 2006; Martins & Martins, 2000; Torkar & Vengus, 2008).

Food commodity	Country
Soy beans	Argentina
Almonds; Brazil nuts	USA
Dried figs	Austria, Switzerland
Nutmeg	Japan
Chilli	Pakistan
Herbs, spices	UK
Spices	Sweden
Peanuts	India, Sudan, Brazil, Egypt, South Africa
Maize	Argentina, India, China, Uganda, Nigeria, USA
Pistachio nuts	Netherlands, USA, Turkey,
Wheat	Uruguay, China, Russia
Rice	Ecuador, China, India
Millet	India
Sunflower oil	China, Russia
Coconut	India
Mustard seed	India

Table 1. Presence of aflatoxins in food commodities (Moss, 2002a; Rustom, 1997).

2.3 Toxicity

Aflatoxins can be both acute and chronic toxins; acute poisoning is usually rare and exceptional, while chronic toxicity is of serious concern and it drives international concern about the occurrence of aflatoxins in food (Moss, 2002a).

AFB1 is toxic for a wide range of animal species. AFB1 is principally a hepatotoxin and hepatocarcinogen (JECFA, 1998), but it can cause a myriad of other effects: immunosuppresion, reduced growth rate, lowered milk and egg production, reduced reproductivity, reduced feed utilization and efficiency and anemia. AFB1 has been shown to induce hepatocellular carcinoma in many animal species including fish, poultry, non-human primates, and rodents (Wogan, 1992).

Species susceptibility to various acute toxic manifestations, as measured by TD50, is also variable (Gold et al., 1984). A wide variation exists in species susceptibility to AFB1 hepatocarcinogenesis.

In humans, acute aflatoxicosis is manifested by vomiting, abdominal pain, pulmonary edema, coma, convulsions, and death with cerebral edema and fatty involvement of the liver, kidneys, and heart (Mwanda et al., 2005). Epidemiological studies have consistently demonstrated that AFB1 is a liver carcinogen in humans (Groopman et al., 1988; Van Rensburg et al., 1985). The International Agency for Research on Cancer has concluded that there is sufficient evidence for the carcinogenicity of AFB1 in humans and hence placed this mycotoxin in group I.

AFB1 is not mutagenically active itself. It is primarily metabolized in the liver and has several metabolites, such as aflatoxicol and AFQ1. AFB1 is mainly activated by cytochrome P450 dependent monooxygenase; most of the metabolic products, such as AFM1 and AFQ1, are less toxic than the parent AFB1, but aflatoxin B1-8-9-expoxide (AFBO) is the most toxic metabolite (Hwan Do & Choi, 2007). The carcinogenic and mutagenic action of AFB1 might be the result of the affinity of the electrophilic and highly reactive AFBO for cellular nucleophiles, such as DNA (Coulombe, 1993). Thus, epoxidation is generally considered in metabolite activation, while hydroxylation, hydration, and demethylation are considered metabolic detoxications. The toxic and carcinogenic effects of aflatoxin B1 are intimately linked to both the rate of activation and the rate of detoxification at the primary and secondary levels of metabolism, in a similar way to chlorinated hydrocarbon (Olaniran et al., 2006).

3. Ochratoxin A

Mycotoxin ochratoxin A (OTA) was discovered in 1965 in South Africa (Van der Merwe et al., 1965): it was isolated as a toxic metabolite of *Aspergillus ochraceus* from corn meal artificially inoculated with the fungus. In 1969, naturally occurring OTA was isolated from a commercial corn sample in the United States (Shotwell et al., 1969). Later, it was recognized as a secondary metabolite of several *Aspergillus* and *Penicillium* spp. which are characterized by widespread occurrence and different behavior which depends on the ecological niches, the products affected and the environment (Duarte et al., 2010).

OTA is one of the most relevant mycotoxins, with great public health and agroeconomic significance, due to the confirmed nephrotoxic, genotoxic, neurotoxic, imunotoxic, embriotoxic and teratogenic effects and its suspected carcinogenicity (JECFA, 2008). OTA has been documented as a global contaminant of a wide variety of commodities and staple food. Humans are directly and indirectly exposed to OTA: it can enter the food chain, through contamination of the ingredients or foodstuffs consumed by humans, or the feed chain, through contamination of the feeds for animals destined for human consumption (Cark & Snedeker, 2006).

The chemical name of OTA is N-[(3R)-(5-Chloro-8-hydroxy-3-methyl-1-oxo-7-isochromanyl)carbonyl]-L-phenylalanine; OTA belongs to a group of metabolites with a similar chemical structure, as shown in Figure 2 and Table 2.

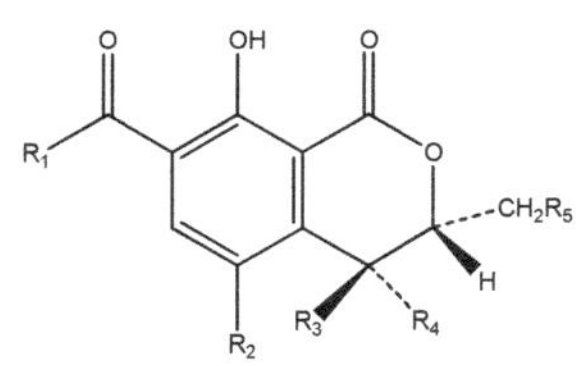

Fig. 2. General structure of OTA and its metabolites (El Khoury & Atoui, 2010).

The biosynthetic pathway for OTA has not yet been completely established; however, the isocoumarin group is a pentaketide skeleton formed from acetate and malonate via a polyketide synthesis pathway with the L-phenylalanine being derived from the shikimic acid pathway (O'Callaghan et al., 2003).

OTA is a weak organic acid (the pKa is 7.1 and the molar mass is 403.8 g mol^{-1}). In acidic conditions, OTA is soluble in polar organic solvents, slightly soluble in water and insoluble in petroleum ethers and saturated hydrocarbons. In alkaline conditions, OTA is soluble in aqueous sodium bicarbonate solutions and in all alkaline ones. It has a melting point of about 90 °C, when crystallized from benzene as a solvate (El Khoury & Atoui, 2010; Keeper-Goodman & Scott, 1989). Due to its resistance to acidic conditions and high temperatures, OTA is characterized by high stability. Thus, it is very difficult to eliminate the molecule: OTA is only partially degraded at normal cooking conditions and after three hours of high pressure steam sterilization at 121 °C, or even at 250 °C, its destruction is not complete (Boudra et al., 1995).

Name	R1	R2	R3	R4	R5
OTA	Phe*	Cl	H	H	H
OTB	Phe	H	H	H	H
OTC	Ethyl-ester, Phe	Cl	H	H	H
OTA methyl-ester	Methyl-ester, Phe	Cl	H	H	H
OTB methyl-ester	Methyl-ester, Phe	H	H	H	H
OTB-ethyl-ester	Ethyl-ester, Phe	H	H	H	H
OTα	OH	Cl	H	H	H
OTβ	OH	H	H	H	H
4-R-hydroxy OTA	Phe	Cl	H	OH	H
4-s-hydroxy OTA	Phe	Cl	OH	H	H
10-hydroxy OTA	Phe	Cl	H	H	OH
Tyr* analog of OTA	Tyr	Cl	H	H	H
Ser* analog of OTA	Ser	Cl	H	H	H
Hyp* analog of OTA	Hyp	Cl	H	H	H
Lys* analog of OTA	Lys	Cl	H	H	H

Table 2. Radicals in OTA metabolites *(Phenylalanine; Tyrosine; Serine; Hydroxyproline; Lysine) (El Khoury & Atoui, 2010).

3.1 Fungi

Ochratoxin A is produced by *Aspergillus* and *Penicillium* species listed in Table 3. These microorganisms differ according to the ecological conditions and commodities that characterize different geographical regions. In general, *Penicillium verrucosum* is responsible for OTA contamination in cool-temperate conditions, whereas *Aspergillus ochraceus* is particularly relevant in hot-tropical regions (Battaccone et al., 2010; Scudamore, 2005).

The major *Aspergillus* producers in food and feeds are *A. alliaceus*, *A. carbonarius*, *A. ochraceus*, *A. steynii* and *A. westerdijkiae*. *A. melleus*, *A. ostianus*, *A. persii* and *A. petrakii* may produce trace amounts of OTA, but since the publication by Ciegler (1972) and Hesseltine et al. (1972) no further confirmation has been found.

In the genus *Penicillium*, *P. verrucosum* and *P. nordicum* are the only species that are able to produce OTA (Abruhnosa et al., 2010; El Khoury & Atoui, 2010). *P. chrysogenum*, *P. brevicompactum*, *P. crustosum*, *P. olsonii* and *P. oxalicum* have been claimed as OTA producers, but a confirmation of these findings is required (Paterson, 2006).

Aspergillus section Circumdati
A. cretensis; A. flocculosus; A. ochraceus ; A. pseudoelegans; A. roseoglobulosus; A. sclerotiorum ; A. steynii; A. sulphureus ; A. westerdijkiae; Neopetromyces muricatus
Aspergillus section Flavi
A. alliaceus; Petromyces albertensis
Aspergillus section Nigri
A. carbonarius; A. lacticoffeatus; A. niger; A. sclerotioniger; A. citricus ; A. fonsecaeus
Penicillium
P. nordicum; P. verrucosum

Table 3. OTA producing fungi (Abrunhosa et al., 2010; El Khoury & Atoui, 2010; Moss, 2002b).

3.2 Food

OTA has a widespread diffusion, and it has in fact been detected in agricultural commodities, livestock products and processed food (Abrunhosa et al., 2010).

The main OTA contamination concerns cereals and their products, listed in Table 4, which include food and beverages for human consumption, but also by-products that are usually utilized as animal feeds. Ochratoxin contamination can occur from temperate to tropical climates, from hot and wet climatic conditions to low temperature environments, and affects numerous countries: Northern America, Northern and Western Europe, African countries, South Asia and South America (Battaccone et al., 2010; Cabanes et al., 2010; El Khoury & Atoui, 2010; Moss, 2002b). Vega et al. (2009) suggests that cereals should be considered a major source of OTA contamination, as 50% of human daily intake of this mycotoxin is due to the consumption of different cereal derived products.

cereals
Corn (grains, gluten); Rice; Wheat; Barley; Oats; Rye; Sorghum; Millet
cereal products for human consumption
Beer; Baby food; Breakfast cereals; Bread
cereal feed products
Cracked grains; Cereal cleanings; Wheat bran; Corn bran; Rice bran

Table 4. Cereals contaminated by OTA (Abrunhosa et al., 2010; Scudamore et al., 2003).

Cereals may be colonized by both *Aspergillus* and *Penicillium*. The two fungal species do not invade the crop in the field, but mainly do in the post-harvest phase. Considering that the main abiotic factors that influence mould growth and OTA production are water availability and temperature, cereals should be dried quickly after harvesting and maintained at a lower moisture content than 14.5% during storage to avoid OTA contamination (Magan & Aldred, 2005). OTA is mainly concentrated in the seed coat, which is often used for animal feeding. Moreover, on-farm production and the storage of barley and wheat with a high moisture content increases the risk of mould growth and toxin production (Scudamore et al., 2003). Some cereal processing, like malting, malt fermentation, bread production and feed extrusion, can contribute significantly to reduce OTA concentration in the final food products (Baxter et al., 2001; Scott et al., 1995; Scudamore et al., 2003). Other practices can increase OTA values; for example, cracked grains and cereal cleanings are often the most contaminated fractions and are usually directed for feed proposes (Scudamore et al., 2003).

Wine is considered the second source of the human consumption of OTA. Many works have highlighted the presence of considerable levels of this toxin in wines, musts and grape juices. This occurrence has been explained by the fact that grapes are contaminated in the vineyard by various ochratoxigenic species, belonging above all to the *Aspergillus* section *Nigri* genus (*A. carbonarius* and *A. niger* aggregates) and that OTA production increases rapidly with the maturation stage. Thus, the date of the grape harvest would have an important effect on the OTA content in grapes and their derived products (Cabanes et al., 2002; El Khoury et al., 2006).

OTA contamination of many other raw agricultural products has been well documented; such a contamination occurs in a variety of food and feeds, such as coffee beans, pulses, spices, meat and cheese products (Wolff et al., 2000).

3.3 Toxicity

OTA can have several effects, such as nephrotoxic, hepatotoxic, neurotoxic, teratogenic and immunotoxic effects on several species of animals, and can cause kidney and liver tumours in mice and rats; OTA toxicity varies depending on the sex, the species and the cellular type of the tested animal (El Khoury & Atoui, 2010).

Nephropathy is the main toxic effect of OTA; it is potentially nephrotoxic in all non-ruminant mammals (Ribelin et al., 1978). OTA plays an important role in the etiology of porcine nephropathy (Elling et al., 1985). This mycotoxin was also associated with human nephropathy and it is suspected to be the cause of the human Balkan Endemic Nephropathy (BEN) and the Tunisian Nephropathy (TCIN) (Hassen et al., 2004; Pfohl-Leszkowicz, 2009).

The administration of OTA at gestation period in rats induced many malformations in the central nervous system. OTA can be regarded as a possible cause of certain lesions as well as damage at the cerebral level. OTA seems to be highly toxic for the nervous cells and able to reach the neural tissue (Soleas et al., 2001).

OTA is a potent teratogen for laboratory animals. It can cross the placenta and accumulate in fetal tissue, causing various morphological anomalies. It has been reported to elicit prenatal dysmorphogenesis in rats, mice, hamsters and chick embryos (El Khoury & Atoui, 2010).

OTA also has an immunosuppressor effect. Necroses of lymphoid tissues has been reported, and humoral and cellular immunity affections have also been described (Creppy et al., 1991; Holmberg et al., 1988). OTA seems to play a role in the inhibition of proliferation of the peripheral T and B lymphocytes and stops the production of interleukin 2 (IL2) and its receptors (Lea et al., 1989). Moreover, it blocks the activity of killer cells as well as the production of interferon (Pfohl-Leszkowicz & Castegnaro, 1999). OTA is taken as an important immunosupressor agent, in fact it is considered to be the cause of lymphopenia, regression of the thymus, and suppression of the immunity response (Petzinger & Weidenbach, 2002).

OTA is anticipated to be a human carcinogen based on sufficient evidence of carcinogenicity in experimental animals. Hepatocellular tumors, renal cell tumors, hepatomas, and hyperplastic hepatic nodules have been observed in male mice (Huff et al., 1992). OTA has been correlated with hepatocellular and renal-cell carcinomas and adenomas in mice and rats (El Khoury & Atoui, 2010). On the other hand there are no adequate studies of the relationship between exposure to OTA and human cancer; incidence and mortality from urothelial urinary tract tumours have been correlated with the geographical distribution of Balkan endemic nephropathy in Bulgaria and Yugoslavia (Feier & Tofana, 2009).

4. Trichothecenes

Trichothecin was first isolated from *Trichothecium roseum* and described by Freeman and Morrison in 1949. The discovery of trichothecin was followed by the isolation and description of other trichothecenes (TCTs), such as diacetoxyscirpenol (DAS), T-2 toxin (T-2), nivalenol (NIV) and deoxynivalenol (DON) (Yazar & Omurtag, 2008).

The Alimentary Toxic Aleukia (ATA) that occurred in Russia during World War II was caused by T-2 toxin and its derivatives; *F. sporotrichioides* was isolated from contaminated grains (Yazar & Omurtag, 2008). DON is the most prevalent toxin associated with Fusarium Head Blight (FHB), and it belongs to the phytotoxic type B trichothecene (Foroud & Eudes, 2009).

TCTs are the most important group of mycotoxins and they are produced above all by various *Fusarium* plant pathogen species (Kimura et al., 2007). They are non-volatile, low-molecular-weight tricyclic sesquiterpenes with a basic 12,13-epoxy-trichothec-9-ene ring system (Figure 3) and are further classified as macrocyclic, or non macrocyclic depending on the presence of a macrocyclic ester or an ester–ether bridge between C-4 and C-15 (Bennett & Klich, 2003; Merhej et al., 2011). Trichothecenes are a family of more than 200 related compounds which are divided into four subclasses (Types A–D), according to their characteristic functional groups. Type A has a functional group other than a ketone at position C-8; Type B has a ketone at position C-8; Type C has a second epoxy group at C-7,8 or C-9,10 and Type D contains a macrocyclic ring between C-4 and C-5 with two ester linkages (Sweeney & Dobson, 1998). The major Type A trichothecenes in *Fusarium* species include T-2 toxin (T-2) and HT-2 toxin (HT-2), both of which have an isovalerate function in C-8. Type B TCTs include Fusarenone-X, deoxynivalenol (DON) and nivalenol.

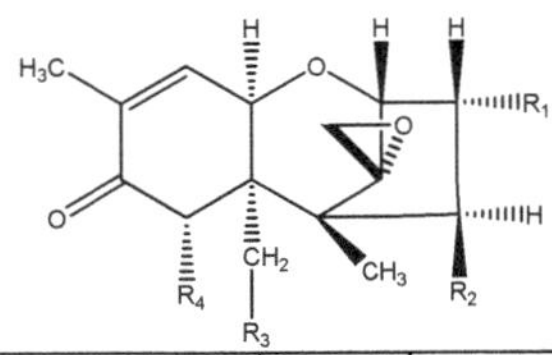

Trichothecene	R1	R2	R3	R4
Deoxynivalenol (DON)	-OH	-H	-OH	-OH
Nivalenol (NIV)	-OH	-OH	-OH	-OH

Fig. 3. Chemical structure of trichothecens.

The trichothecene biosynthetic pathway in *Fusarium* has been reported extensively by Sweeney & Dobson (1998) and Desjardins & Proctor (2007); it begins with the cyclization of the isoprenioid intermediate farnesyl pyrophosphate to trichodiene by the enzyme trichodiene synthase. After this a number of oxygenation, isomeritation, cyclization, and esterification leading from trichodiene to dyacetoxyscirpenol, T-2 toxin and 3-acetyldeoxynivalenol (Huffman et al., 2010). The recent advances concerning the regulation of trichothecene biosynthesis in *Fusarium* and the potential implication of various general regulatory circuits has been reported in the work of Merhej et al. (2011); the knowledge of the role of these regulatory systems might be useful in designing new strategies to reduce mycotoxin accumulation.

Deoxynivalenol (DON) is the most studied mycotoxin produced by *Fusarium*. DON, also known as vomitotoxin, is a polar organic compound, which is soluble in water and polar organic solvents (e.g. aqueous methanol, ethanol, chloroform, acetonitrile and ethyl acetate); it is optically active. The chemical name of DON is 12,13-epoxy-3α,7α,15-trihydroxytrichothec-9-en-8-one, and the molar mass is 296,32 g mol^{-1}. DON is very stable at temperatures within the 170°C to 350°C interval with no reduction in DON concentration after 30 min at 170°C (Sobrova et al., 2010). DON shows great stability during storage/milling and in the processing and cooking of food.

4.1 Fungi

The *Fusarium* genus includes a number of important plant pathogens that produce a wide range of mycotoxins (TCTs, fumonisine and zearalenone) which are mainly found in cereal grains (Vesonder & Golinski, 1989). *Fusarium* is the main genus implicated in the production of the non-macrocyclic TCTs. Many toxigenic *Fusarium* species have been associated with infected grain, and the predominant pathogens are *F. graminearum* and *F. culmorum*. During infection, *F. graminearum* produces various mycotoxins in grains, in particular deoxynivalenol (DON), a type B trichothecene. *F. graminearum* is the most important DON producer, followed by *Fusarium culmorum*, but other species such as *Fusarium sporotrichioides* or *Fusarium langsethiae* have also been reported. The geographical distribution of the species is probably related to temperature requirements (Merhej et al., 2011). From an economic point of view, the most important TCT producers are *Fusarium* species that cause Fusarium Head Blight (FHB) in small-grain cereals and Gibberella Ear Rot (GER) in maize (Bottalico & Perrone, 2002). The first documented FHB-outbreak occurred in England in 1884, where the disease was named "wheat scab". Outbreaks have since been reported in the Americas, Asia, Australia, Europe, and South Africa (Foroud & Eudes, 2009). These diseases are associated with the temperature in the grain growing region: *F. graminearum* (optimal growth range between 24 and 26°C, minimum a_w value 0.90) is more dominant in warmer regions (North America and China), while *F. culmorum* (psychrotrophic strain, optimal temperature growth 21°C) is more dominant in cooler regions (northern Europe) (Sweeney & Dobson, 1998).

The main species responsible for the production of T-2 toxin is *F. sporotrichioides*. The natural occurrence of this species has been reported in Asia, Africa, South America, Europe and North America (CAST, 2003).

Apart from *Fusarium*, several other fungal genera are capable of producing TCTs: *Myrothecium, Phomopsis, Stachybotrys, Trichoderma, Trichothecium*. Macrocyclic TCTs are produced largely by *Myrothecium, Stachybotrys* and *Trichothecium* (Bennett & Klich, 2003).

Fusarium species are pathogens that are found on cereal crops and they produce mycotoxins before, or immediately after, harvesting. Consequently, strategies for the prevention of TCTs from entering human and animal food chain include the elimination of TCTs in grains in the field, detoxification of TCTs that are already present in food and feeds, and inhibition of TCTs absorption in the gastrointestinal tract (He et al., 2010).

To date, the control of *Fusarium* proliferation in the field is not ensured, therefore, generation of resistant varieties of crop plants still remain the best way to reduce grain contamination by *Fusarium* without using chemical fungicides. The discovery of biological or chemical molecules which would be able to specifically block the biosynthetic pathway in order to limit the residual synthesis of toxins is a new challenge (Merhej et al., 2011).

At harvest and during the storage of cereals, the key factor for TCT formation is the presence of conidia and humidity combined with temperature. Minimizing or avoiding conidia contaminated materials, cleaning at an early stage during the harvest and drying the grain at low temperatures will allow cereals to be stored for more than 12 months without increasing TCTs levels (Yazar & Omurtag, 2008).

4.2 Food

TCTs are mainly associated with cereals grown in the temperate regions of Europe, America and Asia: wheat, barley, oats, rye, maize and rice (Yazar & Omurtag, 2008). Their presence has also been reported in soybeans, potatoes, sunflower seeds, peanuts and bananas. TCTs have also been found in processed foods, especially those produced from cereals (bread, breakfast cereals, noodles, and beer). The TCTs that are dominant in grains are deoxynivalenol (DON), nivalenol, and their acetylated derivatives (Foroud & Eudes 2009; Karlovsky, 2011). Corn, wheat, barley, oats, rice, rye and other crops have been reported to contain T-2 toxin (CAST, 2003). Moreover, TCTs can enter the food chain through milk, meat and eggs from livestock and poultry that are fed with contaminated feeds, although the exposure risk to human through the consumption of animal tissue is much less than the direct consumption of contaminated grains (He et al., 2010).

Food and feed contamination by TCT have been associated with chronic and fatal toxicoses of humans and animals, including Alimentary Toxic Aleukia in Russia and Central Asia, Akakabi-byo (red mould disease) in Japan, and swine feed refusal in the central United States (Karlovsky, 2011). The epidemy that occurred in Russia between 1942 and 1948, where at least 100,000 people died, was caused by the ingestion of grain contaminated with T-2 produced by *F. sporotrichoides* or *F. poae* (Foroud & Eudes, 2009) .

4.3 Toxicity

At the cellular level, the main mechanism of TCT mycotoxins appears to be a primary inhibition of ribosomal protein synthesis, which is followed by a secondary disruption of DNA and RNA synthesis (Desjardins & Proctor, 2007; Richard, 2007; Zain, 2011), cytotoxicity, and apoptosis (Rocha et al., 2005; Rotter et al., 1996).

TCTs affect dividing cells, such as those coating the gastrointestinal tract, the skin, and lymphoid and erythroid cells. The toxic action of TCTs results in extensive necrosis of the oral mucous and skin in contact with the toxin, an acute effect on the digestive tract and decreased bone marrow and immune functions (Richard, 2007; Rocha et al., 2005).

In general, acute exposure of animals to DON resultes in decreased feed consumption (anorexia) and vomiting (emesis), while longer exposure causes reduced growth, and adverse effects on the thymus, spleen, heart, and liver (Sobrova et al., 2010).

Nowadays, the real concern is not related to acute exposure, but to a prolonged daily exposure, which leads to chronic toxicity, since it has been demonstrated that DON deregulates the immune response and induces cytokine up regulation (Merhej et al., 2011; Pestka & Smolinskj, 2005). It has been demonstrated that the ingestion of DON with contaminated feeds and food leads to growth retardation, and reproductive disorders in animals (Pestka, 2010; Rocha et al., 2005; Sobrova et al., 2010). To date, all the animal species evaluated have shown a differential level of susceptibility to DON with the pigs being the most susceptible (Pestka & Smolinski, 2005).

Human exposure to DON-contaminated grains has been reported to cause acute temporary nausea, vomiting, diarrhea, abdominal pain, headache, dizziness, and fever (Sobrova et al., 2010).

In general, TCTs are heat-stable molecules and are not fully eliminated during the processes currently used in cereal-based food manufacturing (Hazel & Patel 2004). They are also stable at neutral and acidic pH and consequently, they are not hydrolyzed in the stomach after ingestion (Yazar & Omurtag 2008). Since DON is water soluble, its level is reduced in cooked pasta (Sobrova et al., 2010).

The chemical detoxification of DON by ozone (Young et al., 2006), ammonia, chlorine, hydrogen peroxide (He et al., 2010), sodium bisulfite (Young et al. 1986), sodium carbonate (Abramson et al. 2005), and chlorine dioxide (Wilson et al., 2005) has been demonstrated. Therefore, the best way to prevent contamination would be to limit TCT biosynthesis at the field level during crop cultivation (Merhej et al., 2011).

The enzymes involved in biological detoxification of DON and their application to genetically engineered crops and feed additives have been reviewed in the work by

Karlovsky (2011). Bacterial enzymes that catalyze oxidation, epimerization, and, but to a lesser extent, de-epoxidation of DON as well as of the application of acetylation in plant biotechnology have been described (He et al., 2010).

5. Fumonisins

Fumonisins are a group of non-fluorescent mycotoxins. They were discovered and characterized in 1988 (Bezuidenhout et al., 1988). The predominant fungus isolated from fumonisin contaminated corn, associated with the outbreak of Equine Leukoencephalomalacia (ELEM) in South Africa in 1970 and Porcine Pulmonary Edema (PPE) in Iowa, Illinois, and Georgia in 1989, was *F. verticillioides* (Yazar & Omurtag, 2008).

To date, twenty-eight fumonisins have been isolated and they can be divided into four series (A, B, C and P). FB1, FB2 and FB3 are the principal fumonisins analyzed as natural contaminants of cereals (CAST, 2003; Yazar & Omurtag, 2008). Fumonisin B1 is generally the most abundant member of this mycotoxin family; it comprises about 70 % of the total fumonisin content of *Fusarium* cultures (Reddy et al., 2010). Fumonisins have a similar structure to sphingosine, which forms the backbone of sphingolipids within the cell membrane (Sweeney & Dobson, 1998).

Fumonisins are polyketide metabolites, derived from the repetitive condensation of acetate units or other short carboxylic acids, via a similar enzymatic mechanism to that responsible for fatty acid synthesis (Huffman et al., 2010). The fumonisin biosynthetic pathway in *Fusarium* species begins with the formation of a linear dimethylatedpolyketide and condensation of the polyketide with alanine, followed by a carbonyl reduction, oxygenations, and esterification with two propane-1,2,3-tricarboxylic acids (Desjardins & Proctor, 2007).

Fumonisin biosynthetic genes have been mapped to one locus in the *F. verticillioides* genome (Desjardins & Proctor, 2007).

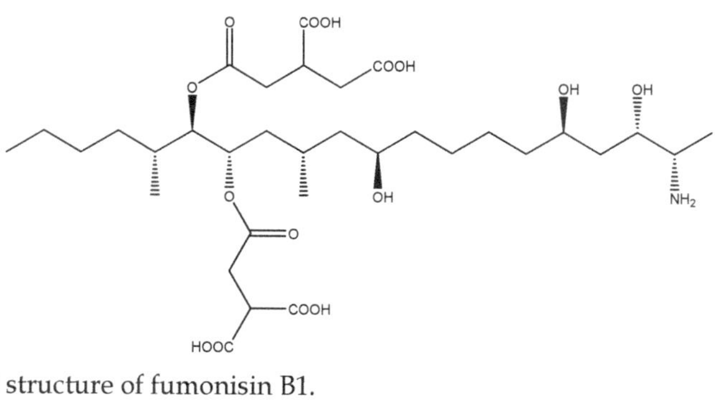

Fig. 4. Chemical structure of fumonisin B1.

The basic chemical structure of fumonisins is a C-20 aliphatic chain with two ester linked hydrophilic side chains (Richard, 2007). The chemical structure of FB1 is 1,2,3-Propanetricarboxylic acid, 1,1N-[1- (12 amino-4,9,11-trihydroxy-2-methyltridecyl)- 2-(1-methylpentyl)-1,2-ethanediyl] Ester (Figure 4). FB2 is the C-10-deoxy analogue of FB1 and FB3 is the C-5-deoxy analogue of FB1(Yazar & Omurtag, 2008). The molecular mass of FB1 is 721 g/mol, while FB2 and FB3 have the same value of molecular mass (705g/mol). FB1 is soluble in water to at least to 20 mg/ml, and in methanol and acetonitrile-water. FB1 and

FB2 are stable in methanol at −18 °C and degrade at 25 °C. However, in a mixture of acetonitrile/water (1:1) and at −25 °C, all fumonisins result to be stable (Wan Norashima et al., 2009).

5.1 Fungi

Fumonisins are produced by a number of *Fusarium* species, notably *F. verticillioides* (formerly *Fusarium moniliforme=Gibberella fujikuroi), F. proliferatum, F. anthophilum, F. nygamai* as well as *Alternaria alternata* f. sp. *lycopersici* (Kumar et al., 2008; Sweeney & Dobson, 1998; Yazar & Omurtag, 2008). Recently, *Aspergillus niger* has also been found to produce fumonisins (i.e., fumonisins B2 and B4), and a new B-series fumonisin (FB6) has been identified from this fungus (Huffman et al., 2010).

Fumonisins found in food are produced mainly in the field; temperature and moisture conditions are important factors that affect *Fusarium* infection and toxin synthesis as is insect damage of corn ears and kernels (Richard, 2007; Yazar & Omurtag, 2008).

At a field level, two approaches are known to reduce infections and mycotoxin accumulation: pre-harvesting control strategies, which consist of crop practices designed to reduce the infection and the development of toxinogenic fungi (Nicholson et al., 2004; Wan Norashima et al., 2009;) and the utilization of genetically resistant hybrids (Munkvold, 2003; Blandino & Rejneri, 2008). Direct fungal control with chemical or biological products (e.g. microbial antagonist or competitor) has only recently been considered (Mazzoni et al., 2011).

Mycotoxin risk can be reduced by enhancing the resistence of insect attack, by inducing the process of detoxification pathway that inhibits the production of mycotoxins and increasing the resistance of the plant to infection by means of genetic engineering. A recent approach to the search for hybrids that are resistant to mycotoxin contamination consists in the obtaining of genetically modified hybrids which create the resistance action through transgenes (Blandino & Rejneri, 2008; Wan Norashima et al., 2009).

An increase in concentrations of fumonisins during storage does not appear to be a major problem. However, grains should be harvested without additional kernel damage, screened to remove broken kernels, dry stored and maintained at moisture concentrations < 14% (Richard, 2007).

5.2 Food

Fumonisins have been found to be a very common contaminant of corn-based food and feeds in Africa, China, France, Indonesia, Italy, the Philippines, South America, Thailand and the USA (Kumar et al., 2008).

In addition to corn or corn-based food and feeds, the occurrence of fumonisins has also been reported in other products, such as: rice and sorghum (CAST, 2003), wheat noodles, curry, beer and corn-based brewing adjuncts (Yazar & Omurtag, 2008).

Fumonisins B1 and B2 have been reported in "black oat" feeds from Brazil and forage grass in New Zealand. FB1 and FB2 have been found in rural areas of South Africa, in home-grown corn produced and consumed by the people living in those areas. Commercial corn based human foodstuff from retail outlets in several countries contains fumonisins (Wan Norashima et al., 2009).

Castelo et al. (1998) have reported that fumonisins found in artificially contaminated cornmeal samples are unstable under roasting conditions, but remain fairly stable during the canning and baking of corn-based foods because the canned and baked products reach lower temperatures than the roasted products. Jackson et al. (1996) indicated that foods that reach greater temperatures than 150 °C during processing may have lower fumonisin levels.

5.3 Toxicity

At a cellular level, the structural similarity between sphinganine and FB1 suggests that the action mechanism of this mycotoxin is mainly via the disruption of the sphingolipid metabolism. This mechanism is reflected in effects on cell growth and differentiation, in cell death (apoptosis) and carcinogenicity (Yazar & Omurtag, 2008). Fumonisins have often been found to be involved in liver and kidney toxicity; they have been shown to be hepatocarcinogenic in male rats and female mice and nephrocarcinogenic in male rats (Mazzoni et al., 2011). Purified FB1 has been shown to cause Equine Leukoencephalomalacia (ELEM) and Porcine Pulmunary Edema (PPE). In most animal species, the main target organs for FB1 are the liver and kidneys (Richard, 2007; Yazar & Omurtag, 2008; Wan Norashima et al., 2009). There is no carryover of fumonisins into milk in cattle and there appears to be little absorption of them in tissues (Richard et al., 2007). The high incidences of esophageal cancer in the Transkei region of South Africa, in northern Italy and in China have been linked to the ingestion of fumonisin contaminated maize; recent findings suggest that fumonisins might increase the risk of neural tube defects in populations that consume large amounts of contaminated maize (Mazzoni et al., 2011; Yazar & Omurtag, 2008).

6. Patulin

Patulin (PAT) was discovered in 1943 in relation to *P. griseofulvum* and *P. expansum*. The molecule was first studied as a potential antibiotic, but the subsequent research demonstrated its toxicological properties (Baert et al., 2007; Birkinshaw et al., 1943). PAT is produced by several species of *Aspergillus*, *Penicillium*, among these, *P. expansum* is the most relevant. In fact, almost all *P. expansum* isolates are PAT producers (Puel et al., 2010). This mycotoxins can be found in different food products and raw materials, but apples and apple by-products are of greatest concern regarding PAT accumulation: the frequency of contamination in other food resources and products is much lower than in apple processing (Moake et al., 2005).

PAT has been reported to be mutagenic and to cause neurotoxic, immunotoxic, genotoxic and gastrointestinal effects in rodents; therefore, there is some concern that similar effects may occur in humans as a consequence of the long-term consumption of contaminated food or beverages (Hopkins, 1993).

PAT, 4-Hydroxy-4H-furo[3,2-c]pyran-2(6H)-one, is a water-soluble unsaturated heterocyclic lactone (Figure 5). The biosynthesis involves a series of condensation and redox reactions. The pathway consists of approximately 10 steps, as suggested from several biochemical studies and from the identification of several mutants that are blocked at various steps in the PAT biosynthetic pathway. A cluster of 15 genes involved in PAT biosynthesis, containing characterized enzymes, a regulation factor and transporter genes, has recently been reported (Puel et al., 2010).

PAT is a colourless and crystalline low-molecular weight compound, which is relatively heat resistant , with a melting point of 110 °C and a maximum UV absorption at 276 nm. It is soluble in water, ethanol, ethyl acetate, chloroform and acetone, while it is weakly soluble in ethyl ether and benzene and insoluble in petroleum ether, pentane and hexane (Pohland et al., 1982). PAT is unstable in a basic solution and stable in acidic media; in sulfurous compound solutions, the instability is accompanied by the loss of biological activity (Harrison, 1989). It is stable at pH values ranging between 3,0 and 6,5: if the pH is higher, the lactone ring is opened and the toxic effect is lost (Janotovà et al., 2011).

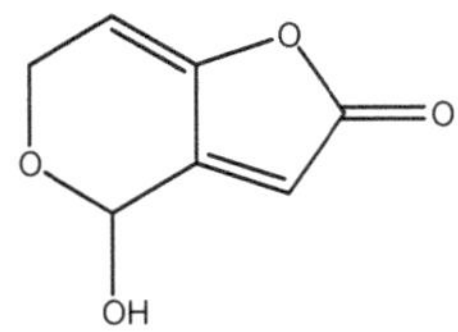

Fig. 5. Chemical structure of patulin.

6.1 Fungi

PAT has been isolated from several species of *Penicillium*, *Aspergillus*, *Paecilomyces* and *Byssochlamys* (Puel et al., 2010). Recent studies based on HPLC-DAD (High Pressure Liquid Chromatography-Diode Array Detector) or LC-MS (Liquid Chromatography-Mass Spectometry) analysis of secondary metabolites, have established the reliable PAT producing species, which are listed in Table 5.

Aspergillus, Clavati group (Varga et al., 2007)
A. clavatus; A. giganteus; A. longivesica
Penicillium (Frisvad et al., 2004)
P. carneum; P. clavigerum; P. concentricum; P. coprobium; P. dipodomyicola; P. expansum; P. glandicola; P. gladioli; P. griseofulvum; P. marinum; P. paneum; P. sclerotigenum; P. vulpinum
Paecylomyces (Samson et al., 2009)
Paecylomyces saturatus
Byssochlamys (Samson et al., 2009)
B. nivea

Table 5. Patulin producing fungi.

6.2 Food

Patulin-producing strains have been isolated from a variety of fruit and vegetables and both pastorized and unpastorized related products, but within the food industry, apples and apple products are of predominant concern as far as PAT contamination (Moake et al., 2005; Sant'Ana et al., 2008). PAT occurs mostly in apples evidently mould-damaged fruit, but sometimes fungal growth can occur internally, as a consequence of various kinds of damage, and can result in the occurrence of PAT in externally undamaged fruit. Therefore, apples must be handled adequately before and during processing to avoid all kinds of damage. It is also fundamental to reduce the possibility of contamination by eliminating mouldy fruit and taking particular care when cleaning containers (Codex, 2003b; Food and Agriculture Organisation of the United Nations [FAO], 2003). In terms of apple storage conditions, in general *P. expansum* shows psychrotrophic characteristics, in fact it is able to growth and produce PAT under refrigerated storage, but different strains show different capacities to produce PAT in different storage conditions (refrigeration and or controlled atmosphere) (Lovett et al., 1975; Paster et al., 1995; Taniwaki et al., 1989). The elimination of mouldy fruit is fundamental during storage because the greater the percentage of damaged fruit in a stored batch, the greater the amount of PAT in the derived products. It has also been shown that the concentration of PAT in deck stored apples increases with storage time (FAO, 2003; Sydenham et al., 1997). In order to improve storage under refrigeration and in a controlled atmosphere against fungal growth and PAT production, additive treatments can be employed, including the use of sanitizers, natural or biological agents or a combination of the two (Chen et al., 2004). Another alternative is the use of polyethylene (PE) packages, with or without a controlled atmosphere, during storage and transport (Moodley et al., 2002). PAT can be reduced in stored apples through a washing stage with tap water, or tap water with active chlorine, or with highly pressurized water; the decrease percent depends on the initial amount of mycotoxin. The use of pressurized water makes it possible to remove the rotten parts of the fruit and also to reduce the fungal population, but it can also suspend and disperse PAT and spores in the washing water because it disturbs the rotten areas (Acar et al., 1998; Marin et al., 2006; Sydenham et al., 1997). Of all the apple products, apple juice is the most important source of PAT in the human diet throughout the world (World Health Organisation [WHO], 1995); the main steps of this production are summarized in Figure 6.

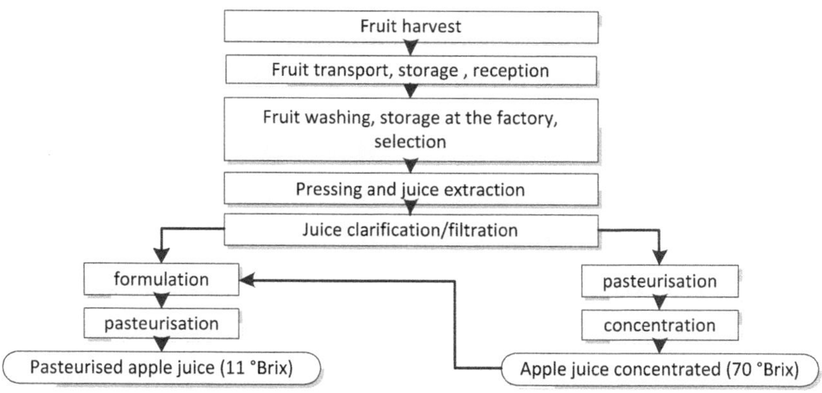

Fig. 6. Apple juice processing steps (modified from Sant'Ana et al., 2008)

PAT can be removed from juice by means of stirring or filtering through granulated activated carbon (Kadakal & Nas, 2002); the obtained percent of PAT reduction depends on the type of carbon, type of activation (physical or chemical), the solid content of the juice and the contact time (Leggott et al., 2001). As far as the heat treatments of juice, it is known that PAT is heat stable in acidic environments; nowadays various research and controversial results exist concerning the effect of the first pasteurization of the juice on the toxin (Kadakal & Nas, 2003). Experimental studies, on various combinations of temperature/time, generally demonstrate the heat stability of PAT to various time/temperature binomials (e.g. 80°C for 30 min, 100°C for 15 min at pH 2.0). Moreover, these studies show that if the contamination is high in the initial processing stages, it will be practically impossible to obtain significant reductions in the level of PAT. On the other hand, despite the studies showing no significant reduction in PAT in apple juice after pasteurization, the destruction of the spores of *P. expansum* reduces the risk of the subsequent production of this mycotoxin (FAO, 2003). Vacuum distillation is usually adopted for the concentration step of the juice and it can allow a reduction in the PAT level because of the time and temperature

exposition. Possibly PAT transformation occurs, while PAT removal to volatile phase is unprobable. As regard that, the results obtained in various studies are controversial, and in some cases show a certain reduction while in other no changes are observed (Kadakal & Nas, 2003; Leggott et al., 2000). The PAT levels in formulated juices may be affected by adding ingredients such as ascorbic acid, thiamine hydrochloride, pyridoxine hydrochloride and calcium pantothenate (Yazici & Velioglu, 2002). Nevertheless, the use of these additives has some limitations; as regard ascorbic acid, its use is influenced by the storage conditions and if it is oxidized, no further degradation of PAT is observable (Drusch et al., 2007). Other possible additives are sulphur dioxide, sodium benzoate and potassium sorbate (Lennox & McElroy, 1984; Roland et al., 1984), but the current demand for healthy food, free of additives, could result in an impediment to the use of such techniques. Thus, it is preferable to use treatments that guarantee the elimination/inactivation of the ascospores of the heat resistant fungi (such as filtration with diatomaceous earth) than to apply these additives.
It can be said that, although the juice manufacturing process stages are capable of reducing the amount of PAT in the final products to a certain extent, the incidence of this mycotoxin throughout the World confirms its stability to some degree; when faced with the techniques currently in use, only the adoption of adequate controls to reduce the incidence of fruit damage and rot, during pre-harvest, harvest and post-harvest, can lead to an important reduction in the final product, whether it is fruit for direct consumption or one of the various fruit products (Sant'Ana et al., 2008).

6.3 Toxicity

The health risks of PAT for humans include acute and chronic symptom and effects at a cellular level.
Some of the acute toxic signs that have consistently been reported are agitation, convulsions, dyspnea, pulmonary congestion, edema, and ulceration, hyperemia and distension of the gastro intestinal tract (WHO, 1995). Sub-acute toxicity has also been indicated: PAT is recognized to mainly induce gastrointestinal disorders; it has mainly been studied in rats, where it has been shown to induce weight loss, gastric and intestinal changes and alterations in the renal function (Puel et al., 2010).
PAT is genotoxic; most assays carried out with mammalian cells have been positive while those with bacteria have mainly been negative. Some studies have indicated that PAT impairs DNA synthesis. These effects might be related to the ability PAT to react with sulphydryl groups and to induce oxidative damage (Liu et al., 2007). The IARC has placed PAT in group 3, as "not classifiable as to its carcinogenicity to humans" (IARC, 1986). PAT can also alter the immune response of the host (Oswald & Comera, 1998). Several in vitro studies have demonstrated that PAT inhibits various macrophage functions. PAT has also been found to reduce the cytokine secretion of IFN-γ and IL-4 by human macrophages and of IL-4, IL-13, IFN-γ, and IL-10 by human peripheral blood mononuclear cells and human T cells (Luft et al., 2008; Wichmann et al., 2002). In vivo studies using mice have indicated variable effects of PAT on the immune system, such as an increased number of splenic T lymphocytes and depressed serum immunoglobulin concentrations (Escoula et al., 1988; Paucod et al., 1990). As regard humans, exposure to PAT, at levels that are consistent with potential human exposure in food, would not be likely to alter immune responses (Llewellyn et al., 1998). When injected into the air cell of chick eggs, PAT is found embryotoxic, depending on the age of the embryo, and teratogenic (Ciegler et al., 1976). PAT can induce a reduction in the protein and DNA content, in the yolk sac diameter, crown rump length, and somite

number count; it can also increase the frequency of defective embryos. Anomalies can include growth retardation, hypoplasia of the mesencephalon and telencephalon, and hyperplasia and/or blisters of the mandibular process (Smith et al., 1993).

At a cellular level, PAT is believed to cause cell damage by forming adducts with thiol-containing cellular components (Barhoumi & Burghardt, 1996); in fact, many enzymes with a sulfhydryl group in their active site are sensitive to PAT. PAT has also been shown to induce intra- and intermolecular protein cross-links (Fliege & Metzler, 1999). Finally, PAT can interact directly with DNA and RNA inhibiting transcription and translation (Lee & Roschenthaler, 1987).

7. References

Abramson, D.; House, J.D. & Nyachoti, C.M. (2005). Reduction of deoxynivalenol in barley by treatment with aqueous sodium carbonate and heat. *Mycopathologia*, Vol. 160, pp. 297–301.

Abrunhosa, L.; Robert R.; Paterson, M. & Venâncio, A. (2010). Biodegradation of Ochratoxin A for food and feed decontamination. *Toxins*, Vol. 2, pp. 1078-1099.

Acar, J.; Gökmen, V. & Taydas, E. E. (1998). The effects of processing technology on the patulin content of juice during commercial apple juice concentrate production. *Zeitschrift fur Lebensmittel-Untersuchung und -Forschung A*, Vol. 207, pp. 328–331.

Baert, K.; Devlieghere, F.; Flyps, H.; Oosterlinck, M.; Ahmed, M.M.; Rajković, A.; Verlinden, B.; Nicolaï, B.; Debevere, J. & De Meulenaer, B. (2007). Influence of storage conditions of apples on growth and patulin production by *Penicillium expansum*. *International Journal of Food Microbiology*, Vol. 119, pp. 170–181.

Barhoumi, R. & Burghardt, R.C. (1996). Kinetic Analysis of the Chronology of Patulin- and Gossypol-lnduced Cytotoxicity in Vitro. *Fundamental and Applied Toxicology*, Vol. 30, pp. 290-297.

Bakirci, I. (2001). A study on the occurrence of aflatoxin M1 in milk and milk products produced in Van province in Turkey. *Food Control*, Vol. 12, 47-51.

Bankole, S.A.; Adenusi, A.A.; Lawal, O.S. & Adesanya, O.O. (2010). Occurrence of aflatoxin B1 in food products derivable from 'egusi' melon seeds consumed in southwestern Nigeria. *Food control*, Vol. 21, pp. 974-976.

Barbiroli, A.; Bonomi, F.; Benedetti, S.; Mannino, S.; Monti, L.; Cattaneo, T. & Iametti, S. (2007). Binding of Aflatoxin M1 to Different Protein Fractions in Ovine and Caprine Milk. *Journal of Dairy Science*, Vol. 90, pp. 532-540.

Battacone, G.; Nudda, A. & Pulina, G. (2010). Effects of Ochratoxin A on Livestock Production. *Toxins*,Vol. 2, pp. 1796-1824.

Baxter, E.D.; Slaiding, I.R. & Kelly, B. (2001). Behavior of ochratoxin A in brewing. *Journal of the American Society of Brewing Chemists*, Vol. 59, pp. 98–100.

Bennett, J.W. & Klich, M. (2003). Mycotoxins. *Clinical Microbiology Review*, Vol. 16, pp. 497–516.

Bezuidenhout, S.C.; Gelderblom, W.C.A.; Gorst-Allman, C.P.; Horak, R.M.; Marasas, W.F.O.; Spiteller, G. & Vleggaar, R. (1988). Structure elucidation of the fumonisins, mycotoxins from *Fusarium moniliforme*. *Journal of the Chemical Society, Chemical Communication*, Vol. 11, pp. 743-745.

Bianchini, A. & Bullerman, L.B. (2010). Biological control of molds and mycotoxins in foods. *In mycotoxin prevention and control in agriculture*. ACS symposium series, American Chemical Society, Washington, DC, 2010, pp 1-16.

Birkinshaw, J.H; Michael, S.E.; Bracken, A. & Raistrick, H. (1943). Patulin in the common cold collaborative research on a derivative of *Penicillium patulum* Bainier. II. Biochemistry and Chemistry. *Lancet*, Vol. 245, pp. 625.

Blandino, M. & Reyneri, A. (2008). Effect of Maize Hybrid Maturity and Grain Hardness on fumonisin and zearalenone contamination. *Italian Journal of Agronomy*, Vol. 2, pp. 107-117.

Blount, W.P. (1960). Disease of turkey poults. *Veterinary Record*, Vol. 72, pp. 786.

Bottalico, A. & Perrone, G. (2002). Toxigenic *Fusarium* species andmycotoxins associated with head blight in small-grain cereals in Europe. *European Journal Plant Pathology*, Vol. 108, pp. 611–624.

Boudra, H.; Le Bars, P. & Le Bars, J. (1995). Thermostability of ochratoxin A in wheat under two moisture conditions. *Applied and Environmental Microbiology*, Vol. 61, pp. 1156–1158.

Bozoglu F. (2009) Different mycotoxin inactivation application and their inactivation mechanisms. *Proceeding for Natural Science* Matica Srpska Novi Sad, Vol. 117, pp. 27-35.

Brackett R.E. & Marth, E.H. (1982). Association of aflatoxin M1 with casein. *Zeitschrift fur Lebensmitteluntersuchung und – Forschung A*. Vol. 174, pp. 439-441.

Bryden, W.L. (2007). Mycotoxins in the food chain: human health implications. *Asia Pacific Journal of Clinical Nutrition*, Vol. 16, pp. 95-101.

Bullerman, L.B. & Bianchini, A. (2007). Stability of mycotoxins during food processing. *International Journal of Food Microbiology*, Vol. 119, pp. 140–146.

Cabañes, F.J.; Accensi, F.; Bragulat, M.R.; Abarca, M.L.; Castella, G.; Minguez, S. & Pons, A. (2002). What is the source of Ochratoxin A in wine? *International Journal of Food Microbiology*, Vol. 79, pp. 213–215.

Cabañes, F.J.; Bragulat, M.R. & Castellá, G. (2010).Ochratoxin A Producing Species in the Genus *Penicillium*. Toxins, Vol. 2, pp. 1111-1120;

Cark, H.A. & Snedeker, S.M. (2006). Ochratoxin A: its cancer risk and potential for exposure, *Journal of Toxicology and Environmental Health B*, Vol. 9 , pp. 265–296.

Carnaghan, R.B.; Hartley, R.D. & O'Kelly, J. (1963). Toxicity and Fluorescence Properties of the Aflatoxins. *Nature*, Vol. 200, pp. 1101.

CAST (Council for Agricultural Science and Technology). Mycotoxins: Economic and Health Risks.Ames, Iowa, USA.Vol. IA, pp. 116, 1989.

CAST(2003). Mycotoxins: Risks in Plant, Animal and Human Systems. Task Force Report N°139, Ames, Iowa, USA.

Castelo, M.M.; Sumner, S.S. & Bullerman, L.B. (1998). Stability of fumonisins in thermally processed corn products. *Journal of Food Protection*, Vol. 161, pp. 1030-1033.

Chen, L.; Ingham, B. H. & Ingham, S. C. (2004). Survival of *Penicilliumexpansum*and patulin production on stored apples after wash treatments. *Journal of Food Science*, Vol. 69, No. 8, pp. 669–675.

Ciegler, A. (1972). Bioproduction of ochratoxin A and penicillc acid by members of the *Aspergillusochraceus* group. *Canadian Journal of Microbiology*, Vol. 18, pp. 631–636.

Ciegler, A.; Beckwith, A.C. & Jackson, L.K. (1976). Teratogenicity of patulin and patulin adducts formed with cysteine. Applied and Environmental Microbiology, Vol. 31, pp. 664-667.

CODEX (Codex Alimentarius Commission). (2003b). Code of practice for the prevention and reduction of patulin contamination in apple juice and apple juice ingredients in other beverages. CAC/RCP, Vol. 50, pp. 1–6.

Coulombe, R. A. (1993). Biological action of mycotoxins. *Journal of Dairy Science*, Vol. 76, pp. 880-891.

Creppy, E.E.; Betbeder, A.M.; Gharbi, A.; Counord, J.; Castegnaro, M.; Bartsch, H.; Moncharmont, P.; Fouillet, B.; Chambon, P. & Dirheimer, G. (1991). Human ochratoxicosis in France. In Mycotoxins, Endemic Nephropathy and Urinary Tract Tumours; Castegnaro, M., Plestina, R., Dirheimer, G., Chernozemsky, I.N., Bartsch, H., Eds.; IARC Scientific Publ.: Lyon, France, 1991.

Creppy, E. E. (2002). Update of survey, regulation and toxic effects of mycotoxins in Europe. *Toxicology Letters*, Vol. 127, pp. 19–28.

Dalvi, R.R. (1986). An overview of aflatoxicosis of poultry: its characteristics, prevention and reduction. *Veterinary Research Communications*, Vol. 10, pp. 429–443.

Desjardins, A.E. & Proctor, R.H. (2007). Molecular biology of *Fusarium*mycotoxins. *International Journal of Food Microbiology* ,Vol. 119, pp. 47–50.

Dragacci, S.; Gleizes, E.; Fremi, J.M. & Candlish, A.A.G. (1995) Use of immunoaffinity chromatography as a purification step for the determination of aflatoxin M1 in cheeses. *Food Additives and Contaminants*, Vol. 12, pp. 59-65.

Drusch, S.; Kopka, S. & Kaeding, J. (2007). Stability of patulin in a juice-like aqueous model system in the presence of ascorbic acid. *Food Chemistry*, Vol. 100, pp. 192–197.

Duarte, S.C.; Pena, A. & Lino, C.M. (2010). A review on ochratoxin A occurrence and effects of processing of cereal and cereal derived food products. *Food Microbiology*, Vol. 27, pp. 187–198.

Elgerbi, A.M.; Aidoo, K.E.; Candlish, A.A.G. & Tester, R.F. (2004). Occurrence of aflatoxin M1 in randomly selected North African milk and cheese samples. *Food Additives and Contaminants*, Vol. 21, pp. 592-597.

El Khoury, A. & Atoui, A. (2010). Ochratoxin A: General Overview and Actual Molecular Status. *Toxins*, Vol. 2, pp. 461-493.

El Khoury, A.; Rizk, T.; Lteif, R.; Azouri, H.; Delia, M.L. & Lebrihi, A. (2006). Occurrence of ochratoxin A and aflatoxin B1-producing fungi in Lebanese grapes and OTA content in musts and finished wines during year 2004. *Journal of Agriculture and Food Chemistry*,Vol. 54, pp. 8977–8982.

Elling, F.; Nelsen, J.P.; Lillehoj, E.B.; Thomassen, M.S. & Stormer, F.C. (1985). Ochratoxin A induced porcine nephropathy: Enzyme and ultrastructural changes after short-term exposure. *Toxicon*, Vol. 23, pp. 247–254.

Escoula, L.; Thomsen, M.; Bourdiol, M.; Pipy, B.; Peuriere, S. & Roubinet, S. (1988). Patulin immunotoxicology: effect on phagocyte activation and the cellular and humoral immune system of mice and rabbits. *International Journal of Immunopharmacology*, Vol. 10, pp. 983-989.

Etzel R. (2002). Mycotoxins. *Journal American Medical Association*, Vol. 287, pp. 425-427.

Fallah, A.A. (2010). Assessment of aflatoxin M1 contamination in pasteurized and UHT milk marketed in central part of Iran. *Food and Chemical Toxicology*, Vol. 48, pp. 988–991.

FAO (Food and Agriculture Organization of the United Nations). (2003). Manual on the application of the HACCP system in mycotoxin prevention and control. FAO Food and Nutrition Paper, Vol. 73, pp. 1-124.

Feier, D. & Tofana, M. (2009). Ochratoxin A–Toxicological Aspects. *Bulletin of University of Agricultural Sciences and Veterinary Medicine Cluj-Napoca. Agriculture*, Vol. 66, pp. 308–312.

Fliege, R. & Metzler, M. (1999). The mycotoxin patulin induces intra- and intermolecular protein crosslinks in vitro involving cysteine, lysine, and histidine side chains, and α-amino groups. *Chemical Biological Interactions*, Vol. 123, pp. 85–103.

Foroud, N.A. & Eudes, F. (2009). Trichothecenes in Cereal Grains. *International Journal of Molecular Science*, Vol. 10, pp. 147-173.

Freeman, G.G. & Morrison, R.I. (1949). The isolation and chemical properties of trichothecin, an antifungal substance from *Trichothecium roseum* Link. *Biochemical Journal* , Vol. 44, pp. 1.

Frisvad, J.C.; Smedsgaard, J.; Larsen, T.O. & Samson, R.A. (2004). Mycotoxins, drugs and other extrolites produced by species in *Penicillium*subgenus *Penicillium*. *Studies in Mycology*, Vol. 49, pp.201–241.

Frobish, R. A.; Bradley, B. D.; Wagner, D. D.; Long-Bradley, P. E. & Hairstone, H. (1986). Aflatoxin residues in milk of dairy cows after ingestion of naturally contaminated grain. *Journal of Food Protection*, Vol. 49, pp. 781–785.

Galvano, F.; Piva, A.; Ritieni, A. & Galvano, G.(2001). Dietary strategies to counteract the effects of mycotoxins: a review. *Journal of Food Protection*, Vol. 64, pp. 120–131.

Gold, L.S.; Sawyer, C.B.; Magaw, R.; Backman, G.M.; de Veciana, M.; Levinson, R.; Hooper, N.K.; Havender, W.R.; Bernstein, L. & Peto, R. (1984). A carcinogenic potency database of the standardized results of animal bioassays. *Environmental Health Perspectives*, Vol. 58, pp. 9–319.

Goto, T.; Wicklow, D.T. & Ito, Y. (1996). Aflatoxin and cyclopiazonic acid production by a sclerotium-producing *Aspergillus tamarii* strain. *Applied and Environmental Microbiology*, Vol. 62, pp. 4036-4038.

Gourama, N., & Bullerman, L. B. (1995). *Aspergillusflavus* and *Aspergillusparasiticus*: aflatoxigenic fungi of concern in foods and feeds: a review. *Journal of Food Protection*, Vol. 58, pp. 1395-1404.

Govaris A.; Roussi V.; Koidis P.A. & Botsoglou N.A. (2001).Distribution and stability of aflatoxin M1 during processing, ripening and storage of Telemes cheese. *Food Additives and Contaminants*, Vol. 18, pp. 437-443.

Groopman, J.D.; Cain, L.G. & Kensler, T.W. (1988). Aflatoxin exposure in human populations: measurements and relationship to cancer. *Critical Reviews in Toxicology*, Vol. 19, pp. 113–145.

Gurbay, A. S.; Engin, A. B.; Çaglayan, A. & Sahin, G. (2006). Aflatoxin M1 levels in Commonly consumed cheese and yoghurt samples in Ankara, Turkey. *Ecology of Food and Nutrition*, Vol. 45, pp. 449-459.

Harrison, M. A. (1989). Presence and stability of patulin in apple products: A review. *Journal of Food Safety*, Vol. 9, pp. 147–153.

Hassen, W.; Abid-Essafi , S.; Achour, A.; Guezzah, N.; Zakhama, A.; Ellouz, F.; Creppy, E.E. & Bacha, H. (2004). Karyomegaly of tubular kidney cells in human chronic interstitial nephropathy in Tunisia: Respective role of Ochratoxin A and possible genetic predisposition. Human & Experimental Toxicology, Vol. 23, pp. 339–346.

Hazel, C.M. & Patel, S. (2004). Influence of processing on trichothecenelevels. *Toxicology Letters*, Vol. 153, pp. 51–59.

He, J.; Zhou, T.; Young, J.C.; Boland, G.J. & Scott, P.M. (2010). Chemical and biological transformation for detoxification of trichothecenemycotoxins in human and animal food chains: a review. *Trends in food & science Technology*, Vol. 21, pp. 67-76.

Hesseltine, C.W.; Vandegraft, E.E.; Fennell, D.I.; Smith, M.L. & Shotwell, O.L. (1972). Aspergilli as ochratoxin producers. *Mycologia*, Vol. 64, pp. 539–550.

Heussner, A.H.; Dietrich, D.R. & O'Brien, E. (2006). *In vitro* investigation of individual and combined cytotoxic effects of ochratoxin A and other selected mycotoxins on renal cells. *Toxicology in Vitro*, Vol. 20, pp. 332–341.

Hill, R.A.; Blankenship, P.D.; Cole, R.J. & Sanders, T.H. (1983). Effects of Soil Moisture and Temperature on Preharvest Invasion of Peanuts by the *Aspergillusflavus* Group and Subsequent Aflatoxin Development. *Applied and Environmental Microbiology*, Vol. 45, pp. 628-633.

Holmberg, T.; Thuvander, A. & Hult, K. (1988). Ochratoxin A as a suppressor of mitogen-induced blastogenesis of porcine blood lymphocytes. *ActaVeterinariaScandinavica*, Vol. 29, pp. 219–223.

Hopkins, J. (1993). The toxicological hazards of patulin. *Food and Chemical Toxicology*, Vol. 31, pp. 455–456.

Huff, W.E.; Kubena, L.F.; Harvey, R.B. & Phillips, T.D. (1992). Efficacy of hydrated sodium calcium aluminosilicate to reduce the individual and combined toxicity of aflatoxin and ochratoxin A. *Poultry Science*, Vol. 71, pp. 64–69.

Huffman, J.; Gerber, R. & Du, L. (2010). Recent advancement in the biosynthetic mechanism for polyketide-derived mycotoxins. *Biopolymers*, Vol. 93, pp. 764-776.

Hwan Do, J. & Choi, D. (2007). Aflatoxins: Detection, Toxicity, and Biosynthesis. *Biotechnology and Bioprocess Engineering*, Vol. 12, pp. 585-593.

IARC (International Agency for Research on Cancer). (1986). Summaries & Evaluations, Patulin. Vol. 40, p. 83.

Jackson, L.S.; Hlywka, J.J.; Senthil, K.R. &Bullerman, L.B. (1996). Effect of thermal processing on the stability of fumonisins. *Advances in Experimental Medicine and Biology*, Vol. 392, pp. 345-353.

Janotová, L.; Cízková, H.; Pivonka, J. & Voldrich, M. (2011). Effect of processing of apple puree on patulin content. *Food Control*, Vol. 22, pp. 977-981.

JECFA. Joint FAO/WHO Expert Committee on Food Additives. Safety evaluation of certain food additives and contaminants. World Health Organization, Geneva,1998.

JECFA. Evaluation of certain additives and contaminants – sixty-eighth report of the Joint FAO/WHO Expert Committee on Food Additives. In: WHO Food Additives Series 59. Joint FAO/WHO Expert Committee on Food Additives. 2008.

http://whqlibdoc.who.int/publications/2007/9789241209472_eng.pdf. Accessed 17 Nov 2008.

Kadakal, C. & Nas, S. (2002). Effect of activated charcoal on patulin, fumaric acid, and some other properties of apple juice. *Nahrung/Food*, Vol. 46, pp. 31–33.

Kadakal, C. & Nas, S. (2003). Effect of heat treatment and evaporation on patulin and some other properties of apple juice. *Journal of the Science of Food and Agriculture*, vol. 83, pp. 987–990.

Kamkar, A. (2008). The study of aflatoxin M1 in UHT milk samples by ELISA. *Journal of Veterinary Research*, Vol. 63, pp. 7–12.

Karlovsky, P. (2011). Biological detoxification of the mycotoxindeoxynivalenol and its use in genetically engineered crops and feed additives. *Applied Microbiology Biotechnology*, Vol. 91, pp. 491–504.

Keeper-Goodman, T. & Scott, P.M. (1989). Risk assessment of the mycotoxinochratoxin. *Biomedical and Environmental Sciences*, Vol. 2, pp. 179–248.

Kimura, M.; Tokai, T.; Takahashi-Ando, N.; Ohsato, S. & Fujimura, M. (2007). Molecular and genetic studies of *Fusarium*trichothecene biosynthesis: pathways, genes, and evolution. *Bioscience Biotechnology Biochemistry*,Vol. 71, pp. 2105–2123.

Kumar, V.; Basu, M.S. & Rajendran, T.P. (2008). Mycotoxins research and mycroflora in some commercially important agricultural commodities. *Crop Protection*, Vol. 27, pp. 891-905.

Lea, T., Steien. K. & Stormer. C. (1989). Mechanism of ochratoxin A induced immunosuppresion. *Mycopathologia*, Vol. 107, pp. 153–159.

Lee, K.S. & Röschenthaler, R. (1987). Strand scissions of DNA by patulin in the presence of reducing agents and cupric ions. Journal of Antibiotics, Vol. 40, pp. 692-696.

Leggott, N. L., Sheppard, G. S., Stockenström, S.; Staal, E. & van Schalkwyk, D. J. (2001). The reduction of patulin in apple juice by three different types of activated carbon. *Food Additives and Contaminants*, Vol. 18, pp. 825–829.

Leggott, N. L.; Vismer, H. F.; Sydenham, E. W.; Shephard, G. S.; Rheeder, J. P. & Marasas, W. F. O. (2000). Occurrence of patulin in the commercial processing of apple juice. *South African Journal of Science*, Vol. 96, pp. 241–243.

Lennox, J. E. & McElroy, L. J. (1984). Inhibition of growth and patulin synthesis in *Penicilliumexpansum* by potassium sorbate and sodium propionate in culture. *Applied and Environmental Microbiology*, Vol. 48, pp. 1031–1033.

Liu, B.; Wu, T.; Yu, F. & Su, C. (2007). Induction of oxidative stress response by the mycotoxin patulin in mammalian cells. *Toxicological Sciences*, Vol. 95, pp. 340-347.

Llewellyn, G.C.; McCay, J.A.; Brown, R.D.; Musgrave, D.L.; Butterworth, L.F.; Munson, A.E.; & White, K.L. Jr. (1998). Immunological evaluation of the mycotoxin patulin in female B6C3F$_1$ mice. *Food and Cosmetics Toxicology*, Vol. 36, pp. 1107-1115.

Lopez, C.; Ramos, L.; Ramada´n, S.; Bulacio, L. & Perez, J. (2001). Distribution of aflatoxin M1 in cheese obtained from milk artificially contaminated.*International Journal of Food Microbiology*, Vol. 64, pp. 211–215.

Lovett, J.; Thompson, R. G. & Boutin, B. K. (1975). Patulin production in apples stored in a controlled atmosphere. *Journal of AOAC*, Vol. 58, pp. 912–914.

Luft, P.; Oostingh, G.J.; Gruijthuijsen,Y.; Horejs-Hoeck, J.; Lehmann, I. & Duschl,A. (2008). Patulin influences the expression of Th1/Th2 cytokines by activated peripheral blood mononuclear cells and T cells through depletion of intracellular glutathione. *Environmental Toxicology*, Vol. 23, pp. 84–95.

Magan, N.; Hope, R.; Cairns, V. & Aldred, D. (2003). Post-harvest fungal ecology: impact of fungal growth and mycotoxin accumulation in stored grain. *European Journal of Plant Pathology*, Vol. 109, pp. 723–730.

Magan, N. & Aldred, D. (2005). Conditions of formation of ochratoxin A in drying, transport and in different commodities. *Food Additives and Contaminants*, Vol. 1, pp. 10–16.

Magan, N. (2006). Mycotoxin contamination of food in Europe: early detection and prevention strategies. *Mycopathologia*, Vol. 162, pp. 245–253.

Magan, N. & Aldred, D. (2007). Post-harvest control strategies: minimizing mycotoxins in the food chain. *International Journal of Food Microbiology*, Vol. 119, pp. 131–139.

Marin, S.; Morales, H.; Hasan, H. A.; Ramos, A. J. & Sanchis, V. (2006). Patulin distribution in Fuji and Golden apples contaminated with *Penicilliumexpansum*. *Food Additives and Contaminants*, Vol. 23, pp. 1316–1322.

Martins M.L. & Martins H.M. (2000). Aflatoxin M1 in raw and ultra-high temperature-treated milk commercialized in Portugal. *FoodAdditives and Contaminants*, Vol. 17, pp. 871-874.

Mazzoni, E.; Scandolara, A.; Giorni, P.; Pietri, A. & Battilani, P. (2011). Field control of *Fusarium* ear rot, *Ostrinianubilalis*(Hübner), and fumonisins in maize kernels. *Pest Management Science*, Vol. 67, pp. 458–465.

Merhej, J.; Richard-Forget, F. & Barreau, C. (2011). Regulation of trichothecene biosynthesis in *Fusarium*: recent advances and new insights. *Applied Microbiology Biotechnology*, Vol. 91, pp. 519–528.

Milićević, D. R.; Škrinjar, M. & Baltić, T. (2010). Real and Perceived Risks for Mycotoxin Contamination in Foods and Feeds: Challenges for Food Safety Control. *Toxins*, Vol. 2, pp. 572-592.

Moake, M.M.; Padilla-Zakour, O.I. & Worobo, R.W. (2005).Comprehensive Review of Patulin Control Methods in Foods. *Comprehensive Reviews in Food Science and Food Safety*, Vol. 1, pp. 8-21.

Molinie´, A.; Faucet, V.; Castegnaro, M. & Pfohl-Leszkowicz, A. (2005). Analysis of some breakfast cereals on the French market for their contents of ochratoxin A, citrinin and fumonisin B1: development of a method for simultaneous extraction of ochratoxin A and citrinin. *Food Chemistry*, Vol. 92, pp. 391–400.

Moodley, R. S.; Goviden, R. & Odhav, B. (2002). The effect of modified atmospheres and packaging on patulin production in apples. *Journal of Food Protection*, Vol. 65, pp. 867–871.

Moss, M.O. (2002a). Risk assessment for aflatoxins in foodstuffs. *International Biodeterioration& Biodegradation*, Vol. 50, pp. 137-142.

Moss, M.O. (2002b). Mycotoxin review - 1. *Aspergillus* and *Penicillium*. *Mycologist*, Vol. 16, No. 3, pp. 116-119.

Munkvold, G.P. (2003). Cultural and genetic approaches to managing mycotoxins in maize. *Annual Review of Phytopathology*, Vol. 41, pp. 99-116.

Mwanda, O.W.; Otieno, C.F. & Omonge, E. (2005). Acute aflatoxicosis: case report. *East Africa Medical Journal*, Vol. 82, pp. 320-324.

Nguefacka, J.; Leth, V.; AmvamZollo, P.H. & Mathur, S.B. (2004). Evaluation of five essential oils from aromatic plants of Cameroon for controlling food spoilage and mycotoxin producing fungi *International Journal of Food Microbiology*, Vol. 94, pp. 329- 334.

Nicholson, P.; Gosman, N.; Draeger, R. & Steed, A. (2004). Control of *Fusarium* and *Aspergillus* species and associated mycotoxins on wheat and maize. In: Barug D., van Egmond H., Lopez-Garcia R., van Osenbruggen T., Visconti A. (eds.): Meeting the mycotoxin menace, pp. 113-132, Wageningen Academic Publishers, The Netherlands.

Niessen, L. (2007). PCR-based diagnosis and quantification of mycotoxin producing fungi. *International Journal of Food Microbiology*,Vol.119, pp. 38-46.

O'Callaghan, J.; Caddick, M.X. & Dobson, A.D.W. (2003). A polyketide synthase gene required for ochratoxin A biosynthesis in *Aspergillusochraceus*. *Microbiology*, Vol.149, pp. 3485-3491.

Olaniran, A.O.; Mfumo, N.H.; Pillay, D. & Pillay, B. (2006). Synergistic utilization of dichloroethylene as sole carbon source by bacterial consortia isolated from contaminated sites in Africa. *Biotechnology and Bioprocess Engineering*, Vol. 11, pp. 205-210.

Oliveira, C. A.; Rosmaninho, J. & Rosim, R. (2006). Aflatoxin M 1 and cyclopiazonic acid in fluid milk traded in São Paulo, Brazil. *Food Additives and Contaminants*, Vol. 23, pp. 196-201.

Oruc, H.H.; Cengiz, M. & Kalkanli, O. (2006). Comparison of aflatoxin and fumonisin levels in maize grown in Turkey and imported from the USA. *Animal Feed Science and Technology*, Vol. 128, pp. 337-341.

Oswald, I.P. & C. Comera. (1998). Immunotoxicity of mycotoxins. *Review in Medical Veterinary*, Vol. 149, pp. 585-590.

Oveisi, M. R.; Jannat, B.; Sadeghi, N.; Hajimahmoodi, M. & Nikzad, A. (2007). Presence of aflatoxin M1 in milk and infant milk products in Tehran, Iran. *Food Control*, Vol. 18, pp. 1216-1218.

Paster, N.; Huppert, D. & Barkai-Golan, R. (1995). Production of patulin by different strains of *Penicilliumexpansum*in pear and apple cultivars stored at different temperatures and modified atmospheres. *Food Additives and Contaminants*, Vol. 12, pp. 51–58.

Paterson, R.R.M. (2006). Identification and quantification of mycotoxigenic fungi by PCR. *Process Biochemistry*, Vol. 41, pp. 1467–1474.

Paterson, R.R.M. & Lima, N. (2010). How will climate change affect mycotoxins in food? *Food Research International*, Vol. 43, pp. 1902–1914.

Paucod, J.C.; Krivobok, S. & Vidal, D. (1990). Immunotoxicity testing of mycotoxins T-2 and patulin on Balb/C mice. *Acta Microbiologica Hungarica, Vol.* 37, pp. 331-339.

Pestka, J.J. & Smolinski, A.T. (2005). Deoxynivalenol: toxicology and potential effects on humans. *Journal of Toxicology and Environmental Health Part B Critical Reviews*, Vol. 8, pp. 39–69.

Pestka, J.J. (2010). Deoxynivalenol: mechanisms of action, human exposure, and toxicological relevance. *Archives of Toxicology*, Vol. 84, pp. 663–679.

Petzinger, E. & Weidenbach, A. (2002). Mycotoxins in the food chain: The role of ochratoxins. *Livestock Production Science*, Vol. 76, pp. 245–250.

Pfohl-Leszkowicz, A. & Castegnaro, M. (1999). L'Ochratoxine A dans: Mycotoxines: Evaluation et gestion du risqué, Lavoisier; Tec & Doc: Paris, France, pp. 249–278.

Pfohl-Leszkowicz, A. (2009). Ochratoxin A and aristolochic acid involvement in nephropathies and associated urothelial tract tumours. *Archives of Industrial Hygiene and Toxicology*, Vol. 60, pp. 465–483.

Pitt, J.I. (2000).Toxigenic fungi and mycotoxins. *British Medical Bulletin*, Vol. 56 , pp. 184-192.

Pohland, A.E.; Schuller, P.L. & Steyn, P.S. (1982). Physicochemical data for some selected mycotoxins. *Pure and AppliedChemistry*, Vol. 54, pp. 2219-2284.

Prandini, A.; Tansini, G.; Sigolo, S.; Filippi, L.; Laporta, M. & Piva, G. (2009). On occurrence of aflatoxin M1 in milk and dairy products. *Food and Chemical Toxicology*, Vol. 47, pp. 984-991.

Puel, O., Galtier, P. & Oswald, I.P. (2010). Biosynthesis and Toxicological Effects of Patulin. *Toxins*, Vol. 2, pp. 613-631.

Reddy, K.R.N.; Nurdijati, S.B. & Salleh, B. (2010). An overview of plant-derived products on control of mycotoxigenic fungi and mycotoxins. *Asian Journal of Plant Science*, Vol. 9, pp. 126-133.

Ribelin, W.E.; Fukushima, K. & Still, P.E. (1978). The toxicity of ochratoxin A to ruminants. *Canadian Journal of Comparative Medicine*, Vol. 42, pp. 172–176.

Richard, J.L. (2007). Some major mycotoxins and their mycotoxicoses- An overview. *International Journal of Food Microbiology*, Vol. 119, pp. 3–10.

ROC (Reports on Carcinogens), (2003). Aflatoxins. Eleventh ed. U.S. Department of Health and Human Services. Date of access 10/02/11, Available from: http://ntp.niehs.nih.gov/ntp/roceleventh/profiles/s006afla.pdf.

Rocha, O.; Ansari, K. & Doohan, F.M. (2005). Effects of trichothecenemycotoxins on eukaryotic cells: a review. *Food Additive Contamination Part A*, Vol. 22, pp. 369–378.

Roland, J. O.; Beuchat, L. R.; Worthington, R. E. & Hitchcock, H. L. (1984). Effects of sorbate, benzoate, sulfur dioxide and temperature on growth and patulin production by *Byssochlamysnivea* in grape juice. *Journal of Food Protection*, Vol. 47, pp. 237–241.

Rotter, B.; Prelusky, D.B. & Pestka, J.J. (1996). Toxicology of deoxynivalenol (vomitoxin). *Journal of Toxicology and Environmental Health*, Vol. 48, pp. 1–34.

Rustom, I.Y.S. (1997). Aflatoxin in food and feed: Occurrence, legislation and inactivation by physical methods. *Food Chemistry*, Vol. 59, pp. 57–67.

Samson, R.A.; Houbraken, J.; Varga, J. & Frisvad J.C. (2009). Polyphasic taxonomy of the eat resistant ascomycete genus *Byssochlamys*and its *Paecilomyces*anamorphs. *Persoonia*, Vol. 22, pp. 14–27.

Sant'Ana, A.; Rosenthal, A. & de Massaguer, P. (2008). The fate of patulin in apple juice processing: A review. *Food Research International*, Vol. 41, pp. 441–453.

Scott, P.M.; Kanhere, S.R.; Lawrence, G.A.; Daley, E.F. & Farber, J.M. (1995). Fermentation of wort containing added ochratoxin A and fumonisins B1 and B2. Food Additives and Contaminants, Vol. 12, pp. 31–40.

Scudamore, K.A. (2005). Prevention of ochratoxin A in commodities and likely effects of processing fractionation and animal feeds. *Food Additives and Contaminants A*, Vol. 22, pp. 17–25.

Scudamore, K.A.; Banks, J. & MacDonald, S.J. (2003). Fate of ochratoxin A in the processing of whole wheat grains during milling and bread production. Food Additives and Contaminants, Vol. 20, pp. 1153–1163.

Shotwell, O.L.; Hesseltine, C.W. & Goulden M.L. (1969). Ochratoxin A: Occurrence as Natural Contaminant of a Corn Sample. *AppliedMicrobiology*, Vol. 17, pp. 765–766.

Shundo, L.; Navas, S.A.; Lamardo, L.C.A.; Ruvieri, V., & Sabino, M. (2009). Estimate of aflatoxinM1 exposure in milk and occurrence in Brazil. *Food Control*, Vol. 20, pp. 655-657.

Smith, E.E.; Duffus, E.A. & Small, M. H. (1993). Effects of patulin on postimplantation rats embryos. *Archives of Environmental Contamination and Toxicology*, Vol. 25, pp. 267-270.

Sobrova, P.; Adam, V.; Vasatkova, A.; Beklova, M.; Zeman, L. & Kizek, R. (2010). Deoxynivalenol and its toxicity. *Interdisciplinary Toxicology*, Vol. 3, pp. 94–99.

Soleas, G.J.; Yan, J. & Goldberg, D.M. (2001). Assay of ochratoxin A in wine and beer by high-pressure liquid chromatography photodiode array and gas chromatography mass selective detection. *Journal of Agriculture and Food Chemistry*, Vol. 49, pp. 2733–2740.

Solomon, P.S. (published online 12 August 2011).Assessing the mycotoxigenic threat of necrotrophic pathogens of wheat. *Mycotoxin Research,* DOI 10.1007/s12550-011-0108-5

Soubra, L.; Sarkis, D.; Hilan, C. & Verger, P. (2009). Occurrence of total aflatoxins, ochratoxin A and deoxynivalenol in foodstuffs available on the Lebanese market and their impact on dietary exposure of children and teenagers in Beirut. *Food Additives and Contaminants*, Vol. 26, pp. 189-200.

Speijers, G.J.A. & Speijers, M.H.M. (2004).Combined toxic effects of mycotoxins. *Toxicology Letters*, Vol. 153, pp. 91-98.

Sweeney, M.J. & Dobson, A.D.W. (1998). Mycotoxin production by *Aspergillus*, *Fusarium* and *Penicillium* species. *International Journal of Food Microbiology,* Vol. 43, pp. 141–158.

Sydenham, E. W.; Vismer, H. F.; Marasas, W. F. O.; Brown, N. L.; Schlechter, M. & Rheeder, J. P. (1997). The influence of deck storage and initial processing on patulin levels in apple juice. *Food Additives and Contaminants*, Vol. 14, pp. 429–434.

Taniwaki, M. H.; Bleinroth, E. W. & De Martin, Z. J. (1989). Boloresprodutores de patulinaemmacã e suco industrializado. *Coletânea do Instituto de Tecnologia de Alimentos*, Vol. 19, pp. 42–49.

Thembo, K.M.; Vismer, H.F.; Nyazema, N.Z.; Gelderblom, W.C.A. & Katerere D.R. (2010). Antifungal activity of four weedy plant extracts against selected mycotoxigenic fungi. *Journal of Applied Microbiology*, Vol. 109, pp. 1479–1486.

Torkar, K.G. & Vengus, A. (2008). The presence of yeasts, moulds and aflatoxin M1 in raw milk and cheese in Slovenia.*Food Control*, Vol. 19, pp. 570–577.

Unusan, N. (2006). Occurrence of aflatoxin M1 in UHT milk in Turkey. Food and Chemical Toxicology, Vol. 44, pp. 1897-1900.

Van der Merwe, K.J.; Steyn, P.S.; Fourie, L.; Scott, De B. & Theron, J.J. (1965). Ochratoxin A, a toxic metabolite produced by *Aspergillusochraceus*Wilh. *Nature*, Vol. 205, pp. 1112–1113.

Van Rensburg, S.J.; Cook-Mozaffari, P.; Van Schalkwyk, D.J.; Van der Watt, J.J.; Vincent, T.J. & Purchase, I.F. (1985). Hepatocellular carcinoma and dietary aflatoxin in Mozambique and Transkei. *British Journal of Cancer*, Vol. 51, pp. 713–726.

Vega, M.; Muñoz, K.; Sepúlveda, C.; Aranda, M.; Campos, V.; Villegas, R. & Villarroel, O. (2009). Solidphase extraction and HPLC determination of ochratoxin A in cereals products on Chilean market. *Food Control*, Vol. 20, pp. 631–634.

Varga, J.; Due, M.; Frisvad, J.C. & Samson, R.A. (2007). Taxonomic revision of *Aspergillus*section *lavati*based on molecular, morphological and physiological data. *Studies in Mycology*, Vol. 59, pp. 89–106.

Vesonder, R.F. & Golinski P. (1989). Metabolites of *Fusarium*, In: *Fusariummycotoxins, taxonomy and pathogenigity*, Chelowski, J., pp.1-40, Elsevier, Amsterdam.

Wan Norashima, W.M.; Abdulamir, A.S.; Abu Bakar, F.; Son, R. & Norhafniza, A. (2009). The health and toxic adverse effects of *Fusarium* fungal mycotoxin, fumonisins, on human population. *American Journal of Infectious Diseases*, Vol. 5, pp. 273-281.

WHO (World Health Organization). (1995). Evaluation of certain food additives and contaminants. WHO. *Forty-fourth report of the joint FAO/WHO expert committee on food additives*, WHO, Geneva: Technical Report Series, Vol. 859, pp. 36–38.

Wichmann, G.; Herbarth, O. & Lehmann, I. (2002). The mycotoxins citrinin, gliotoxin, and patulin affect interferon-? rather than interleukin-4 production in human blood cells. *Environmental Toxicology*, Vol. 17, pp. 211-218.

Wild, C.P. & Gong Y.Y. (2010). Mycotoxins and human disease: a largely ignored global health issue. *Carcinogenesis*, Vol. 31, pp. 71-82.

Wilson, D.M. & Payne, G.A. (1994). Factors affecting *Aspergillusflavus* group infection and aflatoxin contamination of crop. In: Eaton, D.L., Groopman, J.D. (Eds.), The Toxicology of Aflatoxin Human Health, Veterinary and Agricultural Significance. Academic Press, San Diego, pp. 383–406.

Wilson, S.C.; Brasel, T.L.; Martin, J.M.; Wu, C.; Andriychuk, L.; Douglas, D.R.; Cobos L. & Straus, D.C. (2005). Efficacy of chlorine dioxide as a gas and in solution in the inactivation of two trichothecenemycotoxins. *International Journal of Toxicology*, Vol. 24, pp. 181–186.

Wogan, G.N. (1992). Aflatoxins as risk factors for hepatocellular carcinoma in humans. *Cancer Research*, Vol. 52, pp. 2114–2118.

Wogan, G.N. (1999). Aflatoxin as a human carcinogen. *Hepatology*, Vol. 30, pp. 573-575.

Wolff, J.; Bresch, H.; Cholmakov-Bodechtel, C.; Engel, G.; Garais, M.; Majerus, P.; Rosner, H. & Scheuer, R. (2000). Ochratoxin A: Contamination of foods and consumer exposure. *Archiv fur Lebensmittelhygiene*, vol. 51, pp. 181–128.

Yazar, S. & Omurtag, G.Z. (2008). Fumonisins, trichothecenes and zearalenone in cereals. *International Journal of Molecular Science* Vol. 9, pp. 2062-2090.

Yazici, S. & Velioglu, S. (2002). Effect of thiamine hydrochloride, pyridoxine hydrochloride and calcium-d-pantothenate on the patulin content of apple juice concentrate. *Nahrung/Food*, Vol. 46, pp. 256–257.

Young, J.C.; Subryan, L.M.; Potts, D.; McLaren, M.E. & Gobran F.H. (1986). Reduction in levels of deoxynivalenol in contaminated wheat by chemical and physical treatment. *Journal of Agricultural and Food Chemistry*, Vol. 34, pp. 461–465.

Young, J.C.; Zhu, H. & Zhou, T. (2006). Degradation of trichothecenemycotoxins by aqueous ozone. *Food and Chemical Toxicology*, Vol. 44, pp. 417–424.

Yousef, A.E. & Marth, E.H. (1985). Degradation of aflatoxin M1 in milk by ultraviolet energy. Journal of Food Protection, Vol. 48, pp. 697–698.

Yu, J.; Bhatnagar, D. & Ehrlich, K.C. (2002). Aflatoxinbiosynthesis. *Revista Iberoamericana de Micología*, Vol. 19, pp. 191-200.

Zain M.E. (2011). Impact of mycotoxins on human and animals. *Journal of Saudi Chemical Society*, Vol. 15, pp. 129-144.

Zinedine, A.; Brera, C.; Elakhdari, S.; Catano, C.; Debegnach, F.; Angelini, S.; De Santis, B.; Faid, M.; Benlemlih, M.; Minardi V. & Miraglia, M. (2006). Natural occurrence of mycotoxins in cereals and spices commercialized in Morocco. *Food Control*, Vol. 17, pp. 868–874.

11

Microbial Pectic Enzymes in the Food and Wine Industry

Carmen Sieiro[1], Belén García-Fraga[1], Jacobo López-Seijas[1], Abigaíl F. da Silva[1] and Tomás G. Villa[2]
[1]*Department of Functional Biology and Health Sciences, University of Vigo*
[2]*Department of Microbiology and Parasitology, University of Santiago de Compostela*
Spain

1. Introduction

Pectins are polysaccharides ubiquitous in the plant kingdom and constitute the major component of plant cell walls. The pectinases are a group of related enzymes capable of degrading pectin. Therefore, this group of enzymes have been used for decades in the food and winemaking industry for the processing of fruit juices (Mohnen, 2008; Prade et al., 1999; Ribeiro et al., 2010).

The pectinases are synthesized by plants and microorganisms, the latter being used for industrial production. Microorganisms are used to produce many enzymes of industrial interest in processes relatively inexpensive and environmentally friendly. Moreover, enzymatic catalysis is preferred over other chemical methods since it is more specific, less aggressive and generates less toxicity (Hoondal et al., 2002; Lara-Márquez et al., 2011).

Advances in biotechnology, especially in the fields of molecular biology and microbial genetics, have led to major advances in enzyme technology and have allowed, in many cases, the development of new producing strains and microbial enzymes. The production of pectinases may also benefit from these technologies.

This article reviews the characteristics of pectic substances, the types and mode of action of enzymes which degrade them and the main applications of commercial preparations of microbial pectinases in the food and winemaking industry, followed by a review of new microorganisms and pectolytic enzymes, evaluating new approaches to their production, marketing and use.

2. Pectic substances

Pectic substances are polysaccharides of high molecular weight, with a negative charge, appearing mostly in the middle lamella and the primary cell wall of higher plants, found in the form of calcium pectate and magnesium pectate. They are formed by a central chain containing a variable amount although in high proportion of galacturonic acid residues linked through α-(1-4) glycosidic bonds partially esterified with methyl groups (Fig. 1).

This molecule is known as pectin, while the demethylated molecule is known as polygalacturonic acid or pectic acid. Several L-rhamnopyranosyl residues may be attached to the main chain through its C-1 and C-2 atoms. In addition, galacturonate residue may be acetylated at the C-2 and C-3 positions, and side chains of residues of neutral sugars may be linked to the galacturonic acid or to the C-4 of the rhamnose residue in the main chain (Caffall & Mohnen, 2009; Mohnen, 2008; Pilnik & Voragen, 1970; Rombouts & Pilnik, 1980).

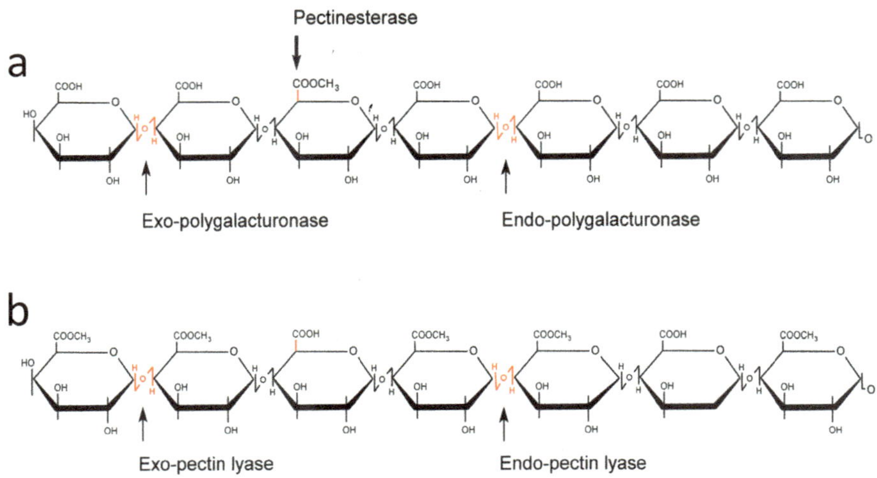

Fig. 1. Structure (main chain) of low (a) and high (b) methylated pectic substances and site of action of enzymes involved in their degradation

The generic name of pectic substances is used for referring to four types of molecules: protopectin (pectic substance in intact tissue), pectinic acids (polygalacturonan containing >0-75% methylated galacturonate units), pectic acids (polygalacturonan that contains negligible amount of methoxyl groups), and pectins (pectinic acid with at least 75% methylated galacturonate units). Protopectines are insoluble in water, while the rest are wholly or partially soluble in water (Alkorta et al., 1998; Kertesz, 1951).

Pectic substances represent between 0.5-4% of fresh weight plant material (Jayani et al., 2005; Sakai et al., 1993). In addition to their role as cementing and lubricating agents in the cell walls of higher plants, they are responsible for the texture of fruits and vegetables during growth, maturation and their storage (Alkorta et al., 1998; Caffall & Mohnen, 2009). Furthermore, pectic substances are involved in the interaction between plant hosts and their pathogens (Collmer & Keen, 1986; Prade et al., 1999).

Pectins have numerous and important applications in the food and pharmaceutical industries. In the food sector, it is primarily used as a gelling agent, replacing sugars and/or fats in low-calorie food and as nutritional fiber (Panchev et al., 1988; Sakai et al., 1993; Thakur et al., 1997). The pharmaceutical industry offers them as preparations to reduce cholesterol or to act as a lubricant in the intestines thus promoting normal peristaltic movement without causing irritation. In addition, these polysaccharides are used as drug delivery systems, which can also reduce the toxicity of these and make their activity longer lasting without altering their therapeutic effects (Morris et al., 2010; Pilnik & Voragen, 1970; Schols et al., 2009; Thakur et al., 1997).

3. Pectolytic enzymes

The enzymes which hydrolyze pectic substances are known as pectic enzymes, pectinases or pectinolytic enzymes (Blanco et al., 1999). Based on its mode of action and substrate preference these enzymes are classified into three types:
I. Protopectinases, which solubilize protopectin forming soluble pectin
II. Esterases (pectin methyl esterases and pectin acetyl esterases), which eliminate methoxyl and acetyl residues from pectin giving rise to polygalacturonic acid
III. Depolymerases, which break the glycosidic α-(1- 4) bonds between galacturonic residues via:
1. Hydrolysis (polygalacturonases)
2. Transelimination (pectin lyases and pectate lyases)
Also, the latter enzymes are subdivided into endo- if its pattern of action is random or exo- if its pattern of action is at the terminal end (Fogarty & Kelly, 1983; Rexova-Bencova & Markovie, 1976; Sakai, 1992; Whitaker, 1990). The detailed classification of these enzymes, their mode of action and final product are shown in Table 1 and in Fig. 1.

Enzyme	EC N°	Main substrate	Mode of action	Product
Esterases				
Pectin methyl esterase	3.1.1.11	Pectin	Hydrolysis	Pectic acid + methanol
Pectin acetyl esterase	3.1.1.6	Pectin	Hydrolysis	Pectic acid + methanol
Depolymerases				
Hydrolases				
Protopectinases		Protopectin	Hydrolysis	Pectin
Endopolygalacturonase	3.2.1.1.5	Pectic acid	Hydrolysis	Oligogalacturonates
Exopolygalacturonase	3.2.1.6.7	Pectic acid	Hydrolysis	Monogalacturonates
Lyases				
Endopectate lyase	4.2.2.2	Pectic acid	Transelimination	Unsaturated oligogalacturonates
Exopectate lyase	4.2.2.9	Pectic acid	Transelimination	Unsaturated oligogalacturonates
Endopectinlyase	4.2.2.10	Pectin	Transelimination	Unsaturated methyl-oligogalacturonates

Table 1. Pectolytic enzymes classified according to its mode of action

4. Pectic enzymes in nature: Microbial pectinases

Pectic enzymes are widely distributed in nature and are produced by bacteria, yeast, fungi (Fig. 2A) and plants. (Lang & Dornenburg, 2000; Whitaker, 1990). In plants, pectic enzymes are very important since they play a role in elongation and cellular growth as well as in fruit ripening (Sakai, 1992; Ward & Moo-Young, 1989; Whitaker, 1990). Pectolytic activity of microorganisms plays a significant role, firstly, in the pathogenesis of plants since these enzymes are the first to attack the tissue (Collmer & Keen, 1986; Whitaker, 1990). In addition, they are also involved in the process of symbiosis and the decay of vegetable residues (Hoondal et al., 2002; Lang & Dornenburg, 2000).

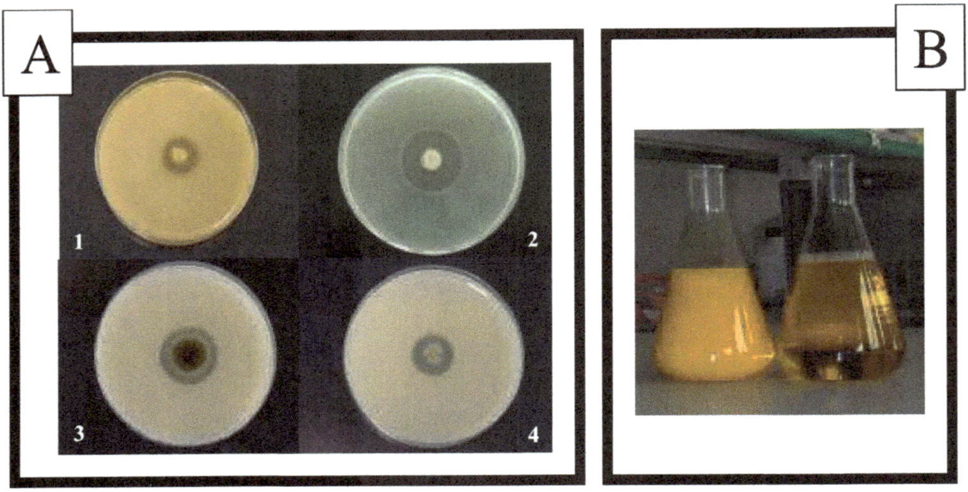

Fig. 2. A: Pectic enzymes produced by different microorganisms growing on minimal medium supplemented with polygalacturonic acid and detected by a clear halo around the colonies. 1, *Xanthomonas campestris* CECT 97; 2, *Kluyveromyces marxianus* CECT 1043; 3, *Aspergillus niger* CECT 2088; 4, *Botryotinia fuckeliana* CECT 20518. B: Cloudy (peach) and clear (apple) juices. CECT: Spanish Type Culture Collection

The microbial world has shown to be very heterogeneous in its ability to synthesize different types of pectolytic enzymes with different mechanisms of action and biochemical properties (Favela-Torres et al., 2005; Gummadi & Panda, 2003). Pectic enzymes are produced by both prokaryotic microorganisms, which primarily synthesize alkaline pectinases, and by eukaryotic microorganisms, mostly fungi that synthesize acid pectinases (Hoondal et al., 2002; Jayani et al., 2005; Kashyap et al., 2001). Furthermore, the production of these enzymes has also been described in yeast (Alimardani-Theuil et al., 2011; Blanco et al., 1999).

There are many studies that have been conducted related to the characterization of different microbial pectic enzymes concerning their mechanisms of action and biochemical properties. The optimal pHs that these enzymes may act range between 3.5-11, while the optimal temperatures vary between 40-75 °C (Gummadi & Panda, 2003; Kashyap et al., 2001). Table 2 shows an example of the diversity of pectic enzymes produced by micro-organisms. Given the features of the substrate on which they act and the effect that is required of them, acidic and depolymerizing pectinases are of great interest for the food industry although some applications such as the extraction of oils requires the alkaline ones (Hoondal et al., 2002).

5. Pectinases in the food industry

5.1 Extraction of fruit and vegetable juices

The main industrial application of pectinases is the extraction and clarification of fruit and vegetable juices. Most of the microbial pectinases produced by the industry are dedicated to this purpose. Pectins are responsible for the turbidity and consistency of the juice causing an increase in their viscosity which hinders its clarification, filtration and concentration (Alkorta et al., 1998). The degradation of pectic substances in mashed fruit purees is achieved through the addition of pectolytic enzymes resulting in an increase in juice yield and its clarification as well as a decrease in viscosity. Treatment with pectinases also provides filtering of the product (de Gregorio et al., 2002; Fernandez-Gonzalez et al., 2004; Ribeiro et al., 2010; Sarioglu et al., 2001; Souza et al., 2003).

Microorganism	Enzyme	Optimal pH	Optimal temperature (°C)	Reference
Bacteria				
Bacillus sp NT-33	Polygalacturonase	10.5	75	Cao et al., 1992
Bacillus sp DT7	Pectin lyase	8	60	Kashyap et al., 2000
Fungi				
Aspergillus ficuum	Pectin lyase	5	50	Yadav et al., 2008
Penicillium frequentans	Endopolygalacturonase	3.5-5	50	Boirin et al., 1996
Sclerotium rolfsii	Endopolygalacturonase	3.5	55	Chane & Shewal, 1995
Penicillium paxilli	Pectin lyase	5	35	Szajer & Szajer, 1982
Yeasts				
Saccharomyces cerevisiae	Endopolygalacturonase	5.5	45	Blanco et al., 1994
Kluyveromyces marxianus	Endopolygalacturonase	4.5	55	Serrat et al., 2002

Table 2. Biochemical properties of some pectinases

The fruit and vegetable juice industry uses mainly acidic pectinases of fungal origin, principally from *Aspergillus* spp. Commercial preparations are mixtures of polygalacturonases, pectate lyases and pectin esterases. Pectate lyases can act on the esterified pectin while the polygalacturonases act on the desesterified pectin thus it might require previous action of the pectin esterases. Pectic enzymes treatments vary depending on the type of juice (Fig. 2B):

5.1.1 Clear juices (i.e., apple, pear, grape)

In this type of product, pectolytic enzymes are added to increase the yield in juice during the pressing and for removing matter in suspension. In the case of apple juice, the most commonly used enzymes are those that can depolymerize the highly esterified apple pectin. Apple juice can be obtained through a two-step process consisting of a first treatment of the crushed apple mush with pectinases to obtain the premium juice followed by pomace liquefaction treatment made with a mixture of different pectinases and cellulases for the complete extraction of the juice (Will et al., 2000).

After washing and crushing, the apples are pressed to obtain the juice. Pectic enzymes are used to facilitate the pressing and juice extraction and to assist in the separation of a flocculant precipitate by means of sedimentation, filtration or centrifugation. If a cloudy product is required, the juice is pasteurised after pressing to inactivate residual enzymes. Centrifugation removes the large-size remains leaving small particles in suspension.

However, if a clear juice is required, these suspended particles have to be withdrawn. In order to do this a treatment with mixtures of commercial enzymes is carried out containing pectinases, cellulases and hemicellulases. Finally, the fluid is centrifuged to obtain the clear juice (Grassin & Fauquembergue, 1996; Kashyap et al., 2001). Although it has been noted that the proper clarification of apple juice requires mixtures of polygalacturonase and pectin methyl esterase (Yamaski et al., 1964), subsequent studies have shown that it can be clarified by only using a pure pectin lyase (Ishii & Vokotsuka, 1973).

5.1.2 Cloudy juices (i.e., citrus, tomato, nectars)

In the case of orange juice, where natural pectin esterases are present, pectin is only partially methylated. Polygalacturonases are the pectic enzymes which are most commonly used and of great interest for this type of fruit juice.

In the process of orange juice extraction, pectinases can be added at the end of the pulp wash extraction to reduce viscosity or, preferably, at the end of the first finisher. This leads to higher yield in juice, a better extraction of soluble solids and to a lower viscosity. The action of these enzymes just reduces the viscosity without attacking the insoluble pectin that maintains the stability of the cloud. Enzyme preparations should lack or have the least possible amount of pectin methyl esterases to avoid the clarification of the product. It has been suggested that the best enzyme might be pure pectin lyase (Kashyap et al., 2001; Rebeck, 1990).

5.2 Maceration products of plant tissues

The enzymatic maceration of plant tissues allows the transformation of these organized tissues in suspensions of intact cells that constitute the pulpy products that are used as a basis for preparing juices, nectars, baby food and some dairy products such as yoghurts. Enzyme preparations for this purpose contain cellulases, hemicellulases and pectic enzymes which should only act on the middle lamella of the plant tissue (Kashyap et al., 2001).

5.3 Extraction of vegetable oil

Vegetable oils of olive, sunflower, coconut, palm or canola are obtained by extraction with organic solvents such as hexane. The use of pectolytic enzymes, in this case preferably alkaline, allows the extraction of vegetable oils in an aqueous process. Enzyme preparations based on cellulases, hemicellulases and pectinases have been used successfully in the extraction of olive oil. The enzyme treatment not only improves oil yield and stability but also increases polyphenols and vitamin E content enhanzing its organoleptic quality (Hoondal et al., 2002; Iconomou et al., 2010; Kashyap et al., 2001; West, 1996).

5.4 Coffee, cocoa and tea fermentation

Traditionally, fermentation of coffee is made with pectolytic microorganisms in order to remove the layer of mucilage from the coffee beans. With the same purpose, commercial enzyme preparations containing pectinase is sprayed onto the beans to ferment. A cheaper alternative is to use, with the same purpose, the filtrate of inoculated fermentations. The enzyme treatment significantly reduces the fermentation time (Amorim & Amorim, 1977; Kashyap et al., 2001; Serrat et al., 2002; Silva et al., 2000). Cocoa fermentation is essential to develop the chocolate flavour. This fermentation is carried out by a succession of different microorganisms, some of them pectolytic. Pectic enzymes allow the degradation of the cocoa pulp and are indispensable for the fermentation process and the good quality of fermented beans (Ouattara et al., 2010; Schwan & Wheals, 2004). Similarly, treatment of tea leaves with pectic enzymes of fungal origin (at a dose adjusted to avoid damaging the leaf), facilitates and accelerates the fermentation (Carr, 1985; Kashyap et al., 2001).

6. Pectinases in the wine industry

Wine is the result of the fermentation of grape juice. Pectinases are the most important enzymes used by the winemaking industry although commercial preparations may contain other enzymatic activities such as hemicellulases, glucanases and glycosidases (Rombouts & Pilnik, 1980).

Pectic enzymes are synthesized naturally by the plant and are present in the grape. However, they have low activity during the wine producing process (Ducasse et al., 2011). Microbial pectolytic enzymes especially of fungal origin are resistant to the conditions of fermentation and can be used to facilitate processes, improve quality and diversify products. So far, commercial enzymes are produced all from fungi, mainly of the genus *Aspergillus*. Although all enzymes are produced by *Aspergillus*, pectinase preparations currently available for the wine market are very different. Both the type of activities as well as their concentration in the preparations depend on the strain of *Aspergillus* used, the fermentation conditions for production, the nature of the fermented substrate and the degree of partial purification.

Various research studies have shown that the addition of pectolytic enzymes leads to increased levels of methanol in wine (Revilla & González-SanJosé, 1998; Servili et al., 1992) due to the activity of pectin methyl esterase. Methanol is toxic and its maximum concentration in wine is regulated. Therefore, pectin methyl esterase activity should be at low concentrations in commercial mixtures.

The functions of pectic enzymes in the winemaking process are to support the extraction process, maximize juice yield, facilitate filtration and intensify the flavour and colour.

6.1 Extraction, clarification and filtration

The addition of pectinase to the must reduces its viscosity and causes the grouping of suspended particles in larger aggregates that can be removed by sedimentation. If the enzymes are added to the pulp before pressing, must yield increases, facilitating the pressing and enhancing the colour. Macerating enzyme preparations for this purpose contain pectinases as well as cellulases (Ribereau-Gayon et al., 2006). A high level of polygalacturonase is very effective for clarification but may require the prior action of pectin lyase activity. For this reason the enzyme preparations with a high content of pectin lyase are desirable when very fast racking is required to prevent any problems related to must oxidation, development of endogenous microbiota and nutrient loss. In addition, wines made with pectic-enzyme-treated grape must significantly reduces filtration times (Blanco et al., 1997; Blunt, 2000).

6.2 Intensification and stabilization of colour

In the case of musts obtained from red grapes, the degradation of cell walls in the skin of the grapes through pectolytic enzyme treatment results in an increased release of phenolic compounds responsible for colour (Busse-Valverde et al., 2011; Pinelo et al., 2006). Early work related to the use of pectinases to enhance the colour of the wines were a bit confusing because Ough et al. (1975) were able to intensify the colour of wine by the enzyme treatment, but Wightman et al. (1997) found that some enzyme preparations containing pectic enzymes reduced red wine colour. Subsequent studies performed with two different enzyme preparations (Watson et al., 1999) confirmed that both allow to produce wines with higher concentrations of anthocyanins and total phenols and have a higher colour intensity. Similarly, in 1994 the Australian Wine Research Institute conducted a study with different commercial preparations of pectic enzymes and in all cases concluded that its use leads to faster and better colour extraction during maceration, pressing and fermentation as well as improve the clarification of the product (van Rensburg & Pretorius, 2000).

6.3 Boosting aroma and flavour

The aromatic profile of wines consists of two components: the varietal aromas characteristic of the variety of grape used and the aromas originated by the yeast during fermentation (Piñeiro et al., 2006; Vilanova & Sieiro, 2006). In many cases the grape variety used completely determines the aroma of the wine, especially in young wines. The volatile compounds of grapes include monoterpenes, C13 nor-isoprenoides, benzene derivatives and aliphatic alcohols. The aromatic components of the grape may appear as free forms, which contribute directly to its scent or bound forms, of greater concentration, to sugars and nonvolatile. Nonodorous glycoside flavour precursors accumulate in the grape especially in the skin during the ripening process (Bayonove, 1993; Williams et al., 1989; Winterhalter & Skouroumounis, 1997). The hydrolysis of these precursors via beta-glucosidases releases the olfactory active aglycones (Günata et al., 1988; Williams et al., 1989). Pectic enzymes help break down the cell walls of the grapes and thus to extract the aromatic precursors. The addition of pectic enzymes during the extraction or fermentation of the must results in an increase in aromatic precursors susceptible to being attacked by beta-glucosidases from the must, those produced by yeasts and bacteria during fermentation or those which are included in commercial enzyme preparations, thereby enhancing the aroma of wines (Comitini et al., 2011; du Toit et al., 2011; Gómez-Plaza et al., 2000; Pinelo et al., 2006).

7. Biotechnology production of microbial pectic enzymes

As discussed until now, there are numerous applications of microbial pectic enzymes in food and wine-making industry. Not surprisingly, the sales volume of these enzymes represents 25% of the enzymes that are commercialized in the food and alcoholic beverages industry. These many applications require one or more types of pectinases that must act in very different condition according to the process in which they are involved. For example, while in the extraction and clarification of fruit juice enzymes can be used at temperatures between 45-95 °C (Kashyap et al., 2001), the wine industry employs temperatures below 15-10 °C (Gómez-Plaza et al., 2000). Although most food applications pectinases have to act in a medium acid, in others, such as oil extraction, they should perform in a medium alkaline (Hoondal et al., 2002).

So far, commercial pectic enzymes are prepared only from cultures of filamentous fungi, mainly of the genus *Aspergillus*. In fact, in the European Union, the OIV (Organisation

Internationale de la Vigne et du Vin) regulates the origin of these enzymes for the wine industry. Commercial products contain mainly a mix of polygalacturonase, pectin lyase and pectin methyl esterase. The use of different strains of *Aspergillus* and modification of substrates and culture conditions can lead to mixtures enriched in one type of enzyme. Companies, including AEB, Gist-Brocades, Novo Nordisk and Lallemand, offer different products with pectic enzymes in different proportions to suit the needs of each process.

However, these commercial pectic preparations are not always optimal for each process, and its use is not without side effects and controversy. Thus, while pectin methyl esterase activity present in the samples may be necessary for the action of polygalacturonase in the case of pectin with a high degree of esterification, its action may lead to an undesirable increase of methanol in the products (Vilanova et al., 2000). Moreover, although the pectic enzymes are the most abundant in commercial mixtures, they can also contain other undesirable activities, such as in the case of making wine, polyphenoloxidases or cinnamyl esterases (Mantovani et al., 2005; van Rensburg & Pretorius, 2000).

At present especially in recent years, the accumulated knowledge of new microbial pectolytic enzymes as well as methodological and technological advances can address the production and use of these enzymes with a different approach. The diversity of applications and conditions in which these enzymes must work also demand a large number of different enzymes capable of acting in such conditions. Even more interesting would be having more robust, broad-spectrum enzymes which allow for a more versatile use in different applications.

Considering the variety of enzymes, traditionally the majority of studies refer to the pectinases of *Erwinia* and *Bacillus* within the bacteria and various fungi especially *Aspergillus* (de Vries and Visser, 2001; de Vries et al., 2002; Gummadi & Panda, 2003; Hoondal et al., 2002; Jayani et al., 2005; Yadav et al., 2009) although there has been a major advance in the description and characterization of pectic enzymes produced by yeast in the last 15 years (Alimardani-Theuil et al., 2011; Blanco et al., 1999; Rodríguez-Gámez & Serrat, 2008). In recent years there has been also a growing interest in studying pectic enzymes with very interesting properties from the point of view of their application. These include thermostable pectinases (Kar & Ray, 2011; Swain & Ray, 2010) or pectinases with optimal activity at low temperatures (Cabeza et al., 2011; Merin et al., 2011; Nakawagua et al., 2004; Padma et al., 2011).

The possibility of producing different types of pectolytic enzymes separately and for later preparation of their mixtures in the proper proportions would allow to provide more suitable commercial preparations for each application and lacking undesirable activities.

In this sense, microorganisms such as some strains of yeast *Saccharomyces cerevisiae* and *Kluyveromyces marxianus* which produce only one type of enzyme (Blanco et al., 1999; Serrat et al., 2002) or that constitutively synthesize it (Serrat et al., 2002) are of great interest. Similarly, obtaining constitutive mutants from producing strains allow the optimization of production and contribute to making cost-effective production processes (Favela-Torres et al., 2006; Trigui-Lahiani et al., 2008). Furthermore, these strains can be used to produce pectinases that accumulate together with other metabolites in the culture broth, which contributes favourable to the overall economy of the process (Serrat et al., 2011). Heterologous expression of enzymes in prokaryotic or eukaryotic systems is a technique of great interest for the production of a single type of enzyme. Table 3 shows some of the many pectic enzymes, both from bacteria, yeast and fungi, which are rightly expressed in

Escherichia coli and different species of yeast. Although *E. coli* is unable to carry out post-translational modifications of proteins, fungal pectolytic enzymes expressed in these bacteria are active (Wang et al., 2011). However, one of the most interesting strategies seems to be the expression of pectinase genes in yeast, particularly in *Pichia pastoris*, in which very high levels of constitutive expression has been achieved (Sieiro et al., 2009). In some cases changes in glycosylation patterns conducted by yeast did not affect the activity and characteristics of recombinant pectinases (Sieiro et al., 2009), while other changes do occur with respect to the characteristics of the native protein, which even lead to enzymes with interesting properties for certain applications (Lang & Looman, 1995; Sieiro et al., 2003).

Microorganism gene origin	Type of enzyme	Host strain	Reference
Xanthomonas campestris	Pectate lyase Polygalacturonase	*E. coli*	Xiao et al., 2008
Streptomyces coelicolor	Polygalacturonase	*E. coli*	Xiao et al., 2008
Pseudoalteromonas haloplanktis ANT/505	Pectate lyase	*E. coli*	Truong et al., 2001
Thermotoga maritima	Exo-polygalacturonase	*E. coli*	Parisot et al., 2003
Burkholderia capacia	Endo-polygalacturonase	*E. coli*	Massa et al., 2007
Phytophthora capsici	Pectate lyase	*E. coli*	Wang et al., 2011
Erwinia chrysantemy	Pectate lyase Polygalacturonase	*S. cerevisiae*	Laing & Pretorius, 1993
Aspergillus niger RH5344	Polygalacturonase	*S. cerevisiae*	Lang & Looman, 1995
Aspergillus aculeatus	Pectin methyl esterase	*S. cerevisiae*	Christgau et al., 1996
S. cerevisiae IM1-8b	Endo-polygalacturonase	*S. cerevisiae*	Blanco et al., 1998
S. cerevisiae IM1-8b	Endo-polygalacturonase	*Schizosaccharomyces pombe*	Sieiro et al., 2003
K. marxianus CECT1043	Endo-polygalacturonase	*P. pastoris*	Sieiro et al., 2009
Bispora sp. MEY-1	Endo-polygalacturonase	*P. pastoris*	Yang et al., 2011

Table 3. Microbial pectic enzymes expressed in different host strains

8. Conclusions

The enzymes that degrade the pectic substances play an essential role in the food and winemaking industries because they are used to degrade the pectins that interfere with the extraction and clarification of fruit juices and oils as well as being important in the fermentation of coffee, cocoa and tea. Also, in the wine industry they play an important role by contributing to the release of the molecules responsible for aroma and colour, two of the major components that characterize a wine.

Traditionally, this industry uses different mixtures of pectolytic enzymes derived from fungi cultures, mainly of the genus *Aspergillus*, not always completely adequate for the processes they must carry out because of the type and concentration of different enzyme activities that they contain, not without undesirable effects due to other non-pectic enzymes that may be present in the mixtures.

The exploration of microbial biodiversity has allowed, especially in recent years, to identify and characterize new pectic-enzyme-producing microorganisms with different biochemical characteristics, some potentially very interesting from the point of view of their application. Also, it has been technically possible, on the one hand, to select wild strains and constitutive mutants that produce a single enzyme, and, on the other hand, the heterologous expression in bacteria and yeast of numerous genes which encode pectic enzymes, obtaining producing strains of interest.

All this opens the possibility of producing different pectic enzymes individually and preparing commercial mixtures of these, adapted to each process. Research focused on protein engineering in order to obtain pectic enzymes more robust and versatile as well as the optimization of production processes with new strains are necessary for the successful completion of this new approach for the production and use of microbial pectinases.

9. References

Alimardani-Theuil, P., Gainvors-Claise, A. & Duchiron, F. (2011). Yeasts: An attractive source of pectinases-From gene expression to potential applications: A review. *Process Biochemistry,* Vol.46, pp. 1525-1537

Alkorta, I., Garbisu, C., Llama, M.J. & Serra, J.L. (1998). Industrial applications of pectic enzymes: a review. *Process Biochemistry*, Vol.33, pp. 21-28

Amorim, H.V. & Amorim, V.L. (1977). Coffee enzyme and coffee quality. In: *Enzymes in food and beverage processing,* R.L. Ori & A.J. st Angelo, (Eds.), pp. 27-56, American Chemical Society, ISBN: 084120375X, Indiana, USA

Bayanove, C. (1993). Les composés terpeniques. In : *Les acquisitions récentes en chromatographie du vin. Applications á l´ analyse sensorielle des vins,* B. Doneche, (Ed.), pp. 99-119, ISBN: 2877773647, Paris, France

Blanco, P., Sieiro, C. & Villa, T.G. (1999). Production of pectic enzymes in yeasts. *FEMS Microbiology Letters*, Vol.175, pp. 1-9

Blanco, P., Sieiro, C., Díaz, A. & Villa, T.G. (1994). Production and partial characterization of an endopolygalacturonase from *Saccharomyces cerevisiae. Canadian Journal of Microbiology*, Vol.40, pp. 974-977

Blanco, P., Sieiro, C., Díaz, A., Reboredo, N.M. & Villa, T.G. (1997). Grape juice biodegradation by polygalacturonases from *Saccharomyces cerevisiae. International Biodeterioration and Biodegradation,* Vol.40, pp. 115-118

Blanco, P., Sieiro, C., Reboredo, N.M. & Villa, T.G. (1998). Cloning, molecular characterization, and expression of an endo-polygalacturonase-encoding gene from *Saccharomyces cerevisiae* IM1-8b. *FEMS Microbiology Letters*, Vol.164, pp. 249-255

Blunt, M.K. (2000). Cellulases and related enzymes in biotechnology. *Biotechnology Advances*, Vol.18, pp. 355-383

Borin, M.D.F., Said, S. & Fonseca, M.J.V. (1996). Purification and biochemical characterization of an extracelular endopolygalacturonase fron *Penicillium frequentans*. *Journal of Agricultural and Food Chemistry*, Vol.44, pp. 1616-1620

Busse-Valverde, N., Gómez-Plaza, E., López-Roca, J.M., Gil-Muñoz, R. & Bautista-Ortín, A.B. (2011). The extraction of anthocyanins and proanthocyanidins from grapes to wine during fermentative maceration is affected by the enological technique. *Journal of Agricultural and Food Chemistry*, Vol.59, pp. 5450-5455

Cabeza, M.S., Baca, F.L., Puntes, E.M., Loto, F., Baigori, M.D. & Morata, V.I. (2011). Selection of psychrotolerant microorganisms producing cold-active pectinases for biotechnological processes at low temperature. *Food Technology and Biotechnology*, Vol.49, pp. 187-195

Caffall, K.H. & Mohnen, D. (2009). The structure, function, and biosynthesis of plant cell wall pectic polysaccharides. *Carbohydrate Research*, Vol.344, pp. 1879-1900

Cao, J., Zheng, L. & Chen, S. (1992). Screening of pectinase producer from alkalophilic bacteria and study on its potential application in degumming of ramie. *Enzyme and Microbial Technology*, Vol.14, pp. 1013-1016

Carr, J.G. (1985). Tea, coffee and cocoa. In: *Microbiology of fermented foods*, B.J.B. Wood, (Ed.), pp. 133-154, Elsevier Applied Science, ISBN: 0751402168, London, UK

Chane, P.S. & Shewal, J.G. (1995). Pectinase production by *Sclerotium rolfsii*. Effects of culture conditions. *Folia Microbiologica*, Vol.40, pp. 111-117

Christgau, S., Kofod, L.V., Halkier, T., Andersen, L.N., Hockauf, M., Dorreich, K., Dalboge, H. & Kauppinen, S. (1996). Pectin methyl esterase from *Aspergillus aculeatus*: Expression, cloning in yeasts and characterization of the recombinant enzyme. *Biochemical Journal*, Vol.319, pp. 705-712

Collmer, A. & Keen, N.T. (1986). The role of pectic enzymes in plant pathogenesis. *Annual Review of Phytopathology*, Vol.24, pp. 383-409

Comitini, F., Gobbi, M., Domizio, P., Romani, C., Lencioni, L., Mannazzu, I. & Ciani, M. (2011). Selected non-*Saccharomyces* wine yeasts in controlled multistarter fermentations with *Saccharomyces cerevisiae*. *Food Microbiology*, Vol.28, pp. 873-882

de Gregorio, A., Mandalani, G., Arena, N., Nucita, F., Tripodo, M.M. & lo Curto, R.B. (2002). SCP and crude pectinase production by slurry-state fermentation of lemon pulps. *Bioresource Technology*, Vol.83, *pp. 89-94*

de Vries, R.P. & Visser, J. (2001). *Aspergillus* enzymes involved in degradation of plant cell wall polysaccharides. *Microbiology and Molecular Biology Reviews*, Vol.65, pp. 497-522

de Vries, R.P., Jansen, J., Aguilar, G., Parenicova, L., Joosten, V., Wulfert, F., Benen, J.A. & Visser, J. (2002). Expression profiling of pectinolytic genes from *Aspergillus niger*. *FEBS Letters*, Vol.530, pp. 41-47

Demir, N., Acar, J., Sarioglu, K., Mutlu, M. (2001). The use of commercial pectinase in fruit juice industry. Part 3: Immobilized pectinase for mash treatment. *Journal of Food Engineering*, Vol.47, pp. 275-280 du Toit, M., Engelbrecht, L., Lerm, E. & Krieger-Weber, S. (2011). *Lactobacillus*: the next generation of malolactic fermentation starter cultures-an overview. *Food and Bioprocess Technology*, Vol.4, pp. 876-906

Ducasse, M.A., Williams, P., Canal-Llauveres, R.M., Mazerollest, G., Cheynier, V. & Doco, T. (2011). Effect of macerating enzymes on the oligosaccharide profiles of Merlot red wines. *Journal of Agricultural and Food Chemistry*, Vol.59, pp. 6558-6567

Favela-Torres, E., Aguilar, C.N., Contreras-Esquivel, J.C. & Viniegra-González, G. (2005). Pectinases. In: *Enzyme Technology*. A. Pandey, C. Webb, C.R. Soccol & C. Larroche, (Eds.), pp. 265-287, Asiatech Publishers Inc., ISBN: 8187680121, New Delhi, India

Favela-Torres, E., Volke-Sepúlveda, T. & Viniegra-González, G. (2006). Production of hydrolytic depolymerizing pectinases. *Food Technology and Biotechnology*, Vol.44, pp. 221-227

Fernández-González, M., Úbeda, J.F., Vasudevan, T.G., Otero, R.R.C. & Briones, A.I. (2004). Evaluation of polygalacturonase activity in *Saccharomyces cerevisiae* wine strains. *FEMS Microbiology Letters*, Vol.237, pp. 261-267

Fogarty, W.M. & Kelly, C.T. (1983). Pectic enzymes. In: *Microbial enzymes and biotechnology*, W.M. Fogarty, (Ed.), pp. 131-182, Applied Science Publishers, ISBN: 0853341850, London, England

Gómez-Plaza, E., Gil-Muñoz, R., López-Roca, J.M. & Martínez, A. (2000). Colour and phenolic compounds of a young red wine, influence of wine-making techniques, storage temperature, and length of storage time. *Journal of Agricultural and Food Chemistry*, Vol.48, pp. 736- 741

Grassin, C. & Fauquembergue, P. (1996). Fruit Juices. In: *Industrial Enzymology*. T. Godfrey & S. West, (Eds.), pp. 225-264. Industrial enzymology, second ed. Stockholm Press, ISBN: 0935859381, New York, USA

Gummadi, S.N. & Panda, T. (2003). Purification and biochemical properties of microbial pectinases: a review. *Process Biochemistry*, Vol.38, pp. 987-996

Günata, Y.Z., Bitteur, S., Brillouet, J.M., Bayonove, C. & Cordonnier, R. (1988). Sequential enzymatic hydrolysis of potentially aromatic glycosides from grapes. *Carbohydrate Research*, Vol.184, pp. 139-149

Hoondal, G.S., Tiwari, R.P., Tewari, R., Dahiya, N. & Beg, Q.K. (2002). Microbial alkaline pectinases and their industrial applications: a review. *Applied Microbiology and Biotechnology*, Vol.59, pp. 409-418

Iconomou, D., Arapoglou, D. & Israilides, C. (2010). Improvement of phenolic antioxidants and quality characteristics of virgin olive oil with the addition of enzymes and nitrogen during olive paste processing. *Grasas y Aceites*, Vol.61, pp. 303-3011

Ishii, S. & Yokotsuka, T. (1973). Susceptibility of fruit juices to enzymatic clarification by pectin lyase and its relation to pectin in fruit juice. *Journal of Agricultural and Food Chemistry*, Vol.21, pp. 269-272

Jayani, R.S. Saxena, S. & Gupta, R. (2005). Microbial pectinolytic enzymes: a review. *Process Biochemistry*, Vol.40, pp. 2931-2944

Kar, S. & Ray, R.C. (2011). Purification, characterization and application of thermostable exo-polygalacturonase from *Streptomyces erumpens* MTCC 7317. *Journal of Food Biochemistry*, Vol.35, pp. 133-147

Kashyap, D.R., Chandra, S., Kaul, A. & Tewari, R. (2000). Production, purification and characterization of pectinase from *Bacillus* sp. DT7. *World journal of Microbiology and Biotechnology*, Vol.16, pp. 277-282

Kashyap, D.R., Vohra, P.K., Chopra S. & Tewari, R. (2001). Applications of pectinases in the commercial sector: a review. *Bioresource Technology*, Vol.77, pp. 215-227

Kertesz, Z.I. (1951). *The pectic substances*. Interscience Publishers, ISBN: 0471377538, New York, USA

Laing, E. & Pretorius, I.S. (1993). Co-expression of an *Erwinia chrysantemy* pectate lyase-encoding gene (*pelE*) and an *E. carotovora* polygalacturonase-encoding gene (*peh1*) in *Saccharomyces cerevisiae*. *Applied Microbiology and Biotechnology*, Vol.39, pp. 181-188

Lang, C. & Dörnenburg, H. (2000). Perspectives in the biological function and the technological application of polygalacturonases. *Applied Microbiology and Biotechnology*, Vol.53, pp. 366-375

Lang, C. & Looman, A. C. (1995). Efficient expression and secretion of *Aspergillus niger* RH5344 polygalacturonase in *Saccharomyces cerevisiae*. *Applied Microbiology and Biotechnology*, Vol.44, pp. 147-156

Lara-Márquez, A., Zavala-Páramo, M.G., López-Romero, E. & Camacho, H.C. (2011). Biotechnological potential of pectinolytic complexes of fungi. *Biotechnology Letters*, Vol.33, pp. 859-868

Mantovani, C.F., Geimba, M.P. & Brandelli, A. (2005). Enzymatic clarification of fruit juices by fungal pectin lyase. *Food Biotechnology*, Vol.19, pp. 173-181

Massa, C., Degrassi, G., Devescovi, G., Venturi, V. & Lamda, D. (2007). Isolation, heterologous expression and characterization of an endo-polygalacturonase produced by the phytopathogen *Burkholderia capacia*. *Protein Expression and Purification*, Vol.54, pp. 300-308

Merin, M.G., Mendoza, L.M., Farias, M.E., & de Ambrosini, V.I.M. (2011). Isolation and selection of yeasts from wine grape ecosystem secreting cold-active pectinolytic activity. *International Journal of Food Microbiology*, Vol.147, pp. 144-148

Mohnen, D. (2008). Pectin structure and biosynthesis. *Current Opinion in Plant Biology*, Vol.11, pp. 266-277

Morris, G., Kök, S., Harding, S. & Adams, G. (2010). Polysaccharide drug delivery systems based on pectin and chitosan. *Biotechnology & Genetic Engineering Reviews*, Vol.27, pp. 257-284

Nakawawa, T., Nagaoka, T., Taniguchi, S., Milyaji, T. & Tomizuka, N. (2004). Isolation and characterization of psychrophylic yeasts producing cold-adapted pectinolytic enzymes. *Letters in Applied Microbiology*, Vol.38, pp. 383-387

Ouattara, H.G., Reberchon, S., Niamke, S.L. & Nasser, W. (2010). Biochemical properties of pectate lyases produced by three different *Bacillus* strains isolated from fermenting cocoa beans and characterization of their cloned genes. *Applied and Environmental Microbiology*, Vol.76, pp. 5214-5220

Ough, C.S., Noble, A.C., & Temple, D. (1975). Pectic enzyme effects on red grapes. *American Journal of Enology and Viticulture*, Vol.26, pp. 195-200

Padma, P.M., Anuradha, K. & Reddy, G. (2011). Pectinolytic yeasts isolates for cold-active polygalacturonase production. *Innovative Food Science & Emerging Technologies*, Vol.12, pp. 178-181

Panchev, I.N., Kirtchev, N.A., Kratchanov, C.G. & Proichev, T. (1988). On the molecular weight of pectic substances and its relation to their gel strengths. *Carbohydrate Polymers*, Vol.8, pp. 257-269

Parisot, J., Laglois, V., Skanyan, V. & Rabiller, C. (2003). Cloning, expression and characterization of a thermostable exopolygalacturonase from *Thermotoga maritima*. *Carbohydrate Research*, Vol.338, pp. 1333-1337

Pilnik, W. & Voragen, A.G.J. (1970). Pectic substances and other uronides. In: *The biochemistry of fruits and their products*, A. C. Hulme, (Ed.), Vol.1, pp. 53-87, Academic Press, ISBN: 0123612012, London, England

Piñeiro, Z., Natera, R., Castro, R., Palma, M., Puertas, B. & Barroso, C.G. (2006). Characterization of volatile fraction of monovarietal wines: Influence of winemaking practices. *Analytica Chimica Acta*, Vol.563, pp. 165-172

Pinelo, M., Arnous, A. & Meyer, A.S. (2006). Upgrading of grape skins: Significance of plant cell-wall structural components and extraction techniques for phenol release. *Trends in Food Science & Technology*, Vol.17, pp. 579-590

Prade, R.A., Zhan, D., Ayouby, P. & Mort, A.J. (1999). Pectins, pectinases and plant-microbe interactions. *Biotechnology & Genetic Engineering Reviews*, Vol.16, pp. 361-391

Rebeck, H. (1990). Processing of citrus juices. In: *Production and packaging of non-carbohydrate fruit juices and fruit beberages*, D. Hick, (Ed.), Van Nosrand Reihold, New York, USA

Revilla, I. & González-SanJosé, M.L. (1998). Methanol release during fermentation of red grapes treated with pectolytic enzymes. *Food Chemistry*, Vol.63, pp. 307-312

Rexová-Benková, L. & Marcovic, O. (1976). Pectic enzymes. *Advances in Carbohydrate Chemistry*, Vol.33, pp. 323-385

Ribeiro, D.S., Henrique, S.M.B., Oliveira, L.S., Macedo, G.A. & Fleuri, L.F. (2010). Enzymes in juice processing: a review. *Interntional Journal of Food Science and Technology*, Vol.45, pp. 635-641

Ribereau-Gayon, P., Dubourdieu, D., Doneche, B. & Lonvaud, A. (2006). *Handbook of Enology, Vol. I. The Microbiology of Wine and Vinifications* (2nd edition), John Wiley & Sons, ltd., ISBN: 0470010347, New York, USA

Rodríguez-Gámez, O. & Serrat, M. (2008). Poligalacturonasas de levaduras: un producto biotecnológico de grandes potencialidades. *Tecnología Química*, Vol.28, pp. 80-90

Rombouts, F.M. & Pilnik, W.L. (1980). Pectic enzymes. In: *Economic microbiology: microbial enzymes and bioconversions*, A. H. Rose, (Ed.), Vol.5, pp. 227-282. Academic Press, ISBN: 0125965559, London, England

Sakai, T. (1992). Degradation of pectins. In: *Microbial degradation of natural products*, G. Winkelmann, (Ed.), pp. 57-81, VCH, ISBN: 3527283544, Weinheim & New York, Germany & USA

Sakai, T., Sakamoto, T., Hallaert, J. & Vandamme, E.J. (1993). Pectin, pectinase and protopectinase: production, properties and applications. *Advances in Applied Microbiology*, Vol.39, pp. 213-294

Sarioglu, K., Demir, N., Acar, J. & Mutlu, M. (2001). The use of commercial pectinase in the fruit juice industry, part 2: Determination of the kinetic behaviour of immobilized commercial pectinase. *Journal of Food Engineering*, Vol.47, pp. 271-274

Schols, H.A., Visser, R.G.F. & Voragen, A.G.J. (2009). Pectin and Pectinases. Wagenigen Academic Publishers, ISBN 978-90-8686-677-9, Wagenigen, Netherlands

Schwan, R.F. & Wheals, A.E. (2004). The microbiology of cocoa fermentation and its role in chocolate quality. *Critical Reviews in Food Science and Nutrition*, Vol.44, pp. 205-221

Serrat, M., Bermúdez, R.C. & Villa, T.G. (2002). Production, purification and characterization of a polygalacturonase from a new strain of *Kluyveromyces marxianus* isolated from coffee wet-processing wastewater. *Applied Biochemistry and Biotechnology*, Vol.97, pp. 193-208

Serrat, M., Rodríguez, O., Camacho, M., Vallejo, J.A., Ajeitos, J.M. & Villa, T.G. (2011). Influence of nutritional and environmental factors on ethanol and endopolygalacturonase co-production by *Kluyveromyces marxianus* CCEBI 2011. *International Microbiology*, Vol.14, pp. 41-49

Servili, M., Begliomini, A.L., Montedoro, G., Petruccioli, M. & Federici, F. (1992). Utilisation of a yeast pectinase in olive oil extraction and red wine making processes. *Journal of the Science of Food and Agriculture*, Vol.58, pp. 253-260

Sieiro, C., Poza, M., Vilanova, M. & Villa, T.G. (2003). Heterologous expression of the *Saccharomyces cerevisiae PGU1* gene in *Schizosaccharomyces pombe* yields an enzyme with more desirable properties for the food industry. *Applied and Environmental Microbiology*, Vol.69, pp. 1861-1865

Sieiro, C., Sestelo, A.B.F. & Villa, T.G. (2009). Cloning, characterization, and functional analysis of the *EPG1-2* gene: a new allele coding for an endopolygalacturonase in *Kluyveromyces marxianus*. *Journal of Agricultural and Food Chemistry*, Vol.57, pp. 8921-8926

Silva, C.F., Schwan, R.F., Dias, E.S. & Wheals, A.E. (2000). Microbial diversity during maturation and natural processing of coffee cherries of *Coffea arabica* in Brazil. *International Journal of Food Microbiology*, Vol.60, pp. 251-260

Souza, J.V.B., Silva, E.S., Maia, M.L.S. & Teixeira, M.F.S. (2003). Screening of fungal strains for pectinolytic activity: endopolygalacturonase production by *Peacilomyces clavisporus* 2A.UMIDA.1. *Process Biochemistry*, Vol.39, pp. 455-458

Swain, M.R. & Ray, R.C. (2010). Production, characterization and application of a thermostable exo-polygalacturonase by *Bacillus subtilis* CM5. *Food Biotechnology*, Vol.24, pp. 37-50

Szajer, I. & Szajer, C. (1982). Pectin lyase of *Penicillium paxilli*. *Biotechnology Letters*, Vol.4, pp. 549- 552

Thakur, B.R., Singh, R.K. & Handa, A.K. (1997). Chemistry and uses of pectin: a review. *Critical Reviews in Food Science and Nutrition*, Vol.37, pp. 47-73

Trigui-Lahiani, H., Ayadi, M., Hadj-Taieb, N., Ben Ali, M. & Gargouri, A. (2008). Genomic organization of a polygalacturonase gene from a hyperpectinolytic mutant strain of *Penicillium ocitanis*. *FEMS Microbiology Letters*, Vol.281, pp. 23-29

Truong, L.V., Tuyen, H., Helmke, E., Binh, L.T. & Schweder, T. (2001). Cloning of two pectate lyase genes from the marine Antarctic bacterium *Pseudoalteromonas haloplanktis* strain ANT/505 and characterization of the enzymes. *Extremophiles*, Vol.5, pp. 35-44 van Rensburg, P. & Pretorius, I.S. (2000). Enzymes in winemaking: Harnessing natural catalysts for efficient biotransformation: a review. *South African Journal of Enology and Viticulture*, Vol.21, pp. 52-73

Vilanova, M. & Sieiro, C. (2006). Determination of free and bound terpene compounds in Albariño wine. *Journal of Food Composition and Analysis*, Vol.19, pp. 694-697

Vilanova, M., Blanco, P., Cortés, S., Castro, M., Villa, T.G. & Sieiro, C. (2000). Use of a *PGU1* recombinant *Saccharomyces cerevisiae* strain in oenological fermentations. *Journal of Applied Microbiology*, Vol.89, pp. 876-883

Wang, H.Z., Fu, L. & Zhang, X.G. (2011). Comparison of expression, purification and characterization of a new pectate lyse from *Phytophthora capsici* using two different methods. *BMC Biotechnology*, (DOI: 10. 1186/1472-6750-11-32)

Ward, O.P. & Moo-Young, M. (1989). Enzymatic degradation of cell wall and related plant polysaccharides. *Critical Reviews in Biotechnology*, Vol.8, pp. 237-274

Watson, B., Goldberg, N., Chen, H.P. & McDaniel, M. (1999). Fermentation processing effects on colour, phenolic profiles, and sensory character of Oregon Pinot noir wines. In: *Proceedings of the 12th International Oenological Symposium*, Montreal, Canada

West, S. (1996). Olive and other oils. In: *Industrial enzymology* (2nd edition), T. Godfrey & S. West, (Eds.), pp. 293-300, Stockholm Press, ISBN: 0333594649, New York, USA

Whitaker, J.R. (1990). Microbial pectinolytic enzymes. *Microbial enzymes and biotechnology* (2nd edition), W.M. Fogarty & C.T. Kelly, (Eds.), pp. 133-176, Elsevier Science Ltd., ISBN: 1851664866, London, England

Wightman, J.D., Price, S.F., Watson, B.T. & Wrolstad, P.E. (1977). Some effects of processing enzymes on anthocyamino and phenolics in Pinot noir and Cabernet Sauvignon wines. *American Journal of Enology and Viticulture*, Vol.48, pp. 39-48

Will, F., Bauckhage, K. & Dietrich, H. (2000). Apple pomace liquefaction with pectinases and cellulases: analytical data of the corresponding juices. *European Food Research and Technology*, Vol.211, pp. 291-297

Williams, P.J., Sefton, M.A. & Wilson, B. (1989). Non volatile conjugated of secondary metabolites as precursors of varietal grape flavor components. In: *Flavor chemistry: trends and developments*, R. Teranishi, R.G. Buttery & F. Shaihidi, (Eds.), pp. 35-48, ISBN: 0841215707, Washington D.C., USA

Winterhalter, P. & Skouroumounis, G.K. (1997). Glycoconjugated aroma compounds: occurrence, role and biotechnological transformation. In: *Biotechnology of aroma compounds. Advances in biochemical engineering/biotechnology*, T. Scheper, (Ed.), pp. 74-105, Springer, ISBN: 3540614826, Berlin, Germany

Xiao, Z., Boyd, J., Grosse, S., Beauchemin, M., Coupe, E. & Lau P.C.K. (2008). Mining *Xanthomonas* and *Streptomyces* genomes for new pectinase-encoding sequences and their heterologous expression in *Escherichia coli*. *Applied Microbiology and Biotechnology*, Vol.78, pp. 973-981

Yadav, S., Yadav, P., Yadav, D. & Yadav, K. (2008). Purification and characterization of an acidic pectin lyase produced by *Aspergillus ficuum* strain MTCC 7591 suitable for clarification of fruit juices. *Annals of Microbiology*, Vol.58, pp. 61-65

Yadav, S., Yadav, P.K., Yadav, D. & Yadav, K.D.S. (2009). Pectin Lyase: A review. *Process Biochemistry*, Vol.44, pp. 1-10

Yamaski, M., Yasui, T. & Arima, K. (1964). Pectic enzymes in the clarification of apple juices. Part I. Study on the clarification reaction in a simplified model. *Agricultural and Biological Chemistry,* Vol.28, pp. 779-787

Yang, Y., Luo, H.I., Li, J.A., Wang, K., Cheng, H.P., Bai, Y.G., Yuan, T.Z., Fan, Y.L. & Yao, B. (2011). Cloning, expression and characterization of an acidic endopolygalacturonase from *Bispora* sp. MEY-1 and its potential application in juice clarification. *Process Biochemistry,* Vol.46, pp. 272-277

12

Dried Probiotics for Use in Functional Food Applications

Karla Bigetti Guergoletto[1*]
[1] *Universidade Estadual de Londrina, Brazil*

1. Introduction

The increasing consumer demand for foods with health benefits has led the food industry to diversify its products. Most of the foods containing probiotic bacteria are dairy products, although there is a rapidly growing demand for incorporating probiotics in other segments of the food industry.

According to Food and Agriculture Organization (FAO, 2002), probiotics are live microorganisms that, when administered in adequate amounts, confer health benefits on the host. In order to produce such health benefits, it is essential that live probiotic bacteria survive after passing through the gastrointestinal tract and reach their site of action intact. Foods containing probiotic microorganisms are expected to have a promising future, but will require the development of new technologies, specifically developed to enable their use in innovative products in a wide range of industries. The challenge for the industry is to produce large enough amounts of viable and stable probiotic cultures for use as inoculums or to be introduced directly into a dried food.

Although specific numbers are not mentioned at the FAO definition, some studies demonstrated that high levels of viable microorganisms are recommended in probiotic foods for efficacy (Meng et al., 2007). However, maintaining such counts throughout preparation, processing and storage requires constant study by the scientific and industrial community.

Drying process are often used as a means to stabilize probiotics and facilitate storage, handling, transportation and subsequent use in functional foods (Santivarangka et al., 2008). Drying by spray-drying and freeze-drying are the most commonly used techniques for dehydrating probiotic cultures and lactic acid bacteria starter cultures (Betoret et al., 2003), although other methods such as vacuum oven drying, fluid bed drying or combinations of these may also be used.

Drying processes are a major cause of loss of viability of probiotics, and in the specific case of freeze-drying, the freezing step causes additional stress on the bacterial cells, making the

*Karla Bigetti Guergoletto[1], Kátia Sivieri[2], Alessandra Yuri Tsuruda[1], Elvis Peboni Martins[1],
Jean Clovis Bertuol de Souza[3], Salvador Massaguer Roig[3], Elisa Yoko Hirooka[1] and Sandra Garcia[1]
[1]*Universidade Estadual de Londrina,*
[2]*Universidade Estadual Paulista,*
[3]*Universidade Norte do Paraná*
Brazil

lipid fraction of the cell membrane more susceptible to damage during the process. Although the exact mechanisms of cell inactivation during drying processes are not yet fully elucidated, it is known that bacterial cells consist of 70 to 95% water, and its removal poses serious physiological obstacles to the survival of cells (Santivarangka et al., 2008).

Survival of microorganisms during drying processes and subsequent storage, depends on factors, such as species and strain, drying conditions, inoculums and culture media, the use of cultures pre-adapted to the stress inherent to the processing conditions and the use of cryoprotective agents (Champagne et al., 2005). Examples of such agents include trehalose, sucrose, glycerol, skim milk powder, whey, betaine, gum acacia (Meng et al., 2007) and exopolysaccharides, the latter of which are produced by the cells themselves as a protective mechanism in response to stress situations (Ruas Madiedo et al., 2002). The use of fibers and prebiotics as multi-functional protective agents and vehicles for probiotics / symbiotics has gained wide interest over the past few years (Saarela et al., 2006).

This chapter describes some methods to increase the viability of probiotics, along with new uses and applications in the functional food industry. Dehydration methods, such as freeze-drying, spray-drying and the use of cell protective additives will be introduced. Finally, the use of probiotics in ice cream after drying by freeze-drying and the use of probiotics in dark chocolate after drying by spray-drying will be discussed.

1.1 Drying processes

There is currently a real interest on the part of the food processing, chemical and pharmaceutical industries and in agriculture in the use of products containing sensitive compounds or living cells, due to beneficial functional properties and effects on the environment and human health (Ivanova et al., 2005). These products are used in dehydrated form to increase stability, and may be either encapsulated or attached to a support for this purpose (Anal & Singh, 2007). They can be developed for therapeutic use by oral administration in capsules/tablets, or as direct starter cultures, as functional supplements or as biocontrol agents. Commercial dried probiotic cultures obtained by processes the cost-benefits of which are interesting may help to expand the market and enhance the use of these microorganisms in several other applications (Fu & Chen, 2011).

Drying is an ancient method of food preservation and the term is generally used to refer to the withdrawal of moisture from a substance. It is used in the drying of particulate solids, pastas, solutions and provides the greatest diversity among the unit operations used in foods (Ratti, 2001). Conventional hot air drying allows to extend the shelf life for long periods of time, however, the quality of the products obtained is drastically reduced when compared to traditional drying. Drying by convection may, however, offer an efficient alternative in terms of costs and transport, although studies on the intrinsic factors and optimization of extrinsic processes should be carried out to minimize losses that may occur during the hot drying process (Fu & Chen, 2011; King & Su, 1994).

During thermal processing, stress caused by heat and by dehydration may be significant to the cells. Both forms of stress at lethal levels cause permanent loss of viability. Excessive heat causes denaturation of the structure of macromolecules or breaks the bonds between monomeric units. Stress caused by dehydration affects principally the cytoplasmic membrane by changing fluidity or the physical state of the membrane, in addition to causing lipid peroxidation. Microorganisms subjected to dehydration by heat would undergo simultaneously the two types of stress in a synergistic manner (Fu & Chen, 2011).

Culturing conditions and the cellular growth phase have great influence on viability during drying, with the cells in the stationary phase being more resistant than in any of the other stages. For that reason, this is the stage of development that has been used in different drying processes, such as: spray drying; fluid bed drying; vacuum drying; as well as freeze drying (Fu & Chen, 2011).

When the water is removed at a fast rate, the microorganism has not the time it needs to adapt itself through genetic expression or by adjusting its metabolism. An osmotic pressure response can be induced by stress conditions and leads to accumulation of compatible solutes, as well as to a cross-response with the effects of the other stress conditions. Pre-adaptation to heat induces the heat-shock response of proteins and enzymes, such as chaperones and proteases. The first help to stabilize the RNA and repair denatured proteins, while the proteases degrade denatured proteins, with both reactions leading to heat stress resistance (Fu & Chen, 2011).

The industrial use of a certain and specific culture sometimes does not allow to replace this same culture by another and, for that reason, pre-adaptation offers the opportunity to increase survival against adverse conditions (Fu & Chen, 2011).

Freeze-drying, also known as lyophilization or cryo-desiccation, was initially used as a commercial technique in the 1930's for the preservation of blood plasma and instant coffee, however, from that time onward, it has been being used for the preservation of other foods and products (Food Today, 2009). It is the best method for removing water and generally the water content falls to levels lower than 4% (Muller et al., 2009).

For the lyophilization, the food or suspension medium was quickly frozen, resulting in the formation of small ice crystals, while at the same time the non-frozen phase becomes concentrated. Large crystals formed during slow freezing might damage the structure of the product or penetrate the cell walls (Stapley, 2009; Food Today, 2009). A number of factors affects the efficacy of cryopreservation of microorganisms, but one of the most important conditions is the composition of the suspension medium for freezing (Hubalek, 2003). The solid state of the water during lyophilization protects the primary structure and shape of the product with minimal volume reduction. In spite of the numerous microbiological advantages and the high quality of the products as compared to other methods, lyophilization is always considered a slow and the most expensive dehydration process (Ratti 2001; Peighambardoust et al., 2011), in addition to also being seen as less favorable for the production of larges masses of dry cells (Fu & Chen, 2011).

In lyophilization, the frozen material is exposed to a head space with partial vapor pressure a slightly below the partial equilibrium vapor pressure of the ice at the temperature of the material. This is achieved by placing close to the food a condensation unit that causes the sublimation of the ice crystals (primary sublimation) and also desorption of non-crystalline water present in the food matrix (secondary drying). The condensation surface temperature should be significantly lower than that of the drying material (Stapley, 2009). In conventional lyophilization, the vacuum in the lyophilization chamber is maintained by a vacuum pump, the function of which is to remove non-condensable gases, since these reduce the transfer rate of the water vapor from the food to the condenser. For being endothermic, in sublimation heat must be applied to the food to maintain and accelerate the process. (Stapley, 2009; Food Today, 2009), without thawing the food and maintaining its rigidity. However, the temperature of the food should always be kept below the temperature of collapse, since the latter is related to the glass transition temperature (T_g). Several researchers have verified that the temperature of collapse lies 2-3K above T_g or between the values of T'_g and the phase transition temperature of the cytoplasmic membrane (T'_m), though it varies as a function of moisture content (Stapley, 2009). The collapsing phenomenon occurs when the solid food matrix can no longer maintain it own weight, leading to a drastic structural change which is marked by a decrease in volume, an increase in stickiness of dry powders, loss of porosity, etc (Chuy & Labuza, 1994, Ratti, 2001).

A comparison between the reduction in volume during lyophilization and conventional drying shows that in the first case this reduction is minimal (5 to 15%), whereas it is excessive in the case of air drying (close to 80%), and that this behavior is observed in several foods. In the case of hot air the reduction is accompanied by wrinkling, deformation or color changes indicating a collapse of the food dehydrated by this method. Also according to data shown by Ratti (2001), in conventional drying the temperature of the product remains above T_g throughout the entire process, reason why collapse and inferior quality of the final product are expected.

With regard to costs, lyophilization is generally seen as an expensive method of preservation (Peighambardoust et al., 2011), which depends on the type of material, the products, the packaging, the capacity of the processing plant, duration of the cycle, etc. The total cost is about 4 to 8 times greater than that of hot air drying. However, studies indicate that, when an analysis is made of the total energy involved in the different processes, the inclusion of the expenses with home storage, lyophilization is advantageous. Furthermore, comparing the several steps involved, the added value and improvement in the quality of raw materials and the differentiated products, such as is the case of probiotics, lyophilization should not be considered prohibitive in terms of cost (Ratti, 2001). Lyophilization involves four main steps: freezing; vacuum; sublimation and condensation, with each contributing to the total energy consumption. Any development in the classical process to reduce energy costs should focus on improving heat transfer in the sublimation phase, reduce time and consequently expenses with vacuum and avoid the use of condensers. Technological developments such as microwave heating, desorption freeze-drying using desiccants, atmospheric fluidized bed freeze-drying, atmospheric spray-freeze drying have not yet been totally incorporated by the industries due to a series of technical problems (Ratti, 2001).

Probiotic lyophilized or atomized yogurts and fermented products have been described by various authors and, in these cases, the viability and activity of the bacteria are important to obtain the beneficial effects, in addition to maintaining the original authentic qualities of the products (Kumar & Mishra, 2004; Peighambardoust et al., 2011). Different approaches can be used for evaluating the survival or activity of the dehydrated cells. Among these, one can cite the determination of cell viability in itself, the determination of the metabolic capacity and the determination of the ability of the microorganisms to tolerate stressing environmental conditions, such as acidity and bile (Fu & Chen, 2011). Lyophilization and drying by atomization are among the most used processes to maintain these properties. Since lyophilization utilizes milder conditions, the survival rates are higher than those obtained with atomization (Wang et al., 2004), with the specific exception of the damage caused by freezing. The physical mechanism of dehydration and cryoinjuries caused by freeze-drying are different from what occurs in heat-drying, as well as the damage on the cell structure (Fu & Chen, 2011). In the case of *L. paracasei*, for example, lower temperatures applied during freezing stage of the lyophilization process lead to lower survival rates probably due to stress during freezing (Bauer et al., 2011). Protection of the cells during the different steps is of fundamental importance (Semyonov et al., 2010) and the effect of protectants should be studied on a case-by-case basis as a function of the lineages used (Carvalho et al., 2004). However, atomization for the commercial production of cultures allows to reduce costs (Peighambardous et al., 2011; King & Su, 1994) and increase the yield. (Zamora et al., 2006). In spite of being economic and effective, spray-drying inflicts high a mortality rate the cells as a result of the simultaneous action of heat, oxygen and dehydration stress (Anal & Singh, 2007; Fu & Chen, 2011).

Atomization (spray-drying) is considered a long-term preservation method for lactic bacteria and dates from 1914 when it was described in a study by Rogers on dehydrated lactic cultures. The concept of spray-drying was patented by Samuel Percy in 1872 and industrial use with milk started in 1920. The drying rate and continuous production capacity are useful for large-scale production. In spite of this, it is commercially less employed due to the low survival rates during drying, low stability during storage and difficulty to rehydrate the product (Peighambardoust et al., 2011). Spray-drying, a convective drying process (Fu & Chen, 2011) is a unique process in that the particles are formed at the same time they are being dehydrated. The final product should be compatible with the quality standards for particle size and particle distribution, residual moisture content, density and shape of the particles. During the process, dry granulated particles are obtained from a paste. By atomization the moist product in the shape of small droplets is passed at high speed through a flow of hot air between 150 a 200ºC. The atomized droplets have a large surface area, which results in a very short drying time when exposed to hot air in the drying chamber. There are several types of atomizers and a classification would probably be based on the kind of energy involved: centrifugal, kinetic, pressure or sound energy (Vega-Mercado et al., 2001). Selecting the configuration of the atomizer depends however, on the nature and viscosity of the feeding flow, as well as on the characteristics of the dehydrated product. In the period of constant drying rate, the temperature of the product and the thermal inactivation are limited to the temperature of the wet bulb and if the evaporation rate is high, the cells are protected against the high temperatures of the air in the chamber (Fu & Chen, 2011 ; Peighambardoust et al., 2011). The falling rate period of the particle is important and the optimal residence time is the time necessary for complete removal of moisture with a minimum increase in the temperature of the dry product (Peighambardoust et al., 2011).

The preparation of dry starter cultures is a long process which involves various steps from culturing up to storage in powdered form (Peighambardoust et al., 2011). During drying by atomization, bacteria may suffer damage to their cytoplasmatic membrane due to the high temperatures and survivors with sub-lethal injuries become more sensitive to the storage conditions, losing viability during storage time (Corcoran et al., 2004). Several investigators have reported that increased outlet temperatures reduce the survival of the microorganisms upon completion of the spray-drying process, with this being one of the most important parameters affecting the viability of the starter cultures (Peighambardoust et al., 2011). Other pre-drying processes, such as emulsification, homogenization and pressure applied prior to atomization may cause injury or inviability by 2 to 3 log cycles. For that reason, it is

suggested that probiotic bacterial or yeast cultures be added only after these processes are completed. In addition to mechanical stress, the high atomization pressures lead to the formation of smaller particles which are subject to inactivation caused by dehydration and heat, thereby emphasizing the importance of selecting the best atomizer (Fu & Chen, 2011).
Another alternative drying method is vacuum drying, which is practiced at low temperatures. This method is applied using temperature ranges between 30 and 80°C and may cause great cell losses due to heat damage. However, by working slightly above the triple point of water, temperatures close to 0°C would be obtained. This process is called Controlled Low-Temperature Vacuum Dehydration (CLTV), and is interesting for sensitive food ingredients and can be used for drying microorganisms such as *L. acidophilus* (King & Su, 1994). Compared to spray-drying, other processes using lower drying temperatures allow for greater and better survival of microorganisms (Fu & Chen, 2011).
Survival rates are species-specific, depend on the drying method used and experiments are essential to attain maximum survival under a given drying condition (Fu & Chen, 2011; Otero et al., 2007). The different behavior of the cells during dehydration may be explained by differences in the cell membrane (Bauer et al., 2011). Indeed, Bauer et al (2011) have demonstrated differences in behavior between probiotics by comparing controlled low-temperature vacuum dehydration (LTVD) and lyophilization (Freeze-drying - FD). For the survival of *Bifidobacterium lactis Bb-12*, FD would be the method of choice, for *L. paracasei* there wasn't any difference between the two methods and for starter culture *L.delbruekii ssp bulgaricus* manufactured using the LTVD, the culture exhibited higher survival rates.
The loss of viability during storage increases with temperature between 4 and 37°C (Sunny-Roberts & Knorr, 2009). The mechanism behind the loss of cell viability during the storage period depends on a series of factors, including temperature, moisture content, protectors and oxidative stress. Since there is generally no heat stress during storage, the mechanism of death must be different from that which occurs in thermal processes (Fu & Chen, 2011).

1.2 Cell protectors

Initiation bacterial cultures (starters) are of great importance for the industry for the production of fermented foods, with drying being the most used method to prepare and store these microorganisms. However, this practice may cause injuries that cause a drop in the number of viable cells and consequent losses in industrial production. In this way, a well-succeeded drying process depends on the capacity to maintain a sufficient number of cells that contains and preserves the same characteristics of the original culture (Santivarangkna et al., 2008).
A method commonly employed to increase the survival of bacterial cultures is the addition of protective agents. These agents have been previously added to drying media to protect the cells against adverse processing and storage conditions. Among these protectors may be cited skim milk powder, whey, trehalose, glycerol, betaine, saccharose, glucose, and lactose, in addition to some polysaccharides (Meng et al., 2007).
Several hypotheses have been proposed to explain the effect of these protectors on the process of cell dehydration. Among these, the presence of compatible solutes, such as sugars, quaternary amines and some amino acids, may equip the cells against hyperosmotic stress, which occurs during the drying process, bearing a direct relationship between the accumulation of these solutes and survival under adverse conditions (Kets et al., 1996). According to Santivarankna et al. (2008), it is believed that the protective effect of these

solutes occurs due to their preferential exclusion, which is the main mechanism of protection of macromolecules against moderate loss of water. Still according to the same authors, the preferential exclusion occurs where a macromolecule has greater affinity with water than with the solute, with the consequence of the latter becoming in excess on the surface of the macromolecule.

Cell growth is the first step in the preparation of dehydrated starter cultures, but its viability during drying and the subsequent storage period may be affected by the growth conditions and growth medium (Santivarankna et al., 2008). It has been suggested that adverse conditions such as heating or the application of osmotic stress during the microbial growth phase may leave the lactic acid bacteria more tolerant to dehydration processes. In the study published by Tymczyszyn et al. (2007), *Lactobacillus bulgaricus* were pre-adapted in a low-water activity medium, containing lactose and polyethylene glycol during the cellular growth phase and it was observed that both conditions at different concentrations brought about an increase in the survival of the cells subjected later on to the vacuum drying process.

In this aspect, the presence of carbohydrates in the growth medium has been shown to be of great importance. Panoff et al. (2000) observed that the growth of *Lactobacillus delbrueckii* subsp. *bulgaricus* in the presence of sugars such as lactose, saccharose, trehalose and cryoprotectants, such as glycerol, causes the cells to adapt themselves to freezing and thawing in the lyophilization process with a consequent increase in cell survival. In the study by Carvalho et al. (2004), supplementation of the growth medium of *L. bulgaricus* with glucose, fructose, lactose and mannose, resulted in increase in the level of protection conferred by the agents used in the drying medium as compared to non-supplementation. Furthermore, the data suggest that cell growth in the presence of several sugars produces cells with different morphological and physiological characters, which represent distinct levels of resistance to the various stress treatments investigated. The osmotic stress caused by the solutes in this growth stage may favor the assimilation of the protectants inside the cells, which in its turn leads to a reduction of the osmotic effects that dehydration causes (Kets et al., 1997), since the protectant accumulated within the cell reduces the osmotic difference between the internal and external environment (Capela et al., 2006).

On the other hand, several lactic acid bacteria (LAB) produce exopolysaccharides (EPS), which may remain either attached to the cell wall (capsular) or be excreted in free form (mucus). The majority of the studies involving the production of EPS by the LAB bacteria focus on their use in the food industry as viscosity agents, stabilizers, emulsifiers or gelling agents (De Vuyest &Degeest, 1999), although some authors have suggested the importance of their role in the mechanism to protect the cells against dehydration processes (Ruas Madiedo et al., 2002). It has been observed that the production of this metabolite requires the utilization of substrates and energy, and since this polysaccharide is not considered essential to the cell, it is believed that its production is connected to an important biological function, such as for example, obtaining a selective advantage in the environment (Looijesteijn et al., 2001). In this manner, the presence of certain carbohydrates such as gum acacia, may favor their production and consequently increase bacterial survival in subsequently adverse conditions (Schiavão-Souza et al., 2007).

Protective agents in the drying medium may be available both in the form of sugars that are not utilized in the growth medium when there is no washing stage, or from the build-up of compatible solutes. However, in most cases they are additionally mixed or used as drying medium to increase the viability of bacterial cultures (Santivarangkna et al., 2008).

The damage caused in biological systems by drying processes should be attributed to: denaturation of critical main components, such as DNA and RNA (Linders et al., 1997); change in the physical state of the lipid membrane; and changes in the structure of more sensitive proteins. The removal of water from region of the polar chains of the phospholipids, causes the closure of these chains, forcing the encounter of acyl groups, favoring the formation of Van der Waals interactions. As a result, the lipid may pass from the crystalline liquid to the gel phase and with rehydration, this change occurs again. For that reason, in the process of the membrane going through these phase changes, there are regions that become defective, turning the membrane leaky (Leslie et al., 1995).

With the addition of protectants, such as the disaccharide trehalose before drying, a reduction in the membrane transition temperature is observed. This occurs due to the replacement of water between the front end of the lipid groups, which prevents the change of phase and consequent leakage during rehydration. In addition, these sugars have the capacity to form a vitreous state, in which they present high viscosity and low mobility, leading to an increase in the stability of the material to be preserved (Crowe et al., 1996). In lyophilization processes, the presence of cryoprotectors may raise the vitreous transition temperature and this way, the viable cells become able to reach this state of stability without nucleation of intracellular ice, which might cause damage inside the cell by the crystals formed (Meng et al., 2007).

The damage inflicted on dehydrated probiotic cultures during drying may be aggravated during their storage, which would make them undesirable or unfit for further industrial use. Teixeira et al. (1996) observed that after storage of dried cells of *L. bulgaricus*, the unsaturated fatty acids/saturated acids ratio of the lipid membrane diminishes considerably, thereby evidencing the damage caused by lipid oxidation. This oxidation may result in some physical changes in the functions and structure of the membrane. The increase in the proportion between saturated fatty acids brings about an increase in the transition temperature and consequently leads to a reduction in the fluidity of the membrane up to a certain temperature and an increase in leakage through the membrane during rehydration. In addition to this, it is known that the consequence of a great number of biological oxidations results in the formation of free radicals and that these may reduce the hydrophobicity of fatty acids due to the introduction of hydrophilic groups, and thus, weaken the hydrophobic interaction with the proteins of the membrane, which may be essential to their activity (Santivarangkna et al., 2008).

Nonetheless, the use of protectants in drying medium may reverse this pattern. Some sugars like sorbitol, maltose and mannitol are also effective protective agents capable of providing cultures with protection against oxidation damage inflicted over long periods of storage. Sunny-Roberts & Knorr (2009) observed an increase in the viability of *Lactobacillus rhamnosus LGG* stored at room temperature in dried form in the presence of trehalose supplemented with monosodium glutamate (MSG), indicating a possible antioxidant action attributed to a combination of compounds. The possible mechanisms suggested behind this antioxidant protection provided by sugars are reported for being free of free radicals and metal quelants, and having capacity to form complexes of hydrogen peroxide and restrict the viscous flow of oxygen diffusion (Santivarangkna et al., 2008).

Additionally, it is suggested that temperature is one of the more critical parameters for microbial survival during storage (Gardiner et al., 2000), and that this effect may be directly related with the speed or rate of the biochemical and physiological reactions that occur inside the microorganisms. Dehydrated probiotics are generally stable when stored at temperatures of 4ºC. However, the stability of dried probiotics kept at non-refrigeration temperatures is strongly recommended when these are used to supplement products such as milk powder, cereals, confectionery and chocolate, in addition to significantly reducing the costs with transportation and storage (Foerst et al., 2011).

Studies have shown that the use of sugars during drying and in the course of subsequent storage may result later in higher survival at higher storage temperatures. Foerst et al. (2011) evaluated the storage at different temperatures of *L. paracasei* F19 vacuum-dried in the presence and absence of protectants and observed that when drying was done in the presence of sorbitol, the microorganism may be stored at room temperature without any significant loss of its viability. Ananta et al. (2005) obtained similar data and information on the survival for *Lactobacillus rhamnosus* LGG atomized with reconstituted skim milk (RSM).

Trehalose (α-D- glucopyranosyl, α-D- glucopyranoside) is a non-reducing disaccharide synthesized by procaryotic and eucaryotic organisms, such as yeasts, bacteria, molds, algae, plants and invertebrate insects (Conrad et al., 2000). Several studies have demonstrated that trehalose is one of the best protein stabilizers as compared to other compatible solutes (Crowe et al., 1996) under the identical conditions of drying and storage, because of its uncommon capacity to stabilize proteins in their native conformation by changing the water environment surrounding the peptide (Magazú et al., 2005). Trehalose may interact directly with the polar chains of the phospholipids of the membrane via a hydrogen bridge between the hydroxyl groups of the sugar and the phosphate groups of the lipids, stabilizing them under adverse conditions (Welsh, 2000). Zayed & Ross (2004) reported a 10% increase in the viability of *L. salivarius* when lyophilization of these microorganisms was performed in the presence of 15% trehalose.

It was observed that the production of trehalose by organisms that synthesize it is directly linked to the cell protection mechanism under adverse conditions. Li et al. (2009), observed that under stressing conditions such as heating, the genes responsible for the codification of its biosynthesis are activated, promoting their formation and release into the environment, consequently providing protection to the cells. The study by Mamose et al. (2010) supports the hypothesis that the presence of trehalose before freezing, in addition to reducing the damage caused by this procedure also favors the formation of several processes for the recovery from post-freezing damage.

In addition to the mono-and disaccharides, polysaccharides have been shown as promising for use as cell protectants. In spite of the size of their molecules, they can be inserted between the phospholipid chains, changing the transition temperatures of the membranes. This polysaccharide–phosphoplipid interaction depends on the flexibility of its structure. In polysaccharides with rigid structure, the interaction is dependent upon size, whereas for those with a flexible structure, the negative steric effects of increasing degrees of polymerization is counterbalanced by the elasticity of the structure (Santivarangkna et al., 2008). The use of gum acacia in the drying medium of probiotic cultures, has been proven promising. Desmond et al. (2002) obtained a survival rate 10 times greater when the microorganism *L. paracasei* NFBC 338 was dried by spray-drying in the presence of 10% (g/v) gum acacia as compared to drying without the polysaccharide. In the study by Lian et al. (2002), four lineages of bifidobacteria were successfully subjected to spray-drying with gum acacia.

Reconstituted skim milk (RSM) is another carrier matrix utilized in the production of dehydrated probiotics. In studies investigating the protective effect of different food systems in the survival of cells subjected to drying, the results obtained with RSM stood out in comparison to those yielded by other compound such as gelatin and starches (Lian et al., 2002). Fu & Chen (2011) suggest that their effective protection may be related to the presence of lactose in the composition of RSM, since this is the disaccharide that interacts with the cell membrane and helps to maintain its integrity in a similar way as non-reducing disaccharides do with trehalose and saccharose. However, this mechanism has not yet been totally elucidated as yet. In another way RSM might create a porous structure in the lyophilized product, which would make its rehydration easilier (Carvalho et al., 2004).

Nonetheless, researchers are satisfactorily using formulations containing RSM and other compounds to increase the survival of probiotics subjected to drying and subsequent storage for later use. Corcoran et al. (2004) evaluated the influence of RSM combined with polydextrose on the lactobacillus viability during spray-drying and observed that these combinations produced an increase in the survival of these cells during storage. Likewise, Ananta et al. (2005) used reconstituted skim milk at a concentration of 20% (w/w) for spray-drying *Lactobacillus rhamnosus* GG, and found that the incorporation of raftilose and polydextrose to this medium, resulted in an increase in the survival of the probiotic after drying.

For that reason, the use of compounds that may protect bacterial cells against drying processes constitutes an important strategy for the industry, which desires to obtain probiotics with high counts to be added to several types of foods, in a way so as to provide and guarantee the benefits this type of product.

2. Perspective of trehalose and gum acacia on the survival of *Lactobacillus casei* subsp. *paracasei* LC-1 culture upon lyophilization and application in ice-cream

2.1 Use the trehalose and gum acacia on the survival of *Lactobacillus casei* subsp. *paracasei* LC-1

Lyophilization has been widely used to produce probiotics in powder form. Lyophilization is a process based on the sublimation of water, in which bacterial cells are first frozen and then dried under high vacuum (Leslie et al., 1995). Since water plays an important role in stabilizing the structure and maintaining the functional integrity of cellular macromolecules, its removal causes damage to the surface of cell walls and membranes, thereby compromising the viability of the bacterial cells (Ruas Madiedo et al., 2002).

For most probiotic bacterial cultures of commercial interest to the dairy industry, skim milk powder is frequently selected as the drying medium because it prevents cellular injury by stabilizing the cell membrane constituents, in addition to creating a porous structure in the freeze-dried product that makes rehydration easier (Carvalho et al., 2004).

The addition of protective compounds such as trehalose, sucrose, betaine and gum acacia is another option for maintaining the viability of microorganisms during freezing, drying and storage (Carvalho et al., 2004). Leslie et al (1995) showed that the survival of freeze-dried *Escherichia coli* DH5a and *Bacillus thuringiensis* HD-1 was respectively 60% and 45% higher in the presence of trehalose compared with its absence. Guergoletto et al (2010) observed a positive and significant effect of trehalose on the survival of *Lactobacillus casei* subsp. *paracasei* LC-1 subjected to vacuum drying. However, in these studies the effect of these compounds was evaluated separately.

Our group evaluated the effect of trehalose and acacia gum on *Lactobacillus casei* subsp. *paracasei* LC-1 culture upon completion of lyophilization. Table 1 shows the different concentration of trehalose and gum acacia, their coded values and the results of cellular viability calculated from the number of colonies on MRS Agar plates, before and after lyophilization using skim milk and whey media. The highest survival rates were obtained with the media containing 5% trehalose and 4.5% gum acacia, which statistically corresponds to the center point. The response surface plot (Figure. 1) shows the experimental data distributed in the intervals studied and clearly shows that the highest survival rates of LC-1 are located in the region around 5% trehalose and 4.5% acacia gum.
Conrad et al. (2000), reported a survival rate of 90% of *L. acidophillus* cells after lyophilization in a drying medium containing 5% (w/v) trehalose. Desmond et al (2002) observed that the use of 10% (w/v) gum acacia in the medium used for drying *L. paracasei* in a spray-dryer increased survival of the probiotic by 10 times compared to the medium without gum acacia.
Schiavão-Souza et al. (2007) evaluated the use of 4.5% and 5.5% gum acacia in the media of skimmed milk powder to optimize EPS production by *Lactobacillus casei* and observed that the addition of 4.5% gum acacia, favored the production of exopolysaccharides (EPS). According to Ruas-Madiedo et al. (2002) the production of these compounds by lactic bacteria may be considered as a protection mechanism against stress situations, such as dehydration processes, which may explain the higher survival rates obtained in our study.

2.2 The use of *L. casei* (LC-1) protected with trehalose and acacia gum in ice cream processes

Ice cream is an appropriate food matrix for the addition of probiotic and prebiotic ingredients. Some studies have demonstrated that it is possible to produce ice cream of high sensory quality with added probiotic microorganisms (Shah, 2000). Among other innovative products as a functional food, ice cream has a promising market and a huge growth potential. However, the addition of the probiotic microorganism should not change the sensory attributes of the product.

Trehalose (x1)	Trehalose (% w/v)	Gum acacia (x2)	Gum acacia (%w/v)	Cellular viability before lyophilization (log cfu)	Cellular viability after lyophilization (log cfu)
-1.0	2.5	-1.0	3.5	10.3	10.2
-1.0	2.5	1.0	5.5	10.5	8.4
1.0	7.5	-1.0	3.5	10.2	8.0
1.0	7.5	1.0	5.5	10.6	9.7
0.0	5.0	0.0	4.5	10.8	11.0
0.0	5.0	0.0	4.5	10.6	10.9
-1.4	1.5	0.0	4.5	10.3	10.0
1.4	8.5	0.0	4.5	10.5	9.6
0.0	5.0	-1.4	3.1	10.7	9.9
0.0	5.0	1.4	5.9	10.4	9.5

Table 1. Variables trehalose and gum acacia: actual (% w/v) and coded values (x1 and x2) and the survival of *Lactobacillus casei* subsp. *paracasei* (LC-1- values in log cfu.g^{-1}) in culture media containing milk solids, before and after lyophilization

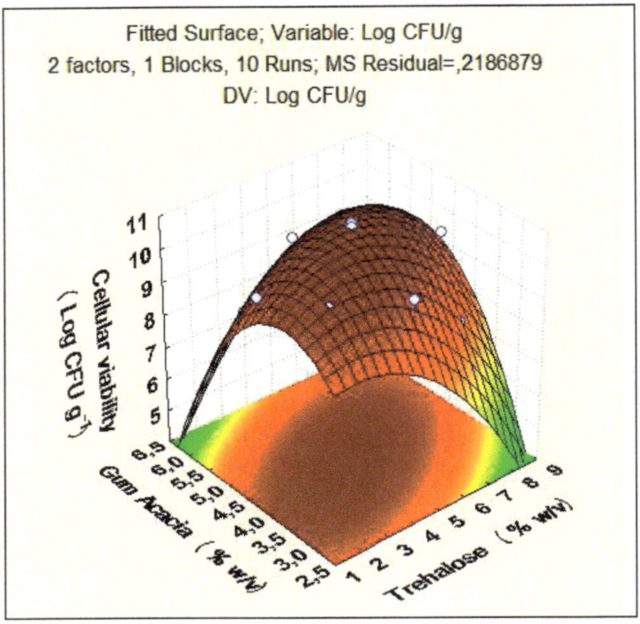

Fig. 1. Response surface plot obtained after lyophilization of *Lactobacillus casei* subsp. *paracasei* LC-1 culture suspended in a skim milk and whey medium, for the percentage of variables trehalose and gum acacia, with the viability of LC-1 expressed in log cfu.g^{-1} as response.

Our group evaluated the viability of Lactobacillus casei (LC-1) in the free form and/or protected with trehalose and gum acacia in ice cream throughout 98 days storage at -20°C, as well as the acceptability from a sensory viewpoint. Figure 2 shows the results of the free and protected L. casei populations over a total 98-day storage period -20ºC. At the beginning of storage, populations of 9.20 and 9.80 log CFU.g-1 were observed, which after 98 days of storage had been reduced to 6.90 and 8.17 log CFU.g-1 for free L. casei and protected L. casei, respectively. There was a significant difference ($p<0.01$) in the L.casei populations of the two treatments tested at the end of the storage period investigated. Hekmat & McMahon (1992) verified a decline by 2 logarithmic log cycles in the L. acidophilus counts in ice creams after 17 weeks storage at -29°C. Similar results were observed in our experiments with free L. casei.

The reduction in the population of probiotic microorganisms in ice creams during shelf life was observed by Homayouni et al., (2008); Vasiljevic & Shah (2008). The drop in the bacterial population of may be the result of freezing, causing cell injuries and eventually cell death (Cruz et al., 2009). One strategy to diminish cell injuries is the use of cell protectants, such as trehalose or gum acacia. Studies such as those by Linders et al., (1997), Conrad et al., (2000) reported the action of these substances on the protection of probiotic microorganisms during dehydration processes such as lyophilization. Our results show that the use of trehalose and that of gum acacia as cell protective agents was effective in maintaining a high population of L. casei, with a reduction by only one logarithmic cycle after 98 days storage at -20 °C in ice cream.

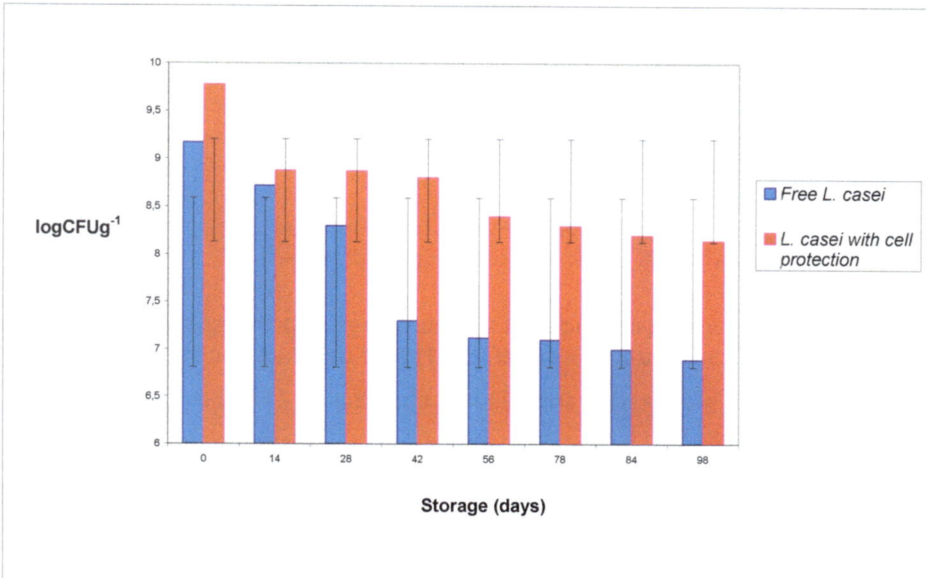

Fig. 2. Means and standard deviations of the viability of *L. casei* in ice cream during 98 days storage at -20°C.

Producing an ice cream containing probiotic of high sensory acceptability is a difficult task, since it requires a great deal of technical knowledge and most often functional ice creams present low sensory performance, when compared to conventional ice cream (Cruz et al., 2009).

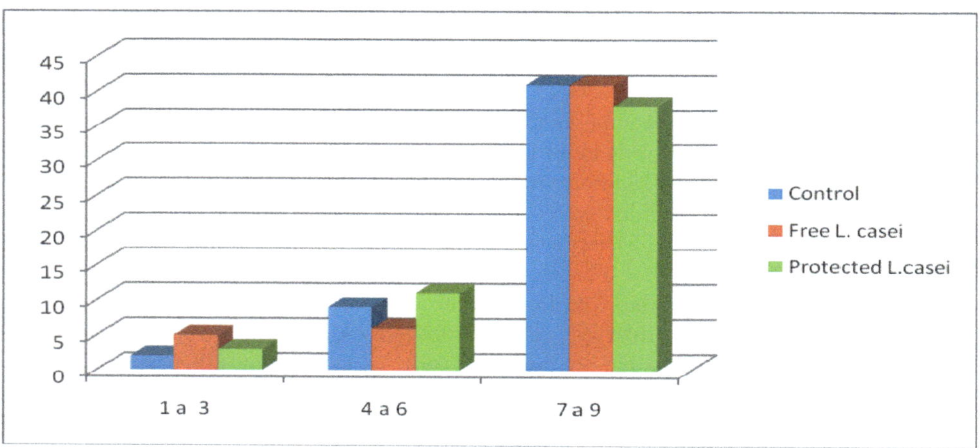

Fig. 3. Frequency hystogram for the attribute taste of the control ice creams, with addition of free and protected *L. casei*.

The results of sensory analysis of the taste of probiotic ice creams protected with trehalose and gum acacia are depicted in Figure 3. The three types of ice creams evaluated (control, added with free *L. casei* and added with protected *L. casei*) exhibited a sensory profile similar in terms of the attribute taste and did not statistically differ from each other ($p<0.05$). Figure 3 shows that the product received liking ratings between 7 and 9 ("liked" to "liked a lot"); thus, the addition of the probiotic microorganism (in the free and in the protected form) did not affect the taste of the finished product. A similar result was observed by Homayouni et al., (2008), who found that the addition of microencapsulated *L.casei* in ice cream did not affect the sensory quality of the product.

In view of all this, one may conclude that trehalose and gum acacia have excellent perspectives as an industrial application as bacterial cell protectants.

3. Perspective of spray-drying and vacuum-drying processes for the preservation of *Lactobacillus casei* (Lc-1) and use in chocolate

3.1 Combination of atomization and vacuum chamber on *Lactobacillus casei* (Lc-1) drying

The group of *Lactobacillus casei* formed by strains of *Lactobacillus casei / paracasei* and *Lactobacillus rhamnosus,* in addition to possessing probiotic properties proven through clinical trials (Buriti & Saad, 2007) are also species that show good survival characteristics to drying processes by atomization. Corcoran et al. (2004) and Ananta et al. (2005) obtained dried cultures from *Lactobacillus rhamnosus* GG, while Gardiner et al. (2000) demonstrated enhanced survival to drying of *Lactobacillus paracasei* NFBC 338 compared to *Lactobacillus salivarius* UCC 118 showing adequacy of this group to this dehydration process.

As discussed on section "Cells Protectors" the growing and drying media have an effect on microbial survival during dehydration and the use of reconstituted skim milk (RSM) as drying vehicle can be inexpensive and effective alternative. Authors such as Corcoran et al. (2004), Gardiner et al. (2002) and Simpson et al. (2005) obtained high survival rates (greater than 70%) of lactic acid bacteria produced by drying by atomization using RSM.

Among the factors that influence the preservation of dried lactic acid bacteria - apart from temperature -, moisture control is indispensable, since according to the studies of Kurtmann et al. (2009), a gain in moisture content results in loss of Lactobacillus during storage. The authors demonstrated that one of the causes of the drop in the live lactobacillus population is lipid oxidation of the plasma membrane constituted by unsaturated phospholipids (Maness et al., 1999). Having this in mind, the aim of the present study was a combined use of spray and vacuum-drying to reduce the final moisture content of fermented powder.

Prior to drying, the lactobacillus were multiplied in three different concentration of skim milk powder (10, 20 and 30%). There were no differences in the amounts of *L. casei* between the fermentations with 20 and 30% reconstituted skim milk powder (RSMP), with mean values of, respectively, 9.36 and 9.41 (log cfu.ml^{-1}). The concentration of the probiotic was lower in fermentation with 10%(w/v) RSMP with 9.07 (log cfu.ml^{-1}). This behavior can be explained by the reduced availability of carbon sources (lactose) and nitrogen sources (milk proteins) necessary to the growth of the microorganisms (Pham & Shah, 2008).

Based on this, an experimental design was conducted evaluating all the variables that can influence the viability of microorganisms at spray drying method. The variables selected and the results obtained after drying are showed in Table 2. With the program Statisca 5.1, it was observed that the main factors to exert influence on the survival of *L. casei* (LC-1) were the inlet temperature (IT), the flow rate (FR) and the percentage of reconstituted skim milk powder (RSMP) (w/v).

Experimental test	X_1	X_2	X_3	X_4	IT (°C)	FR ($ml.h^{-1}$)	% RSMP ($m.v^{-1}$)	FAr	Log ($cfu.g^{-1}$)
1	-1	-1	-1	-1	160	260	20	54	8.5
2	1	-1	-1	1	190	260	20	66	7.7
3	-1	1	-1	1	160	520	20	66	9.6
4	1	1	-1	-1	190	520	20	54	8.7
5	-1	-1	1	1	160	260	30	66	6.6
6	1	-1	1	-1	190	260	30	54	5.9
7	-1	1	1	-1	160	520	30	54	8.3
8	1	1	1	1	190	520	30	66	6.6
9	0	0	0	0	175	390	25	60	8.2
10	0	0	0	0	175	390	25	60	7.4

Table 2. Experimental Design $2^{(4-1)}$ and their viability of *Lactobacillus casei* (Lc-1) after spray drying method for spray drying method. Coded Independents variables: X_1, X_2, X_3, X_4, Original independents variables: Inlet temperature (IT), Flow rate (FR), % of reconstitute skim milk powder (RSMP) and Air Flow (FAr).

With the intention to obtains higher survival of the microorganism and choose the best conditions to dry these materials, a new experiment was conducted applying an experimental displacement with the best conditions. After this new adjustments it was observed that the conditions TE = 144°C; TA = 650 ml/h and %RSMP = 13.25 produced higher mean viabilities (9.95 log $cfu.g^{-1}$) with a variation coefficient (VC) smaller than 2%.

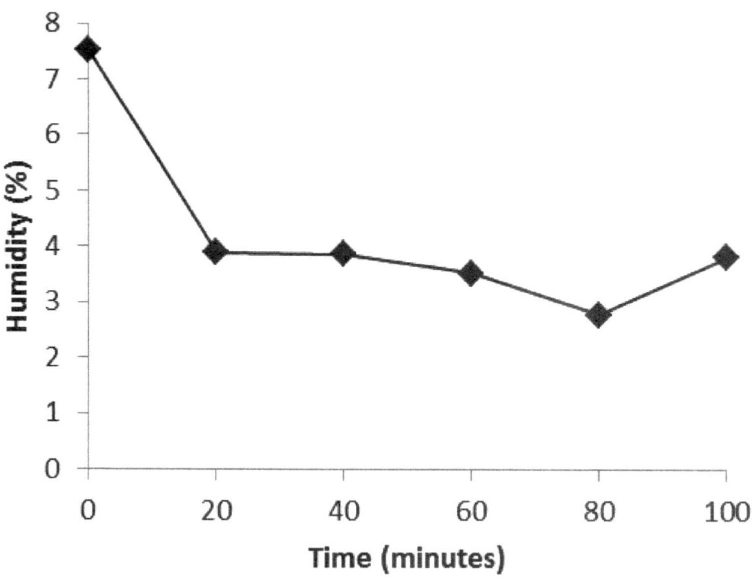

Fig. 4. Humidity values after vacuum chamber drying of dry fermentate (DF) at 45°C.

In spite of the satisfactory results as to the high concentration of viable lactobacillus after drying by atomization, the moisture content in % 7.2 ± 0.4 was greater than the maximum recommended by Corcoran et al. (2004), which is 4% for the preservation of viable lactobacillus during storage in RSMP. An alternative was vacuum drying, which reduced the moisture level of the powders dried by atomization to values lower than 4%. Authors such as Santivarangkna et al. (2007) and Tymczyszyn et al. (2008) demonstrated that it is possible to dry lactobacillus in vacuum chambers and they employed temperatures lower than 45°C which are less aggressive to the microorganisms.

To determine the influence of vacuum dehydration time on viabilities and final humidity, the analyses were conducted every 20 minutes. There was no difference between the counts of *L. casei* (Lc-1) as related to drying times ($p < 0.05$). The mean value after drying was 9.96 ± 0.4 % (Log cfu.g^{-1}). The moisture level (in %) are demonstrated at Figure 4, and varied from 7.2 ± 0.4 at the beginning to 3.9 ± 0.2 (CV) after 20 minutes drying (quick stage). At 80 minutes the percentage of humidity reached 2.8 ± 0.2 (slow stage) and then recapture of humidity from the air occurred after 100 minutes drying, with the moisture content increasing to 3.8 ± 0.1.

Complementary drying of the vacuum-chamber dried fermentate (DF) did not cause loss of viability of the lactobacillus ($p < 0.05$) with a survival rate of 99.7 %. According to studies of Santivarangkna et al. (2007), vacuum drying causes damage to the cell envelop (cell wall and plasma membrane) of the lactobacillus, resulting in losses in viability during drying. These losses were not noted in this study, since the maximum drying time used was 100 minutes, while Santivarangkna et al. (2007) utilized 24 hours, which increases the contact of the bacteria with degradation processes such as: high temperature, enzyme action and lipid oxidation (Kurtman et al., 2009; Maness et al., 1999; Fennema, 1996).

These results makes vacuum drying a technologically feasible post-processing option for the control and standardization of the moisture level of lactobacillus dried by atomization without causing losses in terms of their viability. To produce the DF, a standardization process of the moisture content of the DF was performed in a vacuum chamber for 80 minutes at 45°C, which resulted in the standardized dry fermentate (SDF).

After fermentation and drying processes, 20g of powders were packaged in bi-oriented polypropylene (BOPP) packs measuring 11 x 11.5 cm and stored in a desiccator with silica at 25°C (Simpson et al., 2005).

In addition to being packaged in BOPP, the SDF was also wrapped in polyethylene compound film therephthalate/aluminum/low density (PET/Al/PEBD) with 58 μm thickness and 11 x 11.5 cm in size. The flexible films used and their respective oxygen transmission rates (O_2TR), expressed in cm^3 (CNTP)/m^2day and water vapor transmission rates (WVTR) expressed in g water/m^2 day at 38° C / 90% RH, were: PET/Al/PEBD (O_2TR = 0.32; WVTR = 0.036) according to Lima, Silva and Gonçalves (1999). The bi-oriented polypropylene (BOPP) has a WVTR of 4.6 to 6.2 and a O_2TR in the 1800 - 3120 range according to Teixeira Neto & Vitali (1996). The SDF packaged in two packages was named as compound standardized fermentate (CSDF).

The pronounced drop noted up to the sixth week in the viability of the DF (control, dried only by atomization) can be explained by to the high initial level moisture (7.2%) of the samples and according to Corcoran et al. (2004) moisture levels should be kept below 4% to preserve the lactobacillus during storage. Kurtman et al. (2009) revealed in their studies that the best relative humidity level for storing lactobacillus is 11%, which fits the low moisture levels required by storage. After 6 weeks, the difference in the DF counts (control) and the

SDF was 2.9 logarithmic cycles. The DF with initial moisture content of 7.2% maintained only 2.2×10^6 cfu/g, whereas the SDF (3% initial moisture) preserved 1.9×10^9 cfu.g^{-1} greater than DF ($p < 0.05$) and according to Ishibashi & Shimura (1993), a probiotic level should be 10^7 cfu.g^{-1}.

Since both DF as SDF were kept under the same storage conditions and departed from the same ($p< 0.05$) amount of viable *L. casei* cells (1.2×10^{10} and 1.0×10^{10} cfu.g^{-1}, respectively), it may be stated that the post-processing in a vacuum chamber used to obtain the SDF exhibited the largest number of survivors as compared to the control for 42 days.

Up to the sixth week, the evolution of loss in viability of SFS and CSDF remained the same with losses of 0.7 log cycles for both. However, there was a moisture gain in SDF from 3.1 to 4.1% from the third to the sixth week. Since the 4% moisture limit was exceeded (Corcoran et al., 2004), the loss in viability of SDF was no longer linear and increased between the sixth and the ninth month (Figure 5). Between day 42 and day 63, the loss of viable lactobacillus was 2.3 logarithmic cycles. From the ninth month onwards, there was a tendency toward equality in the counts of *L. casei* between SDF and DF. After 18 weeks, there were no differences in the counts ($p < 0.05$) of DF (control) and SDF, with 6.1 and 6.2 log cfu.g^{-1}, respectively.

The loss in viability of lactic acid bacteria with increasing moisture levels can be better explained by the increase in oxidation and enzyme hydrolysis during storage. According to Fennema (1996), the oxidation process depends on the moisture level and is reduced whenever the BET monolayer value is established. The gain in moisture in excess of the monolayer results in increased oxidation of the microorganism and consequent loss of viability of the lactobacillus (Fennema, 1996; Kurtman et al., 2009).

Acceleration of oxidation driven by the gain in moisture occurs through the increase in the solubility of oxygen and the swelling of the macromolecules exposing more catalytic sites. According to Maness et al. (1999), the bacterial cell membrane is constituted by polyunsaturated phospholipids and is susceptible to oxidation by attack of oxigen reactive species. In addition to oxidation, the gain in moisture favors enzyme hydrolytic action (Fennema, 1996), which explains the greater loss in viability of the lactobacillus in comparison to storage conducted at increasing water activity values, even under low oxygen concentrations (Kurtman, 2009).

The enhances preservation of CSDF compared to SDF can be explained by the lower water vapor transmission rate (WVTR) of the laminated PET/Aluminum/PE film used to package CSDF, whereas SDF was packaged in BOPP which has a higher WVTR. Since the dried lactobacillus are sensitive to the increase in moisture contents as described in studies by Kurtman et al. (2009). After 18 weeks, CSDF kept itself 1.5 logarithmic cycles higher than SDF. At the beginning of storage, SDF was 9.6 log (cfu.g^{-1}), and this value was reduced to 6.2 log (cfu.g^{-1}) after 18 weeks.

In CSDF there was an average loss of 2.2 logarithmic cycles in 18 weeks. Departing from 9.9 at the beginning and falling to 7.7 log cfu.g^{-1} at the end of 126 days. This concentration of viable lactobacillus is greater than the probiotic level established by Ishibashi & Shimura (1993).

These values were higher than those described by Nebesny et al. (2007) who maintained 5 log cfu.g^{1} for 4 to 5 months storage at 30°C in chocolate added with lyophilized lactobacillus.

In the samples kept below 4% as recommended by Corcoran et al (2004), the behavior of loss of viability is linear (SDF up to the sixth week and CSDF). The D_{FSPC} value was 8.2 weeks, with this result being a little below those found by Ananta et al. (2005) who reported a D = 11.5 in storage for 6 weeks at 25°C with controlled relative humidity at 11%. Simpson et al (2005) selected a *Bifidobacterium thermophilum* NCIMB 702554 from among 16 other bifidobacterial strains. The selected strain maintained 7.3 log cfu.g^{-1} at 25°C for 90 days, an amount smaller than that obtained for CSDF in our study, which maintained 8.3 log cfu.g^{-1} viable *L. casei* cells for 84 days at 25°C.

To observe the aspects of the microparticles containing *L. casei* after drying by atomization, a scanning electron microscopy was performed at Microscopy Laboratory of Analysis. Figure 6 shows micrographs obtained at a magnification of 3000X (A) and 6000X (B). These microparticles are spherical in shape and of varying sizes. At various ends or extensions they had the appearance of "flat balls" with well-evident wrinkles on their surface as if they had been severely dehydrated. Lian et al. (2002) observed more spherical microparticles in the drying of *B. Longum* B6 at an inlet temperature of 100°C, indicating that high drying temperatures may be responsible for the deformations observed on the surface of the microparticles. Like with the images taken by our group, Crittenden et al. (2006) also verified in their work absence of free or non-microencapsulated cells in the micrographs at a magnification of 1000X and proposed that the microorganisms may have been effectively encapsulated within the microparticles.

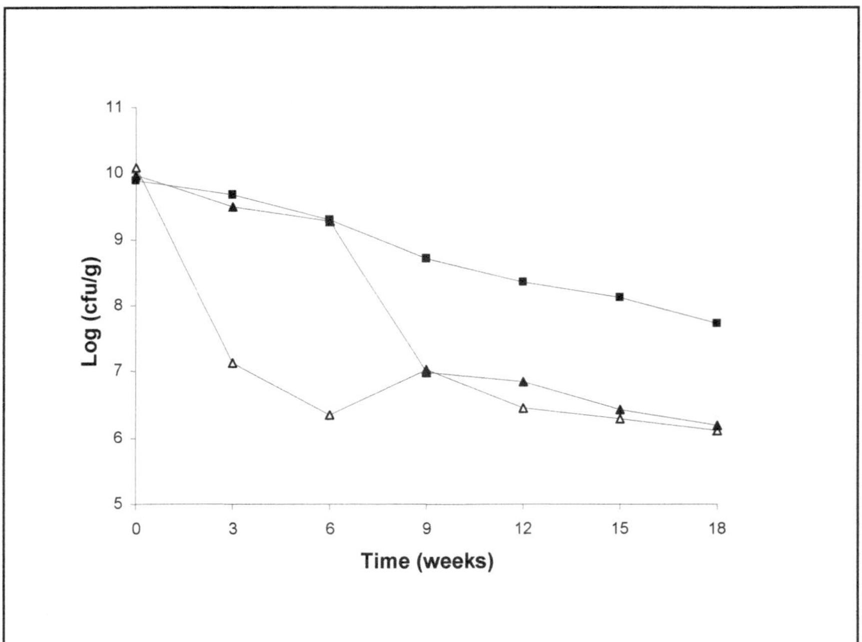

Fig. 5. Loss of viability of L. casei (Lc-1) at 25°C during 18 weeks:(■) Compound standardized dry fermentate (CSDF), (▲) standardize dry fermentate (SDF); (Δ) Dry fermentate (DF)

(A)

(B)

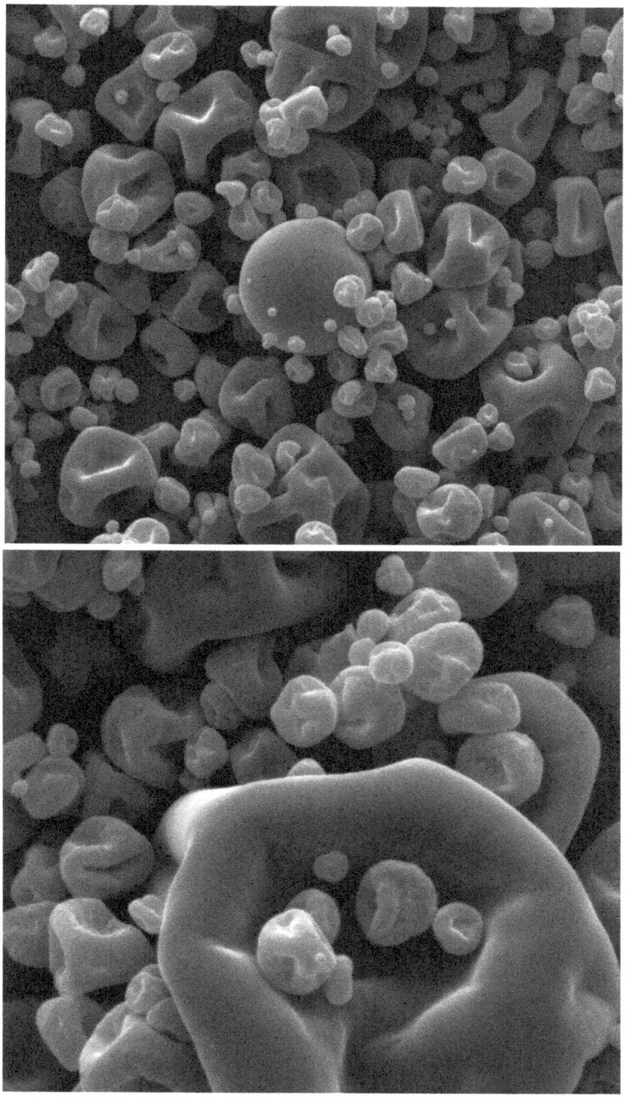

Fig. 6. Scanning electron micrographs at magnifications of (A) 3000x and (B) 6000x, respectively, of the microparticles of fermentate dried by atomization containing *L. casei*. Laboratory of Electron Microscopy and Microanalyses - State University of Londrina.

3.2 The use of *L. casei* (LC-1) microencapsulated by spray-drying in the processing of chocolate bars

The fermented powder obtained could be applied in several food products being chocolate an option that has been exhibited good results in terms of incorporating dried probiotics, since it melts at temperatures that microorganisms support, thereby avoiding loss of viability of the probiotic during the chocolate manufacturing process. Chocolate can further preserve lactic acid bacteria, at high counts (greater than 10^7 cfu.g^{-1}), for 6 months storage at 18°C (Nebesny et al., 2005). Additionally, chocolate has an exclusive taste, aroma and texture which make it one of the most popular foods in the world, in addition to being a source of biologically active substances, such as polyphenols, which have been shown to possess significant antioxidant properties and to provide a beneficial impact on the human health, particularly on the cardiovascular system (Nesbeny, 2004).

The dried fermentate was produced by applying drying conditions that had been optimized with the aid of the experimental design $2^{(4-1)}$ and used to supplement bars of dark chocolate to produce a potentially probiotic product. Bars of dark chocolate 9 cm x 3 cm in size were produced with a standardized weight of 10.5g ±0.2g and the survival of *L. casei* to the production process was 85%, resulting in a concentration of 4.2×10^8 Cfu.g^{-1} chocolate. The consumption of 10g of the product would provide a dose of 4.2×10^9 Cfu, a value above the minimum limit of 10^7 Cfu.g or ml^{-1} for probiotic products (Ishibashi & Shimura, 1993). A desirable characteristic in a functional probiotic food is that the selected microorganism remains viable from the processing of the product until consumption at counts that are adequate to promote the desired benefit. Within this context, our group evaluated the viability of *Lactobacillus casei* (LC-1) microencapsulated by spray-drying in dark chocolate during 90 days storage at 4°C and 25°C. Lactobacillus counts were performed after 0, 30, 60 and 90 days storage. The results are presented in Figure 7.

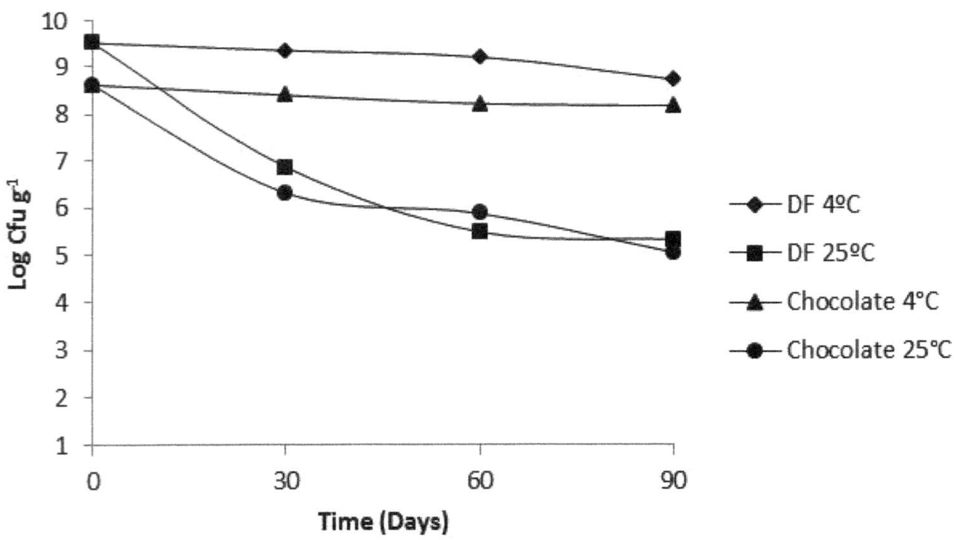

Fig. 7. Viability of *L. casei* in the Dry Fermentate (DF) and in little dark chocolate bars. (♦) DF at 4°C, (▲) Chocolate at 4°C, (■) DF at 25°C and (•) chocolate at 25°C.

After 12 weeks storage at 4°C (refrigerator temperature), there was a slight loss in viability of *L. casei,* but this did not surpass 0.5 log Cfu.g^{-1}. The survival rate was reduced from 8.6 log Cfu.g^{-1} to 8.2 log Cfu.g^{-1}, thus kept at a level above the minimum level of 7 log Cfu.g^{-1} (Ishibashi & Shimura, 1993). In storage at 25°C, the loss in viability was a little more abrupt and greater than 3 log Cfu.g^{-1}, falling back from 8.6 log Cfu.g^{-1} to 5.0 log Cfu.g^{-1}. These data reinforce that the damage to the cell caused by the drying process become evident during storage of the product. The extent of the loss of viability depends on the storage conditions such as temperature, water activity, moisture content and osmotic tension (Nesbeny, 2007). The choice of the food matrix to which the probiotic will be added must be well analyzed, since the constituents of the food may negatively affect the viability of the microorganisms (Possemiers et al., 2010). Chocolate showed to be a good matrix for using probiotic microorganisms since it exhibits a survival profile quite similar to that of the DF.

4. Conclusions

At present, there are numerous products containing probiotic microorganisms, with the main commercial products being fermented milks and yogurts. The great obstacle to the use of these microorganisms in other food systems is the loss of viability during processing and storage. Within this context, the use of cell protectants enables the use of these microorganisms in several processes and formulations. In this chapter, previously unreleased studies conducted by our research group, using cell protectants in *Lactobacillus casei* (Lc1) and their use in different food systems, such as ice cream and chocolate were presented. The results show that the use of cell protective agents enables the addition of probiotics to different food matrices or food systems, thereby opening new possibilities for the development of foods that can positively affect consumer health.

5. Acknowledgment

The authors would like to thank the Conselho Nacional de Desenvolvimento Científico e Tecnológico (CNPq) and the Coordenação de Aperfeiçoamento de Pessoal de Nivel Superior (Capes), Brazil for financial support.

6. References

Anal, A.K. & Singh, H. (2007). Recent advances in microencapsulation of probiotics for industrial application and target delivery. *Trends in Food Sci & Tecnol.*, v.18, pp. 240-251.

Ananta, E.; Volkert, M. & Knorr, D. (2005). Cellular injuries and storage stability of spray-dried *Lactobacillus rhamnosus* GG. *International Dairy Journal,* v. 15, pp. 399-409.

Bauer, S. A W.; Schneider, S.; Behr, J.; Kulozik, U. & Foerst, P. (Available online june 2011). Combined influence os fermentation and drying conditions on survival and metabolic activity of starter and probiotic cultures after low-temperature vacuum drying. *J. Biotechnol.* (2011), doi:10.1016/j.jbiotec.2011.06.016

Betoret, N.; Puente, L.; Diaz, M. J.; Pagan, M. J.; Garcia, M. J.; Gras, M. L.; Martinez-Monzo, J. & Fito, P. (2003). Development of probiotic-enriched dried fruits by vacuum impregnation. *Journal of Food Engineering,* v. 56, pp. 273–277.

Buriti, F. A. C. & Saad, S. M. I. (2007). Bactérias do grupo *Lactobacillus casei*: caracterização, viabilidade como probióticos em alimentos e sua importância para a saúde humana. *Archivos Latinoamericanos de Nutricion,* v. 57, pp. 373-380.

Capela, P.; Hay, T.K.C. & Shah, N.P. (2006). Effect of cryoprotectants, prebiotics and microencapsulation on survival of probiotic organisms in yoghurt and freeze-dried yoghurt. *Food Research International,* v. 39, pp. 203-211.

Carvalho, A. S.; Silva, J.; Ho, P.; Teixeira, P.; Malcata, F. X. & Gibbs, P. (2004). Effects of various sugars added to growth and drying media upon thermotolerance and survival throughout storage of freeze dried *Lactobacillus delbruekii ssp. bulgaricus. Biotechnology Progress*, v.20, pp. 248-254.

Champagne, C. P.; Gardner, N. & Roy, D. (2005). Challenges in the addition of probiotic cultures to foods. *Critical Reviews in Food Science and Nutrition*, v. 45, pp. 61-84.

Conrad, P. B.; Miller, D. P.; Cielenski, P. R., & Pablo de, J. J. (2000). Stabilization and preservation of Lactobacillus acidophilus in saccharide matrices. *Cryobiology*, v. 41, pp. 17−24.

Corcoran, B. M.; Ross, R.P.; Fitzgerald, G. F. & Stanton, C. (2004). Comparative survival of probiotic lactobacili spray-dried in the presence of prebiotic substances. *Journal of Applied Microbiology*, v. 96, pp. 1024-1039.

Chuy, L.E.& Labuza, T.P. (1994). Caking and stickiness of dairy-based food powders as related to glass transition. J. Food Sci., v. 59, pp. 43-46.

Crowe, L. M.; Reid, D. S. & Crowe, J. H. (1996). Is trehalose special for preserving dry biomaterials? Biophysical Journal, v. 71, pp. 2087-2093.

Cruz, A. G. (2009). Ice cream as a probiotic food carrier. *Food Research International*, v.42, pp.1233-1239.

Desmond, C.; Ross, R. P.; O'Callaghan, E.; Fitzgerald, G. & C. Stanton, C. (2002). Improved survival of Lactobacillus paracasei NFBC 338 in spray-dried powders containing gum acacia. *Journal of Applied Microbiology*, v. 93, pp.1003–1011.

De Vuyst, L. & Degeest, B. (1999). Heteropolysaccharides from lactic acid bacteria. *FEMS Microbiology Reviews*, v. 23, pp. 153-177.

Fennema, O. R. (1996). Food Chemistry. 3. ed. p. 54.

Foerst, P.; Kulozik, U.; Schimitt, M.; Bauer, S. & Santivarangkna C. (Available online 15 june 2011). Storage stability of vacuum-dried probiotic bacterium *Lactobacillus paracasei* F19. *Food Bioproducts Process.* doi:10.1016/j.fbp.2011.06.004

Food Today- The European food information Council newsletter, 07/2009

Food and Agriculture Organization (FAO). (2002). Guidelines for evaluation of probiotics in food. Ontario, Canada, 30 april and 01 mat.

Fu, N. & Chen, X. D. (2011). Towards a maximal cell survival in convective thermal drying processes. *Food Research International*, v. 44, pp. 1127-1149.

Gardiner, G. E.; O' Sullivan, E.; Kelly, J.; Auty, M. A.; Fitzgerald, G. F. & Collins, J. K. (2000). Comparative survival rates of human-derived probiotic *Lactobacillus paracasei* and *L. salivarius* strains during heat treatment and spray-drying. *Applied and Environmental Microbiology*, v. 66, pp. 2605-2616.

Guergoletto, K. B; Magnani, M.; Martins, J. S.; Andrade, C. G. T. J. & Garcia, S. (2010). Survival of *Lactobacillus casei* (LC-1) adhered to prebiotic vegetal fibers. *Innovative Food Science and Emerging Technologies*, v. 11, pp. 415–421.

Hekmat, S. & Mcmahon, D. J. (1992). Survival of *Lactobacillus* and *Bifidobacterium* bifidum in ice cream for use as a probiotic food. *Journal of Dairy Science*, v.75, pp. 1415–1422.

Homayouni, A. (2008). Effect of microencapsulation and resistant starch on the probiotic survival and sensory properties of symbiotic ice-cream. *Food Chemistry*, v.111, pp. 50–55.

Ishibashi, N. & Shimura, S. (1993). Bifidobacteria: Research and development in Japan. *Food Technology*, v. 46, pp. 126-135.

Ivanova, E.; Teunou, E. & Poncelet,D. (2005). Encapsulation of water sensitive products effectiveness and assessment of fluid bed dry coating. *J. Food Engineering*, v. 71, pp. 223-230.

Kets, E. P. W.; Teunissen, P. J. M. & Bont, J. A. M. (1996). Effect of compatible solutes on survival of lactic acid bacteria subjected to drying. *Applied and Environmental Microbiology*, v. 62, pp. 259-261.

King, V.A.E.& Su, J.T.(1994). Dehydration of *Lactobacillus acidophilus*. *Process Biochemistry*, v.28, pp.47-52.

Kumar, P.& Mishra, H.N. (2004). Yoghurt powder- A review of process technology, storage and utilization. Trans IChem E, Part C. *Food and Bioproducts Processing*, v.82(C2), pp. 133-142.

Kurtmann, L.; Carlsenl, C. U.; Risbo, J. & Skibsted, L. H. (2009). Storage stability of freeze-dried *Lactobacillus acidophilus* (La-5) in relation to water activity and presence of oxigen and ascorbate. *Cryobiology*, v.10, pp. 1016-1021.

Leslie, S.B.; Israeli, E.; Lighthart, B.; Crowe, J.H. & Crowe, L.M. (1995). Trehalose and sucrose protect both membranes and proteins in intact bacteria during drying. *Applied Environmental Microbiology*, v. 61, pp. 3592-3597.

Li, L.; Ye, Y.; Pan, L.; Zhu, Y.; Zheng, S. & Lin, Y. (2009). The induction of trehalose and glycerol in *Saccharomyces cerevisiae* in response to various stresses. *Biochemical and Biophysical Research Communications*, v. 387, pp. 778–783.

Lian, W. C.; Hsiao, H. C. & Chou, C. C. (2002). Survival of bifidobacteria after spray-drying. *International Journal of Food Microbiology*, v. 74, pp. 79– 86.

Linders, L. J. M.; Jong, G. I. W.; Meerdink, G. & Riet K. V. (1997). Carbohydrates and the Dehydration *Inactivation* of *Lactobacillus plantarum:* The Role of Moisture Distribution and Water Activity. *Journal of Food Engineering*, v. 31, pp. 237-250.

Looijesteijn, P. J.; Trapet, L.; Vries E.; Abee T. & Hugenholtz, J. (2001). Physiological function of exopolysaccharides produced by *Lactococcus lactis*, v.64, pp. 71-80.

Maness, P. C; Smolinski, S.; Blake, D. M.; Huang, Z.; Wolfrume, E. J. & Jacoby, W. A. (1999). Bactericidal Activity of Photocatalytic TiO2 Reaction: toward an Understanding of Its Killing Mechanism. *Applied and Environmental Microbiology*, v. 65, pp. 4094–4098.

Meng, X. C.; Stanton, G. F.; Fitzgerald, C. D., & Roos, R. P. (2007). Anhydrobiotics. *Food Chemistry*, v.106, pp. 1406–1416.

Muller, J.A.; Ross, R.P.; Fitzgerald, G.F. & Stanton, C. (2009). Manufacture of probiotic bacteria. In: *Prebiotic and probiotics, Science and technology*. Eds Charalampopoulos, D. and Rastall, R.A. p.727.

Otero, M.C.; Espeche, M.C.& Nader-Macias, M.E. (2007). Optimization of freeze-drying media and survival throughout storage of freeze-dried *Lactobacillus gasseri* and *Lactobacillus delbruekii* subsp. *delbruekii* for veterinarian probiotic application. *Process Biochemistry*. V.42, pp.1406-1411.

Panoff, J.M.; Thammavongs, B.; Guéguen, M. (2000). Cryoprotectans lead to Phenotypic Adaptation to Freeze-Thaw Stress in *Lactobacillus delbrueckii* ssp. *bulgaricus* CIP 101027T. *Cryobiology*, v. 40, pp.264-269.

Peighambardoust, S.H.; Golshan, T. & Hesari, J. (2011). Application of spray drying for preservation of lactic acid starter cultures: a review. *Trends in Food Science &technology*, v.22, pp. 215-224.

Ratti,C. (2001). Hot air freeze-drying of high- value foods: a review. *Journal of Food Engineering*, v.49, pp. 319.

Ruas-Madiedo, P.; Hugenholtz, J., & Zoon, P. (2002). An overview of the functionality of exopolisaccharides produced by lactic acid bacteria. *International Dairy Journal*, v. 12, pp. 163–171.

Saarela, M.; Virkajarvi, I.; Nohynek, L.; Vaari, A. & Matto, J. (2006). Fibres as cariies for *Lactobacillus rhamnosus* during freeze-drying and storage in Apple juice and chocolate-coated breakfast cereals. *International Journal of Food Microbiology,* v. 112, pp. 171–178.

Santivarangkna, C.; Wenning, M.; Foerst P. & Kulozik. U. (2007). Damage of cell envelope of *Lactobacillus helveticus* during vacuum drying. *Journal of Applied Microbiology,* v. 102, pp. 748–756.

Santivarangkna, C., Higl, B., & Foerst, P. (2008). Protection mechanisms of sugars during different stages of preparation process of dried lactic acid starter cultures. *Food Microbiology,* v. 25, pp. 429–441.

Semyonov, D.; Ramon, O.; Kaplun, Z.; Levin- Brener, L.; Gurevich, N. & Shimoni, E. (2010) Microencapsulation of *Lactobacillus paracasei* by spray freeze drying. *Food research International,* v. 43, pp.193-202.

Shah, N. P. (2000). Probiotic bacteria: selective enumeration and survival in dairy foods. *Journal of Dairy Science,* v.83, pp.894-907.

Simpson, P. J.; Stanton, C.; Fitzgerald, G. F. & Ross, R. P. (2005). Intrinsic tolerance of *Bifidobacterium* species to heat and oxygen and survival following spray drying and storage. *Journal of Applied Microbiology,* v. 99, pp. 493-501.

Stapley, A. (2009). Freeze drying. In: *Frozen Food Science and Technology* (ed J.A. Evans) Blakwell, Publishing Ltd, Oxford, UK doi:10.1002/9781444302325.ch12

Sunny-Roberts, E. O. & Knorr, D. (2009). The protective effect of monosodium glutamate on survival of *Lactobacillus rhamnosus* GG and *Lactobacillus rhamnosus* E-97800 (E800) strains during spray-drying and storage in trehalose-containing powders. *International Dairy Journal,* v. 19, pp. 209–214.

Teixeira Neto, R.O. & Vitali, A.A. (1996). Reações de transformação e vida-de-prateleira de alimentos processados. 2. ed. n. 6, Campinas: ITAL, pp. 74.

Tymczyszyn, E. E.; Gómez-Zavaglia, A., & Disalvo, E. A. (2007). Effects of sugars and growth media on the dehydration of Lactobacillus delbrueckii ssp. bulgaricus. *Journal of Applied Microbiology,* v. 102, pp. 845−851.

Tymczyszyn, E. E.; Díaz, R.; Pataro, A.; Sandonato, N.; Gómez-Zavaglia, A. & Disalvo, E. A. (2008). Critical water activity for the preservation of *Lactobacillus bulgaricus* by vacuum drying. *International Journal of Food Microbiology,* v. 128, pp. 342-347.

Valsiljevic, T. & Shah, N. P. (2008). Probiotics-from metchnikoff to bioactives. *International Dairy Journal,* v.18, pp.714-728.

Vega-Mercado, H.; Gongora-Nieto, M.& Barbosa-Canovas, G.V. (2001). Advances in dehydration of foods. *J. Food Engineering,* v. 49, pp. 271-289.

Wang, Y. C.; Yu, R. C. & Choy, C. C. (2004). Viability of lactic acid bacteria and bifidobacteria in fermented soymilk after drying, subsequent rehydration and storage. *International Journal of Food Microbiology,* v. 93, pp. 209-217.

Welsh, D. T. (2000). Ecological significance of compatible solute accumulation by micro-organisms: from single cells to global climate. *FEMS Microbiology Reviews,* v. 24, pp. 263-290.

Zamora, L.M.; Carretero, C. & Pares, D. (2006). Comparative survival rates of lactic acid bacteria isolated from blood, following spray-drying and freeze-drying. *Food Science and Technology International,* v.12, pp.77-84.

Zayed, G., & Roos, Y. H. (2004). Influence of trehalose and moisture content on survival of *Lactobacillus salivarius* subject to freeze-drying and storage. *Process Biochemistry,* v. 39, pp. 1081−1086.

13

Electrochemical Biosensors for Food Quality Control

Margarita Stoytcheva[1], Roumen Zlatev[1], Marcela Ovalle[2], Zdravka Velkova[3], Velizar Gochev[4] and Benjamin Valdez[1]
[1]*Universidad Autónoma de Baja California, Instituto de Ingeniería*
[2]*Universidad Autónoma de Yucatán, Facultad de Química*
[3]*Plovdiv University of Food Technology*
[4]*Plovdiv University "Paisii Hilendarski*
[1,2]*México*
[3,4]*Bulgaria*

1. Introduction

The electrochemical biosensors are analytical devices designed by coupling biological recognition elements and electrochemical transducers. The transducer converts the analytical signal produced as a result of the biochemical and electrochemical interactions into measurable electrical one (Thévenot et al., 1999).

The electrochemical biosensors are self-contained, simple to handle, and able to provide specific, sensitive, accurate and cost-effective *in situ* and *on line* measurements in real time, without or with a minimum sample preparation. Because of these advantages over the conventional analytical methods, they are well suited for the detection of a large spectrum of compounds, entering food and subjects of analytical control.

The present work is intended to demonstrate the applicability of the electrochemical biosensors for arsenic determination in beverages.

2. Arsenic content in food and beverages

Arsenic is a chemically active, toxic, and carcinogenic element (Moore & Ramamoorthy, 1984). It is among the 129 priority pollutants of the environment and among the 25 hazardous substances representing a significant potential threat to human health (EPA: Toxic and priority pollutants). It occurs naturally in soil and groundwater, but additionally enters the environment in a large quantity because of the human industrial and agricultural activities. The most affected by arsenic pollution are fishes and other aquatic organisms, since they accumulate it. High arsenic concentrations in plants are registered when using for irrigation arsenic-rich groundwater or contaminated water because of the industrial discharges and the treatment of soils with fertilizers and pesticides. Lead arsenate insecticides were extensively used in some countries until 1981 (Peryea, 1998). Arsenic content in food from plant and animal origin, with the exceptions of seafood and animal and poultry offal, does not habitually exceed 0.25 mg kg^{-1}, according to WHO data (Arsenic.

WHO Food Additives Series 18). The average daily arsenic intakes for various countries are summarized in Fig. 1. Arsenic concentration in food and beverages, as evaluated by the US Food and Drug Administration (FAD, 2010) in its annual Total Diet Study, is shown in Table 1.

Product	As, mg kg^{-1}
Cheese, American, processed	0.002
Beef roast, chuck, oven-roasted	0.001
Turkey breast, oven-roasted	0.004
Liver (beef/calf), pan-cooked w/oil	0.001
Fish sticks or patty, frozen, oven cooked	0.527
Peanut butter, creamy	0.013
Peanuts, dry roasted, salted	0.014
Rice, white, enriched, cooked	0.065
Oatmeal, plain, cooked	0.002
Cream of wheat (farina), enriched, cooked	0.001
Corn, fresh/frozen, boiled	0.001
Bread, white, enriched	0.001
Bread, whole wheat	0.002
Muffin, fruit or plain	0.007
Corn/tortilla chips	0.001
Fruit-flavoured cereal, presweetened	0.013
Raisin bran cereal	0.006
Crisped rice cereal	0.135
Granola w/raisins	0.021
Oat ring cereal	0.028
Pear, raw (w/peel)	0.001
Strawberries, raw/frozen	0.001
Fruit cocktail, canned in light syrup	0.002
Grapes (red/green), raw	0.003
Cantaloupe, raw/frozen	0.008
Raisins	0.014
Avocado, raw	0.001
Apple juice, bottled	0.005
Prune juice, bottled	0.004
Spinach, fresh/frozen, boiled	0.001
Collards, fresh/frozen, boiled	0.003
Tomato, raw	0.001
Tomato sauce, plain, bottled	0.001
Cucumber, peeled, raw	0.011
Brownie	0.006
Syrup, chocolate	0.001
Jelly, any flavour	0.002
BF, cereal, rice, dry, prepared w/water	0.041
Beef steak, loin/sirloin, broiled	0.001
Chicken thigh, oven-roasted (skin removed)	0.009
Catfish, pan-cooked w/oil	0.012
Fruit juice blend (100% juice), canned/bottled	0.005

Lettuce, leaf, raw	0.002
Beef w/vegetables in sauce, from Chinese carry-out	0.004
Potato, baked (w/peel)	0.002
Chili con carne w/beans, canned	0.003
Quarter-pound hamburger on bun, fast-food	0.001
Meatloaf, beef, homemade	0.001
Chicken potpie, frozen, heated	0.001
Soup, tomato, canned, cond., prepared w/water	0.003
Cake, chocolate w/ icing	0.013
Sweet roll/Danish pastry	0.001
Gelatine dessert, any flavour	0.001
Wine, dry table, red/white	0.010
BF, beef and broth/gravy	0.001
BF, macaroni, tomato and beef	0.002
BF, peaches	0.001
BF, juice, apple	0.022
BF, vanilla custard/pudding	0.002
BF, fruit dessert/pudding	0.003
Chicken breast, oven-roasted (skin removed)	0.004
Shrimp, boiled	0.265
Bread, cracked wheat	0.003
Bagel, plain, toasted	0.001
English muffin, plain, toasted	0.001
Crackers, graham	0.004
Grape juice, frozen conc., reconstituted	0.007
Mushrooms, raw	0.073
Eggplant, fresh, peeled, boiled	0.001
Okra, fresh/frozen, boiled	0.001
Beef stroganoff w/noodles, homemade	0.012
Tuna noodle casserole, homemade	0.164
Fish sandwich on bun, fast-food	0.380
Egg, cheese, and ham on English muffin, fast-food	0.002
Clam chowder, New England, canned, cond., prepared w/ whole milk	0.128
Coffee, from ground	0.0002
BF, teething biscuits	0.004
Salmon, steaks/fillets, baked	0.288
BF, cereal, rice w/apples, dry, prepared w/water	0.033
Chicken breast, fried, fast-food (w/skin)	0.013
Chicken leg, fried, fast-food (w/skin)	0.013
Tuna, canned in water, drained	1.00
Cranberry juice cocktail, canned/bottled	0.004
Sweet potatoes, canned	0.001

Table 1. Arsenic occurence in food (US Food & Drug Administration - Total Diet Study - Market Baskets 2006-1 through 2008-4).

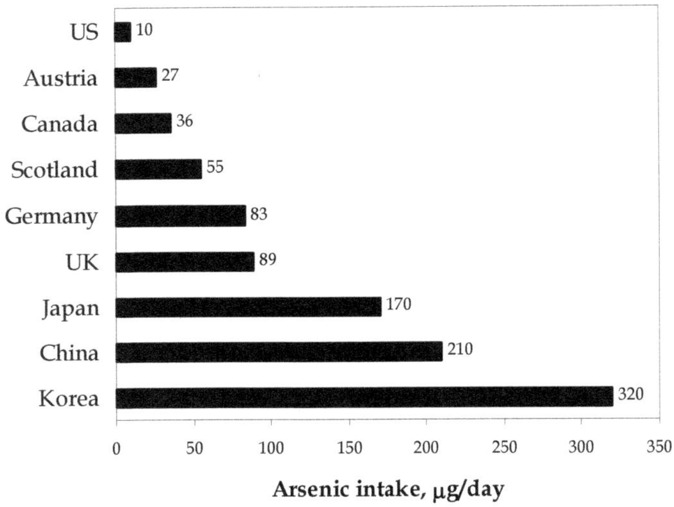

Fig. 1. Average daily arsenic intakes for various countries (Arsenic. WHO Food Additives Series 18)

3. Sample collection, preparation and treatment for arsenic determination in food and beverages

Sample collection, preparation and treatment for arsenic determination in food and beverages are performed according to the established procedures (WHO, 2011). These include: collection of samples, representative of the food consumed in a population; sample conservation in acid washed plastic containers; freezing of samples if necessary, to -80°C; food preparation or cooking in a manner similar to those that would be used at home, if appropriate; sample homogenization and digestion, applying various techniques guided by the subsequent analysis technique.

4. Methods for inorganic arsenic determination in food and beverages

Inorganic arsenic determination could be performed applying a number of methods (WHO, 2011). Some of them, such as the spectrophotometric analysis with silver diethyldidhiocarbamate and certain modifications of the atomic absorption spectrometry (AAS) and the inductively coupled plasma (ICP) are standardised (DIN 38405-D12; APHA/AWWA/WPCF 3500-As C; AOAC 33.125-33.132 in combination with 25.041 and 25.042; EPA 7061; DIN 38405-D1; APHA/AWWA/WPCF 3500-As B: 3114 B; APHA/AWWA/WPCF 3500-As E: 3120 B; DIN 38406-E22) and are among the mostly applied for arsenic determination in food and beverages (Bingöl et al., 2010; Conklin, 2010; Husáková et al., 2007; Karadjova et al., 2005; Niu Jianjun & Wang Bingwu, 1992; Roberge et al., 2009; Stafilov et al., 2004; Syr-Song Chen et al., 2003; Tašev et al., 2005). Nevertheless, arsenic is one of the few elements for which AAS is not enough sensitive. Using special supplies such as arsine generators and electrothermal analysers allows lowering the detection limit, but causes difficulties in the routine analysis. The other advanced instrumental methods such as ICP, neutron activation analysis (NAA), and X-ray fluorescence permit the determination of arsenic at trace levels, but they require expensive and sophisticated equipment. The spectrophotometric methods, although simple and cost effective, do not provide the required sensitivity.

The electrochemical methods for inorganic arsenic determination (Cavicchioli et al., 2004), including mainly anodic stripping voltammetry and differential pulse polarography, in spite of their limited application in food quality control, could be considered as an alternative to the above mentioned analytical techniques. For instance, their sensitivity is similar to this of mass spectrometry and NAA, but they are much more simple, require low costing equipment, and allow distinguishing the electro-active As(III) and the electro-inactive As(V), in contrast to the enumerated techniques. As(III) and As(V) have different toxicity, biological activity, and physiological action. The toxicity of As(III) is known to be greater of that of As(V). Thus, the distinction between the two forms is of primary importance.

The further development of the electrochemical methods is associated with the appearance, during the 1960s, of the so-called electrochemical biosensors. They combine the high sensitivity, accuracy and reproducibility of the electrochemical analysis with the substrate specificity and catalytic activity of the biological molecules. A number of them found an application in food industry, namely in food safety and quality control, and in the control of the fermentation processes (Mutlu, 2010; Scott, 1998; Prodromidis & Karayannis, 2002; Wagner & Guilbault, 1994).

5. Acylcholinesterase based sensor for arsenic determination in wine

Arsenic determination in wine, using the suggested in this work acetylcholinesterase electrochemical sensor, is based on the following reactions:

$$\text{acetylthiocholine} + H_2O \xrightarrow{ACh} \text{thiocholine} + CH_3COOH$$

$$\text{thiocholine} \rightarrow \text{dithio-bis-choline} + 2H^+ + 2e^-$$

The acetylcholinesterase Ach (EC 3.1.1.7) catalysed hydrolysis of acetylcholine generates the electroactive product thiocholine. The current of its oxidation is recorded amperometrically at a potential of +0.80 V/SCE. In the presence of As(III), because of the enzyme inhibition that it provokes, the quantity of the produced thiocholine decreases. Thus, the current of its oxidation also decreases as a function of As(III) concentration under similar conditions.

The acetylcholinesterase based electrochemical sensor was prepared as described in our previous works (Stoytcheva et al., 1998a, 1998b), i. e.: acetylcholinesterase was covalently immobilized onto the surface of a rotating disc electrode elaborated from spectrally pure graphite (Ringsdorf Werke, Germany). The analysis was carried out in an electrolysis cell of conventional type, at a temperature of 25°C, with a rotation speed of the working electrode of 1000 rpm. The auxiliary electrode was a glassy carbon electrode. A saturated calomel electrode was used as a reference.

The response of the biosensor was measured for various acetylthiocholine iodide concentrations in the presence of different amounts of As(III) in the form of AsO_3^{3-} in a Britton-Robinson buffer solution with pH 7. The obtained results are presented in Fig. 2, where ΔI is the difference between the registered steady-state currents of thiocholine oxidation in the absence and in the presence of inhibitor (to note that iodide oxidation to iodine occurs, too).

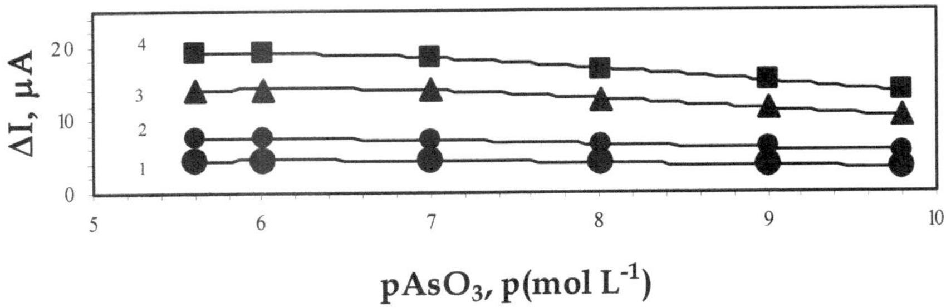

Fig. 2. Calibration curves for AsO_3^{3-} determination using different substrate concentrations: 1) 0.2 mmol L^{-1}; 2) 0.4 mmol L^{-1}; 3) 0.6 mmol L^{-1}; 4) 1 mmol L^{-1}; 5) 1.2 mmol L^{-1}. $pAsO_3$ is the negative decimal logarithm of the AsO_3^{3-} concentration.

As shown, the linear dynamic range of the calibration curves suitable for AsO_3^{3-} determinations varies from 0.2 nmol L^{-1} to 0.02 μmol L^{-1}. AsO_3^{3-} concentrations superior to 10 μmol L^{-1} caused an increase of the sensor response, due to the following concurrent reactions:

$$3I^- - 2e^- = I_3^-$$

$$H_3AsO_3 + I_3^- + H_2O = H_3AsO_4 + 3I + 2H^+$$

The sensitivity of the determinations, as shown in Table 2, increased with the increase of the acetylthiocholine iodide concentration until the enzyme saturation with 1.0 mmol L^{-1} acetylthiocholine iodide.

Substrate concentration, mmol L^{-1}	Sensitivity, μA/p(mol L^{-1})
0.2	0.39
0.4	0.65
0.6	1.25
1.0	1.65

Table 2. Sensitivity of As(III) determination

The method allows As(III) and As(V) differentiation, due to the fact that AsO_4^{3-} does not inhibit the acetylcholinesterase.

These preliminary results served for the development of a method for As(III) determination in wine. As known, arsenic content in some type of wines exceeds 0.1 mg L^{-1} (Crecelius, 1997). Arsenic concentration in contaminated illicit whiskey (moonshine) was found to be more than 0.4 mg L^{-1} (Gerhardt et al., 1980).

For the purposes of the analysis, commercially available wine was artificially contaminated with AsO_3^{3-} 0.0133 mmol L^{-1} (0.001 mg L^{-1}). The sample, without any pretreatment, was analysed according to the following protocol: (i) registration of the amperometric response of the electrochemical biosensor for a substrate concentration of 1.0 mmol L^{-1}, for which the sensitivity of the AsO_3^{3-} determination is maximal (25°C, Britton-Robinson buffer 0.1 mol L^{-1}, pH 7, 1000 rpm, +0.80 V/SCE) in the presence of no contaminated wine; (ii) registration of the amperometric response of the electrochemical biosensor in similar conditions, but in the presence of the contaminated wine sample; (iii) ΔI calculation and AsO_3^{3-} concentration evaluation using a preliminary constructed calibration curve. The relative error of the analysis was found to be inferior to 3%.

6. Conclusion

The modern food analysis requires sensitive, accurate, and express methods for food safety, food quality, and food technology control. The growing field of the biosensors in food industry represents an answer to this demand. Thus, the method for As(III) determination in wine described in this work is an example demonstrating the viability of the electrochemical biosensors in food quality control.

7. References

Arsenic (WHO Food Additives Series 18). http://www.inchem.org/documents/jecfa/jecmono/v18je17.htm (accessed on June, 29, 2011)

Bingöl, M.; Yentür, G.; Er, B. & Öktem, A. B. (2010). Determination of some heavy metal levels in soft drinks from Turkey using ICP-OES method. *Czech J. Food Sci.*, 28, 3, 213–216

Cavicchioli, A.; La-Scalea, M. A. & Gutz, I. G. R. (2004). Analysis and speciation of traces of arsenic in environmental, food and industrial samples by voltammetry: A review. *Electroanalysis*, 16, 9, 698-711

Conklin, S. (2010). Elemental analysis manual: Section 4.10: High Performance Liquid Chromatography-Inductively Coupled Plasma-Mass Spectrometric determination of four arsenic species in fruit juice. U.S. Food and Drug Administration. http://www.fda.gov/Food/ScienceResearch/LaboratoryMethods/ElementalAnalysisManualEAM/ucm219640.htm (accessed on July, 2, 2011)

Crecelius, E. (1997). Arsenite and arsenate levels in wine. *Bull. Environ. Contamination and Toxicology*, 18, 2, 227-230

EPA: Toxic and priority pollutants. http://water.epa.gov/scitech/methods/cwa/pollutants.cfm

FAD. 2010. Total Diet Study Statistics on Element Results 2006-2008 http://www.fda.gov/downloads/Food/FoodSafety/FoodContaminantsAdulteration/TotalDietStudy/UCM184301.pdf (accessed on June, 29, 2011).

Gerhardt, R. E.; Crecelius, E. A.; Hudson J. B. (1980). Moonshine-related arsenic poisoning. *Arch Intern Med.*, 140, 2, 211-3 http://www.inchem.org/documents/jecfa/jecmono/v18je17.htm (accessed on June, 29, 2011)

Husáková, L.; Černohorský, T.; Šráková, J. & Vavrušov, L. (2007). Direct determination of arsenic in beer by electrothermal atomic absorption spectrometry with deuterium background correction (D2-ET-AAS). *Food Chemistry*, 105, 286–292

Karadjova, I.; Lampugnani, L.; Onor, M.; D'Ulivo, A. & Tsalev, D. (2005). Continuous flow hydride generation-atomic fluorescence spectrometric determination and speciation of arsenic in wine. *Spectrochimica Acta Part B: Atomic Spectroscopy*, 60, 6, 816-823

Moore, J. & Ramamoorthy, S. (1984). Heavy metals in natural waters: Appl. Monitoring and impact assessment. Springer, N. Y.

Mutlu, M. (2010). *Biosensors in food rrocessing, safety, and quality control*. CRC Press, ISBN: 1439819858, 9781439819852, Boca Raton

Niu Jianjun & Wang Bingwu (1992). Total arsenic determination and its species and valent state analysis in beverages. *Chinese Journal of Analysis Laboratory*, 1992-03

Peryea Francis J. Historical use of lead arsenate insecticides, resulting soil contamination and implications for soil remediation. Proceedings, 16th World Congress of Soil Science (CD Rom), Montpellier, France. 20-26 Aug. 1998. http://soils.tfrec.wsu.edu/leadhistory.htm (accessed on June, 29, 2011).

Prodromidis, M. & Karayannis, M. (2002). Enzyme based amperometric biosensors for food analysis. *Electroanalysis*, 14, 4, 241-261

Roberge, J.; Abalos, A.; Skinner, J.; Kopplin, M. & Harris, R. (2009). Presence of arsenic in commercial beverages. *Am. J. Environ. Sciences*, 5, 6, 688-694

Scott, A. (1998). *Biosensors for food analysis*. Royal Society of Chemistry, ISBN: 0854047506, 9780854047505, Great Britain

Stafilov, T.; Karadjova, I.; Arpadjan, S.; Bundaleska, J. & Tašev, K. (2004). Speciation analysis of arsenic in natural waters and beverages by HG-AAS. Adnan Menderes University, 4th AACD Congress, 29 Sept-3 Oct. 2004, Kuşadası-AYDIN, Turkey, Proceedings Book

Stoytcheva, M.; Sharkova, V. & Magnin, J-P. (1998b). Electrochemical approach in studying the inactivation of immobilized acetylcholinesterase by arsenate(III). *Electroanalysis*, 10, 14, 994-998

Stoytcheva, M.; Sharkova, V. & Panayotova, M. (1998a). Electrochemical approach in studying the inhibition of acetylcholinesterase by arsenate(III): analytical characterisation and application for arsenic determination. *Anal. Chim. Acta*, 364, 195-201

Syr-Song Chen, Chieu-Chen Cheng & Shin-Shou Chou (2003). Determination of arsenic in edible oils by Direct Graphite Furnace Atomic Absorption Spectrometry. *Journal of Food and Drug Analysis*, 11, 3, 214-219

Tašev, K.; Karadjova, I. & Stafilov, T. (2005). Determination of inorganic and total arsenic in wines by Hydride Generation Atomic Absorption Spectrometry . *Microchimica Acta*, 149, 1-2, 55-60

Thévenot, D. R.; Tóth, K.; Durst, R. A. & Wilson, G. S. (1999). Electrochemical biosensors: recommended definitions and classification. *Pure Appl. Chem.*, 71, 12, 2333-2348

Wagner, G. & Guilbault, G. (1994). *Food biosensor analysis*. M. Dekker, ISBN: 0824791509, N.Y.

WHO, 2011. Arsenic and arsenic compounds. Environmental health criteria 224. http://www.inchem.org/documents/ehc/ehc/ehc224.htm (accesssed on July, 2, 2011)

Part 3

Industrial Processes

14

Maillard Reaction Products in Processed Food: Pros and Cons

Deborah Markowicz Bastos, Érica Monaro,
Érica Siguemoto and Mariana Séfora
Nutrition Department, School of Public Health, São Paulo University
Brazil

1. Introduction

The Maillard reaction was first reported in 1912 by Louis-Camille Maillard, who described that upon gently heating sugars and amino acids in water, a yellow-brown color developed. The reaction that leads to these colorful compounds, firstly described from a simple observation, is actually the result of a complicated pathway of chemical reactions. The Maillard reaction is often described in food systems but it also occurs in living organisms, and in this case, it is called glycation. In biological systems, the ramifications of the Maillard reaction have been observed and analyzed, as this reaction has become important in the field of food science and medicine (Finot, 2005; Gerrard, 2002a).

The consumption of Maillard Reaction Products (MRPs) has increased in recent decades and there are evidences that these substances are absorbed and may participate in pathological processes such as, cataract, diabetes, degenerative diseases, atherosclerosis and chronic renal failure. On the other hand, these compounds are responsible for essential sensory attributes of thermally processed food products, contributing to their appearance, flavor, aroma and texture.

This chapter will cover the chemistry of Maillard reaction products generation, the role of these products in food acceptability, the analysis of these compounds both in food products and in the human body and the biological activities attributed to these compounds, since this is a contemporary and controversy subject in food science and nutrition field.

2. The chemistry of Maillard reaction products generation

Since the first description of a browning reaction of glycine with glucose by Louis Maillard, the knowledge on chemical structures derived from the reaction of carbonylic and amino compounds has considerably increase (Nass *et al.*, 2007).

Amino-carbonyl and related interactions of food constituents comprise those changes commonly termed as "non-enzymatic browning reactions". Specifically, reactions of amines, amino acids, peptides, and proteins with reducing sugars and vitamin C (Maillard reaction, caramelization, ascorbic acid degradation) and quinones (enzymatic browning) cause deterioration of food during storage and processing (Friedman, 1996).

Non-enzymatic browning reactions depend on many parameters (Table 1), such as, temperature, water activity (a_w), pH, moisture content and chemical composition. In general, maximum browning occurs at a_w between 0.60 and 0.85 and the browning rate increases with increasing pH, up to a pH of around 10 (Gerrard, 2002a; Morales & Van Boekel, 1997).

Mechanism	Oxygen requirement	NH_2 requirement	pH optimum	Temperature	a_w
Maillard reaction	No	Yes	Basic/acid	Medium	Medium/high
Caramelization	No	No	Basic/acid	High	Low
Ascorbic acid degradation	Yes/No	No	Slightly acid	Medium	Medium/high

Table 1. Main differences between non-enzymatic browning reactions (based on Finot, 2005)

From 1940, amino-carbonyl reactions and the resulting browning pigments have been investigated by many chemists. In 1953 Hodge proposed a three stages (initial, advanced and final) scheme for the Maillard reaction (Figure 1) (Nursten, 2005). The initial stage starts from the sugar amine systems leading to browning pigments generation. Amadori colorless compounds are formed in this stage, and an increased content of unsaturated carbonyl compounds is observed. During the intermediate stage, both fluorescence and radiation absorbing properties of the system increase due to the formation of small molecules with chromophores. Aldehydes formed by the Strecker degradation of amino acids can condense either with themselves, sugar fragments, furfurals, or with other dehydration product forming brown pigments. Although Strecker's pathway is not the major color-producing reaction, it is responsible for the origin of off-flavours usually associated with Maillard browning. The final stage is characterized by the formation of unsaturated, brown nitrogenous polymers (melanoidins) which may also be generated from the condensation reaction of furfurals or dehydro reductones (Finot, 2005; Hodge, 1953; Morales & Van Boekel, 1997).

Browning development occurs after an induction period, characterized by the production of fluorescent uncolored intermediates. Fluorophores are considered precursors of brown pigments and allow detecting the progress of the reaction before any visual change occurs. Fluorescence from the Maillard reaction is attributed to molecular structures with complex bonds between carbon and nitrogen, and the contribution of sugar caramelization to global fluorescence is insignificant in amino-acid containing systems (Matiacevich & Buera, 2006; Rozycki *et al.*, 2010).

Maillard reaction ratio is proportional to the heat-treatment severity during food processing, when temperatures range from 100 to 250 °C (baking, grilling, frying, extruding and roasting) and/or during storage for long periods at room temperature. This reaction is of most importance for roasted products such as coffee, chocolate and peanuts. In the medical arena, several authors describe the role of the Maillard reaction during ageing and chronic diseases, as diabetes and renal failure (Gerrard, 2002a; Nguyen, 2006; Nursten, 2005).

Maillard reaction occurs in biological systems and the final products are refered as Advanced Glycation End Products (AGEs). AGEs are a heterogeneous group of compounds that arise non-enzymatically by the reaction of reducing sugar and other α-carbonylic compounds with amino groups on proteins, lipids and nucleic acids. Actually, glycation in living organisms have other pathways that are linked to glucose metabolism and lipid peroxidation, whose products are termed Advanced Lipoxidation End Products (ALEs) (Goldberg *et al.*, 2004; Nass *et al.*, 2007).

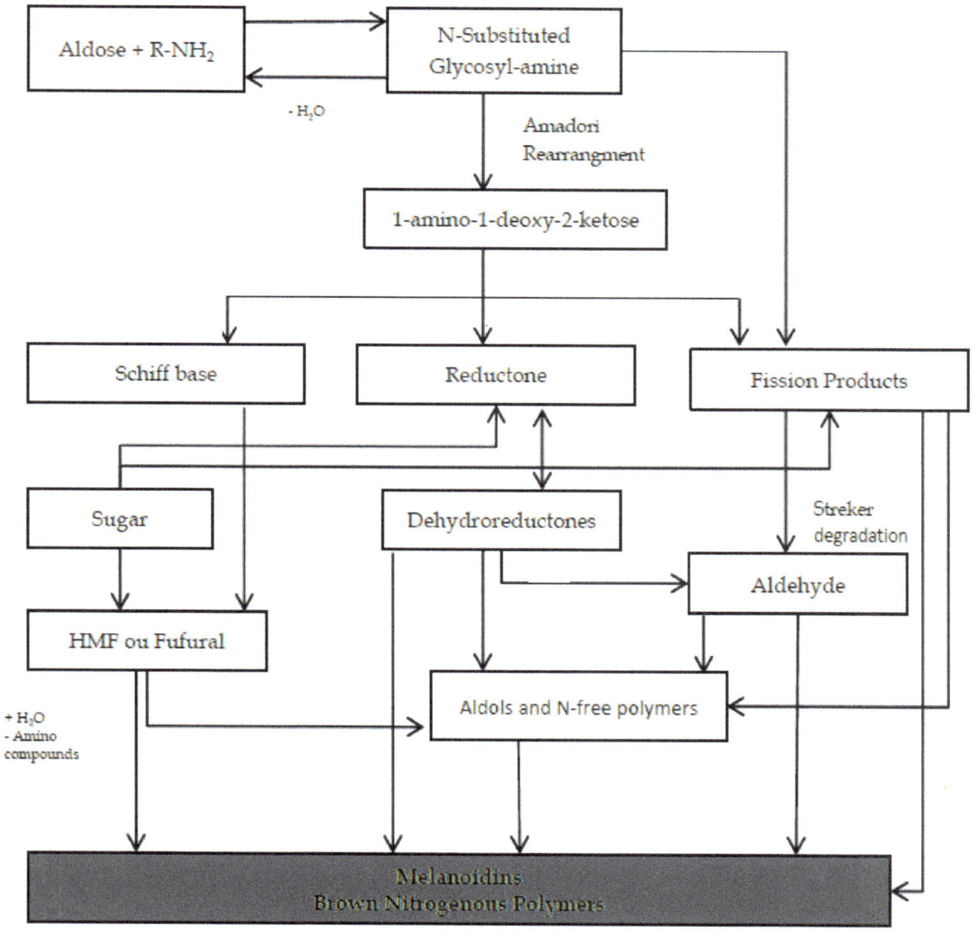

Fig. 1. Main stages in Maillard reaction proposed by Hodge (adapted from Nursten, 2005)

MRPs/AGEs generated in food and biological systems are shown in Figures 2 and 3. Nguyen (2006) describes the MRP/AGE content of selected popular foods, such as roasted almonds (66.5 kU/g), oil (120.0 kU/g), butter (94.0 kU/g), mayonnaise (265.0 kU/g), broiled chicken for 15 minutes (58.0 kU/g), fried chicken for 15 minutes (61.0 kU/g), homemade pancakes (10.0 kU/g), bread (0.5 kU/g). It is noteworthy that temperature and cooking process are more relevant for the formation of MRPs than time of cooking or other parameters (Nguyen, 2006).

Fluorescence/crosslinked

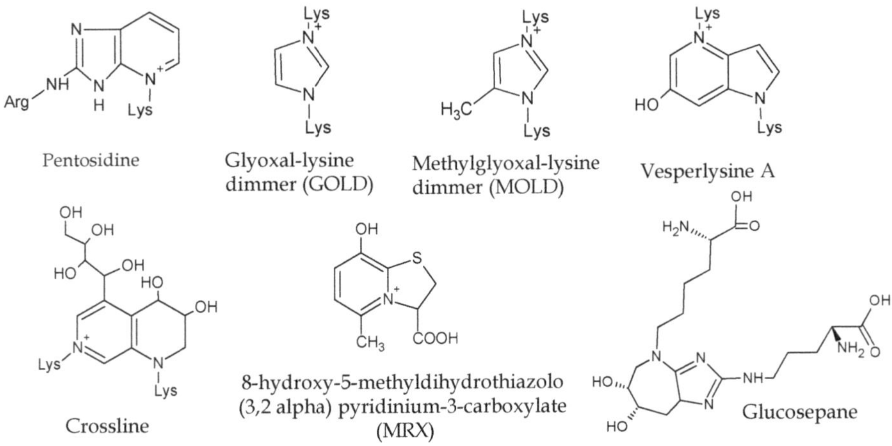

Fig. 2. Main chemical structures of Fluorescence MRPs/AGEs

Non-fluorescence/Non-crosslinked

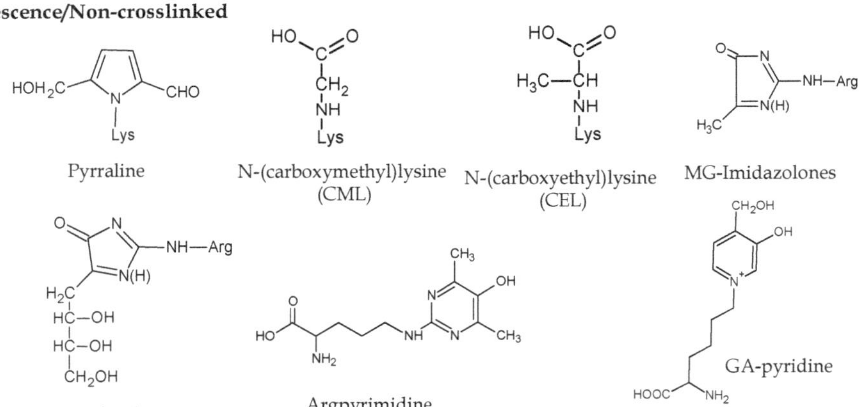

Fig. 3. Main chemical structures of non-fluorescence MRPs/AGEs

Sato *et al.* (2006a) proposed a scheme of formation of six distinct AGEs *in vivo* (AGE-1, AGE-2, AGE-3, AGE-4, AGE-5 and AGE-6). AGE-1 is formed from glucose through Schiff base and Amadori products, AGE-2 from glyceraldehyde, AGE-3 from glycoaldehyde, AGE-4 from methylglyoxal, AGE-5 from glyoxal, and AGE-6 from 3-deoxyglucosone. AGE-2 and AGE-3 are considered toxic AGEs by contributing to the neuronal cell toxicity. They also proposed that pentosidine, Nε-(carboxymethyl)lysine, pryrraline and crossline are nontoxic AGEs, however, other studies are needed to validate these conclusions (Nguyen, 2006).

AGEs formation occurs slowly and naturally in the body of healthy people, however, this process is accelerated under certain conditions, such as hyperglycemia and oxidative stress. MRPs dietary are added to the intra and extracellular AGEs produced, contributing to the accumulation of these compounds in tissues and, thus favoring the onset and progression of metabolic complications (Barbosa *et al.*, 2008).

Up to 90% of pyrraline free and Amadori products excreted by the kidneys are from food source. Free pentosidine, for example, presents about 60 to 80% bioavailability (Forster *et al.*, 2005). Somoza *et al.* (2006) administered casein-linked lysinoalanine (LAL), Nε-fructoselysine (FL) and Nε-(carboxymethyl)lysine (CML) to rats and revealed that the kidneys are the main organs in which AGEs are accumulated and excreted. In this study, the dietary LAL, FL and CML excreted in the urine was 5.6, 5.2 and 29%, respectively.

In other words, the rate of absorption and renal excretion of MRPs depends on dietary intake or the presence of pathologies as well as the amount and type of compound ingested.

3. The role of Maillard reaction products in food acceptability

Maillard reaction is one of the most important reaction which results from food processing. Maillard reaction products (MRPs) greatly influence essential food quality attributes such as flavor, aroma, color and texture. Actually, this reaction can be used to design foods that present sensory attributes demanded by the consumer (Ames, 1990; Yu & Zang, 2010).

3.1 Color

Color formation is the primary characteristic of the Maillard reaction. In the last decade, efforts have been driven to detect Maillard reaction kinetics and the formation ratio of colored compounds, mainly with the use of model systems. Brown color development during processing and storage is desirable for many products such as baked foods, coffee, cookies while undesirable in some kinds of food products orange juice, white chocolate, milk and powder egg. Predicting and controlling food color development are particularly important for companies to satisfy consumer preference, since a complex array of melanoidins produced by the Maillard reaction is strongly dependent on the food matrix composition as well as the technological conditions of the reaction (Wang *et al.*, 2011). Melanoidin can also be formed by sugar caramelization without the participation of amino groups.

The presence of melanoidins, brown nitrogen-containing high molecular weight pigments, responds for the characteristic color of roasted foods such as coffee, cocoa, bread and malt. Although the chemical structures and health effects of these compounds produced both in food and model systems have been investigated for over 30 years, the health effects are not well understood, mainly because their bioavailability depends on several parameters that include gut microbiota metabolism. Despite of the lack of general knowledge, the positive correlation between melanoidins content in food and antioxidant activity is well documented in the literature.

3.2 Flavor and aroma

Flavor and aroma development due to the Maillard reaction depends on the reaction temperature, time, pH, water content and on the type of sugars and amino acids involved (Yu & Zhang, 2010; Van Boekel, 2006). In most cases, the first factor mentioned influences the kinetics parameters, while the second factor determines the type of flavor compounds formed.

The intermediate and final stages of the Maillard reaction are the most important to flavor development, especially the so-called Strecker degradation step, in which amino acids are degraded by dicarbonyls formed previously in the reaction,leading to the aminoacids deamination and decarboxylation (Ames, 1990; Rizzi, 2008).

The volatile products of the Maillard reaction can be classified according to the sugar dehydration/fragmentation products as furans, pyrones, cyclopentenes, carbonyls and acids; the amino acid degradation products as aldehydes and sulfur compounds; and the volatiles produced by further interactions as pyrroles, pyridines, imidazoles, pyrazines, oxazoles, thiazoles, and others. Pyrazines and alkylpyrazines are associated with the flavor and aroma of cooked (roasted) and nutty, respectively. Alkylpyridines confer to foods flavor and aroma of green, bitter, astringent and burnt, and furans, furanones and pyranones of sweet, burnt, pungent and caramel-like flavors/aromas.

Compounds that are essential to the characteristic flavor and aroma of food products are generally present at trace levels. The oxygen-containing aroma compounds 2,3-butanedione, 2,3-pentanedione, methylpropanal, 3-methylbutanal, phenylacetaldehyde, 3-hydroxy-4,5-dimethyl-2(3*H*) furanone and 2,5-dimethyl- 4-hydroxy-3(2*H*)-furanone occur in concentration ranging from 1μg/kg up to 100 mg/kg. The nitrogen-containing aroma compounds 2-ethyl-3,5dimethylpyrazine, 2,3-diethyl-5-methylpyrazine and 2-acetyl-1-pyrroline are present in food in an order of magnitude of 0.001–10 mg/kg. On the whole, sulfur containing Maillard odorants constitute the most powerful aroma compounds and often play, although at trace levels, a dominant role in the flavor of cooked meats. These volatile compounds are responsible for the flavor and aroma to stewed beef juice, boiled trout, french fries, bread crust, cooked chicken, roasted chicken, boiled beef, cocoa powder, peanuts, pilsner, roasted beef, popcorn and coffee (Cerny, 2008).

The meat-related flavor compounds are mainly sulphur containing compounds, derived from cysteine and ribose (coming from nucleotides), while the amino acid proline gives rise to typical bread, rice and popcorn flavors. Cysteine-containing mixtures seem to have the most intense meat-like and sulphur smell. The other amino acid-containing sulphur, methionine, generates a highly intense smell of potatoes and it is employed, in the food industry, to enhance the soft flavor of potatoes. Mixtures containing amino acids other than cysteine or methionine in a combination with reducing sugars are characterized mostly by caramel and jammy smell (Stanimirova *et al.*, 2011; Van Boekel, 2006).

The food industry invests great effort trying to create synthetic flavors and aromas by reconstituting combinations of these compounds. The process of creating synthetic flavors is limited since the subtleties of flavor perception are many and varied, and although Electronic Noses may detect these compounds, human sensory perception is considered essential to validate instrumental data, (Gerrard, 2002a; Schaller et al., 1998).

3.3 Texture

Texture definition is complex and a general agreement has been reached which evolved from the efforts of a number of researchers. According to Szczesniak (2002), "texture is the sensory and functional manifestation of the structural, mechanical and surface properties of foods detected through the senses of vision, hearing, touch and kinesthetics".

Maillard reaction influences the texture of food via protein cross-linking. Manipulation of the extent and nature of such protein cross-linking during food processing offers a means by which the food industry can modify the functional properties of food. Despite of this, the extent of how protein cross-linking affects food texture in processed foods and how to control this parameter to maximize food quality is not yet known (Gerrard, 2002b). Protein cross-linking by the Maillard reaction will affect not only texture, but the protein digestibility as well.

Although Maillard reaction effects on food color, flavor and aroma are well understood and used by the food industry, its effects on food texture has attracted less attention from the scientific community. However this is a promising tool for texture development.

4. Analyses of Maillard reaction products

MRPs are present in the diet and many authors have highlighted their health benefits and risks. For that reason, it is of most importance characterizing and quantifying MRPs in common foods, to get the best balance between benefits and potential risks, and then to establish guidelines for food health (Delgado-Andrade *et al.*, 2009).

Development of sophisticated analytical techniques made it possible to isolate, characterize, and quantify several specific non browning reaction compounds formed *in vitro* and *in vivo,* both at the early and advanced stages of Maillard reaction. Among them the most common are: Amadori compounds (indirectly analyzed as furosine), Nε-(carboxymethyl)lysine (CML) and some intermediate derivatives of the reaction, such as hydroxymethylfurfural. Measurement of fluorescent compounds formed during the reaction is also a reliable tool to evaluate the extension and ratio of nutritional loss due to thermal processing of foods (Delgado-Andrade *et al.*, 2009; Friedman, 1996).

4.1 Fluorescent compounds

Traditionally, Maillard reaction monitoring in food processing was based on the spectrophotometric evaluation of color development at 420 nm. More recently, the evaluation of fluorescent compounds generated by the Amadori rearrangement product undergoing dehydration and fission has become usual. Besides its use in food systems, fluorescence measurement is also employed to evaluate Maillard reaction at physiological conditions, meaning, AGEs generation, and also to access AGEs correlated pathologies development (Delgado-Andrade *et al.*, 2006).

Fluorescent compounds (FC) are precursors for the brown pigments formed in the Maillard reaction, which present different chemical structures. Evaluation changes in fluorescence intensity helps evaluating the extent of the Maillard reaction in food products (Morales & Van Boekel, 1998; Rufiàn-Henares & Delgado-Andrade, 2009) and biological systems. Fluorescence was first used to evaluate the formation of MRPs in milk and now it is used to monitor the processing of cereals, cookies, soybeans, infant formula, cooked salmon and bakery products.

Fluorescent compounds may be free in the matrix or linked to te protein fraction. Total FC (free + linked to protein) determination demands a previous enzymatic hydrolysis, which requires the use of a nonspecific protease (pronase) (Delgado-Andrade *et al.*, 2008). Free and total FC have been tested in foods such as milk, breakfast cereals, cooked salmon, roasted soy and enteral formula (Delgado-Andrade *et al.*, 2008; Rufiàn-Henares *et al.*, 2002).

Fluorescence of Advanced Maillard products and Soluble Tryptophan (FAST) is a well established method used to evaluate the nutritional and lysine damages. FAST is based on the quantification of protein denaturation using fluorescence: (1) fluorescent of the advanced Maillard products (FAMP), such as pyrrole and imidazole derivatives, at excitation/emission 330/420 nm; and (2) tryptophan fluorescence (FTrp) at excitation/emission 290/340 nm at pH 4.6. The FAST index is calculated as follows: (100*FAMP/FTrp) (Birlouez-Aragon *et al.*, 2002).

4.2 Hydroxymethyfurfural

Hydroxymethylfurfural (HMF) is an intermediate compound formed during the Maillard reaction and by the degradation of hexoses at high temperatures at acid conditions (Figure 4) (Arribas-Lorenzo & Morales, 2010).

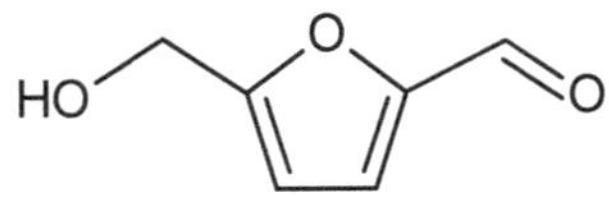

Fig. 4. Chemical structure of HMF

Spectrophotometric (colorimetric) methods, which are the most usual methods for HMF determination in food, are of limited accuracy since other chromophores in foods may absorb radiation in the same wavelength region, interfering with the results. In addition, colorimetric methods have low sensitivity. Chromatographic methods (liquid or gas high resolution chromatography) are more accurate and sensitive for this purpose, and one of the major advantage of the use of chromatographic methods is the individual determination of HMF and furfural, what can not be achieved by spectrophotometric methods (Erbersdobler & Somoza, 2007; Morales *et al.*, 1997; Rufiàn-Henares *et al.*, 2001).
HMF formation is directly linked to the heat intensity applied to food, and because is not usually present in raw and fresh foods, it is considered a thermal damage marker for products containing high carbohydrate concentrations. Moreover, it can be used to monitor the thermal process applied to several different food products such as: breakfast cereals containing dried fruits; caramel and honey; pasta and bakery products (Rufiàn-Henares & Delgado-Andrade, 2009; Rufiàn-Henares *et al.*, 2006).

4.3 Furosine

Amadori compounds are measured as furosine (ε-N-2-furoylmethyl-L-lysine) (Figure 5). The content of furosine present in foods is influenced by the kind of heat treatment and/or the storage time. Levels of furosine tend to decline after prolonged storage or after overheating to give rise to other compounds such as CML (Delgado-Andrade *et al.*, 2005; Friedman, 1996; Rufiàn-Henares *et al.*, 2009).

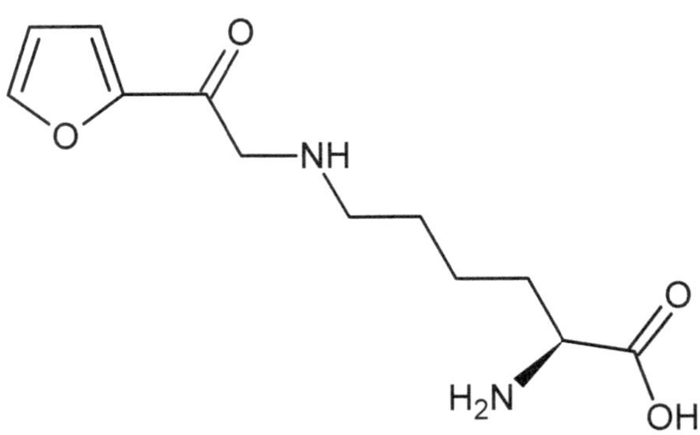

Fig. 5. Chemical structure of furosine

Furosine is the most specific and important indicator of the initial phase of the Maillard reaction. It is widely used in the analysis of cereal products, since lysine is the limiting amino acid of this product and, thus the presence of furosine is an important marker of protein biological value loss. Monitoring furosine formation and contents helps tailor the processing conditions in order to guarantee the maintenance of the nutritional value of food products (Rufiàn-Henares *et al.*, 2004, 2006; Resmini *et al.*, 1990).
Regarding analytical procedures, in 1992, an ion pairing HPLC based methodology was proposed and successfully applied in a series of studies. In 1996, when furosine became commercially available, Reverse Phase-HPLC became the method of choice for furosine analysis. Due to the possible transformation of CML into furosine during heating it is necessary for the acid hydrolysis to be performed in an inert atmosphere, which impairs furosine degradation (Erbersdobler & Somoza, 2007).

4.4 Nε-(carboxymethyl)lysine

Nε-(carboxymethyl)lysine (CML) is a stable, low reactivity advanced Maillard product (Figure 6). CML can be produced by Amadori compound degradation, such as Nε-(fructosyl)lysine (FL). FL undergoes oxidation to form, Nε-(carboxymethyl)lysine (CML). R-dicarbonyls such as glyoxal (GO), formed during the oxidation of the sugar or the Amadori rearrangement products are immediate precursors of CML. Lipid peroxidation is another route to CML, and GO has been suggested as an intermediate. CML is one of the more important markers of bioactive Maillard products and its content is usually correlated to the health risk of ingestion of heat-treated foods (Charissou *et al.*, 2007).

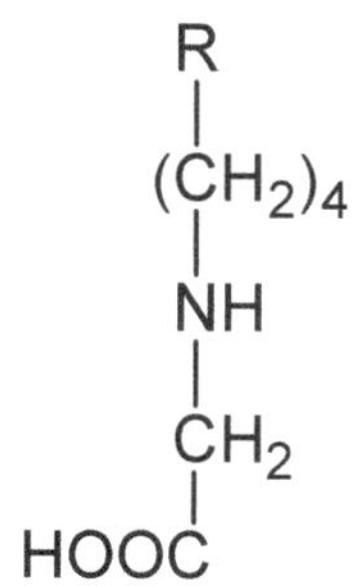

Fig. 6. Chemical structure of CML

The three main metodologies proposed to evaluate CML in foods are: a) RP-HPLC (Reverse Phase/ High Performanc Liquide Chromatography), b) GC/MS (Gas Chromatography connected to Mass Spectrometry) following methylation of the carboxylic group and acylation of the amine group and c) enzyme- linked immunosorbent assay based on a monoclonal anti-CML antibody (Charissou *et al.*, 2007).
Currently, CML analyses in foods are performed by specific immunosorbent (AGE-ELISA-Enzyme Linked Immuno Sorbent Assay). This test is suitable for quantitative CML analysis both in biological samples and food (Goldberg, 2004).
Among the Maillard reaction products, CML is the best characterized end product. It is employed as an advanced glycation end products/advanced lipoxidation end products (AGEs/ALEs) markers in research (Goldberg, 2004).

5. Biological activity of the Maillard Reaction Products (MRPs)

The Maillard reaction is of outstanding importance for the formation of flavor and color of heated foods. Despite the unquestionable beneficial effects on the sensory quality of food products, the Maillard reaction products may be harmful for human health.

Protein nutritional impairment, consequence of the destruction of essential amino acids or the decrease in their bioavailability, is one of the oldest known nutritional implication of this reaction (Delgado-Andrade *et al.*, 2007a; Seiquer *et al.*, 2006). The decrease in the availability of several amino acids, mainly lysine, which may correspond to 50%, is noticeable (Rerat *et al.*, 2002). Total lysine contents in breakfast cereal products can decrease from 20-54% as a result of processing (Rutherfurd *et al.*, 2007; Torbatinejad *et al.*, 2005). Comparing the effects of diets with different MRPs contents on dietary protein utilization in in male adolescents indicated that the consumption of a diet rich in MRPs resulted in 47% higher nitrogen fecal excretion, 12% lower apparent nitrogen and 6% lower nitrogen digestibility. Therefore, the protein apparent absorption absorption and digestibility were significantly lower (Seiquer *et al.*, 2006).

The bioavailability of minerals can also be affected by MRPs, since these compounds are able to chelate minerals interfiring, therefore, with their solubility (Delgado-Andrade *et al.*, 2011). The effects of diets with different Maillard reaction products contents on iron biological utilization showed that significant decrease on dietary iron availability occurred when diets rich in MRP were consumed (García *et al*, 2009a). It was also observed a negative influence on dietary phosphorus absorption in male adolescents exposed to a rich-MRPs diet, as a result of an increased phosphorus fecal excretion concomitant to a decrease in its apparent absorption (Delgado-Andrade *et al.*, 2011). Magnesium and calcium bioavailability can also be affected by the presence of Maillard reaction products in the diet (Delgado-Andrade *et al.*, 2007b; García *et al.*, 2009b).

The results of studies on the genotoxic and mutagenic potential of the MRPs are controversial. While there are studies indicating that the MRPs can cause mutations, no association between MRPs and genotoxicity was found by Wagner et al (2007). In *in vitro* systems, melanoidins mixtures have negligible mutagenic effects (Somoza, 2005), while 5-hydroxymethylfurfural (HMF) is considered a potentially carcinogenic to humans and some epidemiological studies have found an association between acrylamide intake and the occurence of tumors the occurrence of tumors (Capuano & Fogliano, 2011).

The formation of advanced glycation end products (AGEs) is observed during normal ageing and occurs inside as well as outside of cells (Nass *et al.*, 2007). These compounds, when cross linking with proteins profoundly affect protein functionality and irreversibly modify chemical properties and functions of diverse biological structures (Barbosa *et al.*, 2008), which seems to be implicated in inflammatory processes and diabetic complications, such as nephropathy and vascular disease (Jakus & Rietbrock, 2004; Linden *et al.*, 2008; Mostafa *et al.*, 2007).

AGEs accumulate in various tissues during aging, including skin, neural, vascular, renal and cardiac tissues, collagens and crystalline lens. In the skin, glycation is involved in many metabolic processes and, along with aging, affects the functionality of certain cells, such as the synthesis of fibroblasts, enzyme activation of matrix degradation (metalloproteinases) and the organization of the matrix (Hartog *et al.*, 2007; Pageon, 2010; Pawlak *et al.*, 2008). It is proposed that the accumulation of the advanced glycation end products (AGEs) and the activation of the receptor for AGEs in the retina could play a significant role in the initiation and progression of age-related macular degeneration and cataracts (Pawlak *et al.*, 2008). Kalousová *et al.* (2002) and most recently, Mostafa *et al.* (2007) showed that AGEs level in plasma proteins are elevated in patients with diabetes. The high blood glucose levels favor the occurrence of spontaneous reactions (glycation) between glucose and proteins, resulting in the formation and excessive deposition of AGEs (Magalhães *et al.*, 2008). In patients with renal failure AGEs accumulation occurs due to the decrease in the extent of degradation and elimination from the body and, also, to increased exposure to oxidative stress. On the other hand, the AGEs and products derived from the process of oxidation promote damage in the renal tissue, leading to greater accumulation of AGEs, creating a vicious cycle (Hartog *et al.*, 2007). The increase in consumption of heated, cooked or roasted food of AGEs accumulation.

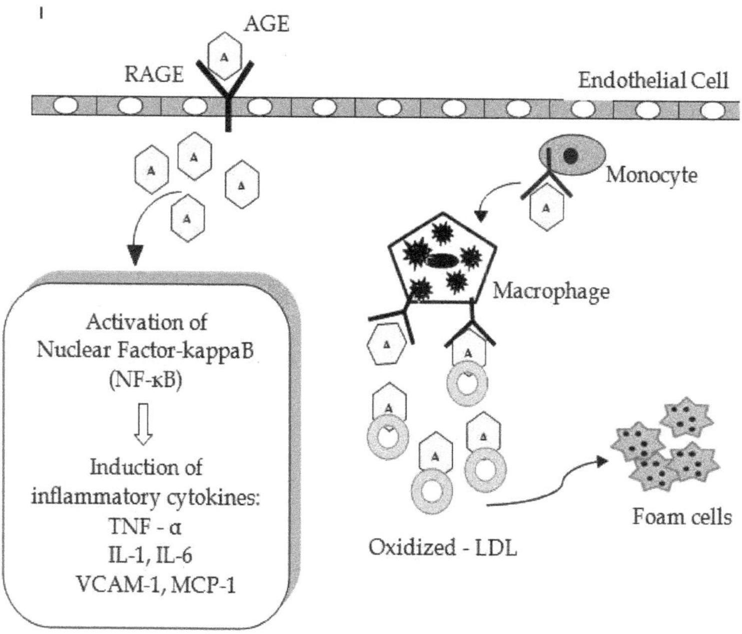

Fig. 7. AGE-RAGE interaction and its association with atherosclerosis (Based on Hartog et al., 2007).

Among the mechanisms by which AGEs may contribute to the development and progression of vascular complications of diabetes, is the interaction of these compounds with receptors on the surface of various cell types, such as RAGEs (Receptors for Advanced Glycation End Products) (Marchetti, 2009). The AGE-RAGE interaction in the endothelial cells activates the transcription of nuclear factor-kappaB (NF-κB), with the induction of proinflammatory cytokines, such as the tumor necrosis factor (TNF), interleukin-1 (IL-1), interleukin-6 (IL-6), monocyte chemotactic protein-1 (MCP-1) and enhances the expression of the vascular cell adhesion molecule-1 (VCAM-1) (Basta, 2008; Magalhães *et al.*, 2008; Méndez *et al.*, 2010; Muscat *et al.*, 2007). In addition, this interaction in monocytes induces their activation to macrophages and promotes monocyte chemotaxis, and in smooth muscle cells, is associated with increased cellular proliferation (Marchetti, 2009). Besides, some studies demonstrate that AGEs may promote atherogenesis by oxidizing low density lipoproteins (LDL) (Basta *et al.*, 2004). Indeed, AGE form crosslinks with low-density lipoprotein (LDL), which become more atherogenic and less susceptible to absorption and subsequent clearance. In addition, LDL modified by AGEs is more easily captured by receptors located on macrophages, generating foam cells (cells with fat droplets and cholesterol) (Hartog *et al.*, 2007; Vasdev *et al.*, 2007) (Figure 7).

It has also been suggested that AGEs are involved in neurodegenerative diseases, such as Alzheimer and Parkinson (Grillo & Colombato, 2008; Sato *et al.*, 2006b), arthritis *(Vytásek et al., 2010),* loss of bone mass (Ding *et al.*, 2006) and promotion of changes in the function and/or structure of DNA and RNA molecules (Li *et al.*, 2008).

Considering AGEs as important mediators of pathological processes, investigations aiming to verify the action of chemical compounds against the synthesis of AGEs and its possible use in therapy of patients with several metabolic complications are in course of development. Substances present in foods, such as pyridoxamine, allyl cysteine (component of garlic extract), phenolic compounds, taurine and carnosine, showed significant anti-AGE effects, but, at present, there is no indication of food components able of reducting AGEs generation *in vivo* (Barbosa *et al.*, 2008). Despite of this, dietary therapies also appear to be an effective alternative in the control of diseases associated with accumulation of AGEs.

Restricting the consumption of fried, grilled or baked foods seems to be the most effective way of decreasing AGEs endogenous *pool*. Dietary AGEs restriction seems to be a successful strategy in suppression of inflammatory molecules in diabetes, implying eventual prevention or delay of atherosclerosis (Vlassara *et al.*, 2002). Several studies indicate that Maillard reaction products from the diet increase the endogenous AGEs pool and, whether this might become a health problem is yet controversial. There are many gaps that must be evaluated before conclusions can be drawn as, for example, the fate of MRPs in the organism. Notwithstanding, several researchers advocate towards the decrease of MRPs ingestion and, therefore, food industry has an important role by considering processes towards the production of foods with lower contents of MRPs.

On the other hand, there are authors who advocate for MRPs as substances that may promote benefits, such as increases in immunity and decreases in the toxicity of some nitrosamines. Figure 8 is a summary of the main biological effects attributed to Maillard Reaction Products (MRPs) / Advanced Glycation End Products (AGEs).

Some studies suggest the MRPs exert positive influence on the gut microbiota (Tuohy *et al.*, 2006). It has been shown that anaerobic bacteria, particularly *Bifidobacteria* strains, are able to use bread melanoidins as carbon source (Borrelli & Fogliano, 2005). Maillard Reaction products in roasted cocoa beans, for example, were able to inhibit the growth of *E. coli* spp. and *Enterobactercloaceae* (Summa *et al.*, 2008).

Furthermore, MRPs exhibit antioxidant activity (Açar *et al.*, 2009; Chang *et al.*, 2011; Chawla *et al.*, 2009; Rao *et al.*, 2011; Summa *et al.*, 2008). Melanoidins from roasted coffee and biscuits exerted protective effects against oxidative stress on human hepatoma HepG2 cells (Goya *et al.*, 2007; Martín *et al.*, 2009). Regarding coffee antioxidant activity, more than 50% of the observed antioxidant activity is due to the low molecular weight compounds linked to the melanoidin skeleton, promoting a chelating ability that is also involved in the shelf life of the product (Delgado-Andrade & Morales, 2005). *In vivo*, a MRPs rich diet was able to suppress lipid peroxidation and to increase antioxidant activity of plasma, although it has not modified the antioxidant enzymes activity (superoxide dismutase, glutathione peroxidase and catalase) (Seiquer *et al.*, 2008).

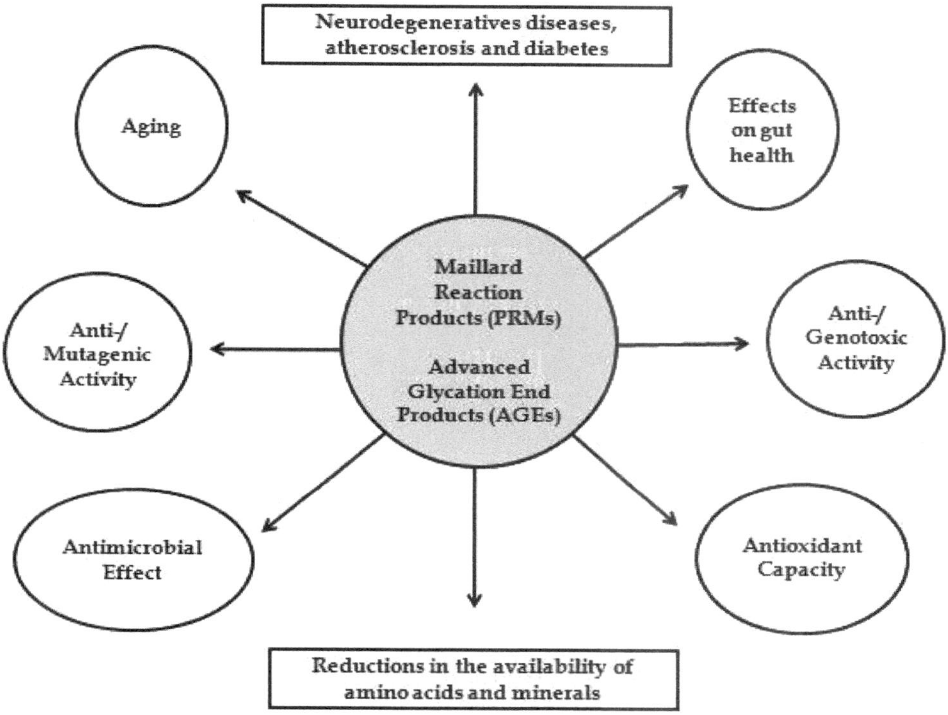

Fig. 8. Possible biological effects of the Maillard Reaction Products (MRPs) and Advanced Glycation End Products (AGEs) (Based on Somoza 2005).

6. Conclusion

For food industry, coping with the Maillard reaction and the effects of the reaction products in food and in health is important to the improvement and development of food products. This chapter discussed the positive and negative aspects of the Maillard reaction in food products. The positive contributions of the Maillard reaction are sensory attributes generation, such as color, flavor, aroma and texture. The negative aspects are off-flavor development, flavor loss, discoloration, and loss of protein nutritional value. In the food industry, the role of flavor and color either desirable or undesirable is the key in the manufacture of products of consistent sensory quality. Contradictory knowledge about the effects of Maillard Reaction Products on health indicates that studies are needed to further expand the AGEs and MRPs database as well as development of methods for reducing MRPs generation during home cooking and food processing. Understanding the chemical, nutritional and toxicological consequences of browning reactions and related transformations, *in vitro* and *in vivo,* can lead to better and safer foods and improved human health.

7. References

Açar, O.C.; Gökmen, V.; Pellegrini, N. & Fogliano, V. (2009). Direct evaluation of the total antioxidant capacity of raw and roasted pulses, nuts and seeds. *European Food Research and Technology*, Vol.229, No.6, October, pp. 961-969.

Ames, J.M. (1990). Control of the Maillard reaction in food systems. *Trends in Food Science and Technology*, Vol.1, No.1, July, pp. 150-154, ISSN 09242244.

Arribas-Lorenzo, G. & Morales, F. J. (2010). Estimation of dietary intake of 5-hydroxymethylfurfural and related substances from coffee to Spanish population. *Food and Chemical Toxicology*, Vol. 48, No.2, December, pp. 644-649.

Barbosa, J.H.P.; Oliveira, S.L. & Seara, L.T. (2008). The Role of Advanced Glycation End-Products (AGEs) in the Development of Vascular Diabetic Complications. *Brazilian Archives of Endocrinology and Metabolism*, Vol.52, No.6, August, pp. 940-950.

Basta, G.; Schmidt, A.M. & Caterina, R. (2004). Advanced glycation end products and vascular inflammation: implications for accelerated atherosclerosis in diabetes.*Cardiovascular Research*, Vol.63, No.4, September, pp. 582- 592, ISSN 0008-6363.

Basta, G. (2008). Receptor for advanced Glycation end products and atherosclerosis: From basic mechanisms to clinical implications. *Atherosclerosis*, Vol.196, No.1, January, pp. 9-21, ISSN 0021-9150.

Birlouez-Aragon, I.; Sabat, P. & Gouti, N. (2002). A new method for discriminating milk heat treatment. *International Dairy Journal*, Vol.12, No.1, August, pp.59-67.

Borrelli, R. C. & Fogliano, V. (2005). Bread crust melanoidins as potential prebiotic ingredients. *Molecular Nutrition & Food Research*, Vol.49, No.7, July, pp. 673-678, ISSN 1613-4125.

Capuano, E. & Fogliano, V. (2011). Acrylamide and 5-hydroxymethylfurfural (HMF): A review on metabolism, toxicity, occurrence in food and mitigation strategies. *Food Science and Technology*, Vol.44, No.4, May, pp.793 - 810, ISSN 1082-0132.

Cerny , C. (2008). The aroma side of the Maillard reaction. *Annals of the New York Academy of Sciences*. Vol.1126, No.1, April, pp. 66-71, ISSN 0077-8923.

Chang, H.L.; Chen, Y.C. & Tan, F.J. (2011). Antioxidative properties of a chitosan-glucose Maillard reaction productand its effect on pork qualities during refrigerated storage. *Food Chemistry*, Vol. 124, No.2, January, pp. 589-595, ISSN 0308-8146.

Charissou, A.; Ait-Ameur, L. & Birlouez-Aragon, I. (2007). Evaluation of gas chromatography/mass spectrometry method for the quantification of carboxymethyllysine in food samples. *J. Chromatogr.*,Vol. 1140, No.1, December, pp.189-194, ISSN 0021-9673.

Chawla, S.P.; Chander, R. & Sharma, A. (2009). Antioxidant properties of Maillard reaction products obtained by gamma-irradiation of whey proteins. *Food Chemistry*, Vol. 116, No.1, September, pp.122-128, ISSN 0308-8146.

Delgado-Andrade, C. & Morales, F. J. (2005). Unraveling the Contribution of Melanoidins to the Antioxidant Activity of Coffee Brews. *J. Agric. Food Chem.*, Vol.53, No.5, February, pp.1403-1407, ISSN 0021-8561.

Delgado-Andrade, C.; Rufiàn-Henares, J. & Morales, F. (2005). Fast method to determine furosine in breakfast cereals by capillary zone electrophoresis. *Eur. Food Res. Technol.*, Vol. 221, No.5, October, pp.707-711, ISSN 1438-2385.

Delgado-Andrade, C.; Rufiàn-Henares, J. & Morales F.J. (2006). Study on fluorescence of Maillard reaction compounds in breakfast cereals. *Molecular Nutrition & Food Research,* Vol. 50, No.9, September, pp. 799-804.

Delgado-Andrade, C.; Rufián-Henares, J.A. & Morales, F.J. (2007a). Lysine availability is diminished in commercialfibre-enriched breakfast cereals. *Food Chemistry,* Vol.100, No.2, December, pp.725–731, ISSN 0308-8146.

Delgado-Andrade, C.; Seiquer, I. & Navarro, M.P. (2007b).Effects of Consumption of Maillard Reaction Products on Magnesium Digestibility and Tissue Distribution in Rats. *Food Science and Technology International,* Vol.13, No.2, April, pp.109–116, ISSN 1082-0132.

Delgado-Andrade, C.; Rufian-Henares, J.; Morales, F.J. (2008). Optimised procedure to analyse Maillard reaction associated fluorescence in cereal-based products. Czech Journal of Food Science, Vol. 26, No. 5, September, pp. 339-346.

Delgado-Andrade, C.; Rufian-Henares, J. & Morales, F. (2009). Colour and fluorescence measurement as unspecific markers for the Maillard reaction, In: *Assessing the generation and bioactivity of neo-formed compounds in thermally treated foods,* Delgado-Andrade, C. & Rufian-Henares, J., 1° ed., Granada: Atrios.

Delgado-Andrade, C.; Seiquer, I.; García, M.M.; Galdo, G. & Navarro, P. (2011). Increased intake of Maillard reaction products reduces phosphorous digestibility in male adolescents. *Nutrition,* Vol. 27, No.1, January, pp.86-91.

Ding, K.H.; Wang, Z.Z.; Hamrick, M.W.; Deng, Z.B.; Zhou, L.; Kang, B.; Yan, S.L.; She, J.X.; Stern, D.M.; Isales, C.M. & Mi, Q.S. (2006). Disordered osteoclast formation in RAGE-deficient mouse establishes an essential role for RAGE in diabetes related bone loss. *Biochem.Biophys. Res. Commun.,* Vol. 340, No. 4, February, pp.1091–1097.

Erbersdobler, H. & Somoza, V. (2007). Forty years of using Maillard reaction products as indicator the nutritional quality of foods. *Mol. Nutr. Food Res.,* Vol.51, No.4, April, pp.423-430, ISSN 1613-4125.

Finot, P. (2005). Historical perspective of the Maillard reaction in food science. *Ann. N.Y. Acad. Sci.,* Vol. 1043, No.1, June, pp.1-8.

Forster, A.; Kuhne, Y. & Henle, T. (2005). Studies on absorption and elimination of dietary maillard reaction products. *Ann N Y Acad Sci.,* Vol.1043, No.1, June, pp.474-481.

Friedman, M. (1996). Food browning and its prevention: an overview. *Journal Agricultural And Food Chemistry,* Vol.44, No.3, January, pp. 631-653, ISSN 0021-8561.

García, M.M.; Seiquer , I.; Delgado-Andrade, C.; Galdó, G. & Navarro, M.P. (2009a). Intake of Maillard reaction products reduces iron bioavailability in male adolescents. *Molecular Nutrition & Food Research,* Vol.53, No.12, December, pp. 1–10, ISSN 1613-4125.

García, M.M.; Seiquer, I. & Navarro, M.P. (2009b). Influence of Diets Rich in Maillard Reaction Products on Calcium Bioavailability.Assays in Male Adolescents and in Caco-2 Cells.*Journal of agricultural and food chemistry,*Vol.57, No.20, October, pp. 9532–9538, ISSN 0021-8561.

Gerrard, J. (2002a). New Aspects of an AGEing Chemistry - Recent developments Concerning the Maillard Reaction. *Australian Journal of Chemistry,* Vol.55, No.5, June, pp.299-310, ISSN 0004-9425.

Gerrard, J. A. (2002b). Protein-protein crosslinking in food: methods, consequences, applications. *Trends Food Science and Technology,* Vol. 13, No. 12, December, pp. 391-399.

Goldberg, T.; Cai, W.; Peppa, M.; Dardaine, V.; Baliga, B.; Uribarri, J. &Vlassara, H. (2004). Advanced Glycoxidation End Products in commonly consumed foods. *J. Am. Diet. Assoc.*, Vol.104, No.8, August, pp. 1287-1291, ISSN 0002-8223.

Goya, L.; Delgado-Andrade, C.; Rufían-Henares, J.A.; Bravo, L. & Morales, F.J. (2007). Effect of coffee melanoidin on human hepatoma HepG2 cells.Protection against oxidative stress induced by tert-butylhydroperoxide. *Molecular Nutrition & Food Research*, Vol.51, No.5, May, pp. 536–545, ISSN 1613-4125.

Grillo, M. A. & Colombatto, S. (2008). Advanced glycation end-products (AGEs): involvement in aging and in neurodegenerative diseases. *Amino Acids*, Vol.35, No.1, June, pp. 29–36, ISSN 0939-4451.

Hartog, J.H.W.; Voors, A.A.; Bakker, S.J.L.; Smit, A.J. & Van Veldhuisen, D.J. (2007). Advanced glycation end-products (AGEs) and heart failure: Pathophysiology and clinical implications. *European Journal of Heart Failure*, Vol.9.No.12, December, pp. 1146–1155, ISSN 1388-9842.

Hodge, J. (1953). Chemistry of browning reactions in model systems.*J. Agric. Food Chem.*, Vol. 1, No.15, October, pp. 928-943, ISSN 0021-8561.

Jakus, V. & Rietbrock, N. (2004). Advanced Glycation End-Products and the Progress of Diabetic Vascular Complications. *Physiol. Res.*,Vol.53, No.2, June, pp. 131-142, ISSN 0862-8408.

Kalousová, M.; Krha, J. & Zima, T. (2002). Advanced Glycation End-Products and Advanced Oxidation Protein Products in Patients with Diabetes Mellitus. *Physiol. Res.*,Vol.5, No.6, April, pp. 597-604, ISSN 0862-8408.

Li, Y.; Cohenford, M.A.; Dutta, U. & Dain, J.A. (2008). The structural modification of DNA nucleosides by nonenzymaticglycation: an in vitro study based on the reactions of glyoxal and methylglyoxal with 2′-deoxyguanosine. *Anal BioanalChem*, Vol.390, No.2, January, pp.679–688.

Linden, E.; Cai, W.; He, J.C.; Xue, C.; Li, Z.; Winston, J.; Vlassara, H. & Uribarri, J. (2008). Endothelial Dysfunction in Patients with Chronic Kidney Disease Results from Advanced Glycation End Products (AGE)-Mediated Inhibition of Endothelial Nitric Oxide Synthase through RAGE Activation. *Clin. J. Am. Soc. Nephrol.*, Vol.3, No.3, May, pp. 691-698.

Magalhães, P.M.; Appell, H.J. & Duarte, J.A. (2008). Involvement of advanced glycation end products in the pathogenesis of diabetic complications: the protective role of regular physical activity. *Eur. Rev. Aging Phys. Act.*, Vol.5, No.1, February, pp. 17–29.

Manzocco, L.; Calligaris, S.; Mastrocola, D.; Nicoli, M. C. & Lerici, C. R. (2001). Review of nonenzymatic browning and antioxidant capacity in processed foods. *Trends in Food Science & Technology*, Vol.11, No.9, September, pp. 340–346.

Marchetti, P. (2009). Advanced glycation end products (AGEs) and their receptors (RAGEs) in diabetic vascular disease. *Medicographia*, Vol.31, No.3, pp.257-265, ISSN 0243-3397.

Martín, M.A.; Ramos, S.; Mateos, R.; Rufián-Henares, J.A.; Morales, F.J.; Bravo, L. & Goya, L. (2009). Biscuit Melanoidins of Different Molecular Masses Protect Human HepG2 Cells against Oxidative Stress. *J. Agric. Food Chemistry*, Vol.57, No.16, July, pp. 7250–7258, ISSN 0021-8561.

Matiacevich, S.B. & Buera, M. P. (2006). A critical evaluation of fluorescence as a potential marker for the Maillard reaction. *Food Chemistry,* Vol. 95, No.3, January, pp. 423-430.

Méndez, J. J.; Xie, J.; Aguilar-Hernández, M. & Méndez-Valenzuela, V. (2010). Trends in advanced glycation end products researchin diabetes mellitus and its complications. *Mo.l Cell. Biochem.,*Vol. 341, No.1, March, pp. 33–41.

Morales, F; Romero, C. & Jimenez-Perez, S. (1997). Chromatographic determination of bound hydroxymethylfurfural as an index of milk protein glycasylation. *Journal of Agricultural and Food Chemistry,* Vol. 45, No. 5, May, pp. 1570-1573, ISSN 0021-8561.

Morales, F. & Van Boekel, M. (1997). A study on advanced Maillard reaction in heated casein/sugar solutions: fluorescence accumulation. *International Dairy Journal,* Vol. 7, No.11, October, pp. 675-683, ISSN 0958-6946.

Morales, F. & Van Boekel, M. (1998). A study on advanced Maillard reaction in heated casein/sugar solutions: colour formation. *International Dairy Journal,* Vol. 8, No.10, January, pp.907-915, ISSN 0958-6946.

Mostafa, A.A.; Randell, E.W.; Vasdev, S.C.; Gill, V.D.; Han, Y.; Gadag, V.; Raouf, A.A. & El Said, H. (2007). Plasma protein advanced glycation end products, carboxymethyl cysteine, and carboxyethyl cysteine, are elevated and related to nephropathy in patients with diabetes. *Mol. Cell Biochem.,* Vol.302, No.1, August, pp. 35–42.

Muscat, S.; Pelka, J.; Hegele, J.; Weigle, B.; Muench, G. & Pischetsrieder, M. (2007). Research Article Coffee and Maillard products activate NF-kbin macrophages via H_2O_2 production. *Molecular Nutrition & Food Research,* Vol.51, No.5, May, pp.525-535, ISSN 1613-4125.

Nass, N.; Bartling, B.; Santos, A.N.; Scheubel, R.J.; Börgermann, J.; Silber, R.E. & Simm, A. (2007). Advanced glycation end products, diabetes and ageing. *Z. Gerontol. Geriat.,* Vol.40, No.5, October, pp.349–356.

Nguyen, C. (2006). Toxicity oh the AGEs generated from the Maillard reaction: On the relationship of food-AGEs and biological-AGEs. *Molecular Nutrition & Food Research,* Vol.50, No.12, December, pp. 1140-1149, ISSN 1613-4125.

Nursten, H. (2005). *The Maillard reaction: chemistry, biochemistry and implications,* Royal Society of Chemistry, ISBN 0-85404-964-9, London, UK.

Pageon, H. (2010). Reaction of glycation and human skin: The effects on the skin and its components, reconstructed skin as a model. *Pathologie Biologie,* Vol.58, No.3, June, pp. 226–231.

Pawlak, A.M.; Glenn, J.V.; Beattie, G.R.; McGarvey, J.J. & Stitt, A.W.(2008). Advanced Glycation as a Basis for Understanding Retinal Aging and Noninvasive Risk Prediction. *Annals of the New York Academy of Sciences,* Vol.1126, No.1, April, pp. 59-65.

Rao, M.S.; Chawla, S.P.; Chander, R. & Sharma, A. (2011). Antioxidant potential of Maillard reaction products formed by irradiation of chitosan–glucose solution. *Carbohydrate Polymers,* Vol.83, No.2, January, pp. 714–719, ISSN 0144-8617.

Rerat, A.; Calmes, R.; Vaissade, P. & Finot, P.A. (2002). Nutritional and metabolic consequences of the early Maillard reaction of heat treated milk in the pig. Significance for man. *European Journal of Nutrition,* Vol.4, No.1, February, pp.1–11, ISSN 1436-6207.

Resmini, P.; Pellegrino, L. & Battelli, G. (1990). Accurate quantification of furosine in milk and dairy products by a direct HPLC method. *Italian Journal of Food Science,* Vol.2, No.3, pp. 173-183. ISSN 1120-1770.

Rizzi, G. (2008). The strecker degradation of amino acids:newer avenues for flavor formation. *Food Reviews International,* Vol. 24, pp. 416–435, 2008, ISSN: 1525-6103.

Rozycki, S.D.; Buera, M.P.; Piagentini, A.M.; Costa, S.C. & Pauletti, M.S. (2010). Advances in the study of the kinetics of color and fluorescence development in concentrated milk systems. *Journal of Food Engineering,* Vol. 101, No.1, June, pp.59–66, ISSN 0260-8774.

Rufiàn-Henares, J. & Delgado-Andrade, C. (2009). Effect of digestive process on Maillard reaction indexes and antioxidante properties of breakfast cereals. *Food Research International,* Vol.42, No.3, April, pp.394-400, ISSN 0963-9969

Rufiàn-Henares, J. A.; Guerra-Hernández, E. & García-Villanova, B. (2002). Maillard reaction in enteral formula processing: furosine, loss of o-phthaldialdehyde reactivity, and fluorescence. *Food Research International,* Vol.35, No.6, April, p.527-533, ISSN 0963-9969.

Rufiàn-Henares, J.; Delgado-Andrade; C. & Morales, F. (2006). Aplication of a fast high-performance liquid chromatography method for simultaneous determination of furanic compounds and glucosylisomaltol in breakfast cereals. *J. AOAC Int.,* Vol. 89, No.1, January/February, pp.161-165, ISSN 1060-3271.

Rufiàn-Henares, J.; Garcia-Villanova B.; & Guerra-Hernandez, E. (2004). Generation of furosine and color in infant/enteral formula-resembling systems. *Journal of Agricultural and Food Chemistry,* Vol.52, No.17, August, pp.5354-5358, ISSN 0021-8561.

Rufiàn-Hernares, J.; Garcia-Villanova, B. & Guerra-Hernandez, E. (2001). Determination of furosine in enteral formula. *J. Liq. Chromatogr.,* Vol.24, No.19, pp. 3049-3061, ISSN 1082-6076.

Rutherfurd, S.M.; Torbatinejad, N.M. & Moughan, P.J. (2007).Available (ileal digestible reactive) lysine in selected cereal-based food products.*Journal of agricultural and food chemistry,* Vol.54, No.25, November, pp.9453-7, ISSN 0021-8561.

Sato, T.; Iwaki, M.; Shimogaito, N.; Xuegang, W. (2006a). AGEs formation *in vivo. Curr. Mol. Med.,* Vol. 6, pp.351-358, ISSN. 1567-2026.

Sato, T.; Shimogaito, N.; Wu, X.; Kikuchi, S.; Yamagishi, S. & Takeuchi, M. (2006b).Toxic Advanced Glycation End Products (TAGE) Theory in Alzheimer's Disease. *American Journal of Alzheimer's Disease & Other Dementias,* Vol.21, No.3, June/July, pp.197-208.

Schaller, E.; Bosset, J.O. & Escher, F. (1998). 'Electronic Noses' and their application to food. *Lebensmittel-Wissenschaft und-Technologie,* Vol.31, No.4, April, pp. 305-316.

Seiquer, I.; Díaz-Alguacil, J.; Delgado-Andrade, C.; López-Frías, M.; Hoyos, A.M.; Galdó, G. & Navarro, M.P. (2006). Diets rich in Maillard reaction products affect protein digestibility in adolescent males aged 11–14 y. *American Journal of Clinical Nutrition,* Vol.8, No.5, May, pp.1082– 8, ISSN 0002-9165.

Seiquer, I.; Ruiz-Roca, B.; Mesías, M.; Muñoz-Hoyos, A.; Galdó, G.; Ochoa, J.J. & Navarro, M.P. (2008). The antioxidant effect of a diet rich in Maillard reaction products is attenuated after consumption by healthy male adolescents. *In vitro* and *in vivo* comparative study. *Journal of the Science of Food and Agriculture,* Vol.88, No.7, March, pp. 1245–1252, ISSN 0022-5142.

Somoza, V. (2005). Five years of research on health risks and benefits ofMaillard reaction products: An update. *Molecular Nutrition & Food Research*, Vol.49, No.7, July, pp.663 - 672, ISSN 1613-4125.

Somoza, V.; Wenzel, E.; Weiss, C.; Clawin-Rädecker, I.; Grübel, N. & Erbersdobler, H. (2006). Dose-dependent utilisation of casein-linked lysinoalanine, N(epsilon)-fructoselysine and N(epsilon)-carboxymethyllysine in rats. *Molecular Nutrition & Food Research*, Vol.50, No.9, September, pp.833-841, ISSN 1613-4125.

Stanimirova, I.; Boucon, C. & Walczaka, B. (2011). Relating gas chromatographic profiles to sensory measurements describing the end products of the Maillard reaction. *Talanta*, Vol. 83, No. 4, January, pp. 1239–1246, ISSN 0039-9140.

Summa, C.; McCourt, J.; Cammerer, B.; Fiala, A.; Probst, M.; Kun, S.; Anklam, E. & Wagner, K. (2008). Radical scavenging activity, anti-bacterial and mutagenic effects of Cocoa bean Maillard Reaction products with degree of roasting. *Molecular Nutrition & Food Research*, Vol.52, No.3, March, pp. 342 - 351, ISSN 1613-4125.

Szczesniak, A. S. (2002). Texture is a sensory property. *Food Quality and Preference,* Vol.13, No.4, June, pp. 215–225, ISSN 0950-3293.

Torbatinejad, N.M.; Rutherfurd, S.M. & Moughan, P.J. (2005). Total and Reactive Lysine Contents in Selected Cereal-Based Food Products. *Journal of agricultural and food chemistry*, Vol.5, No.11, June, pp.4454–4458, ISSN 0021-8561.

Tuohy, K.M.; Hinton, D.J.; Davies, S.J.; Crabbe , M.J.; Gibson, G.R. & Ames, J.M. (2006). Review Metabolism of Maillard reaction products by the human gut microbiota - implications for health. *Molecular Nutrition & Food Research*, Vol.50, No.9, September, pp. 847 - 857, ISSN 1613-4125.

Van Boekel, M. A. (2006). Formation of flavor compounds in the Maillard reaction. *Biotechnology Advances*, Vol.24, No.2, April, pp.230-233, ISSN 0734-9750.

Vasdev, S.; Gill, V. & Singal, P. (2007). Role of Advanced Glycation End Products in Hypertension and Atherosclerosis: Therapeutic Implications. *Cell Biochem.Biophys.*, Vol.49, No.1, July, pp. 48–63.

Vlassara, H.; Cai, W.; Crandall, J.; Goldberg, T.; Oberstein, R.; Dardaine, V.; Peppa, M. & Rayfield, E.J. (2002). Inflamatory mediators are induced by dietary glycotoxins, a major risk factor for diabetic angiopathy. *Proc. Natl. Acad. Sci.*, Vol.99, No.24, January, pp.15596-601.

Vytásek, R.; Sedová, L. & Vilím, V. (2010). Increased concentration of two different advanced glycation end-products detected by enzyme immunoassays with new monoclonal antibodies in sera of patients with rheumatoid arthritis. *BMC Musculoskelet Disordens*, Vol.83, No.11, May, pp.1-11.

Wagner,K.; Reichhold, S.; Koschutnig, K.; Chériot, S. & Billaud, C. (2007). The potential antimutagenic and antioxidant effects of Maillard reaction products used as "natural antibrowning" agents. *Molecular Nutrition & Food Research*, Vol.51, No.4, April, pp.495 - 504, ISSN 1613-4125.

Wang, He-Ya; Qian, H. & Yao, Wei-Rong. (2011). Melanoidins produced by the Maillard reaction: Structure and biological activity. *Food Chemistry*, Vol.128, No.3, October, pp.573–584, ISSN 0308-8146.

Yu, Ai-Nong & Zhang, Ai-Dong. (2010). The effect of pH on the formation of aroma compounds produced by heating a model system containing L-ascorbic acid with

L-threonine/L-serine. *Food Chemistry*, Vol.119, No.1, March, pp.214–219, ISSN 0308-8146.

Zeng, Y.; Zhang, X.; Guan, Y. & Yuanxia, S. (2011). Characteristics and Antioxidant Activity of Maillard Reaction Products from Psicose-Lysine and Fructose-Lysine Model Systems. *Journal of Food Science*, Vol.76, No.3, April, pp.398-403, ISSN 0022-1147.

15

Membrane Separation Process in Wastewater Treatment of Food Industry

Claudia Muro[1], Francisco Riera[2] and María del Carmen Díaz[1]
[1]Institute Technological of Toluca,
[2]University of Oviedo,
[1]México
[2]Spain

1. Introduction

Wastewater derived from food production is highly variable, depending on the specific types of food processing operations (e.g., fruit, vegetable, oils, dairy, meat, and fish).
Advances in membranes technology have showed many advantages for wastewater treatment of food industry. By implementing membranes, the separated substances and clean water are often recoverable in a chemically unchanged form and are therefore easily re-used. Maximum benefits are obtained when one or both the output streams from the membrane system are recycled or re-used, thereby reducing process materials requirement and minimizing waste disposal costs.
This chapter reviews the development and applications of membrane processes in wastewater treatment of food industry. Particular focus is given to membrane's special abilities to wastewater treatment for water regeneration and various re-uses purposes. Influence of engineering aspects is analyzed, specially operating conditions near critical flux to improve processes in wastewater treatment. Detailed discussions are provided with respect to constituents of concern in water reuse applications including recovery of other products with value for food industry.

2. Wastewater of food industry volume and quality

The types of food production processes (e.g., fruit, vegetable, oils, dairy, meat, fish, etc.) vary widely, with associated differences in the specific wastewater contaminants. The characteristics and generation rates of food wastewater are highly variable, depending on the specific types of food processing operations, including wastewater from of activities of food cleaning (sanitizing, peeling, cooking, and cooling); mechanically activities (conveyor medium to transport food materials throughout the process) and clean production equipment between operations. In addition, one important attribute is the general scale of the operations, since food processing extends from small, local operations.
Food processing can be divided into four major sectors: Meat, poultry and seafood; fruit and vegetables; dairy and beverage. Table 1 shows the wastewater volume and pollution charge of some food industries.

Feed processing	Wastewater (m^3ton^{-1}) of product	COD ($mgO_2.L^{-1}$)	BOD_5 ($mgO_2.L^{-1}$)
Meat processing			
- Scalding tube	0.3	1800	1 400
-Chiller showers	1.7	150	140
- Cooling tanks	0.7	550	500
Fruit juice			
-Orange	5.0	11200	8 100
-Apple	1.2	2000	1 400
-Tomato wastewater[1]		1200	
-Fruit Juice (general)[2]		2500-7000	
Vegetable processing			
-frozen carrots	30	5000	4 500
-Olive mill[3]		100 000-200 000	
Potato starch			
-Shower	0.7	3 000	2 500
Starch rinsing	1.5	7 800	6 500
Beer production	4.2	2 500	1 800
Alcohol plant		900-1 200	
Fish industry			
- Unloading fish[5]		5 000-7 000	
-Brines[6]		4 000-14 000	
-Cooked fish[7]		4 000-20 000	
Dairy industry			
-Whey	90	65 000	42 000
-End pipe wastewater	1.5	1 800	860
-Flash cooler condensates		100-570	
-Bottle rinsing[8]		50-1000	
-Caustic solutions[9]		8 000-10 000	

*adapted from [1]Iaquinta et al., 2009; [2]Noronha et al., 2002; [3]Mantzavinos & Kalogerakis, 2005; [4]Madaeni & Mansourpanah, 2006; [5] Matthiasson, 1983; [6]Kuca & Szaniawska, 2009; [7]Walha et al., 2009; [8]Scharnagl et al., 2000; [9]Gésan-Guiziou et al., 2007

Table 1. Wastewater from food industry

Primary and secondary treatments are often used to decompose the high organic contents of wastewater of food industry by aerobic and anaerobic fermentation processes.
After of traditional treatment of wastewater, general requirements are covered by regulations of each country, usually complemented by consent limits based on avoidance of pollution. Discharge licenses may include maxima for flow, temperature, suspended solids, dissolved solids, BOD_5, nitrogen, phosphorous and turbidity. According at quality of water, in most cases, final disposal of treated waste water is into a water course where it will be diluted by the existing flow. However, subsequently one advanced process of effluent treating can be an option desirable to recycle water within a factory of food processing.

3. Membrane process

Membrane filtration is a process used to separate dissolved substances and fine particles from solutions. Membrane acts as a semipermeable and selective barrier that separates particles based on molecular or physical size. Solutes smaller of solution than the membrane pore size are able to pass through the membrane as permeate flux while particles and

molecules larger than the membrane pore size are retained. The two fluxes at outlet of membrane are important because this process has a high efficiency in the separation.
The majority of commercial membranes are made usually of organic polymers (polysulfones and polyamides) and inorganic materials (ceramic membranes based on oxides of zirconium, titanium, silicium and aluminum).
The membranes are implemented in several types of modules. The membrane configuration determines the manner in which the membrane is packed inside the modules. Four main types of membrane configurations are used in the industry. These are: plate-and-frame, spiral wound, tubular and hollow-fiber configurations. The membrane geometry is planar in the first two and cylindrical in the two others. Figure 1 shows schematically a typical hollow fiber module (Okokchina, 2010).

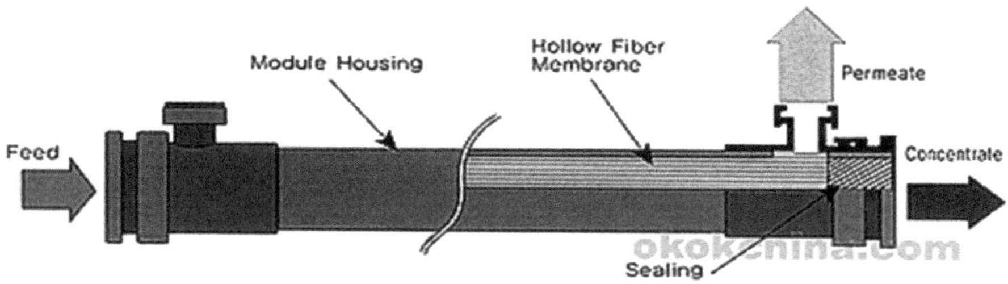

Fig. 1. Scheme of the hollow fiber membrane module with crossflow. A large surface/volume ratio is expected for these modules.

The membrane system is operated in a cross-flow feed mode. The concentrated stream passes parallel to the membrane surface as opposed to perpendicular flow that is used traditionally in filtration. This operating mode allows that accumulation of solute molecules at the membrane surface decreases and the permeate flux remains constant for a long time due to decreased hydrodynamic resistance at the membrane surface by cross-flow induced hydraulic turbulence. Flow direction is usually inside-out, i.e. the concentrate flux inside the fibers and the permeate flux is collected at the shell-side. It is often possible to reverse the flow (outside-in) for cleaning and unclogging of the membrane. Cylindrical configuration provides the possibility of maintaining high tangential velocity in the feed stream and is therefore particularly suitable for applications where the feed contains a high proportion of suspended solids or must be strongly concentrated.
The choice for a certain kind of membrane system is determined by a great number of aspects, such as costs, risks of plugging of the membranes, packing density and cleaning opportunities. The effects of the feed properties, the membrane properties, and the filtration conditions are obviously very important for the success of a membrane filtration process. Principal limitation of membrane lies in membrane fouling which is mainly associated with the deposition of a biosolids cake layer onto the membrane surface (McCutcheon & Elimelech, 2006; Mi & Elimelech, 2008). However, everal alternatives have been implemented to enhance this problem (Al-Akoum et al., 2002; Jaffrin et al., 2004).

3.1 Membrane applications in food industry for wastewater treatment

Membrane separation process has special recognition in food wastewater treatment, applied to the end of conventional treatment systems (Vourch et al., 2008). The process is used primarily to reduce the volume of the food wastewater that is achieved by recovering of two fluxes: permeate water flux having the majority of the original volume, and concentrated flux in a lesser volume (constituents of effluents retained).

The membranes used in food wastewater treatment differ widely in their structure and function. Mainly they are operated in four membrane processes: microfiltration (MF), ultrafiltration (UF), nanofiltration (NF) and reverse osmosis (RO). Solvent permeability and separation selectivity are the two main factors characterizing at these membranes. Transport mechanisms and operating membrane conditions can also explain the pass of species through membranes. Particle size is practically the sole criterion for describer the permeation or rejection of membranes. However, microporous membranes (NF and RO) have ability of separate particles at molecular level and their selectivity is mainly based on the chemical nature of the species.

Several works have been focused on these factors to explain separation selective of residues of food wastewater. Effluents treatment of dairy industry by RO and NF membranes are reported in many investigations, however, a strong development and growth of membrane technology can be observed in the results from the other food industries (Turano et al., 2002). Food industry standards specify that, spent process water intended for reuse (even for cleaning purposes) must be at least of drinking quality. Regulations for other applications, such as boiler make-up water or warm cleaning water, are even more stringent. There has been a study on the possibilities for reuse of vapor condensate in a milk processing company (dried milk production) as boiler make-up water (Hafez et al., 2007), and the reuse of chiller shower water in a meat processing company (sausage production) as warm cleaning water (Mavrov & Bélières, 2000).

3.2 Membrane characteristics

Generally membranes are characterized by pore flow or molecular weight of particle that is retained or is filtered by the membrane. However, important membrane properties such as structure, porosity, thickness, wettability surface and operating conditions, are also studied because affect rejection of solutes. The electrostatic repulsion between the membrane surface and the contaminant may be particularly analyzed to enhance waste solute retention and to increase water flux.

The smallest particle size present in the feed is very important for the selection of membrane pore size. However, currently the feed properties can be changed by pretreatments such as pH adjustment, thermal treatment, addition of chemicals, and pre-filtration. The pH adjustment (Luo et al., 2010) and thermal treatment can decrease the precipitation of certain substances. In addition, chemicals can be added to the feed to increase the particle size through aggregation, and the retention of specific substances can be enhanced through micellation or complexation (Wu et al., 2007). The salt concentration of the feed and the valence of the salt present can also be important to select membrane type (Muro et al., 2009; Lefebvre & Moletta, 2006)

3.2.1 Pore-flow and material membranes

Membrane pore flow is differentiated by the size of particles diameter that they can separate (micrometers, μm) and nominal molecular weight cutoff MWCO (kilo Daltons), which is a performance-related parameter, defined as the lower limit of a solute molecular weight for which the rejection is 95-98% (Boerlage et al., 2004). In theory, compounds having a molecular weight greater than the molecular weight cut off (MWCO) will be retained by the membrane and compounds with molecular weights less than the MWCO will pass through the membrane as permeate. Table 3 shows size range of particles retained with range of MWCO membranes for treatment of wastewater of food industry.

Membrane Process	MWCO membrane (kilo Daltons range)	Retained diameters particle (μm range)	Retained solutes	Application in effluents treatment of food industry
MF	100-500	10^{-1}-10	Bacteria, fat, oil, grease, colloids, organics microparticles	Oil, Cereal, Dairy, Beverage
UF	20-150	10^{-3} - 1	Proteins, pigments, oils, sugar, organics microparticles	Dairy, Cereal, Oil, Tomato puree, Beer, Wine, Fish, Meat, Pickled vegetables
NF	2-20	10^{-3} - 10^{-2}	Pigments, sulfates, divalent cations, divalent anions, lactose, sucrose, sodium chloride	Olive oil, Dairy, Beverage, Meat canning, Pickled vegetable
RO	0.2-2	10^{-4}- 10^{-3}	Salts, sodium chloride and inorganic ions	Dairy, Cereal, Fish, Meat, Pickled vegetables

Table 3. Typical range of application of MWCO, diameter particle and retained solutes type by membrane process in wastewater treatment of food industry.

Retention is obviously affected by the pore size due to the sieving effect, especially when using MF and UF membranes. With tighter (NF and RO) membranes retention will be governed more and more by the electrostatic forces as well as by other interactions between membranes and solutes. Thus MWCO is only a rough indication of the membrane's ability to remove a given compound as molecular shape because polarity and interaction with the membrane affect rejection (Guizard & Amblard, 2009).

Respect to pore diameter, it has frequently been seen that the membrane with the most open pores does not usually give the highest permeate flux in filtration process. Porosity (ratio of void space to total membrane volume in porous membrane) and pore size distribution may influence the apparent size of particles retained. Typical microporous membranes have average porosities in the range 30%-70%. Porosity can also be measured by analyzing processed images obtained from microscopic analyses such as scanning electron microscopy (SEM). Figure 2 shows SEM image of an asymmetric porous structure of a ceramic membrane. It may be noted that the membrane has fine pores through which raw water is filtered (Figure 2a). The most of ceramic membrane elements are constructed from supported multiple ceramic layers constituting an asymmetric porous structure.

Carbon macroporous material is used as support for ceramic membrane deposition (Figure 2b and 2c). Multiple layers are usually resulting from residual spaces created between ceramic particles during sintering. The bottleneck geometry is representative of pores resulting from sintering of almost spherical particles, for example, this is the case of porous structures obtained with titania, zirconia (Guizard et al., 2002; Guizard & Amblard, 2009). The porous sites are uniformly distributed in the membrane and effective diameter of the membrane pore can be determined assuming pores are circular in shape. However, pore geometry (tortuosity; τ) can also affect the retention of molecules by a membrane. Tortuosity reflects the length of the average pore compared to the membrane thickness. Cylindrical pores at right angles to the membrane surface have a tortuosity of one, that is, the average length of the pore is the membrane thickness (Cho et al., 2000; Zhao et al., 2000; Vrijenhoek et al., 2001).

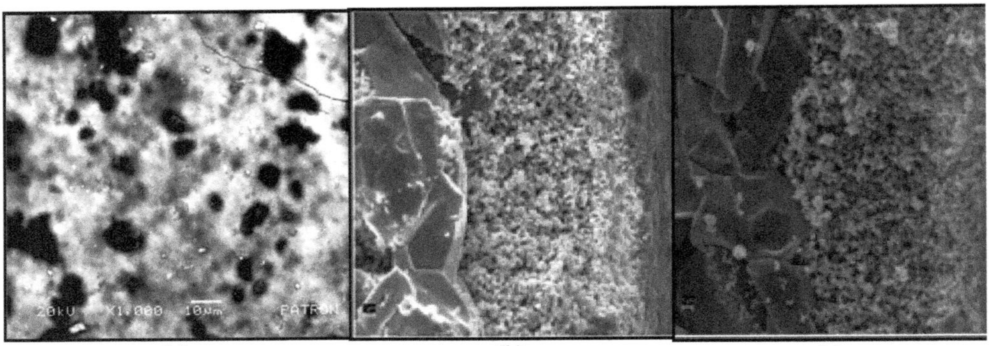

(a) (b) (c)

Fig. 2. SEM image of cross section of a ceramic membrane porous structure of MF, with cut off of 300 kDa and 5 µm pore size, used in wastewater treatment of food industry (From Escobar. PhD thesis, Institute Technological of Toluca, México, 2010)

Chemical composition, hydrophilicity/hydrophobicity, charge, and morphology have also significantly effect on permeability and stability of the membrane (Khayet et al., 2005). Particularly, ceramic membranes have a composite structure, which is used to increase the permeability for small pore size membranes by decreasing the overall hydraulic resistance (Peng et al., 2005; Yu et al., 2006) while polymeric membranes can be modified to make them more hydrophilic and achieve less fouling and better cleaning efficiency.

3.2.2 Surface pore charge. Isoelectric point

Membrane charge affects membrane efficiency in food wastewater treatment, particularly when low cutoff membranes are used for treatment effluents with high salts concentration. The charging occurs due to, for instance, dissociation of functional groups, adsorption of ions from solution, and adsorption of polyelectrolytes, ionic surfactants, and charged macromolecules. Generally, membrane materials carry a negative charge or are modified to have a negative charge because natural organic matter in water is negatively charged at neutral pH, due to phenolic and carboxylic functional groups (Kaeselev et al., 2002). A negatively charge membrane, therefore, prevents rapid deposition of foulants on the membrane surface by charge repellence An increase in the flux of a relatively dense membrane at a high pH may result from an increase in membrane hydrophilicity due to the dissociation of the functional groups in the membrane structure (Schaep & Vandecasteele, 2001; Zhao et al., 2005). Many polymeric membranes are amphoteric, having both negatively and positively charged functional groups in the polymer matrix. Ceramic membranes can also show in water amphoteric behavior and thus their surface charge is pH dependent (Cho et al., 2000).

Membrane charge, as well as hydrophilicity property, can be predicted based on known membrane chemical structure. However, membrane surface/pore charge can be measure by electrical potential (Martín et al., 2003). When the membrane contains strongly acidic groups, the dissociation of the groups occurs immediately at a low pH, and the zeta potential can be expected to be strongly negative even at low pH values (pH 2–3); while when the membrane contains weakly acidic groups, the zeta potential can be expected to become more negative from the point the groups start to dissociate to the point where the groups are totally dissociated. Similarly, strongly basic groups give positive potentials in most of the pH range, while weakly basic groups have no positive charges at pH values higher than 8 (Kim et al., 2005).

The isoelectric point (IP) (pH where net charge is zero) of a membrane is also a referent to determinate the behavior of their surface charge, depending on the pH of the wastewater in contact with the membrane. (Cheng et al., 2008). For example, typically NF polimeric membranes are negatively charged at neutral pH, with IP around pH 3-4, while ceramic membranes have a IP around pH 6-7

The IP of a membrane can be evaluated from the pH dependence of the zeta potential (Martín et al., 2003). However other experiments can also describe this parameter. Figure 3 shows isoelectric point of a ceramic membrane of zirconium and titanium oxide.

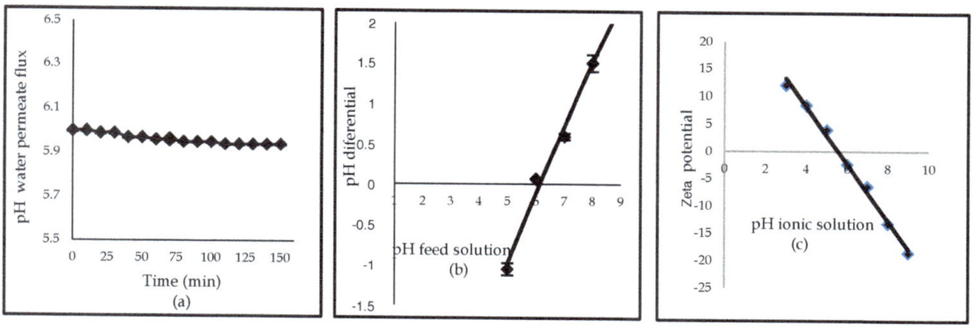

Fig. 3. Isoelectric point of a ceramic membrane of UF. (a) pH of permeate water is measured during an operation time range. (b) pH differential is determinate when pH feed solution is adjustment at 4-8 range, intersection of line with horizontal axe denotes isoelectric point at 6.2. (c) The values of zeta potential are measured in dependence with pH of ionic solution.

Figure 3a, denotes pH determination of pure water during water filtration by membrane of UF. In this membrane the pH value of permeate does not change with operation time. It shows that isoelectric point is around pH 6. Figure 3b shows the pH differential permeate water dependent of pH feed solution. Intersection of line with horizontal axe denotes isoelectric point at pH 6.2 for the same ceramic membrane. Figure 3c shows the values of zeta potential and pH of ionic solution; the values were measured through the pores of ceramic membrane. Intersection line with pH axis, indicates that isoelectric point is 6.2.

4. Influence of engineering aspects in food industrial wastewater treatment

Systematic studies of membrane applications in food wastewater treatment are focused on membrane functionality and performance filtration, under different operation condition. Several researches are specifically focused on optimizing crossflow hydrodynamics and/or membrane filter geometry to increment performance of water flux and maximum rejection or recovering species from effluents. Hydrodynamic factors affecting the membrane functionality, are cross-flow velocity (v) and transmembrane pressure (TMP). Permeate flux can increase or decrease due to simultaneous influence of these variables. Temperature, dilution and pH are also variables involved in the membrane efficiency in membranes filtration. Permeate flux increases with increasing feed temperature due to a decrease in viscosity and/or due to an increase in solubility of suspended solids (Galambos et al., 2004). The exception is the presence of calcium and magnesium salts that might precipitate when temperature is increased. This problem can be avoided at least in some cases through feed pretreatment (Sarkar et al., 2006). The pH has a significant influence on the permeation rate especially around the isoelectric point of certain colloids where they tend to destabilize and precipitate. It also has an effect because of the changes in surface charge of the membrane either due to the amphoteric nature of the surface or due to the specific adsorption of species as presented earlier (Vourch et al., 2008).

4.1 Cross-flow velocity

A hydrodynamic variable of membranes in cross-flow filtration systems is essentially the velocity at which the feed flow is passed across the surface of the membrane. Crossflow velocity (ν) is linear velocity (m/s^{-1}) of the feed flow circulating tangentially across the membrane. This parameter is described by relation of feed flow rate (Qw; m^3/s^{-1}) and the cross sectional area of feed membrane (As; m^2).
Turbulent flow conditions are recommended to maintain the flow tangential to the membrane, thereby reducing the phenomenon of concentration polarization and, consequently, the accumulation of solute near the membrane and inducing acceptable permeate flux for long time. Shear effects induce hydrodynamic filtration of the particles from the boundary layer back into the bulk, with a positive effect on the permeate flux. However, as feed concentration increases, it becomes more difficult to maintain a high recirculation velocity due to an increase in feed viscosity (Muro et al., 2009). In addition, if foods waste water containing macromolecular solutions with flexible solutes, thus also a high velocity can cause deformation of the polymer chains, which favors certain macromolecules that pass through the pores.
The hydrodynamics flow can also be characterized by calculating the Reynolds (Re) number by equation (1).

$$\mathrm{Re} = \nu \frac{d_h}{\mu} \tag{1}$$

Where ν is crossflow velocity, d_h hydraulic diameter of membrane module and μ the dynamic viscosity of fluid.
Normally, Re>2100 guarantees a turbulent flow in the module and a minimum thickness for the concentration polarization layer. Prevention of reversible fouling layer formation is sufficiently achieved by a crossflow velocity of around 2.0 ms^{-1} in UF membranes (McKeown et al., 2005). In practical applications, one has to keep in mind that the permeate flux will be determined by the combination of crossflow velocity and TMP (See Figure 5).

4.2 Transmembrane pressure

The driving force for transport behind membrane process MF, UF, NF and RO, is the pressure difference between feed and permeate flux of the membrane (TMP; bar, psi). TMP is defined as the difference in pressure between the filtrate side of the membrane and the permeate side of the membrane. The average TMP is in general calculated as follows:

$$TMP = \frac{P_i + P_0}{2} - P_p \qquad (2)$$

Where P_i is pressure at the inlet of the membrane module; P_0 is pressure at the outlet of the membrane module and P_p is permeate pressure.

The permeate flux depends directly on the applied TMP for a given surface area under uniform operational conditions. The flux of the pure water is linearly pressure dependent. However when food wastewater is treated by membrane system the flux is more complex. The behavior depends of wastewater composition, membrane type and crossflow velocity.

In food wastewater treatment, one has to keep in mind that the permeate flux will be determined by the combination of crossflow velocity and TMP, due to contaminants (Sarkar et al., 2006; Blöcher et al., 2002; Oktay et al., 2007; Avula et al., 2009).

Figure 4a and 4b show the effect of crossflow velocity and TMP on permeate flux using two membranes of different MWCO (300 kDa and 15 kDa). The experiments were performed by Escobar, 2010. The results indicated that the flux enhancement caused by increasing crossflow velocity was particularly pronounced at range values of the TMP (3-5 bar) and crosflow velocity of 3 ms^{-1}. Fouling occurred over a range of TMPs of 5-6 bar and crossflow velocities at 3.5 ms^{-1}. The permeate flux decreased with time during the development of the fouling layer, but once the fouling layer was established, the permeate flux became constant for a given set of experimental conditions. Therefore these results indicate that at moderate values of TMPs and high flow rates at the membrane surface are operating conditions that conduce at high permeate fluxes in these experiments. Besides, figure 4c shows an overall positive effect of enhanced flow hydrodynamic conditions (TMP = 4 bar) on the average permeate flux, although in the turbulent regime (Re>3,000) a weaker correlation and more data scattering were observed. Therefore a clear correlation between the 3 h flux and Re in the transient regime (Re<3000) could be expected.

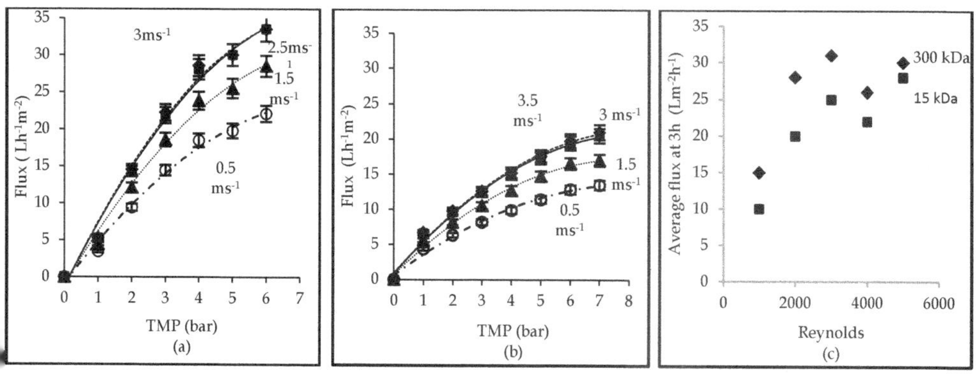

Fig. 4. Effect of crossflow velocity and TMP on the 3 h permeate flux in wastewater treatment of a cereal industry using membranes of MF and UF (a) 300 kDa. (b) 15 kDa. (c) The interdependence between average flux and hydrodynamic conditions for two membranes in a wide range of Re numbers at TMP = 4 bar (From Escobar 2010. PhD thesis, Institute Technogical of Toluca).

Particularly, operational membrane conditions in wastewater treatment show moderate TMP and high flow rates at the membrane surface are conducive of high permeate fluxes in

the MF and UF. An increase in TMP is required to maintain a particular water flux (constant-flux operation) independently of the membrane type and MWCO. However, an increasing flux could lead to an increase in polarization and fouling, which will limit the permeate flux (Abbasi et al., 2011; Simate et al., 2011).
High pressure can also allow membrane compaction, ultimately resulting in the formation of a denser membrane with smaller pores, or one possible enlargement of membrane pores with time, which enables particles to penetrate through the membrane matrix. Choi et al., (2005) showed clearly that pore sizes are modified in the membrane matrix increased with increasing TMP.

4.3 Permeate flow rate

The functionality of a membrane in wastewater treatment is determined by water permeation capacity and retention of solutes. Although permeate flux depends of the characteristics of the membrane and quality of wastewater, the average pore size and pore-size distribution is important since it will give an indication of which transport mechanism can be expected to be dominant for a given specie mixture in a defined material and at given process conditions. There are two theory models to describe the mechanism of permeation in membrane process; one is the solution-diffusion model, in which permeants are diffused through the membrane down a concentration gradient. The other model is the pore-flow model, in which permeants are transported by pressure-driven convective flow through tiny pores. Separation in this case, occurs by excluding of some particles of the pores in the membrane. Fick´s law describes the mass flux through an area perpendicular to the flow direction (Miyoshi, 1998):

$$\frac{dV_i}{Adt} = J_{pi} = -D_i K_i \frac{dC_i}{dx} \tag{3}$$

Where J_{pi} is the linear fluid velocity (ms^{-1}) of component (i) or permeability flux ($Lm^{-2}\,h^{-1}$). The diffusion coefficient D_i (ms^{-1}) reflects the mobility of individual molecules in membrane material and the molecule sorption coefficient K_i reflects the number of molecules dissolved in the membrane material. The product D_iK_i is membrane permeability and is a measure of the membrane's ability to permeate species. dC_i/dx is the concentration gradient ($molL^{-1}$) for component (i) over the length x (m). V_i is the volume of substance (i) transferred (L), t is time (h) and A is perpendicular area (m^2).
Permeability flux $J_{pi} = V_i/At$ is obtained by equation integration (3) and applied for dx = x (membrane thickness or membrane resistance for the pure water transport). C_{i0} and C_{if} are the concentration of component (i) on the feed side and concentration of component (i) on the permeate side respectively. Solution-diffusion model is often used to describe the transport in RO membranes.

$$\frac{V_i}{At} = -L_p \frac{\Delta P}{x} \tag{4}$$

L_p is the hydraulic permeability coefficient (Lm^{-2} bar h^{-1}); ΔP is gradient pressure TMP (bar) in membrane system. Information about porous structure and viscosity of the filtrated liquid is contained in L_p factor.
Membrane resistance (x) is a measure of the hydraulic resistance to flow through a pore channel. However, when wastewater is fed, increment of TMP can cause a decreasing of

membrane permeability because of hydraulic resistance increment by the fouling phenomena. Increment of crossflow velocity, dilution of wastewater, change of temperature of feed and using turbulence promoters such as backflow techniques, feed pulsation and rotation of filter elements, are hydrodynamic methods to increment permeate flux and reduce the hydraulic resistance due to fouling (Jaffrin et al., 2004; Luo et al., 2010).

4.4 Selectivity factor

The best measure of the ability of a membrane to separate molecules (i) of wastewater, is the ratio of their permeability α_j, called the membrane selectivity, which can be written in terms of the apparent sieving coefficient:

$$\alpha_{\mathrm{i}} = \frac{C_{ip}}{C_{if}} \tag{5}$$

C_{ip} is concentration of specie (i) in the permeate flux and C_{if} is the concentration of specie (i) in the feed flow.

The selectivity of a membrane depends on its ability to transmit different species to different extents. Factors that affect solute transmission are solute type, membrane type, solution pH, solution ionic strength, the permeate flux, and the hydrodynamic conditions on the feed side. Membrane selectivity is most often expressed as the membrane retention, R, toward the species to be separated. R is dimensionless parameter, with variation range of 0-100 %.

$$\mathrm{R} = \frac{C_{if} - C_{ip}}{C_{if}} = (1 - \alpha_{\mathrm{i}}) \tag{6}$$

Membrane/Cutt off (kDa)	Ions concentration (mgL^{-1}) in water permeate							
	Na^{1+}		K^{1+}		Ca^{2+}		Fe^{3+}	
	TMP (bar)							
	4	5	4	5	4	5	4	5
MF/150	140	148	23.0	25.1	16.2	16.2	6.4	9.9
UF/15	133	135	22.2	20.1	13.1	13.0	5.7	2.1

Table 4. Effect of (TMP) on the permeability of some ions by MF y UF membranes

Rejection of neutral organic solutes generally increases with the molecular weight (or diameter) of the solute. Species will be retained by the membrane according to their size (sieving effect). For a mixture of multivalent and monovalent co-ions in the feed, mulitivalent co-ions are retained due to their higher electrical charge, while a part of monovalent co-ions pass through the membrane with counter ions to fulfill charge equilibrium criterion on both sides of the membrane (Lefebvre et al., 2003). However, the absolute values of the salt rejection vary over a wide range; the ranking for the different salts is the same for all membranes (Rautenbach & Albrecht, 1989). A high TMP value also affects the selectivity of some ions species. Table 4 shows the effect of TMP conditions on permeability of some ions by two ceramic membranes (Muro et al., 2009). Ions were identified in wastewater of a food industry. The experiments were performed to determine the effect of pressure increment on selectivity of these membranes for these ions. The results indicate that for all PTM values, the ions Fe^{3+} and Ca^{2+} were slightly declined, while ions Na^{+} and K^{+}, were filtered by both membranes.

For other hand, exceptional selectivity for a number of important separations in wastewater treatment of food industry are mentioned in several reports (Vourch et al., 2008; Muro et al., 2010, Escobar et al., 2011; Simate et al., 2011).

Figure 4a and 4b show the difference between selectivity of two ceramic membranes of MF (300 and 150 kDa) and one of UF (50 and 15 kDa) for various TMP values. The data were obtained by experimental study of organics species in micelles with two colorants (a) Brilliant blue. (b) Tartrazine. Membranes denote a low selectivity for the colorants and a high permeability for water. Particularly, membrane of 15 kDa shows the lowest selectivity for two colorants for all TMP values. SEM image denotes, particles deposited on membrane surface, showing a low selectivity of a membrane of 300 kDa for tartrazine colorant.

5. Critical flux conditions

During membrane filtration process are identified three regimens in accordance to the critical flux theory (Field, 1995). Figure 6 shows a typical flux profile by three membranes.

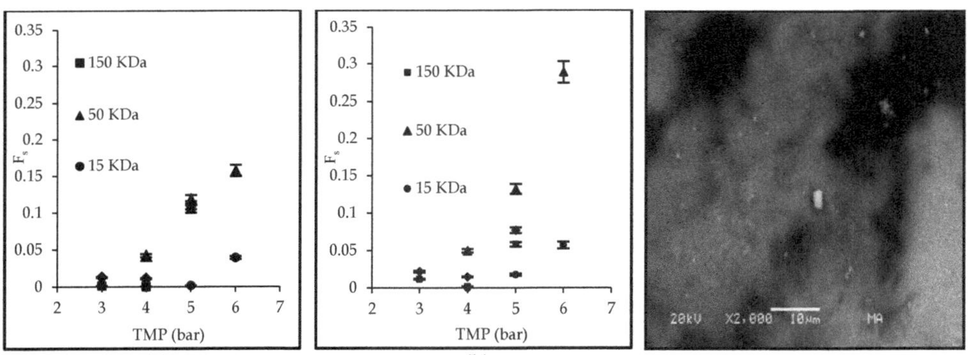

(a) (b) (c)

Fig. 5. Difference between selectivity of ceramic membranes for two colorants from wastewater of a food industry (Muro et al., 2009). (a) brillant blue. (b) tartrazine. (c) SEM image of a ceramic membrane of MF. White small particles of tatrazine may be seen on membrane surface.

Subcritical regime is the first stage of filtration, where flux varies linearly and reversibly with TMP, a high crossflow velocity is employed to increase capacity of permeation and a critical pressure is achieved in the end of this regime Processes where high water purity is required are carried out regime I, because membrane selectivity is optimal. The flux in regime II is independent of TMP, which can be described by an equilibrium stage, where the transport of particles toward the membrane is balanced with the transport of particles toward the bulk flow. At high TMP values, the permeate flux is not significantly affected by increases in pressure. This limiting flux or critical flux increases with increasing crossflow velocity, because materials deposited on the membrane by mass transport are removed by the wall shear force. For soluble species and fine colloids, the critical flux can be considered as the flux below which the wall concentration does not initiate fouling (Cho & Fane, 2000). Choi et al., 2005). High capacity of the concentration of species from wastewater can also be achieved in this regime and the critical flux may either be identical to the clean water flux at the same TMP (Hwang et al., 2006). However, outside the limiting flux, operation at sustained permeability and selectivity is not possible due to the accumulation and compaction of the fouling layer on the membrane. Finally flux decline in time-dependent with high pressure above the critical TMP, are identified in regime III due to increment membrane fouling. Their removal is necessary for stable membrane operation (Espinasse et al., 2002).

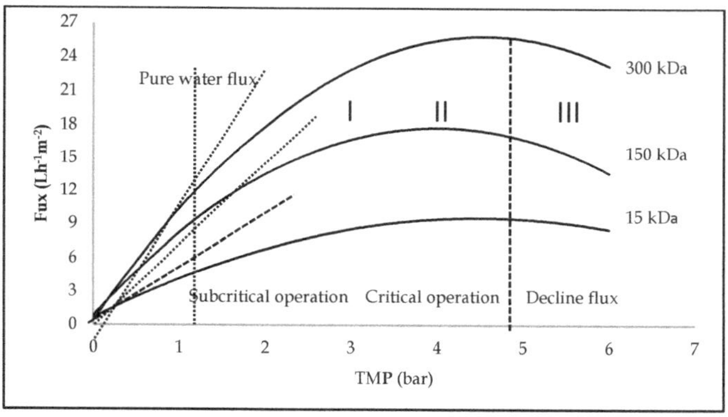

Fig. 6. Critical flux regimes in membranes of 300, 150 and 15 kDa: (I) Subcritical operation (II) Critical operation (III) Decline flux.

The critical flux value depends largely on the hydrodynamic conditions in the process, the membrane pore size, and the feed physicochemical condition (Mänttäri & Nyström, 2000). Appropriate manipulation of these parameters, specifically the hydrodynamic condition, may lead to increment of flux and the reduction or even the elimination of both reversible and irreversible fouling of the membrane. The critical flux can be experimentally identified through constant flux filtration experiments by incrementing the flux until the TMP is no longer steady.

6. Membrane fouling control in food industry for wastewater treatment

Fouling is the most important issue affecting the development of membrane filtration-as it worsens membrane performance and shortens membrane life (Boerlage et al., 2004). Membrane fouling by food wastewater filtration is attributed to depositation of species from effluents onto the membrane surface or within membrane porous, it causes a permeate flux decline with time because the filtration resistance is significant increased (Foley, 2006). Fouling studies on membranes are based in proteins depositation and their interaction in membranes surface. Polydispersity of naturally occurring macromolecules such as polysaccharides and humic substances, have also added a particular complexity on investigation to the fouling membrane mechanisms. Advances in understanding fouling of other species such as bacteria, yeast, emulsions, suspensions, salts and colloids from food wastewater have occurred in microfiltration and ultrafiltration literature (Chan et al., 2002; Foley et al., 2005; Hughes & Field, 2006; Cheng et al., 2008).

There are two form of membrane fouling: the fouling layer that is readily removable from the membrane, it is often classified as polarization phenomena or reversible fouling and is removed by physical procedures. Internal fouling caused by adsorption of dissolved matter into the membrane pores and pore blocking is considered irreversible, which can be removed by chemical cleaning and other methods (Hughes & Field, 2006).

Several aspects such as pretreatment of feed solution (example add flocculants before filtration), membrane surface modification, operating conditions and heavy cleaning procedures such as high temperature, while using caustic, chlorine, hydrogen peroxide, ozone, and strong inorganic acids are carried out on the membrane plant in operation to decrement fouling problem. Hydrodynamic methods used for performance enhancement of membrane filtration as back-pulsed (permeate flow reversal technique), creation of pulsed flow in membrane module, TMP pulsing, creation of oscillatory flow, generation of Dean vortices in membrane module, generation of Taylor vortices in membrane module and use of gas-sparging, have also been developed to reduce membranes fouling (Parck, 2002; Choi et al., 2005; Luo et al., 2010). Specifically, rapid accumulation of foulants, is usually referred to the critical flux (Chan et al., 2002). For single particles depositation, the critical flux occurs at a particular hydrodynamic condition (Espinasse et al., 2002). Critical flux condition can be determined by adsorption process, a slow increase in membrane resistance is always detected by the kinetics of this adsorption, particularly for proteins (Hughes & Field, 2006; Vyas et al., 2002; Ognier et al., 2002). For complex fluid systems, one common practice to experimentally determine the critical flux value is to incrementally increase the flux for a fixed duration. This leads to relatively stable TMP at low fluxes (indicating little fouling), and an ever increasing rate of TMP rises at fluxes beyond the critical flux values (Knutsen and Davis, 2006). In fluids with both macromolecules and particulates, membrane fouling takes place even at low flux rates, but changes dramatically when critical flux is reached. Although rigorous mathematical expressions to determinate membrane fouling, have been reported (Rögener et al., 2002b; Lefebvre et al.,2003), experimental critical flux determination remains an efficient approach to assess the fouling behavior of a given filtration system and to compare different operating conditions (Clech et al., 2006).

7. Optimization membrane process in food industry for wastewater treatment

In order to use membranes filtration as an efficient separation technique and economically interesting, the process optimization is essential. The purpose of the optimization process is the achievement of the highest possible flux production for a long period of time, with acceptable pollution levels.

A well chosen wastewater pretreatment and a proper selection of membrane in relation to the species properties from effluents can be used to assess and predict the optimal flux during filtration. However, the control of the feed pH, ionic strength and temperature is often necessary in order to maximize removal of food production residues.

Optimization methods and statistical designs are widely employed in various field of science from chemistry to engineering to enhance the membrane processes. Particularly, Response Surface Methodology (RSM) is a sequential form of experimentation used to help predict or optimize process. The variables are integrated in a mathematical-statistical model to express the possible simultaneous influence of membrane characteristics, fed composition and operating conditions on water flux performance. Several membrane processes and operating conditions have been reported in the treatment of food wastewater (Stoller and Chianese, 2006; Iaquinta et al., 2009; Escobar et al., 2011)

Table 5, summarize some results that describe the treatment wastewater optimization from production of these food. The permeate water fluxes are different in optimization process, due to membrane type used, membrane area and fed wastewater quality.

Reference	Food wastewater	Membrane process/membrane area (m^2)	Optimum conditions	Maximal permeate flux ($Lh^{-1}m^{-2}$)
Stoller and Chianese, (2006)	Olive oil	UF/32 NF/32	Oil concentration, feed flow velocity, temperature, critical flux, membrane type	415.8 222.0
Iaquinta et al., (2009)	Tomato puree	NF/2.51	Fed concentration, conductivity, temperature, , feed flow velocity, transmembrane pressure	8.21
Escobar et al. (2011)	Cereal	UF/0.56	Transmembrane pressure, membrane type, dye concentration (brilliant blue and tartrazine), flow velocity, filtration time	19.5

Table 5. Membrane conditions in treatment food wastewater optimization

8. Recovery of food industrial effluents by membrane process and water reuse

The drivers for implementation of water reuse practices in food industries is essential due to increasing demands on declining freshwater supplies, severe water shortages and dry periods, and the fact that water quality discharge regulations have become stricter. In addition, environmental and economical aspects are incentives to treat food wastewater with water reuse purpose (Casani et al., 2005).

Food industry looks at membrane processes for wastewater treatments to produce purified water for recycle or reuse due to their characteristics as techniques that can be implemented in any food plant and because they can be combined with other unit operations (hybrid processes (Sarkar et al, 2006). Table 6 summarizes some important results of recycling water and cleaning effluents by membrane technology.

Typical wastewaters in food industries come from different parts of the plant and they are submitted to a wide fluctuation in flow and composition depending on the type of food industry and size and even, on the moment in which the plant is working (different steps of "cleaning in place", heating, sterilization, etc.). They do not contain toxic compounds (except in wastewater from washing fruits and vegetables in which pesticides can be a water contaminant) but they are characterized by high values in biological oxygen demand (BOD) and chemical oxygen demand (COD) as well as total dissolved solids (TSS) in some cases. Those high contents come from organic (proteins, carbohydrates, fats) and inorganic (salts, additives, dyes) compounds.

Reference	Industry/wastewater source	Combined membrane treatments	Water recycling
Chmiel at al., (2003)	Dairy/Vapor condensates from concentration and drying steps	Cartridge filtration-NF-RO-UV-oxidation	Drinking
Mavrov et al., (1997), (2000); Chmiel et al., (2000); Čuda et al., (2006); Vourch et al., (2008)		Two NF steps	Water use in boilers
Koo et al., (2011)	Dairy/Flash coolers	Cartridge filtration-NF-UV	Boiler make up water
Rögener et al., (2002a), (2002b), (2002c); Tay & Jeyaseelan, (1995)	Milk/Bottles machines, chess processing	UF and RO	Unspecified
Scharnagl et al., (2000); Muro et al., (2010)		MF, UF, NF	
Mavrov and Bélières, (2000); Braeken et al., (2004); Simate et al., (2011); Cornelissen, (2002); Blöcher et al., (2002)	Beverage/bottle rinsing, brewing room , bright beer reservoir	MBR-NF, RO	Unspecified
Rajkumar et al., (2010); Muro et al., (2009)	Fruit and vegetable processing/rinsing beans, cereal processing	MF, UF, NF, RO	Rinsing beans
Iaquinta et al., (2006), (2009); Mänttäri & Nyström, (2000)	Tomato/cleaning, sorting and moving the processed	NF	Unspecified
Noronha et al. (2002); Blöcher et al., (2002)	Fruit juices/ bottle washing, fruit processing, juice production and cleaning of tanks, pipes	NF	Drinking
Turano et al. (2002) Mohammadi & Esmaeelifar, (2004); Galambos et al., (2004); Akdemir & Ozer, (2009); Mantzavinos & Kalogerakis, (2005); Rajkumar et al., (2010)	Vegetable oil/olive mill, washing,	MF, UF, NF, RO	Drinking
Fähnrich et al., (1998); Cui & Muralidhara, 2010; Cheryan, 1998; Afonso and Bórquez (2002a); Bohdziewicz et al. (2002), (2003), Bohdziewicz & Sroka, (2005a), (2005b), (2006), Kuca & Szaniawska, (2009); Walha et al., (2009); Dumay et al., (2008).	Meat and seafood/ slaughterhouse fish and crustaceans and tuna cooking	Two NF steps-UV SBR, MBR, UF and RO in different combinations	Drinking

Table 6. Promising applications of membranes in wastewater treatment of food industry

8.1 Recovery of cleaning-in-place solutions

Special attention can give at the recovery cleaning solutions from wastewater of food industries. A large amount of acids and alkalis in cleaning and sanitizing steps are used in dairy industry. The consumption of NaOH, HNO_3 and detergents/disinfectants in a dairy industry that processes 1.5 million liters of milk per day is around 3 500, 1 000 and 1 000 kg per day respectively (Fernández et al., 2010). More than 40% of the total pollution caused by a dairy industry comes from their cleaning in place units (Henk, 1993). Particularly, the cleaning in place (CIP) used in food industries consists in a number of steps that depends on the type of product, but the final waste streams collected from each of these stages are usually treated together and show COD values of 400-600 $mgO_2.L^{-1}$ (Daufin et al., 2001).
There are a number of works describing how to recover contaminated cleaning solutions by membranes (Choe et al., 2005; Fernández et al., 2010; Gésan-Guiziou et al., 2002, 2007; Merin et al., 2002; Räsänen et al., 2002). Dresch et al. (2001) pointed out the NF technology as a promising technique compared to decantation, centrifugation and microfiltration (0.1μm cut-off) for the regeneration of waste NaOH solutions from an industrial CIP system. However, Gésan-Guiziou et al. (2007) reported that MF could be more adequate operation based on that the surfactant contained in the spent detergent is only slightly rejected by the membrane and costs of MF operation are much lower (lower TMP) compared to UF and NF costs, in spite of that the COD permeate when using MF was much higher and its possible uses can be limited.
When using NaOH or HNO_3 solutions in alkaline and acid cleaning steps, their recovery in the permeate is not very difficult, because the rejection of these compounds on an ultrafiltration or even in a NF are very low, obtaining a permeate stream than can be reused in the CIP and being the rest of foulants retained by the membrane. However, when the cleaning agent is composed by other chemicals (antiscalants, anionic/cationic detergents, antifoaming compounds, surfactants, etc.) their recovery in the permeate stream is not so evident (Wendler et al., 2002). The use of MF, UF or NF techniques depends on if surfactants want to be recovered in the permeate o in the concentrate streams. If surfactants are below their critical micelle concentration (CMC) they will not be retained by any of these techniques, but if they are above CMC, MF and UF techniques retain these components and the permeate stream will lose its cleaning properties. Some works based on NF processes with the aim of surfactants recovery in the permeate stream have been published in the last years (Boussu et al., 2007, Forstmeier et al., 2002; Kaya et al., 2006, 2009). In those cases permeate flux and surfactant rejection are strongly dependent on the membrane material (membrane isoelectric point - IEP) and feed conditions (pH, concentration, etc.) due to that NF processes are not only governed by steric reasons and charge interaction between solutes and membrane surface plays an important role in transmission and membrane selectivity.
Diluted caustic and acidic washing solutions (showing COD between 8 000 and 10 000 $mgO_2.L^{-1}$) can be recovered by NF membranes with molecular weight cut off (MWCO) between 150 and 300 Da. Permeate flow rates are moderate (between 7 and 12 $Lh^{-1}m^{-2}$) at pressures around 0.9 MPa (Räsänen et al., 2002). NF shows robust performance for the recovery of caustic solutions when faced with large variations of solution composition, as it happens at industrial CIPs (Dresch et al., 1999; Gésan-Guiziou et al., 2002). In some published research, transmission of NaCl higher than 99% was measured when variable feed composition (COD between 100 and 11 000 $mgO_2.L^{-1}$) and suspended matter between 0.4 and 5.6 gL^{-1} was nanofiltered with ceramic membranes of 1 000 MWCO obtaining high permeate flow rates (40–110 $Lh^{-1}m^{-2}$) at 70°C and 0.4 MPa transmembrane pressure.

Regarding to the acidic detergents used in food industries CIPs, some results have been published (Novalic et al., 1998). Two HNO_3 spent solutions were investigated with NF. Higher COD cleaning solution of 18 500 $mgO_2.L^{-1}$ was obtained after a cleaning step without previous alkaline step. The other solutions was lower in COD (1 800 mg $O_2.L^{-1}$) and was obtained after a previous alkaline cleaning step. Two effluents were nanofiltered at 50 °C and 3.0 MPa and at maximum recovery rate of 75%.

In other studies, several salts ($Ca(NO_3)_2$ and ($Mg(NO_3)_2$)) were analyzed in the cleaning solution. However low COD solution essayed was nanofiltered at a rate of 40 $Lh^{-1}m^{-2}$ and final COD was low (450 $mgO_2.L^{-1}$). Kaya et al. (2009) used NF (1 000 Da cut-off) to treat a detergent composed by anionic and nonionic surfactants, dyes and salts from a dishwasher detergent. Maximum fluxes (around 120 $Lh^{-1}m^{-2}$, 25 °C, 1.2 MPa) were obtained at pH of 5, near to the membrane IEP. However, surfactants have hydrophobic interactions with anionic dyes (tartrazine) what explains higher rejection than expected (Kartal & Akbas, 2005; Zahrim et al., 2011). Authors found also strong influence of temperature and pH on the flux decay along the experiments. Initial higher fluxes at higher temperatures (40°C) rapidly decay due to pores blocking by surfactant monomers and rejections reduces with temperatures due to an increase in solutes diffusion or expansion of membrane structure a higher temperatures (organic membranes).

For other hand, large dairy companies (food companies in general) are changing the conventional cleaning agents for those novel single-phase detergents. These new formulations are expensive but CIP steps are shorter and only have one or two steps (cleaning and disinfection). Single-phase detergents are designed by detergent companies and formulations are not available but alkalis or acids, surfactants, complexant agents and de-foamers usually are included. Recovery of these detergents is not easy because all the components should be permeate through the membrane and to should separate from the rest of foulants, what might be retained. Some authors have been studied the recovery of these detergents by NF processes using a spent single-detergent from a milk company (Fernández et al., 2010). In spite of that NF membrane (200 Da cut-off) maintains constant permeate flux rate (around 45 $Lh^{-1}m^{-2}$) at 0.9 MPa, 70°C and 75% recovery rate after 1800 hours running, infrared studies demonstrated that some compounds present in the fresh single phase detergent are partially retained by the membrane.

8.2 Recovering of the other valuables constituents of wastewater of food industry

An overview of types and applications of membrane separation techniques to recover of proteins and functional compounds from wastewater cheese and fish processing are showed in this section.

Chollangi & Hossain (2007) evaluated the fractionation of dairy wastewater into lactose-enriched and protein-enriched streams using ultrafiltration membrane technique. Three membranes of MWCO of 3, 5 and 10 kDa of regenerated cellulose material were used to determine the efficiency of the process. The performance was determined under various processing conditions that include the operating temperature and TMP across the membrane and the concentration of lactose in the feed solution. It was found that the 3, 5 and 10 kDa membranes provided 70–80%, 90–95% and 100% recovery of lactose in permeate, respectively from made-up solution of pure lactose. The 10 kDa membrane results showed a 100% recovery of lactose from wastewater sample. Muro et al. (2010) worked with residual whey from a cheese industry, it was fractioned to recover proteins, lactose and

minerals by membranes process in filtration stages: UF and NF. The results of membrane process to treatment of whey depended on the operating conditions, but the temperature effect was greater in the ultrafiltration process. 80% of proteins from whey were recovered with the membrane of 15 kDa operating to 2.4 Lh^{-1} to 30°C and 1.5 bar. The NF process showed that the transmembrane pressure affect lactose rejection, obtaining itself 70% of yield with the membrane of 0.150 kDa, using a flow of Lh^{-1} to 25 °C and 1.8 bar.
Respect to wastewaters from fish processing, effluents contain a large amount of potentially valuable proteins. These proteins can be concentrated by means of ultrafiltration (UF) and recycled into the fish meal process, improving its quality and the economic benefits from the raw material, whereas the treated water can be discharged into the sea or reused in the plant. An extensive review of the application of pressure-driven membrane separation processes in the treatment of seafood processing effluents and recovery of proteins therein was presented by Alfonso & Bórquez, (2002b). Two effluents from a fish meal plant located in Talcahuano, Chile, were characterized. A mineral tubular membrane, Carbosep M2 (MWCO = 15 kDa) was used in the UF experiments. The operating conditions were optimized in total recirculation mode, and the subsequent concentration experiments were carried out at 4 bar pressure, 4 ms^{-1}, crossflow velocity, ambient temperature and natural pH. The results show that UF reduces the organic load from the fish meal wastewaters and allows the recovery of valuable raw materials comprising proteins. Dumay´s work focuses on the treatment of washing waters coming from surimi manufacturing using ultrafiltration technology at a laboratory scale. Four membrane materials (poly-ether sulfone, polyacrilonytrile, poly vinylidene fluoride and regenerated cellulose) and 5 MWCO (from 3 to 100 kDa) were studied at bench laboratory scale using the pilot Rayflow® 100, commercialised by Rhodia Orelis. The investigation deals with the ability for membranes to offer a high retention of biochemical compounds (proteins and lipids) (Dumay et al., 2008).

9. Conclusions

Wastewaters produced in the food industry depend upon the particular site activity. Animal processors and rendering plants will generate effluents with different characteristics to those from fruit/vegetable washers and edible oil refiners (suspended/colloidal and dissolved solids, organic pollution and oil and greases as well as microbial contamination).
MF and UF systems can reduce suspended solids and microorganisms, whilst UF/RO combinations can also remove dissolved solids and provide a supply of process water and simultaneously reducing waste streams. UF systems can get more than 90% reduction in BOD and less than 5 $mg.L^{-1}$ in residual solids and less than 50 $mg.L^{-1}$ in grease and oil. NF systems are being used in a number of applications thank to the quick development in new membrane materials. In case of RO process, BOD removal rate of 90-99% is possible providing a low cost controlled source of bacteria-free water.
The favourable characteristics (modular) of membrane technologies allow to use different techniques as it has be seen all along this chapter. These hybrid processes can include traditional techniques as centrifugation, cartridge filtration, disinfection and different membrane techniques building a "cascade design" very used in many of the applications reviewed. The risk of membrane damage due to the contact with particles, salt conglomerates, chemicals or others substances must be minimized to prevent short

membrane life. Operation parameters must be carefully selected to obtain good results, especially not to overpass maximum temperature and transmembrane pressures recommended by membrane manufacturers. From the point of view of each particular process, to work at permeate flow rates below critical flux will assure longer runs. Membrane operating optimization is another aspect of paramount importance.

It seems likely that the application of membrane systems in the food industry will continue growing rapidly. In particular, wastewater treatments will become more important in the next years because of the increasing cost of mains water and effluent sewer disposal. A membrane wastewater treatment system can be a major contribution to a food sector and its introduction may feature as part of the continuous improvement plans within an environmental management system.

10. References

Abbasi, M.; Sebzari, M.; Salahia, A.; Abbasi, S. & Mohammadi, T. (2011). Flux Decline and Membrane Fouling in Cross-Flow Microfiltration of Oil-In-Water Emulsions. *Desalination and Water Treatment*, Vol.28, No.1-3, (April 2011), Pp. 1–7, ISSN 00119164

Afonso, M. & Bórquez, R. (2002a). Review of the Treatment of Seafood Processing Wastewaters and Recovery of Proteins Therein by Membrane Separation Processes - Prospects of the Ultrafiltration of Wastewaters from the Fish Meal Industry. *Desalination*. Vol. 142, No. 1, Pp. 29-45, ISSN 0011-9164

Afonso, M. & Bórquez, R. (2002b). Nanofiltration of Wastewaters from the Fish Meal Industry. *Desalination*. Vol.151, No. 2, Pp. 131-138, ISSN 0011-9164

Akdemir, E.O. & Ozer, A. (2009). Investigation of Two Ultrafiltration Membranes for Treatment of Olive Oil Mill Wastewater. *Desalination*, Vol.249, No. 2, (December 2009), Pp. 660-666, ISSN 00119164

Al-Akoum, O., Mercier-Bonin M., Ding, L.; Fonade, C.; Aptel, P. & Jaffrin M. (2002). Comparison of Three Different Systems Used for Flux Enhancement: Application to Yeast Crossflow Filtration of Yeast Suspensions. *Desalination*, Vol.147, (September 2002), Pp. 31–36, ISSN 00119164

Avula, R.Y.; Nelson, H.M. & Singh, R.K. (2009). Recycling of Poultry Process Wastewater by Ultrafiltration. *Innovative Food Science and Emerging Technologies*, Vol.10, No.1, (January 2009), Pp. 1-8, ISSN 14668564

Blöcher, C.; Noronha, M.; Fünfrockcn, L.; Dorda, J.; Mavrov, V.; Janke, H.D. & Chmiel, H. (2002). Recycling of Spent Process Water in the Food Industry by an Integrated Process of Biological Treatment and Membrane Separation. Desalination, Vol.144, No.1-3, (September 2002), Pp. 143-150, ISSN 00119164

Boerlage, S. F. E.; Kennedy, M.; Tarawneh, Z.; Faber, R. D.; & Schippers, J. (2004). Development of the MFI-UF in Constant Flux Filtration. *Desalination*, Vol.161, No.2, (Februay 2004), Pp. 103-113, ISSN 00119164

Bohdziewicz, J. & Sroka, E. (2005a). Treatment of Wastewater from Meat Industry Applying Integrated Membrane Systems. *Process Biochemistry* Vol., 40, No. 3-4, (March 2005), Pp. 1339-1346, ISSN 00329592

Bohdziewicz, J. & Sroka, E. (2005b). Integrated System of Activated Sludge-Reverse Osmosis in the Treatment of the Wastewater from Meat Industry. *Process Biochemistry*, Vol. 40, No. 5, (April 2005), Pp. 1517-1523, ISSN 13595113

Bohdziewicz, J. & Sroka, E. (2006). Application of Hybrid Systems to the Treatment of Meat Industry Wastewater. *Desalination*, Vol., 198, No. 1-3, (October 2006), Pp. 33-40, ISSN 00119164

Bohdziewicz, J.; Sroka, E. & Korus, I. (2003). Application of Ultrafiltration and Reverse Osmosis to the Treatment of the Wastewater Produced by the Meat Industry. *Polish Journal of Environmental Studies*, Vol.12, No. 3, Pp. 269-274, ISSN 12301485

Bohdziewicz, J.; Sroka, E. & Lobos, E. (2002). Application of the System Which Combines Coagulation, Activated Sludge and Reverse Osmosis to the Treatment of the Wastewater Produced by the Meat Industry. *Desalination*, Vol. 144, No. 1-3, (September 2002), Pp. 393-398, ISSN 00119164

Boussu, K.; Kindts, C.; Vandecasteele, C. & Van Der Bruggen, B. (2007). Surfactant Fouling of Nanofiltration Membranes: Measures and Mechanism. *A European Journal of Chemical Phyics and Phyical Cheistry*, Vol.8, No.12, (August 2007), Pp. 1836-1845, ISSN 14394235

Braeken, L.; Van Der Bruggen, B. & Vandecasteele, C. (2004). Regeneration of Brewery Waste Water Using Nanofiltration. *Water Research*, Vol. 38, No. 13, (July 2004), Pp. 3075-3082, ISSN 00431354

Casani, S.; Rouhany, M. & Knøchel, S. (2005). A Discussion Paper On Challenges and Limitations to Water Reuse and Hygiene in the Food Industry. *Water Research*, Vol.39, No.6, (March 2005), Pp. 1134-1146, ISSN 00431354

Chan, R.; Chen, V. & Bucknall, M.P. (2002). Ultrafiltration of Protein Mixtures: Measurement of Apparent Critical Flux, Rejection Performance, and Identification of Protein Deposition, *Desalination*, Vol.146, No. 1-3, (September, 2002), Pp. 83–90, ISSN 00119164

Cheng J.; Li Y.; Chung T.; Chen S. & Krantz W. (2008). High-Performance Protein Separation by Ion Exchange Membrane Partitioned Free-Flow Isoelectric Focusing System. *Chemical Engineering Science*, Vol.63, No.8, (April 2008), Pp. 2241-2251, ISSN 00092509

Cheryan, M. (1998). Ultrafiltration and Microfiltration Handbook, Technomic, ISBN 1-56676-598-6, Lancaster, USA

Chmiel, H.; Kaschek, M.; Blöcher, C.; Noronha, M. & Mavrov V. (2003). Concepts for the Treatment of Spent Process Water in the Food and Beverage Industries. *Desalination*, Vol.152, No.1-3, (February 2003), Pp. 307-314, ISSN 00119164

Chmiel, H.; Mavrov, V. & Bélières, E. (2000). Reuse of Vapour Condensate from Milk Processing Using Nanofiltration. Fitration&Separation, Vol.37, No.3, (April 2000), Pp.24-27, ISSN 00151882

Cho, B.D. & Fane, A.G. (2000). Fouling Transient in Nominally Sub-Critical Flux Operation of A Membrane Bioreactor. *Journal of Membrane Science*, Vol.209, No.2, (November 2002), Pp. 391-403, 03767388

Cho, J.; Amy, G., & Pellegrino, J. (2000). Membrane Filtration of Natural Organic Matter: Factors and Mechanism Affecting Rejection and Flux Decline With Charged (UF) Membrane. *Journal of Membrane Science*, Vol.164, No.1-2, (January 2000), Pp. 89-110, ISSN 03767388

Choe, E.K.; Son, E.J.; Lee, B.S.; Jeong, S.H.; Shin, H.C. & Choi, J. S. (2005). NF Process for the Recovery of Caustic Soda and Concentration of Disodium Terephthalate from

Alkaline Wastewater from Polyester Fabrics. *Desalination*, Vol. 186, No. 1-3, (December 2005), Pp. 29-37, ISSN 00119164

Choi, K. H.; Zhang, D.D.; Dionysiou, D.; Oerther, D & Sorial G.A. (2005). Effect of Permeate Flux and Tangential Flow On Membrane Fouling for Wastewater Treatment, *Separation and Purification Technology*, Vol. 45, No.1, (September 2005), Pp. 68–78, ISSN 13835866

Chollangi, A. & Hossain, M. (2007). Separation of Proteins and Lactose from Dairy Wastewater. *Chemical Engineering and Processing*, Vol.46, No.5, (May 2007), Pp. 398-404, ISSN 02552701

Clech, P.; Jefferson, B.; Chang, I.S. & S.J. Judd. (2006). Critical flux Determination by the Flux-Step Method in A Submerged Membrane Bioreactor, *Journal membrane Science* 227 (2006) 17–53.ISSN:03767388

Cornelissen E.R.; Janse, W. & Koning, J. (2002). Wastewater treatment with the internal MEMBIOR. *Desalination*, Vol. 146, No. 1-3, (September 2002), Pp. 463-466, ISSN 00119164.

Čuda, P.; Pospíšil, P. & Tenglerová, J. (2006). Reverse Osmosis in Water Treatment of Boilers. *Desalination*, Vol.198, No.1-3, (October 2006), Pp. 41-46, ISSN 00119164.

Cui, Z.F. & Muralidhara, H.S. (Eds.). (2010). Membrane Technology. Elsevier, ISBN 978-1-85617-632-3, Langford Lane, UK

Daufin, G.; Escudier, J-P.; Carrère, H.;Bérot, S.; Fillaudeau, L. & Decloux, M. (2001). Recent and Emerging Applications of Membrane Processes in the Food and Dairy Industry. *Trans. Icheme.*, Vol. 79, No. 2, (June 2001), Pp. 89-102, ISSN 09603085

Dresch, M.; Daufin, G. & Chaufer, B. (1999). Membrane Processes for the Recovery of Dairy Cleaning in Place Solutions. *Lait*, Vol. 79, No. 2, (March 1999), Pp. 245-259, ISSN 00237302

Dresch, M.; Daufin, G. & Chaufer, B. (2001). Integrated Membrane Regeneration Process for Dairy Cleaning-In-Place. *Separation and Purification Technology*, Vol. 22-23, No. 1, (March 2001), Pp. 181-191, ISSN 13835866

Dumay, J.; Radier, S.; Barnathan, G.; Bergé, J. & Jaouen, P. (2008). Recovery of Valuable Soluble Compounds from Washing Waters Generated During Small Fatty Pelagic Surimi Processing by Membrane Processes. *Environmental Technology*, Vol. 29, No. 4, Pp. (451-461), ISSN 0959-3330, London, England

Escobar, J. (2010) Optimización Del Proceso de Ultrafiltración de Efluentes de Una Industria de Cereales. Thesis Doctor, Institute Technological of Toluca, México

Escobar, J.; Muro, C. & Estupiñan J. (2011). Optimización Del Proceso de Ultrafiltración de Efluentes de Una Industria de Cereales. *Afinidad: Revista de Química Teórica Y Aplicada*, Vol.68, No. 551, (January 2011), ISSN 0001-9704

Espinasse, B.; Bacchin, P. & Aimar, P. (2002). On an Experimental Method to Measure Critical Flux in Ultrafiltration. *Desalination*, No.1-3, (September 2002), Pp. 146 91–96, ISSN 00119164

Fähnrich, A.; Mavrov, V. & Chmiel, H. (1998). Membrane Processes for Water Reuse in the Food Industry. *Desalination*, Vol. 119, No. 1-3, (September 1998), Pp. 213-216, ISSN 00119164

Fernández, P.; Riera, F.A.; Álvarez, R. & Álvarez, S. (2010). Nanofiltration Regeneration of Contaminated Single-Phase Detergents Used in the Dairy Industry. *Journal of Food Engineering*, Vol. 97, No.3, (April 2010), Pp. 319-328, ISSN 02608774

Field, R.W.; Wu, D.; Howell, J.A. & Gupta, B.B. (1995). Critical flux concept for microfiltration fouling. *Journal Of Membrane Science*, Vol. 100, No. 3, (abril 1995), Pp. 259-272, ISSN 03767388.

Foley, G. (2006). A Review of Factors Affecting Filter Cake Properties in Dead End Microfiltration of Microbial Suspensions, *Journal of Membrane Science*, Vol. 274, No.1-2, (Abril 2006), Pp. 38–46, ISSN 03767388

Foley, G., Mccarthy, A.A. & Walsh, P.K. (2005). Evidence for Shape-Dependent Deposition in Crossflow Microfiltration of Microbial Cells, *Journal of Membrane Science*, Vol.250, No.1-2, (March 2005), Pp. 311–313, ISSN 03767388

Forstmeier, M.; Goers, B. & Wozuy, G. (2002). Product Recovery and Reduction of Wastewater Disposal Costs in A Detergent Production Facility. *Engineering in Life Sciences*, Vol.3, No.4, (April 2002), Pp. 181-185, ISSN 0009286X

Galambos, I.; Molina, J.; Járay, P.; Vatai, G. & Molnár, E. (2004). High Organic Content Industrial Wastewater Treatment by Membrane Filtration. *Desalination*, Vol.162, (March 2004), Pp.117–120, ISSN 00119164

Gésan-Guiziou, G.; Alvarez, N.; Jacob, D. & Daufin, G. (2007). Cleaning in Place Coupled With Membrane Regeneration for Re-Using Caustic Soda Solutions. *Separation and Purification Technology*, Vol.54, No.3, (May 2007), Pp. 329-339, ISSN 13835866

Gésan-Guiziou, G.; Boyaval, E. & Daufin, G. (2002). Nanofiltration for the Recovery of Caustic Cleaning-In-Place Solutions: Robustness Towards Large Variations of Composition. *Journal of Dairy Research*, Vol. 69, No. 4, (November 2002), Pp. 633-643, ISSN 00220299

Guizard, C. & Amblard, P. (2009). Chapter 6. Current Status and Prospects for Ceramic Membranes Applications in *Handbook of Membrane Separation: Chemical, Pharmaceutical, Food and Biotechnological Applications*. Ed. Taylor Francis Group. Pabby, K.; Rizvi, S. & Sastre A., Pp. 139-179, Boca Ratón, USA.

Guizard, C.; Ayral, A. & Julbe A. (2002). Potentiality of Organic Solvents Filtration With Ceramic Membranes. A Comparison With Polymer Membranes. *Desalination*, Vol.147, (Septembre 2002) Pp. 275–280, ISSN 00119164

Hafez, A.; Khedr, M. & Gadallah H. (2007). Wastewater Treatment and Water Reuse of Food Processing Industries. Part II: Techno-Economic Study of A Membrane Separation Technique. *Desalination*, Vol.214, No.1-3, (August 2007), Pp. 261-272, ISSN 00119164

Henk, M.A. (1993). Recycling of Caustic Cleaning Solutions Using Cross-Flow Filtration in the Dairy Industry. Ph.D. Thesis, University of Zurich, Switzerland

Hughes, D. & Field, R.W. (2006). Crossflow Filtration of Washed and Unwashed Yeast Suspensions At Constant Shear Under Nominally Sub-Critical Conditions. *Journal of Membrane Science*, Vol.280, No.1-2, (September 2006), Pp. 89–98, ISSN 03767388

Hwang, K.J.; Chou, F.Y. & Tung, K.L. (2006). Effects of Operating Conditions On Performance of Cross-Flow Microfiltration of Fine Particle Protein Binary Suspension, *Journal Membrane Science*, 274, 183, 2006.ISSN:03767388

Iaquinta, M.; Stoller, M. & Merli, C. (2006). Development of Synthetic Wastewater from the Tomato Industry for Membrane Processing Purposes. *Desalination*, Vol. 200, No. 1-3, Pp. (314-320), ISSN 0011-9164, Amsterdam, Netherlands

Iaquinta, M.; Stoller, M. & Merli, C. (2009). Optimization of A Nanofiltration Membrane Process for Tomato Industry Wastewater Effluent Treatment. *Desalination*, Vol.245, No.1-3, (September 2009), Pp. 314-320, ISSN 00119164

Jaffrin, M.; Ding, L.; Akoum, O. & Brou A. (2004). A Hydrodynamic Comparison Between Rotating Disk and Vibratory Dynamic Filtration Systems. *Journal of Membrane Science*, Vol.24 No.1-2, (October 2004), Pp. 155–167, ISSN 03767388

Kaeselev, B., Kingshott, P. & Jonsson, G. (2002). Influence of the Surface Structure On the Filtration Performance of UV-Modified PES Membranes. *Desalination*, Vol.146, No.1-2, (September 2002), Pp. 265-271, ISSN 00119164

Kartal, C. & Akbas, H. (2005). Study On the Interaction of Anionic Dye-Noninic Surfactants in A Mixture of Anionic and Non-Ionic Surfactants by Adsorption Spectroscopy. *Dyes Pigments*, Vol. 65, No. 3, (June 2005), Pp. 191-195, ISSN 01437208

Kaya, Y., Aydiner, C.; Barlas, H. & Keskinler, B. (2006). Nanofiltration of Single and Mixture Solutions Containing Anionics and Noionic Surfactants Below Their Critical Micelle Concentration (Cmcs). *Journal of Membrane Science*, Vol. 282, No.1-2, (October 2006), Pp. 401-412, ISSN 03767388

Kaya, Y.; Barlas, H. & Arayici, S. (2009). Nanofiltration of Clearing in Place (CIP) Wastewater in A Detergent Plant: Effects of Ph, Temperature and Transmembrane Pressure On Flux Behaviour. *Separation and Purification Technology*, Vol. 45, No.2, (February 2009), Pp.117-129, ISSN 13835866

Khayet, M.; Matsuura, T. & Mengual, J. (2005). Porous Hydrophobic/Hydrophilic Composite Membranes: Estimation of the Hydrophobic-Layer Thickness. *Journal of Membrane Science*, Vol.266, No.1-2, (Decembre 2005), Pp. 68-78, ISSN 03767388

Kim, Y.; Ahn, W.; Kim, J. & Kim, Y. (2005). in Situ Fabrication of Self-Transformable and Hydrophilic Poly(Ethylene Glycol) Derivative-Modified Polysulfone Membranes. *Biomaterials*, Vol.26, No.16, (Juny 2005), Pp. 2867-2875, ISSN 01429612

Knutsen, J.S. & Davis R.H. (2006). Deposition of Foulant Particles During Tangential Flow Filtration, *Journal of Membrane Science*, Vol.271, No.1-2, (March 2006), Pp. 101–113, ISSN 03767388

Koo, C.G.; Mohammad, A., W. & Suja, F. (2011). Recycling of Oleochemical Wastewater for Boiler Feed Water Using Reverse Osmosis Membranes. A Case Study. *Desalination*, Vol. 271, No. 1-3, (April 2011), Pp. 178-186, ISSN 00119164

Kuca, M. & Szaniawska, D. (2009). Application of Microfiltration and Ceramic Membranes for Treatment of Salted Aqueous Effluents from Fish Processing. *Desalination*, Vol.241, No.1-3, (May 2009), Pp. 227-235, ISSN 00119164

Lefebvre, O. & Moletta, R., (2006). Treatment of Organic Pollution in Industrial Saline Wastewater: A Literature Review. *Water Research*, Vol.40, No.20, (December 2006), Pp. 3671-3682, ISSN 00431354

Lefebvre, X.; Palmeri, J.; Sandeaux, J; Sandeaux, R.; David, P.; Maleyre, B.; Guizard, C.; Amblard, P.; Diaz, J. & Lamaze B.(2003). Nanofiltration Modelling: A Comparative Study of the Salt Filtration Performance of A Charged Ceramic Membrane and an Organic Nanofilter Using the Computer Simulation Program NANOFLUX. *Separation and Purification Technology*, Vol.32, No.1–3, Pp. 117–126, ISSN 13835866

Luo, J; Ding, L.; Wan, Y.; Paullier, P. & Jaffrin M. (2010). Application of NF-RDM (Nanofiltration Rotating Disk Membrane) Module Under Extreme Hydraulic Conditions for the Treatment of Dairy Wastewater. *Chemical Engineering Journal*, Vol.163, No.3, (October 210), Pp. 307-316, ISSN 00219592

Madaeni, S.S. & Mansourpanah, Y. (2006). Screening Membranes for COD Removal from Diluted Wastewater. *Desalination*, Vol.197, No.1-3, (October 2006), Pp. 23-32, ISSN 00119164

Mänttäri, M. & Nyström, M. (2000). Critical Flux in NF of High Molar Mass Polysaccharides and Effluents from the Paper Industry. *Journal of Membrane Science*, Vol. 170, No. 2, (May 2000), Pp. 257-273, ISSN 03767388

Mantzavinos, D. & Kalogerakis, N. (2005). Treatment of Olive Mill Effluents: Part I. Organic Matter Degradation by Chemical and Biological Process-An Overview. *Environmental International*, Vol.31, No.2, (February 2005), Pp. 289-295, ISSN 01604120

Martín, A.; Martínez, F.; Malfeito, J.; Palacio, L.; Prádanos, P.; & Hernández, A. (2003). Zeta Potential of Membranes As A Function of Ph: Optimization of Isoelectric Point Evaluation. *Journal of Membrane Science*, Vol.213, No.1-2, (March 2003), Pp. 225-230, ISSN 03767388

Matthiasson, E. (1983). the Role of Macromolecular Adsorption in Fouling of Ultrafiltration Membranes. *Journal of Membrane Science*, Vol.16, Pp. 23-26, ISSN 03767388

Mavrov, V.; Fähnrich, A. & Chmiel, H. (1997). Treatment of Low-Contaminated Waste Water from the Food Industry to Produce Water If Drinking Quality for Reuse. *Desalination*, Vol.113, No.2-3, (November 1997), Pp. 197-203, ISSN 00119164

Mavrov,V. & Béliéres, E. (2000). Reduction of Water Consumption and Wastewater Quantities in the Food Industry by Water Recycling Using Membrane Processes. *Desalination*, Vol.131, No.1-3, (December 2000), Pp. 75-86, ISSN 00119164

McCutcheon, J. & Elimelech, M. (2006). Influence of Concentrative and Dilutive Internal Concentration Polarization On Flux Behavior in Forward Osmosis. *Journal of Membrane Science*,Vol.284, (November 2006), Pp. 237–247, ISSN 03767388

McKeown, N.B.; Budd, P.M.; Msayib, K J.; Ghanem, B.S.; Kingston, H.J.; Tattershall, C.E.; Makhseed, S., Reynolds, K.J. & Fritsch, D. (2005). Polymers of Intrinsic Microporosity (Pims): Bridging the Void Between Microporous and Polymeric Materials. *Chemistry - A European Journal*, Vol.11, No.9, (April 2005), Pp. 2610–2620, ISSN 00092665

Merin, U.; Gésan-Guiziou, G.; Boyaval, E. & Daufin, G. (2002). Cleaning-In-Place in the Dairy Industry: Criteria for Reuse of Caustic (Naoh) Solutions. *Lait*, Vol. 82, No. 3, Pp. 357-366, ISSN 00237302

Mi, B,& Elimelech M. (2008). Chemical and Physical Aspects of Organic Fouling of Forward Osmosis Membranes. *Journal of Membrane Science* , Vol.320, No.1-2, (July 2008), Pp. 292–302, ISSN 0367388

Miyoshi, H. (1998). Diffusion Coefficients of Ions Through Ion Exchange Membrane in Donnan Dialysis Using Ions of Different Valence. *Journal of Membrane Science*, Vol.141, No.1, Pp. 101–110, (April 1998), ISSN 03767388

Mohammadi, T. & Esmaeelifar, A. 166 (2004). Wastewater Treatment Using Ultrafiltration At A Vegetable Oil Factory, *Desalination*, Vol.166, (August 2004) Pp.329–337, ISSN 00119164

Muro, C.; Díaz, C.; García, B.; Zavala, R.; Ortega, R.; Álvarez, R. & Riera, R. (2010). Recuperación de Los Componentes Del Lactosuero Residual de Una Industria Elaboradora de Queso Utilizando Membranas. *Afinidad: Revista de Química Teórica Y Aplicada*, Vol.67, No.547, (Mayo 2010), Pp. 212-220, ISSN 0001-9704

Muro, C.; Escobar, J.; Zavala, R.; Esparza, M.; Castellanos, J.; Gómez R. & García M. (2009). Evaluación Del Proceso de Microfiltración En Un Efluente Residual de Una Industria Alimentaria Para Su Reuso. *Revista Internacional de Contaminación Ambiental*, Vol.25, No.4, (November 2009), Pp. 229-238, ISSN 01884999

Noronha, M.; Britz, T.; Mavrov., V.; Janke, H.D. & Chmiel, H. (2002). Treatment of Spent Process Water from A Juice Company for Purpose of Reuse: Hybrid Process Concept and On-Site Test Operation of A Pilot Plant. *Desalination*, Vol., 143, No.2, (May 2002), Pp. 183-196, ISSN 00119164

Novalic, S.; Dabrowski, A. & Kulbe, D. (1998). Nanofiltration of Caustic and Acidic Cleaning Solutions With High COD. Part 2. Recycling of HNO_3. *Journal of Food Engineering*, Vol.38, No.2, Pp. 133-140, ISSN 02608774

Ognier, S., Wisniewski, C. & Grasmick, A. (2002). Influence of Macromolecule Adsorption During Filtration of A Membrane Bioreactor Mixed Liquor Suspension, *Journal of Membrane Science*, Vol.209, No.1, (November 2002), Pp. 27–37, ISSN 03767388

Okokchina. (2010). Hollow Fiber Module. 05.07.2011, Available from Http://Www.Okokchina.Com Pall.Com/Pdf/Mtcpaper.Pdf

Oktay, S.; Iskender, G.; Babuna, F.; Kutluay, G. & Orhon D. (2007). Improving the Wastewater Management for A Beverage Industry With In-Plant Control. *Desalination*, Vol.211, No. 1-3, (June 2007), Pp. 138-143 ISSN 00119164

Park Y.G. (2002). Effect of Ozonation for Reducing Membrane-Fouling in the UF Membrane. *Desalination*, Vol.147 No. 1-3, (September 2002), Pp. 43–48, ISSN 00119164

Peng, P., Fane, A. & Li, X. (2005). *Desalination* by Membrane Distillation Adopting A Hydrophilic Membrane. *Desalination*, Vol.173, No.1, (March 2005), Pp. 45-54, ISSN 00119164

Rajkumar, K.; Muthukumar, M. & Sivakumar R. (2010). Novel Approach for the Treatment and Recycle of Wastewater from Soya Edible Oil Refinery Industry—An Economic Perspective. *Resources, Conservation and Recycling*, Vol.54, No.10, (August 2010), Pp. 752-758, ISSN 09213449

Räsänen, E.; Nyström, M.; Sahlstein, J. & Tossavainen, O. (2002). Purification and Regeneration of Diluted Caustic and Acidic Washing Solutions by Membrane Filtration. *Desalination*, Vol. 149, No. 1-3, (September 2002), Pp. 185-190, ISSN 00119164

Rautenbach, R. & Albrecht R. (1989). Membrane Processes, John Wiley & Sons, Inc. Chichester ISSN 19816286

Rögener, F.; Mavrov, V. & Chmiel, H. (2002b). Modeling of Impurity Concentrations in the Cleaning Zones of Bottle Washing Machines in the Treatment of Alkaline Solutions and Rinsing Water. *Chemical Engineering and Technology*, Vol. 25, No. 1, (January 2002), Pp. 41-47, ISSN 09307516

Rögener, F.; Mavrov, V. & Chmiel, H. (2002c). Treatment of Rinsing Water from A Bottle Washing Machine by Membrane Filtration With the Goal of Re-Use. *Chemie-Ingenieur-Technik*, Vol. 74, No. 4, (April 2002), Pp. 517-524+373, ISSN 0009286X

Rögener,F.; Willems, M.; Mavrov, V. & Chmiel, H. (2002a). the Influence of Cleaning Additives On Rejection and Permeability in Nanofiltration and Ultrafiltration of Bottle Washing Solutions. *Separation and Purification Technology*, Vol. 28, No. 3, (September 2002), Pp. 207-217, ISSN 13835866

Sarkar, B.; Chakrabarti, P.P.; Vijaykumar, A. & Kale, V. (2006). Wastewater Treatment in Dairy Industries-Possibility of Reuse. *Desalination*, Vol.195, No.3-4, (March 2009), Pp. 141-152, ISSN 00119164

Schaep, J. & Vandecasteele, C. (2001). Evaluating the Charge of Nanofiltration Membranes. *Journal of Membrane Science*, Vol.188, No.1, (Junmy 2001), Pp. 129-136, ISSN 03767388

Scharnagl, N.; Bunse, U. & Peinemann, K.-V. (2000). Recycling of Washing Waters from Bottle Cleaning Machines Using Membranes. *Desalination*, Vol.131, No.1-3, (December 2000), Pp. 55-63, ISSN 00119164

Simate, G.S.; Cluett, J.; Iyuke, S.E., Musapatika, E.T.; Ndlovu, S.; Walubita,L.F. & Alvarez, A.E. (2011). the Treatment of Brewery Wastewater for Reuse: State of the Art. *Desalination*, Vol.273, No.2-3, (June 2011), Pp. 235-247, ISSN 00119164

Stoller, M. & Chianese, A. (2006). Optimization of membrane batch processes by means of the critical flux theory. *Desalination*, Vol. 191, No. 1-3, (May 2006), Pp. 62-70, ISSN 00119164.

Tay, J.-H. & Jeyaseelan, S. (1995). Membrane Filtration for Reuse of Wastewater from Beverage Industry. Resources, *Conservation and Recycling*, Vol., 15, No 1, Pp. 33-40, ISSN 09213449

Turano, E.; Curcio, S.; de Paola, M.G.; Calabro, V. & Dorio, G. (2002). an Integrated Centrifugation-Ultrafiltration System in the Treatment of Olive Mill Wastewater. *Journal of Membrane Science*, Vol.209, No.2, (November 2002), Pp. 519-531, ISSN 03767388

Vourch, M.; Balannec, B.; Chaufer, B. & Dorange, G. (2008). Treatment of Dairy Industry Wastewater by Reverse Osmosis for Water Reuse. *Desalination*, Vol.219, No.1-3, (January 2008), Pp. 190-202, ISSN 00119164

Vrijenhoek, E., Hong, S., & Elimelech, M. (2001). Influence of Membrane Surface Properties On Initial Rate of Colloidal Fouling of Reverse Osmosis and Nanofiltration Membranes. *Journal of Membrane Science*, Vol.188, No. 1, (June M2001), Pp. 115-128, ISSN 03767388

Vyas, H.K.; Bennett, R.J. & Marshall, A.D. (2002). Performance of Crossflow Microfiltration During Constant Transmembrane Pressure and Constant Flux Operations, *International Dairy Journal*, Vol.12, No.5, Pp. 473–479, ISSN 0250-5118

Walha, K.; Ben Amar, R.; Bourseau, P. & Jaouen, P. (2009). Nanofiltration of Concentrated and Salted Tuna Cooking Juices. *Process Safety and Environmental Protection*, Vol.87, No.5, (September 2009), Pp. 331-335, ISSN 09575820

Wendler, B.; Goers, B. & Wozny, G. (2002). Regeneration of Process Water Containing Surfactants by Nanofiltration. Investigation and Modelling of Mass Transport. *Water Science and Technology*, Vol. 46, No. 1-3, Pp. 287-292, ISSN 02731223

Wu, T.; Mohammada, A.; Jahim, J. & Anuar, N. (2007). Palm Oil Mill Effluent (POME) Treatment and Bioresources Recovery Using Ultrafiltration Membrane: Effect of Pressure On Membrane Fouling. *Biochemical Engineering Journal*, Vol.35, No.3, (August 2007), Pp. 309–317, ISSN 1369703X

Yu, H.; Ziegler, C.; Oszcipok, M.; Zobel, M. & Hebling, C. (2006). Hydrophilicity and Hydrophobicity Study of Catalyst Layers in Proton Exchange Membrane Fuel Cells. *Electrochimica Acta*, Vol.51, No.7, (January 2006), Pp.1199-1207, ISSN 02510790

Zahrim, A.Y.; Tizaoui,C. & Hilal, N. (2011). Coagulation With Polymers for Nanofiltration Pre-Treatment of Highly Concentrated Dyes: A Review. *Desalination*, Vol. 244, No. 1-3, (January 2011), Pp. 1-16, ISSN 00119164

Zhao, C., Zhou, X., & Yue, Y. (2000). Determination of Pore Size Distribution On the Surface of Hollow-Fiber Filtration Membranes: A Review of Methods. *Desalination*, No. 129, Vol.2, (July 2000) Pp. 107-123, ISSN 00119164

Zhao, Y., Xing, W., Xu, N., & Wong, F. (2005). Effects of Inorganic Electrolytes On Zeta Potentials of Ceramic Microfiltration Membranes. *Separation and Purification Technology*, Vol.42, No.2, (March 2005), Pp. 117-121, ISSN 13835866

16

Advanced Oxidation Processes in Food Industry Wastewater Treatment – A Review

Anne Heponiemi[1] and Ulla Lassi[1,2]
[1]*Department of Chemistry, University of Oulu, Oulu*
[2]*Kokkola University Consortium Chydenius, Kokkola*
Finland

1. Introduction

The food industry uses large amounts of water for many different purposes including cooling and cleaning, as a raw material, as sanitary water for food processing, for transportation, cooking and dissolving, as auxiliary water etc. In principle, the water used in the food industry may be used as process and cooling water or boiler feed water (EC, 2006). In 2008, for example, the total industrial water consumption in Finland was 7600 million m^3 of which 34.5 million m^3 was used by the food processing industry (Finnish Food and Drink Industries` Federation, 2010).

As a consequence of diverse consumption, the amount and composition of food industry wastewaters varies considerably. Characteristics of the effluent consist of large amounts of suspended solids, nitrogen in several chemical forms, fats and oils, phosphorus, chlorides and organic matter (Finnish Food and Drink Industries` Federation, 2005). Generally, the BOD (biochemical oxygen demand) and COD (chemical oxygen demand) of food industry wastewater is 10 or even 100 times higher than those of domestic wastewater (EC, 2006). Unpleasant odours are also a typical problem in food industry wastewaters. These odours are usually the result of gases (hydrogen sulphide, indole) produced by the anaerobic decomposition of organic matter (Metcalf & Eddy, 2003).

2. Characteristics of food industry wastewaters

Considering the legislation of wastewater purification, total suspended solids (TSS), organic matter, total nitrogen (N_{tot}) and phosphorus (P_{tot}) must be removed from both municipal and industrial wastewaters before being released into the watercourse (Metcalf & Eddy, 2003).

In 2008 in Finland, the amount of food processing industry wastewater was 4.1 million m^3 and the total amount of industrial wastewaters was 1130 million m^3 (Table 1) (Finnish Environment Institute, 2009).

In 2008, the amount of total suspended solids was 83 tonnes, P_{tot} 4 tonnes, N_{tot} 87 tonnes whilst the proportion of food industry wastewater of the total industry wastewaters was around 0.4 %. However, the volume of organic matter in food industry wastewater was notable: 0.024 kg m^{-3} when for example, compared to BOD_7 in the pulp and paper industry (0.015 kg m^{-3}) (Finnish Environment Institute, 2009).

Industry	1000 m^3 Wastewater	TSS	P_{tot}	t/a N_{tot}	BOD_7	COD_{Cr}
Pulp and paper industry	677058	14787	161	2347	10148	160813
Metal industry	208613	493	2	224	4	70
Chemical industry	115192	1193	14	435	217	2174
Mining and quarry industry	20449	431	1	69	7	686
Food industry	4067	83	4	87	99	290
Others	104981	628	9	42	272	1819
Total	**1130360**	**17615**	**191**	**3204**	**10747**	**165852**

Table 1. Industrial discharge into the watercourse in 2008 in Finland (Finnish Environment Institute, 2009).

2.1 Total suspended solids

The most important physical factor of wastewater is its total solid content which is comprised of floating, settleable and colloidal matter, and matter in a solution. In the characterisation of various solids in wastewater, samples can be classified in ten different fractions. The most important fraction of these is total suspended solids (TSS) which is one of the two universally used effluent standards (along with BOD) to follow the performance of wastewater treatment plants (Metcalf & Eddy, 2003). The TSS of wastewater is determined according to APHA standard 2540 D. "Total Suspended Solids Dried at 103-105 °C" (APHA, 1998).

The solid matter of food industry wastewater can also vary considerably. For example in the slaughterhouse and meat industry wastewaters, the solid matter is composed of hairs, feathers, bowels and piece of tissues (Hiisvirta, 1976) compared to potato and vegetable industry wastewaters whose suspended solids are soil, peels and other vegetable parts (Lehto et al., 2007). Typically, food industry wastewaters contain lots of floating suspended solids which have to be removed since releasing it directly to the watercourse increases sediment (Metcalf & Eddy, 2003).

2.2 Organic matter

Organic compounds consist mainly of carbon, hydrogen and oxygen. The organic matter in wastewaters is typically a mixture of proteins and carbohydrates as well as oils and fats. In the slaughterhouse and meat industry wastewaters, for example, the organic content is mainly composed of grease which can be solid, suspended or emulsified (Hiisvirta, 1976). The low solubility of fats and oils reduces the rate of their biological decomposition and in the wastewater treatment plant, fats can block up the treatment devices of wastewaters. If grease is not removed before the discharge of treated wastewater, it can interfere with the biological life in surface waters. Wastewater also contains urea and small quantities of very large number of simple and extremely complex synthetic organic molecules (Metcalf & Eddy, 2003).

There are a number of different analyses to determine the organic content of wastewater. The analyses can be divided into those used to measure aggregate organic matter content of wastewater and those analyses used to quantify individual organic compounds (APHA, 1998). Typically, only aggregate organic matter content is measured. Those are: biochemical oxygen demand (BOD), chemical oxygen demand (COD) and total organic carbon (TOC) (Metcalf & Eddy, 2003).

2.2.1 Biochemical oxygen demand

The most widely used parameter for the determination of organic matter in wastewater is BOD. In this method, the biodegradable organic matter of wastewater is measured. The biochemical oxygen demand is the amount of oxygen which organic matter (solid or dissolved) in the water is consumed when biodegradation occurs in biological oxygen containing states (SFS-EN 1899-1). There are numerous standards for the determination of BOD such as SFS-EN 1899-1 (SFS-EN 1899-1), APHA standard 5210 B (APHA, 1998) and OECD 301 F-guide (OECD, 1992). For wastewater samples, the standard measuring time is five days (BOD_5) at 20 °C, but other lengths of time and temperatures can also be used. In Finland, for example, the typical measuring time is seven days (BOD_7) (Karttunen, 2003).

Although there is a high content of organic matter in food industry wastewaters, the organic compounds, such as fats and proteins, are usually easily biodegradable. Furthermore, the large amount of micro-organisms, for example in the slaughterhouse and meat processing industry wastewaters, facilitates the decomposition of organic compounds (Hiisvirta, 1976). However, there are also some exceptions such as food industry wastewaters containing salts, disinfectants and cleaning agents (Finnish Food and Drink Industries` Federation, 2005).

2.2.2 Chemical oxygen demand

Chemical oxygen demand describes the number of chemically oxidising organic compounds of wastewater (SFS 3020). The COD value can be determined according to APHA standard 5220 "Chemical oxygen demand (COD)". For wastewaters, the oxidising agent is dichromate in an acid solution (APHA, 1998). The ratio of the BOD and COD can provide more information about the wastewater sample. Usually, for industrial wastewaters, COD is higher than BOD because many organic substances which are difficult to oxidise biologically can be oxidised chemically. If the COD value is much bigger than the BOD value, the organic compounds in wastewater are slowly biodegradable (Hiisvirta, 1976). In food industry wastewaters, the COD and BOD values are often closely matched to each other due to the easily biodegradable organic compounds of the effluent (Finnish Food and Drink Industries` Federation, 2005).

2.2.3 Total organic carbon

Total organic carbon (TOC) describes the amount of organic compounds in wastewater and is used as a more convenient and directs expression of the total organic content than either BOD or COD. TOC analysis provides different information to BOD or COD because the unit of the TOC value is [mg C L^{-1}] while measuring BOD and COD uses the unit of mg O_2 L^{-1}. TOC is also independent of the oxidation state of the organic matter and does not measure other organically bound elements, such as nitrogen and hydrogen, and inorganic compounds that can contribute to the oxygen demand measured by BOD and COD (APHA, 1998). The analysing methods for TOC utilises heat and oxygen, UV radiation, chemical oxidants, or a combination of these to decompose the organic compounds of the sample to carbon dioxide which is measured with an infrared analyser, conductivity or by some other method (Metcalf & Eddy, 2003). The inorganic carbon content of the wastewater sample can be many times greater than the TOC fraction. Therefore, the inorganic carbon (CO_2, carbonates) must first be eliminated by acidifying samples to pH 2 or less to convert inorganic carbon species to CO_2. Alternatively, the inorganic carbon interference may be compensated for by separately measuring total carbon (TC) and inorganic carbon (IC). The TOC can be calculated from the difference between TC and IC. There are different methods available for the determination of TOC such as APHA standard 5310 (APHA, 1998) and SFS-EN 1484 (SFS-EN 1484, 1997). Nowadays, the TOC analysis is more favourable since its measuring time is quite short (5 to 30 minutes) compared to BOD determination which takes several days before the results are known (Metcalf & Eddy, 2003). For food industry wastewaters, TOC measurement provides practical information about the water sample because the organic matter content is usually quite high.

2.3 Nitrogen

Nitrogen is an important nutrient for microbes and other biological organisms. The chemistry of nitrogen is complex, because of the existence of several oxidation states in the element. The most common and important forms of nitrogen in wastewater are ammonia (NH_3), ammonium (NH_4^+), nitrogen gas (N_2), nitrite ion (NO_2^-) and nitrate ion (NO_3^-). Overall, total nitrogen N_{tot} in wastewater is composed of organic nitrogen, ammonia, nitrite and nitrate. Organic nitrogen is determined using the Kjeldahl method (APHA, 1998) where the aqueous sample is first boiled to remove any ammonia, and then wet combusted. During wet combustion, the organic nitrogen is converted to ammonium. In aqueous solution, ammonia nitrogen exists as either ammonia gas or ammonium ion, depending on the pH of the solution according to the equilibrium reaction:

$$NH_3 + H_2O \leftrightarrow NH_4^+ + OH^- \tag{1}$$

At pH 7, over 98% of the ammonia nitrogen is ammonium ion and when the pH is increased, the equilibrium is displaced to the left. Ammonia is determined by raising the pH, distilling off the ammonia with the steam produced during sample boiling, and condensing the steam which absorbs the gaseous ammonia. The measurement can be made colourimetrically, titrimetrically or with specific ion electrodes (APHA, 1998).

Nitrite nitrogen is determined colourimetrically (APHA, 1998). It is relatively unstable and is easily oxidised to the nitrate. The amount of nitrite in wastewaters is seldom above 1 mg L^{-1}. Although present in low concentration, it is important to determine the amount of nitrite because of its extreme toxicity to most fish and other aquatic species (Metcalf & Eddy, 2003).

Nitrate nitrogen, which can also be determined colourimetrically (APHA, 1998), is the most oxidised form of nitrogen found in wastewaters. The typical range (as nitrogen) detected in wastewaters is from 15 to 20 mg L^{-1} (Metcalf & Eddy, 2003).

Ammonium nitrogen, also found in wastewater, can oxidise microbiologically in to a nitrate form (nitrification) and consumes vital oxygen in water systems. The nitrification consumes a relative high amount of oxygen: 1 g ammonium nitrogen needs 4.3 g oxygen for the oxidation process, a reason why ammonium nitrogen has to be converted to nitrate and for the removal of total nitrogen from wastewaters before discharging in to the watercourse (Metcalf & Eddy, 2003).

In food industry wastewaters, the amount of nitrogen is typically bigger than the amount of phosphorus whilst the total nitrogen content can be even ten folds compared with municipal wastewater (Finnish Food and Drink Industries` Federation, 2005). In slaughterhouse and meat industry wastewaters, for example the decomposition of proteins raises the amount of nitrogen in the effluent (Hiisvirta, 1976).

2.4 Phosphorus

Phosphorus is also an essential nutrient to the growth of biological organisms. Due to noxious algal blooms occurring in surface waters, domestic and industrial waste discharges may contain 1 - 2 mg L^{-1} of phosphorus (P) (Metcalf & Eddy, 2003).

The usual forms of phosphorus found in aqueous solutions include orthophosphate (e.g. PO_4^{3-}, HPO_3^{2-}, $H_2PO_4^-$, and H_3PO_4), condensed phosphates (pyro-, meta-, and other polyphosphates) and organic phosphate. Phosphorus analyses include the conversion of a phosphorus form to dissolved orthophosphate and the colourimetric determination of this dissolved orthophosphate. Orthophosphate can be determined directly by adding an ammonium molybdate (forming a coloured complex), while condensed and organic phosphates must first be converted to orthophosphates by digestion before being determined as an orthophosphate (APHA, 1998).

In food industry wastewaters, phosphorus occurs as an organic phosphate that originates from proteins and some detergents used by machine washing which may contain phosphorus. However, the nitrogen content of food industry wastewaters is more significant than phosphorus (Hiisvirta, 1976).

2.5 Odours

Unpleasant odours in food industry wastewater are usually caused by gases produced by anaerobic decomposition of organic matter. The most common odour causing compound is hydrogen sulphide whose characteristic odour is that of rotten eggs (Metcalf & Eddy, 2003).

Sulphur is required in the synthesis of proteins and is released in to the degradation process. Under anaerobic conditions, sulphate is reduced biologically to sulphide which can further combine with hydrogen forming hydrogen sulphide (H_2S). This gas is readily soluble in water, colourless and inflammable, but also toxic. Although hydrogen sulphide is the most common gas formed during the anaerobic decomposition of organic matter when considering odours, other volatile compounds, such as indole, skatole and mercaptans, may cause odours far more unpleasant than H_2S (Metcalf & Eddy, 2003).

In recent years, the control of odours has become more important in the designing and operating of wastewater collection, treatment and disposal plants. Odours are the foremost concern of the public in wastewater treatment processes. Quite often, the psychological stress causing by odours is far more important rather than the harm they do to the health of humans (Droste, 1997).

Unpleasant odours are detected by the olfactory system, but which the precise mechanism is not well known. One of the difficulties in developing a global theory has been the insufficient explanation of why compounds with different molecular structures may have similar odours. Nowadays, some agreement has been achieved that the odour of a molecule has to be related to the molecule as a whole. Malodorous compounds in untreated wastewater are listed in Table 2. All these compounds can be found or may be developing in wastewaters, depending on ambient conditions (Metcalf & Eddy, 2003).

Odorous compound	Chemical formula	Odour quality
Amines	CH_3NH_2, $(CH_3)_3NH$	Fishy
Ammonia	NH_3	Ammonia
Diamines	$NH_2(CH_2)_4NH_2$, $NH_2(CH_2)_5NH_2$	Dead body
Hydrogen sulphide	H_2S	Rotten eggs
Mercaptans (1-2 carbon)	CH_3SH, $CH_3(CH_2)SH$	Decayed cabbage
Mercaptans (over 2 carbon)	$(CH_3)_3CSH$, $CH_3(CH_2)_3SH$	Skunk
Organic sulphides	$(CH_3)_2S$, $(C_6H_5)_2S$	Rotten cabbage
Skatole	C_9H_9N	Faecal matter

Table 2. Malodorous compounds in untreated wastewater (Metcalf & Eddy, 2003).

In the complete characterisation of odour, four independent factors can be classified: intensity, character, hedonics and detectability. Odours can be measured by sensory methods and a specific odorant concentration can be measure by instrumental methods, such as GC-MS analysis (Metcalf & Eddy, 2003). In the sensory method, a panel of human subjects is exposed to odour-free air diluted odours and the minimum detectable threshold odour concentration (MDTOC) is noted. This procedure can be performed according to the Standard Method 2150B Threshold Odour Test (APHA, 1998).

3. Advanced oxidation processes

When selecting the most suitable wastewater treatment method for the specific effluent, both the feasibility of the treatment as well as the economics of the process need to be considered. There are multiplicities of different kinds of techniques available such as physical, chemical and biological wastewater treatments and their combinations.

Advanced oxidation processes (AOPs) belong to the chemical treatment category and are used to oxidise organic compounds found in wastewater which are difficult to handle biologically into simpler end products. Advanced oxidation processes involve the generation of free hydroxyl radical ($HO^{\cdot}$), a powerful, non-selective chemical oxidant (Table 3) (Munter, 2001).

Oxidising agent	Relative oxidation activity
Positively charged hole on titanium dioxide, TiO_2^+	2.35
Hydroxyl radical	2.05
Atomic oxygen	1.78
Ozone	1.52
Hydrogen peroxide	1.31
Permanganate	1.24
Hypochlorous acid	1.10
Chlorine	1.00

Table 3. Relative oxidation activity of some oxidising agents (Munter, 2001).

Organic compound	Rate constant [M^{-1} s^{-1}]	
	O_3	HO
Alcohols	10^{-2}-1	10^8-10^9
Aromatics	1-10^2	10^8-10^{10}
Chlorinated alkenes	10^3-10^4	10^9-10^{11}
Ketones	1	10^9-10^{10}
N-containing organics	10-10^2	10^8-10^{10}
Phenols	10^3	10^9-10^{10}

Table 4. Reaction rate constants for ozone and hydroxyl radical for organic compounds (Munter, 2001).

Hydroxyl radical is one of the most active oxidising agents known. It acts very rapidly with most organic molecules with rate constants in the order of 10^8 - 10^{11} M^{-1} s^{-1} (Table 4) (Munter, 2001). Depending upon the nature of the organic species, generated hydroxyl radical can attack organic radicals by radical addition, hydrogen abstraction, electron transfer and radical combination.

Radical addition. Reaction of the hydroxyl radical and unsaturated or aliphatic organic compound produces organic radical which can further oxidise by oxygen or ferrous iron to form stable oxidised end products.

$$R + HO^{\cdot} \rightarrow ROH \tag{2}$$

Hydrogen abstraction. Generated hydroxyl radical can be used to remove hydrogen from an organic compound forming an organic radical and initiating a chain reaction where the organic radical reacts with oxygen. This produces a peroxyl radical, which can react with another organic compound, and so on.

$$R + HO^{\cdot} \rightarrow R^{\cdot} + H_2O \tag{3}$$

Electron transfer. Electron transfer results in the formation of ions with a higher valence. Oxidation of a monoatomic negative ion will result in the formation of an atom or a free radical.

$$R^n + HO^{\cdot} \rightarrow R^{n-1} + HO^- \tag{4}$$

Radical combination. Two radicals form a stable product.

$$HO^{\cdot} + HO^{\cdot} \rightarrow H_2O_2 \tag{5}$$

Generally, the reaction of hydroxyl radicals and organic compounds will produce water, carbon dioxide and salts (SES, 1994). However, the attack of the $HO^{\cdot}$ radical, in the presence of oxygen, generates a complex series of oxidation reactions in which the exact routes of these reactions to complete mineralisation of the organics are still not quite clear. Chlorine containing organic compounds, for example, are oxidised first to intermediates, such as aldehydes and carboxylic acids, and finally to carbon dioxide and water, and to chlorine ions (Munter, 2001).

A very important point, which has to be considered in the case of natural waters, is the presence of carbonates. Efficient trapping of $HO^{\cdot}$ radicals by bicarbonate (equation 6) and carbonate (equation 7), radical scavengers, can significantly reduce the efficiency of the abatement of pollutants.

$$HO^{\cdot} + HCO_3^- \rightarrow H_2O + CO_3^{\cdot -} \tag{6}$$

$$HO^{\cdot} + CO_3^{2-} \rightarrow HO^- + CO_3^{\cdot -} \tag{7}$$

However, the generated carbonate radical anion is also an oxidant itself, but its oxidation power is less positive compared to a $HO^{\cdot}$ radical (Legrini et al, 1993).

The destruction rate of contaminants is approximately proportional to a constant rate for the pollutant with a $HO^{\cdot}$ radical. As we can see from Table 4, chlorinated alkenes decompose fastest because the double bond is very prone to a hydroxyl attack. Saturated molecules, such as alkanes, are more difficult to oxidise because of a slower reaction rate (Table 4). The powerfulness of the hydroxyl radical gives advanced oxidation processes the ability to achieve oxidative destruction of compounds refractory to conventional hydrogen peroxide or ozone oxidation. AOPs have been used successfully for example, destroying pesticides by photochemical degradation (UV/O_3 and UV/H_2O_2) (Andreozzi et al., 2003), photocatalysis (TiO_2/UV, Fenton and photo-Fenton process) (Legrini et al., 1993; Fallman et al., 1999) and chemical oxidation processes (O_3, O_3/H_2O_2 and H_2O_2/Fe^{2+}) (Masten & Davies, 1994; Benitez et al., 2002), decomposing of organics from textile wastewater, such as surfactants and dyes, by photo-Fenton and H_2O_2/UV-C treatment (García-Montaño et al., 2006), photocatalysis with immobilised TiO_2 (Harrelkas et al., 2008) and also for the destruction of organics in different kind of effluents, such as paper mill wastewaters by photocatalysis (Pérez et al., 2001), landfill leachates by the Fenton process (Lopez et al, 2004; Gotvajn et al., 2009), olive mill wastewaters by wet air oxidation (Gomes et al., 2007) etc.

Advanced oxidation methods can be split into "cold" and "hot" oxidation. Cold oxidation methods work near to ambient temperature and pressure compared to hot oxidation at elevated temperatures and pressure (Verenich, 2003). Suitable applications of cold oxidation methods include effluents containing relatively small amount of COD (≤ 5.0 g L^{-1}). Higher COD contents would require the consumption of large amounts of expensive reactants, such as O_3 and H_2O_2 (Andreozzi et al., 1999). For wastewaters with higher COD values (≥ 5.0 g L^{-1}), hot oxidation techniques are more convenient (Mishra et al., 1995).

3.1 Ozone water processes

Ozonation at elevated values of pH. Ozone is an effective oxidising agent (Table 3) which reacts with most compounds containing multiple bonds, such as C=C, C=N, N=N, but not with species containing single bonds (C-C, C-O, O-H) at high rates (Gogate & Pandit, 2004a). At higher pH values, ozone reacts almost unselectively with all inorganic and organic compounds present in the solution (Staehelin & Hoigne, 1982). Rising the pH of the aqueous solution increases the decomposition rate of the ozone that generates the super-oxide anion radical $O_2^{\cdot -}$ and hydroperoxyl radical $HO_2^{\cdot}$. For example, the ozonide anion $O_3^{\cdot -}$ is formed by the reaction between O_3 and $O_2^{\cdot -}$. The ozonide anion further decomposes to a $HO^{\cdot}$ radical, such that, three ozone molecules will produce two $HO^{\cdot}$ radicals (equation 8) (Munter, 2001):

$$3O_3 + HO^- + H^+ \rightarrow 2HO^{\cdot} + 4O_2 \tag{8}$$

The rate constants of the hydroxyl radicals are typically 10^6 - 10^9 times higher than the corresponding reaction rate constants of molecular ozone (Table 4). The oxidation of organic compounds may also occur due to the combination of reactions with molecular ozone and reactions with hydroxyl radicals (Munter, 2001).

Ozone with hydrogen peroxide. The addition of hydrogen peroxide to the aqueous solution of ozone enhances the decomposition of O_3 with the formation of hydroxyl radicals. To summarise: two ozone molecules will produce two hydroxyl radicals (equation 9) (Munter, 2001):

$$2O_3 + H_2O_2 \rightarrow 2HO^{\cdot} + 3O_2 \quad (9)$$

The action of both ozone molecules and the generated hydroxyl radicals results in a significant improvement in the rates of decomposition of pollutants in aqueous solutions.

Ozone and catalyst. Catalytic ozonation is another opportunity to accelerate ozonation with compounds which are weakly reactive with ozone, such as atratzine. Several homogeneous catalysts, such as zinc and copper sulfates, silver nitrate, chromium trioxide (Abdo et al., 1988) and also heterogeneous catalysts, Ru/CeO_2, (Delanoë et al., 2001), MnO_2 (Ma & Graham, 1997), TiO_2/Al_2O_3 (Beltrán et al., 2004) and Pt/Al_2O_3 (Chang et al., 2009) have been studied. According to these studies, both homogeneous and heterogeneous catalysts are able to improve the efficiency of ozone for the removal of different organic compounds in an aqueous solution.

3.2 Photolysis

In a photo-oxidation reaction, UV radiation (photon) excites an electron of an organic molecule (C) from the ground state to the excited state (C*) (equation 10). The excited organic molecule excites further molecular oxygen (equation 11) with a subsequent recombination of the radical ions or hydrolysis of the radical cation, or homolysis (equation 12) to form radicals which can react with oxygen (equation 13) (Legrini et al., 1993).

$$C \overset{h\nu}{\rightarrow} C^{*} \quad (10)$$

$$C^{*} + O_2 \rightarrow C^{\cdot +}\ O_2^{\cdot -} \quad (11)$$

$$R\text{-}X \overset{h\nu}{\rightarrow} R^{\cdot} + X^{\cdot} \quad (12)$$

$$R^{\cdot} + O_2 \rightarrow RO_2 \quad (13)$$

The rate of the photo-oxidation reaction depends on the adsorption cross section of the medium, the quantum yield of the process, the photon rate at the wavelength of excitation and the concentration of dissolved molecular oxygen (Legrini et al, 1993). However, to achieve the complete mineralisation of the treated effluent, photolysis is usually combined with oxidising compounds (hydrogen peroxide, ozone) or semiconductors (such as titanium dioxide).

3.2.1 UV/ozone, UV/H_2O_2 and UV/O_3/H_2O_2 processes

The combination of UV light and ozone/hydrogen peroxide or both significantly enhances the rate of generating free radicals. Ozone adsorbs UV radiation at a wavelength of 254 nm (equation 14) producing hydrogen peroxide as an intermediate, which decomposes further to hydrogen peroxide radicals (equation 15) (Munter, 2001):

$$O_3 + hv \rightarrow O_2 + O(^1D) \quad (14)$$

$$O(^1D) + H_2O \rightarrow H_2O_2 \rightarrow 2HO^{\cdot} \quad (15)$$

The mechanism for the photolysis of hydrogen peroxide is the cleavage of the molecule into two hydroxyl radicals (equation 16) (Munter, 2001):

$$H_2O_2 \overset{hv}{\rightarrow} 2HO^{\cdot} \quad (16)$$

Depending on the pH value of the aqueous H_2O_2 solution, HO_2^- also absorbs UV radiation at 254 nm to form a hydroxyl radical (equations 17, 18):

$$H_2O_2 \leftrightarrow HO_2^- + H^+ \quad (17)$$

$$HO_2^- \overset{hv}{\rightarrow} + HO^{\cdot} + O^{\cdot -} \quad (18)$$

The combination of UV photolysis and ozone/hydrogen peroxide will be beneficial only for contaminants which require a relatively higher level of oxidation conditions (higher activation energy) (Gogate & Pandit, 2004b).

3.2.2 Photocatalysis

In the photocatalytic process, semiconductor material (often TiO_2) is excited by electromagnetic radiation possessing energy of sufficient magnitude, to produce conduction band electrons and valence band holes (equation 19) (Andreozzi et al, 1999):

$$TiO_2 \overset{hv}{\rightarrow} e^- + h^+ \quad (19)$$

Formed electrons can reduce some metals and dissolved oxygen to produce a superoxide radical ion $O_2^{\cdot -}$ (equations 20, 21):

$$M + e^- \rightarrow O_2^{\cdot -} \quad (20)$$

$$O_2 + e^- \rightarrow O_2^{\cdot -} \quad (21)$$

Remaining holes then oxidise and adsorbed H_2O or HO^- to reactive hydroxyl radicals (equations 22, 23):

$$TiO_2(h^+) + H_2O_{ad} \rightarrow TiO_2 + HO^{\cdot}_{ad} + H^+ \quad (22)$$

$$TiO_2(h^+) + HO^-_{ad} \rightarrow TiO_2 + HO^{\cdot}_{ad} \quad (23)$$

Formed hydroxyl radicals may also react with organic compounds in water as described in the equations (2)-(5).

Several catalytic materials have been studied in photocatalysis although TiO_2 in the anatase form seems to possess the best photocatalytic performance (Andreozzi et al, 1999). TiO_2 in its anatase form has an energy bandgap of 3.2 eV and can be activated by UV radiation with a wavelength up to 387.5 nm. Therefore, many researchers have focused on examining the use of sunlight in photocatalytic processes. Unfortunately, only a few percent of solar energy reaches the surface of the earth that could in principle be utilised as a direct exciter to TiO_2 (Munter, 2001).

Degussa P-25 TiO_2 catalyst is probably the most active catalyst in photocatalytic reactions however its optimum effective will always be strongly dependent on the type and concentration of the treated pollutant (Gogate & Pandit, 2004a). In several studies, the doping of TiO_2 with metals, such as, platinum (Hufschmidt et al, 2002), silver, zirconium and iron (Kment et al, 2010) as well as, sulphur, carbon and nitrogen (Menendez-Flores et al, 2011; Wang et al, 2011) has been proven to enhance the activity of the catalyst.

3.3 Fenton processes

Fenton´s reagent consists of H_2O_2 and ferrous iron, which generates hydroxyl radicals (equation 24) (Munter, 2001):

$$Fe^{2+} + H_2O_2 \rightarrow Fe^{3+} + HO^{-} + HO^{\cdot} \qquad (24)$$

The generated ferric ion decomposes H_2O_2 forming hydroxyl radicals (equations 25, 26):

$$Fe^{3+} + H_2O_2 \rightarrow H^{+} + Fe\text{-}OOH^{2+} \qquad (25)$$

$$Fe\text{-}OOH^{2+} \rightarrow HO_2^{\cdot} + Fe^{2+} \qquad (26)$$

After the reaction (26), the formed ferrous iron again decomposes H_2O_2 (24) etc. The hydrogen peroxide decomposition is an iron salt catalyzed reaction and in reactions (25, 26) iron is regenerated again to iron(II).

The most important operating parameter in the Fenton process is the pH of the solution. According to the majority of researchers, the optimum operating pH to be observed is 3. However, the Fenton process effectively generates hydroxyl radicals, it consumes one molecule of Fe^{2+} for each $HO^{\cdot}$ radical produced, which results in a high concentration of Fe(II) (Munter, 2001).

3.3.1 Photoassisted Fenton processes

In photoassisted Fenton process, Fe^{3+} ions are added to the H_2O_2/UV process. In acidic pH, a $Fe(OH)^{2+}$ complex is formed (equation 27) (Munter, 2001):

$$Fe^{3+} + H_2O \rightarrow Fe(OH)^{2+} + H^{+} \qquad (27)$$

The photolysis of Fe^{3+} complexes allows Fe^{2+} regeneration and formation of hydroxyl radicals (equation 28) (Munter, 2001):

$$Fe(OH)^{2+} \xrightarrow{h\nu} Fe^{2+} + HO^{\cdot} \qquad (28)$$

The combination of H_2O_2/UV and iron salt produces more hydroxyl radicals compared with a conventional Fenton process or photolysis, thus the technique enhances the degradation of treated pollutants (Gogate and Pandit, 2004b).

3.3.2 Electro-Fenton processes

In the Electro-Fenton process, Fe^{2+} and H_2O_2 are generated electrochemically, either separately or concurrently. Hydrogen peroxide can be electrogenated by the reduction of dissolved oxygen (equation 29), and ferrous iron by the reduction of ferric iron (equation 30) or by oxidation of a sacrificial Fe anode (equation 31) (Szpyrkowicz et al., 2001):

$$O_2 + 2H^+ + 2e^- \rightarrow H_2O_2 \tag{29}$$

$$Fe^{3+} + e^- \rightarrow Fe^{2+} \tag{30}$$

$$Fe \rightarrow Fe^{2+} + 2e^- \tag{31}$$

The reaction between H_2O_2 and Fe^{2+} produces hydroxyl radicals (equation 24).

3.4 Wet oxidation processes

The main differences in wet oxidation processes compared with "cold" oxidation techniques described earlier are the operating temperature and pressure. Typically, wet oxidation processes are operating at temperatures from 90 °C (wet peroxide oxidation) to even over 600 °C (supercritical wet air oxidation). The operating conditions of different wet oxidation techniques are described in Table 5.

Operating parameter	Wet peroxide oxidation (WPO)	Catalytic wet air oxidation (CWAO)	Wet air oxidation (WAO)	Supercritical wet air oxidation (SCWO)
Temperature [°C]	100-140	130-250	125-320	> 374
Pressure [bar]	3-5	20-50	5-200	> 221

Table 5. Operating conditions of different wet oxidation techniques (Debellefontaine et al., 1996; Mishra et al., 1995).

The wet oxidation processes are suitable for wastewaters and sludges which are both too diluted to incinerate and too concentrated for biological treatment. The COD level of the wastes appropriate to be treated by WAO techniques is typically between 5 and 200 g L^{-1} (Kolaczkowski et al., 1999).

3.4.1 Wet peroxide oxidation

WPO process is adapted from the Fenton process (Section 3.3) but it operates at temperatures above 100 °C. The oxidation mechanism is the same as for the Fenton´s reaction (equations 24-26), but as a consequence of a higher operating temperature more efficient TOC removal can be obtained (Debellefontaine et al., 1996). A typical catalyst in the WPO is iron but other catalytic materials have also been used successfully in the process such as copper (Caudo et al, 2008), activated carbon (Gomes et al., 2010), ruthenium (Rokhina et al., 2010) etc.

3.4.2 Wet air oxidation

In the WAO process organic and oxidisable inorganic compounds of the liquid phase are oxidised at elevated temperatures and pressures (Table 5) using oxygen containing gas (air, molecular oxygen) to intermediates (short-chain organic molecules), CO_2 and water. The degree of oxidation is dependent on temperature, oxygen partial pressure, operating time and, of course, the oxidisability of the compounds under consideration. To summarise, it can be said that, the higher the operating temperature the higher is the extent of oxidation achieved (Mishra et al. 1995).

According to Lixiong et al. (1991) the WAO reaction starts with the reaction of oxygen and the weakest C-H bonds of the oxidised organic compound (R denotes the organic functional group) forming free radicals:

$$RH + O_2 \rightarrow R^{\cdot} + HO_2^{\cdot} \quad (32)$$

More organic radicals are formed with the reaction of $HO_2^{\cdot}$ and organic molecule:

$$RH + HO_2^{\cdot} \rightarrow R^{\cdot} + H_2O_2 \quad (33)$$

At the high operating temperature of WAO, H_2O_2 decomposes rapidly in the homogeneous or heterogeneous species (term M) to hydroxyl radicals:

$$H_2O_2 + M \rightarrow 2HO^{\cdot} + M \quad (34)$$

The chain reaction continues with the oxidation of organic compounds by hydroxyl radicals following a hydrogen abstraction mechanism described earlier (equation 3).

3.4.3 Catalytic wet air oxidation

Compared with conventional WAO, catalytic wet air oxidation (CWAO) has lower energy requirements. Due to the presence of homogeneous or heterogeneous catalysts, lower operating conditions (air/oxygen pressure and temperature) can be used to achieve much higher oxidation rates. Various heterogeneous catalysts have been synthesised and tested in CWAO reactions, based either on metal oxides or supported noble metals (Levec & Pintar, 2007). Mixtures of metal oxides such as Cu, Zn, Co and Al, are reported to exhibit good activity, but leaching of these metals has been detected (Mantzavinos et al., 1996a). Supported noble metal catalysts, such as Pt, Pd, Rh, Re, Ru on Al_2O_3 (Mantzavinos et al., 1996b) and ceria including doped-ceria supported Pt and Ru (Keav et al., 2010) are less prone to leaching and are also more effective for oxidising organic compounds than metal oxide catalysts. CWAO processes have also been commercialised. In Japan, several companies have developed catalytic wet air oxidation technologies relying on heterogeneous supported noble metal catalysts (Harada et al., 1987; Ishii et al., 1994). In Europe, homogeneous CWAO processes such as Ciba-Geigy, LOPROX, ORCAN and ATHOS have already been developed in the 1990s (Luck, 1999).

3.4.4 Supercritical wet air oxidation

In a supercritical water oxidation process the reaction temperature is above 374 °C and the oxygen containing gas is at a pressure of 221 bar, which is the critical point of water. Above this critical point, water is an excellent solvent for both organic compounds and gases; therefore oxidisable compounds and oxygen can be mixed in a single homogeneous phase. This is the great advantage of supercritical water oxidation and in general, the destruction efficiencies of pollutants are in the order of 99-99.9% at 400-500 °C for a residence time interval of 1-5 minutes (Mishra et al., 1995).

4. AOP applications in food industry wastewater treatment

Food industry wastewaters can generally be treated biologically, in both aerobic and anaerobic reactors. However, as a consequence of diverse consumption, the forming effluents may contain compounds which are poisonous to the micro-organisms in the biological treatment plant. The pre-treatment of the effluent by chemical oxidation, especially with AOPs, can oxidise biorefractory pollutants into a more easily biodegradably form. In the following Sections, there are several examples of AOP applications in different sectors of food industry wastewater treatment.

4.1 Winery and distillery wastewater

The winery industry generates strong organic wastewater whose quality is highly dependent on the production activities. A typical COD value of the effluent containing sugars, ethanol, organic acids, aldehydes, other microbial fermentation products, soaps and detergents, is between 800 and 1200 mg L^{-1} but can easily increase to over 25000 mg L^{-1}. Winery wastewater is quite acidic (pH 3-4) and it usually contains large amounts of phosphorus, but not nitrogen and other trace minerals, which are important for biological treatment (Oller et al., 2010).

There are some studies that have considered the ozonation of winery wastewaters. Lucas et al. (2009a) for example, have treated winery wastewaters by ozonation in a bubble column reactor. During a three hour reaction period the degradation of aromatic and polyphenol content was found to be significant, thus the biodegradability of the wastewater was improved and therefore the ozonation may be considered as pre-treatment to further biological treatment. In addition, Beltrán et al. (1999) have noticed the same BOD/COD enhancement in winery effluent after ozonation whereas Lucas et al. (2010) have combined UV and UV/H_2O_2 with ozone. According to the results, O_3/UV/H_2O_2 combination was identified as the most economical process compared with O_3 and O_3/UV to the treated winery wastewater.

Winery wastewaters were also treated by solar photo-Fenton integrated with activated sludge treatment in a pilot-plant scale (Mosteo et. al., 2008) and by UV and UV/TiO_2 at lab-scale (Agustina et. al., 2008). With both techniques the efficient removal of organics was successfully achieved whereas photo-Fenton combined with biological treatment showed higher mineralisation rates and a significant toxicity decrease of the treated effluent.

Sources of distillery wastewater, much in common with winery wastewater, are stillage, fermenter and condenser cooling water, and fermenter wastewater. These effluents contain high concentrations of organic material (COD 100-150 g L^{-1} and BOD 40-200 g L^{-1}) and fertilisers such as potassium, phosphorus and nitrogen. In addition, the molasses wastewater from ethanol fermentation has a typical brown colour which is difficult to remove by traditional biological treatment (Oller et al., 2010).

Several AOP methods have been studied in the treatment of distillery wastewaters. Beltrán et al. (1997b), Benitez et al. (1999; 2003), Sangave et al. (2007) and Sreethawong & Chavadej (2008) have all used ozonation and its combination with UV, H_2O_2 or Fe-catalyst in the degradation of organics from distillery wastewaters. All these researchers agree that the removal of organic matter improved with the simultaneous presence of UV radiation, hydrogen peroxide or Fe-catalyst in addition to ozone, due to the contribution of hydroxyl radicals generated in these combined processes. In addition, the combined process of ozone

pre-treatment followed by an activated sludge step provides enhancement in the removal of substrate obtained in relation to that obtained in the single aerobic treatment without ozonation, i.e. from 28% to 39 % (Benitez et al., 1999; 2003). The integrated process (ozone pre-treatment-aerobic biological oxidation-ozone post-treatment) achieved almost 80% COD reduction in the treatment of distillery wastewater along with the decolouration of the effluent compared with 35% COD removal for non-ozonated samples (Sangave et al., 2007).
UV/H_2O_2 (Beltrán et al., 1997a) and electro-Fenton (Yavuz, 2007) processes have also been used for the treatment of distillery wastewaters. The EF process (Yavuz, 2007) seems to be a promising technique with the COD removal over 90% compared with a UV radiation and hydrogen peroxide combination whose COD reduction is only 38% (Beltrán et al., 1997a).
Belkacemi et al. (1999; 2000) investigated wet oxidation and catalytic wet oxidation for the removal of organics from distillery liquors. The initial TOC of the effluent was 22500 mg L^{-1} while in the AOPs described earlier the total organic carbon was 10 or even 100 times lower. In the temperature and oxygen partial ranges of 180-250 °C and 5-25 bar respectively, the highest TOC removal (around 60%) was achieved with Mn/Ce oxides and Cu(II)NaY catalysts. These catalysts were found to be very effective for short contact times, while for prolonged exposures catalysts deactivation by fouling carbonaceous deposit was shown to be the prime factor responsible for the loss of catalysts activity (Belkacemi et al., 2000). In the supercritical water oxidation of alcohol distillery wastewater (Goto et al., 1998) almost complete colour, odour and TOC removal was attained when more than stoichiometric amount (over 100%) of oxidant (H_2O_2) was used in temperatures between 200-600 °C.

4.2 Olive industry wastewater

Wastewaters from olive oil extraction plants, also called olive mill wastewaters, and wastewaters generated by table olive production, contain high concentration of phenolic compounds. In olive oil production, an oily juice is extracted from the fruit through milling or centrifugation. Table olive production requires the same treatment in order to eliminate the bitterness of the fruit, due to the presence of polyphenolic compounds (Bautista et al., 2008).
Olive mill wastewater contains polysaccharides, sugars, polyphenols, polyalcohols, proteins, organic acids, oil etc. and therefore, the COD of the effluent may be as high as 220 g L^{-1} and even 190 g L^{-1} for the amount of suspended solids (Oller et al., 2010).
For several years, olive mill wastewater has been the most polluting and troublesome waste produced by olive mills in all the countries surrounding the Mediterranean. Thus, the management of this liquid residue has been investigated extensively and the efficiency of AOPs for treating olive mill effluents has been studied widely (Mantzavinos & Kalogerakis, 2005). Many researchers have also investigated Fenton processes in the treatment of olive mill effluents (Table 6).
Olive mill wastewater has also been treated by several other AOPs such as ozonation or ozone/UV (Lafi et al., 2009) which have increased the biodegradability of the effluent. Minh et al. (2008) and Gomes et al. (2007) have been successful in decreasing the TOC and phenolic content of olive mill wastewater by CWAO. At reaction conditions of 190 °C and 70 bar of air using Pt and Ru supported on titania and zirconia carriers, the toxicity and phytotoxicity of the effluent decreased to a suitable level for anaerobic treatment (Minh et al., 2008). Gomes et al. (2007) reported that with the carbon supported Pt catalyst TOC and the colour of olive mill wastewater were completely removed after 8 h of reaction at 200 °C

at 6.9 bar of O_2. Caudo et al. (2008) have also tested copper-pillared clays as catalysts in wet hydrogen peroxide catalytic oxidation (WHPCO) of olive mill wastewater. According to the research, copper pillared clays are effective and stable catalysts for WHPCO of wastes in water whilst this treatment decreases the toxicity of the olive mill wastewater.

Fenton process	Conclusions	References
Fenton	COD removal 80-90%, followed by biological treatment	Bressan et al. (2004)
	COD and aromatics removal 40%	Ahmadi et al. (2005)
	Considered as a pre-treatment (COD removal 40-50%)	Dogruel et al. (2009)
	COD removal 70%	Lucas & Peres (2009b)
Coagulation and Fenton/photo-Fenton	COD removal 85% by F and 95% by PF	Rizzo et al. (2008)
Coagulation-flocculation-Fenton	COD removal 60% and decrease of phytotoxicity	Ginos et al. (2006)
Fenton with zero-valent iron	Considered as a pre-treatment (COD removal 75-20%) before classical biological process	Kallel et al. (2009)
Solar photo-Fenton	COD removal 85%, phenolic compounds degradation even 100%	Gernjak et al. (2004)
Electro-Fenton	Considered as a pre-treatment (COD removal 66%) before anaerobic digestion	Khoufi et al. (2006)
	Considered as a pre-treatment (COD removal 53%) before anaerobic digestion and ultrafiltration resulting in a complete detoxify of effluent. Pilot plant.	Khoufi et al. (2009)

Table 6. Fenton processes for the treatment of olive mill wastewater.

The organic content of wastewater from a table olive process is quite similar to olive mill wastewater containing phenols, polyphenols, sugars, acids, tannins, pectins and oil residues, with a COD of several grams per litre. The inorganic fraction consists of high concentrations of NaCl and NaOH which are used for debittering and fermentation, as well as trace amounts of various metals. As a consequence of the complexity of these effluents, they are unsuitable for conventional aerobic and anaerobic processes (Oller et al., 2010). Recently, it has been possible to enhance the biodegradability of the table olive processing wastewater by different AOPs. Kyriacou et al. (2005) have scaled this treatment method up from a lab-scale to a pilot-scale for green table olive processing wastewater, which combines biological treatment with an electro-Fenton system. In the pilot plant, 75% COD removal was achieved and the post-treatment by coagulation finally gave an overall 98% COD removal for the treated effluent. Photocatalytic treatment and WAO alone (Chatzisymeon et al., 2008; Katsoni et al., 2008) as well as O_3, O_3/H_2O_2, O_3/UV, UV, UV/H_2O_2, Fenton, photo-Fenton and WAO processes combined with aerobic biological treatment have been studied for organic matter removal from table olive processing wastewaters (Beltran-Heredia et al., 2000; Benitez et al., 2001a, b; Rivas et al., 2000, 2001).

4.3 Meat processing industry wastewater

Meat processing industry wastewaters constitute one of the greatest concerns of the agro-industrial sector, as approximately 62 Mm^3/year of water is used worldwide. However, only a small amount of this becomes a component of the final product. Meat processing industry wastewater contains high concentrations of fat, dry waste, sediments and total suspended matter as well as nitrogen and chlorides whilst possessing high biological and chemical oxygen demand (Sroka et al., 2004). Traditionally, meat processing industry wastewaters are treated by anaerobic or aerobic biological systems (Johns, 1995) but recently few studies concerning AOPs have been published. In the publication of Sena et al (2009), dissolved air flotation (DAF) followed by photo-peroxidation (UV/H_2O_2) and photo-Fenton reactions were evaluated in the treatment of meat processing industry wastewater. According to the results, DAF connected with photo-Fenton treatment achieved the best removals of COD, colour, turbidity and total solids of the treated effluent. WAO have also been used for the removal of organic compounds from meat processing industry wastewater (Heponiemi et al., 2009). After catalytic wet air oxidation treatment, the biodegradability of the wastewater sample has improved.

4.4 Vegetable and fruit processing wastewater

Various factors, such as seasonal and source variations, unit operations etc. affect the composition of vegetable and fruit processing industry wastewater. Typically, this effluent contains high organic loads, e.g. from peeling and blanching, cleaning agents and suspended solids such as fibres, dissolved solids, salts, nutrients etc. Furthermore, residual pesticides, which are difficult to degrade during wastewater treatment, may be a concern (EC, 2006). In some studies, AOPs have been used for the removal of organics from fruit and vegetable processing industry wastewaters. In the research of Beltran et al. (1997a, b), wastewater from a tomato processing plant was treated by UV, UV/H_2O_2, O_3, O_3/H_2O_2 and O_3/UV. According to results, an ozone-UV radiation system achieved the highest degradation rates (90% removal of COD). Due to the improved biodegradability of the treated effluent, Beltran et al. (1997a, b) recommended the combination of this process with biological oxidation. Caudo et al. (2008) have studied copper-pillared clays catalysed wet peroxide oxidation of citrus juice production wastewater. This effluent contains various phenolic compounds with a chemical oxygen demand of over 4000 mg L^{-1}. After 4 h of oxidation reaction, the TOC had decreased 50% and the biodegradability of the effluent (BOD_5/COD index) had increased from 0.05 to 0.4.

4.5 Miscellaneous wastewater

Coffee industry wastewater is another example of a highly polluted food industry wastewater. The coffee industry uses large amounts (around 40-45 L per kilogram of coffee) of water during the various stages of the production process. The forming effluent contains e.g. caffeine, fat and peptic substances, as well as many different macromolecules such as lignins, tannins and humic acids, which are difficult to handle by conventional biological treatment processes. Recently, Zayas et al. (2007) studied the combination of the chemical coagulation-flocculation process with various advanced oxidation processes (UV/H_2O_2, UV/O_3, $UV/H_2O_2/O_3$) for the treatment of coffee industry wastewater. Among the AOPs tested, $UV/H_2O_2/O_3$ process was the most effective in the reduction of COD, colour and turbidity of the treated effluent.

Baker´s yeast is a commercial product of molasses (the end product of sugar manufacture) which constitutes a solution of sugar, organic and inorganic material in water. Baker`s yeast industry wastewater has a high BOD and COD values which contains significant amount of nitrogen and non-biodegradable organic pollutants. In addition, the effluent has a typical dark colour and, therefore the possible decolourisation of the effluent has been investigated by the Fenton process (Pala & Erden, 2005). Fenton oxidation was applied to the biologically pre-treated baker´s yeast industry wastewater. In the optimum operating conditions, 99% colour removal and 88% COD reduction was achieved. Photo-Fenton and UV/H_2O_2 processes have also been studied in the removal of colour and organics from baker´s yeast effluents (Çatalkaya & Şengül, 2006).

Palm oil effluent is a colloidal dispersion of biological origin which has a typical unpleasant odour. The total solids content of the effluent is 5-7% and it constitutes of dissolved, organic and inorganic solids, a reason why it is extremely difficult to treat by conventional wastewater treatment methods (Zinatizadeh et al., 2006). In the study of Babu et al. (2010) a palm oil effluent was treated by a combined electro-Fenton-biological oxidation process. After 2 h of EF and 5 d of biological treatment 86% COD removal was achieved. The treated water can be reused for general purpose in an industrial application.

Dairy industry wastewater has a typical white colour and a high nutrient level as well as organic matter content. It is usually treated by biological methods such as the activated sludge process and anaerobic filters although aerobic biological processes have high energy requirements whilst anaerobic biological methods require additional treatment (Kushwaha et al., 2010). Recently, solar photocatalytic oxidation has been used after anaerobic sludge blanket reactor for the removal of COD from dairy industry wastewater (Banu et al., 2008). The combination of anaerobic process and solar photocatalytic oxidation using TiO_2 as a catalyst resulted in 95% removal of COD from dairy industry wastewater. This integrated system may be a promising alternative for the treatment of dairy industry effluents. In addition, Inamdar & Singh (2008) have applied photocatalysis in the treatment of dairy industry effluent.

5. Conclusion

The characteristics and treatment of food industry wastewaters by different advanced oxidation processes were considered. Typically, the amount and composition of the effluent varies considerably. The high organic matter content is a basic problem in food industry wastewaters but the organic compounds are usually easily biodegradable and the effluents can be treated by conventional anaerobic or aerobic biological methods. However, as a consequence of diverse consumption, the forming effluents may contain compounds which are poisonous to micro-organisms in the biological treatment plant. The pre-treatment of the effluent by chemical oxidation, especially with AOPs, can oxidise biorefractory pollutants to a more easily biodegradable form. Thus, the combination of AOP and biological treatment may be a possible solution for the treatment of variable food industry wastewaters.

6. Acknowledgements

This review has been carried out with the financial support of the Maj and Tor Nessling Foundation and the Finnish Funding Agency for Technology and Innovation (within project 70023/08) from the European Regional Development Fund.

7. References

Abdo, M.S.E.; Shaban, H. & Bader, M.S.H. (1988) Decolorization by ozone of direct dyes in presence of some catalysts. *Journal of Environmental Science and Health. Part A,* Vol. 23, No. 7, pp. 697-710, ISSN 1077-1204

Ahmadi, M.; Vahabzadeh, F.; Bonakdarpour, B.; Mofarrah, E. & Mehranian, M. (2005) Application of the central composite design and response surface methodology to the advanced treatment of olive oil processing wastewater using Fenton's peroxidation. *Journal of Hazardous Materials,* Vol. 123, No. 1-3, pp. 187-195, ISSN 0304-3894

Andreozzi, R.; Caprio, V.; Insola, A. & Marotta, R. (1999) Advanced oxidation processes (AOP) for water purification and recovery. *Catalysis Today,* Vol. 53, No, 1, pp. 51-59, ISSN 0920-5861

Andreozzi, R.; Caprio, V.; Marotta, R. & Radovnikov, A. (2003). Ozonation and H_2O_2/UV treatment of clorifibric in water: a kinetic investigation. *Journal of Hazardous Materials,* Vol. 103, No.3, pp. 233-246, ISSN 0304-3894

APHA (1998) Standard Methods for the Examination of Water and Wastewater, 20th ed., American Public Health Association, Washington, D.C, USA, ISSN 55-1979

Babu, B.R.; Meera, K.S.; Venkatesan, P. & Sunandha, D. (2010) Removal of fatty acids from palm oil effluent by combined electro-Fenton and biological oxidation process. *Water, Air, and Soil Pollution,* Vol. 211, No. 1-4, pp. 203-210, ISSN 0049-6979

Banu, J.R.; Anandan, S.; Kaliappan, S. & Yeom, I.-T. (2008) Treatment of dairy wastewater using ananerobic and solar photocatalytics methods. *Solar Energy,* Vol. 82, No. 9, pp. 812-819, ISSN 0038-092X

Bautista, P.; Mohedano, A.F.; Casas, J.A.; Zazo, J.A. & Rodriguez, J.J. (2008) Review An overview of the application of Fenton oxidation to industrial wastewater treatment. *Journal of Chemical Technology and Biotechnology,* Vol. 83, No. 10, pp. 1323-1338, ISSN 0268-2575

Belkacemi, K.; Larachi, F.; Hamoudi, S. & Sayari, A. (2000) Catalytic wet oxidation of high-strength alcohol-distillery liquors. *Applied Catalysis A: General,* Vol.199, No. 2, pp. 199-209, ISSN 0926-860X

Belkacemi, K.; Larachi, F.; Hamoudi, S.; Turcotte, G. & Sayari, A. (1999) Inhibition and deactivation effects in catalytic wet oxidation of high-strength alcohol-distillery liquors. *Industrial & Engineering Chemistry Research,* Vol. 38, No. 6, pp. 2268-2274, ISSN 0888-5885

Beltrán, F.J.; Encinar, J.M. & González, J.F. (1997b) Industrial wastewater advanced oxidation. Part 2. Ozone combined with hydrogen peroxide or UV radiation. *Water Research,* Vol.31, No.10, pp. 2415-2428, ISSN 0043-1354

Beltrán, F.J.; García-Araya, J. & Álvarez, P.M. (1999) Wine distillery wastewater degradation. 1. Oxidative treatment using ozone and its effect on the wastewater biodegradability. *Journal of Agricultural and Food Chemistry,* Vol. 47, No. 9, pp. 3911-3918, ISSN 0021-8561

Beltrán, F.J.; González, M. & González, J.F. (1997a) Industrial wastewater advanced oxidation. Part 1. UV radiation in the presence and absence of hydrogen peroxide. *Water Research,* Vol.31, No.10, pp. 2405-2414, ISSN 0043-1354

Beltrán, F.J.; Rivas, F.J. & Montero-de-Espinosa, R. (2004) A TiO_2/Al_2O_3 catalyst to improve the ozonation of oxalic acid in water. *Applied Catalysis B: Environmental,* Vol. 47, No. 2, pp. 101–109, ISSN 0926-3373

Beltran-Heredia, J.; Torregrosa, J.; Dominguez, J.R. & Garcia, J. (2000) Aerobic biological treatment of black table olive washing wastewaters: effect of an ozonation stage. *Process Biochemistry,* vol. 35, No.10, pp. 1183–1190, ISSN 0032-9592

Benitez, F.J.; Acero, J.L. & Real, F.J. (2002) Degradation of carbofuran by using ozone, UV radiation and advanced oxidation processes. *Journal of Hazardous Materials,* Vol. 89 No. 1, pp. 51-65, ISSN 0304-3894

Benitez, F.J.; Acero, J.L.; Gonzalez, T. & Garcia, J. (2001a) Ozonation and biodegradation processes in batch reactors treating black table olives washing wastewaters. *Industrial & engineering chemistry research,* Vol. 40, No. 14, pp. 3144-3151, ISSN 0888-5885

Benitez, F.J.; Acero, J.L.; Gonzalez, T. & Garcia, J. (2001b) Organic matter removal from wastewaters of the black olive industry by chemical and biological procedures. *Process Biochemistry,* Vol. 37, No. 3, pp. 257–265, ISSN 0032-9592

Benitez, F.J.; Beltran-Heredia, J.; Real, F.J. & Acero, J.L. (1999) Enhancement of the ozonation of wine distillery wastewaters by an aerobic pretreatment. *Bioprocess Engineering,* Vol. 21, No. 5, pp. 459-464, ISSN 0178-515X

Benitez, F.J.; Real, F.J.; Acero, J.L.; Garcia, J. & Sanchez, M. (2003) Kinetics of the ozonation and aerobic biodegradation of wine vinasses in discontinuous and continuous processes. *Journal of Hazardous Materials,* Vol. 101, No. 2, pp. 203–218, ISSN 0304-3894

Bressan, M.; Liberatore, L.; D álessandro, N.; Tonucci, L.; Belli, C. & Ranalli, G. (2004) Improved combined chemical and biological treatments of olive oil mill wastewaters. *Journal of Agricultural and Food Chemistry,* Vol. 52, No. 5, pp. 1228-1233, ISSN 0021-8561

Çatalkaya, E.Ç. & Şengül, F. (2006) Application of Box–Wilson experimental design method for the photodegradation of baker's yeast industry with UV/H_2O_2 and $UV/H_2O_2/Fe(II)$ process. *Journal of Hazardous Materials,* Vol. 128, No. 2-3, pp. 201-207, ISSN 0304-3894

Caudo, S.; Genovese, C.; Perathoner, S. & Centi, G. (2008) Copper-pillared clays (Cu-PILC) for agro-food wastewater purification with H_2O_2. *Microporous and Mesoporous Materials,* Vol. 107, No. 1-2, pp. 46-57, ISSN 1387-1811

Chang, C.-C.; Chiu, C.-Y.; Chang, C.-Y.; Chang, C.-F.; Chen, Y.-H.; Ji, D.-R.; Yu, Y.-H. & Chiang, P.-C. (2009) Combined photolysis and catalytic ozonation of dimethyl phthalate in a high-gravity rotating packed bed. *Journal of Hazardous Materials,* Vol. 161, No. 1, pp. 287–293, ISSN 0304-3894

Chatzisymeon, E.; Stypas, E.; Bousios, S.; Xekoukoulotakis, N.P. & Mantzavinos, D. (2008) Photocatalytic treatment of black table olive processing wastewater. *Journal of Hazardous Materials,* Vol. 154, No. 1-3, pp. 1090–1097, ISSN 0304-3894

Debellefontaine, H.; Chakchouk, M.; Foussard, J.N.; Tissot, D. & Striolo, P. (1996) Treatment of organic aqueous wastes: Wet air oxidation and wet peroxide oxidation. *Environmental Pollution*, Vol. 92, No. 2, pp.155-164, ISSN 0269-7491

Delanoë, F.; Acedo, B.; Karpel Vel Leitner, N. & Legube, B. (2001) Relationship between the structure of Ru/CeO_2 catalysts and their activity in the catalytic ozonation of succinic acid aqueous solutions. *Applied Catalysis B: Environmental*, Vol. 29, No. 4, pp. 315–325, ISSN 0926-3373

Dogruel, S.; Olmez-Hanci, T.; Kartal, Z.; Arslan-Alaton, I. & Orhon, D. (2009) Effect of Fenton's oxidation on the particle size distribution of organic carbon in olive mill wastewater. *Water Research*, Vol. 43, No. 13, pp. 3974-3983, ISSN 0043-1354

Droste R.L. (1997) *Theory and practice of water and wastewater treatment*, John Wiley & Sons, Inc. New York, USA, ISBN 0-471-12444-3

European Commission (August, 2006) *Integrated Pollution Prevention and Control: Reference Document on Best Available Techniques in the Food, Drink and Milk Industries*, date of access June 15th 2011, Available from: ftp://ftp.jrc.es/pub/eippcb/doc/fdm_bref_0806.pdf

Fallman, H.; Krutzlere, T.; Bauer, R.; Malato, S. & Blanco, J. (1999) Applicability of the photo-Fenton method for treating water containing pesticides. *Catalysis Today*, Vol. 54, No. 2-3, pp. 309-319, ISSN 0920-5861

Finnish Environment Institute (30.9.2009): *The industry emissions into waterways 2008*, date of access 27th June 2011, Available from http://www.ymparisto.fi/default.asp?node=24770&lan=fi

Finnish Food and Drink Industries` Federation (13.12.2010) *Yhteenveto Elintarviketeollisuusliiton vuonna 2010 toteuttamasta ympäristövastuun kyselystä*, date of access June 27th 2011, Available from: http://www.etl.fi/www/fi/julkaisut/Julkaisut/Elintarvteoll_ymparistovastuun_raportti_2010.pdf

Finnish Food and Drink Industries` Federation (2005). *Elintarviketeollisuuden ympäristövastuun raportti 2005*, Laatuketju, Helsinki, Finland

García-Montaño J.; Torrades, F.; García-Hortal, J.A.; Doménech X. & Peral, J. (2006) Degradation of Procion RED H-E78 reactive dye by coupling a photo-Fenton system with a sequencing batch reactor. *Journal of Hazardous Materials*, Vol. 134, No. 1-3, pp. 220-229, ISSN 0304-3894

Gernjak, W.; Maldonado, M.I.; Malato, S.; Cáceres, J.; Krutzler, T.; Glaser, A. & Bauer, R. (2004) Pilot-plant treatment of olive mill wastewater (OMW) by solar TiO_2 photocatalysis and solar photo-Fenton. *Solar Energy*, Vol. 77, No. 5, pp. 567–572, ISSN 0038-092X

Ginos, A.; Manios, T. & Mantzavinos, D. (2006) Treatment of olive mill effluents by coagulation–flocculation–hydrogen peroxide oxidation and effect on phytotoxicity. *Journal of Hazardous Materials*, Vol. 133, No. 1-3, pp. 135–142, ISSN 0304-3894

Gogate, P.R. & Pandit, A.B. (2004a) A review of imperative technologies for wastewater treatment I: oxidation technologies at ambient conditions. *Advances in Environmental Research*, Vol. 8, No. 3-4, pp. 501-551, ISSN 1093-0191

Gogate, P.R. & Pandit, A.B. (2004b) A review of imperative technologies for wastewater treatment II: hybrid methods. *Advances in Environmental Research*, Vol. 8, No. 3-4, pp. 553-597, ISSN 1093-0191

Gomes, H.T.; Figueiredo, J.L. & Faria, J.L. (2007) Catalytic wet air oxidation of olive mill wastewater. *Catalysis Today*, Vol. 124, No. 3-4, pp. 254-259, ISSN 0920-5861

Gomes, H.T.; Miranda, S.M.; Sampaio, M.J.; Silva, A.M.T. & Faria, J.L. (2010) Activated carbons treated with sulphuric acid: Catalysts for catalytic wet peroxide oxidation. *Catalysis Today*, Vol. 151, No. 1-2, pp. 153–158, ISSN 0920-5861

Goto, M.; Nada, T.; Ogata, A.; Kodama, A. & Hirose, T. (1998) Supercritical water oxidation for the destruction of municipal excess sludge and alcohol distillery wastewater of molasses. *Journal of Supercritical Fluids*, Vol. 13, No. 1-3, pp. 277-282, ISSN 0896-8446

Gotvajn, A.Z.; Tisler, T. & Zagorc-Kongan, J. (2009) Comparison of different treatment strategies for industrial landfill leachate. *Journal of Hazardous Materials*, Vol. 162, No. 2-3, pp. 1446-1456, ISSN 0304-3894

Harada, Y.; Nakashiba, A.; Matuura, H.; Okina, T.; Fujitani, H.; Yamasaki, K.; Doi, Y. & Yurugi, S. (1987) *Process for treating wastewater by wet oxidations*. EP: 0224905 (A2)

Harrelkas, F.; Paulo, A.; Alves, M.M.; El Khadir, I.; Zahraa, O.; Pons, M.N. & van der Zee, F.P. (2008) Photocatalytic and combined anaerobic-photocatalytic treatment of textile dyes. *Chemosphere*, Vol. 72, No. 11, pp. 1816-1822, ISSN 0045-6535

Heponiemi, A.; Rahikka, L.; Lassi, U. & Kuokkanen, T. (2009) Catalytic oxidation of industrial wastewaters - a comparison study using different analyzing methods. *Chemical Engineering Transactions*, Vol. 17, pp. 209-214, ISBN 978-88-95608-01-3

Hiisvirta, L. (1976) *Teurastamojen ja lihanjalostuslaitosten jätevesikuormitus ja jätevesien käsittelymahdollisuudet*. Vesihallituksen projekti 7406 B, Oy Vesi-hydro AB, Kyriiri Oy, Helsinki, Finland

Hufschmidt, D.; Bahnemann, D.; Testa, J.J.; Emilio, C.A. & Litter, M.I. (2002) Enhancement of photocatalytic activity of various TiO_2 materials by platinization. *Journal of Photochemistry and Photobiology, A: Chemistry*, Vol. 148, No. 1-3, pp. 223-231, ISSN 1010-6030

Inamdar, J. & Singh, S.K. (2008) Photocatalytic detoxification method for zero effluent discharge in dairy industry: Effect of operational parameters. *International Journal of Chemical and Biological Engineering*, Vol. 1, No. 4, pp. 160-164, ISSN 2010-3867

Ishii, T.; Mitsui, K.; Sano, K.; Shishida, K. & Shiota, Y. (1994) *Catalyst for treating wastewater process for producing it, and process for treating wastewater with the catalyst*. U.S. Patent 5,374,599

Johns, M.R. (1995) Developments in wastewater treatment in the meat processing industry: A review. *Bioresource Technology*, Vol. 54, No. 3, pp. 203-216 ISSN 0960-8524

Kallel, M.; Belaid, C.; Boussahel, R.; Ksibi, M.; Montiela, A. & Elleuch, B. (2009) Olive mill wastewater degradation by Fenton oxidation with zero-valent iron and hydrogen peroxide. *Journal of Hazardous Materials*, Vol. 163, No. 2-3, pp. 550-554, ISSN 0304-3894

Karttunen, E. (2003). *RIL 124-1 Vesihuolto I*, Finnish Association of Civil Engineers RIL, Vammalan Kirjapaino Oy, Helsinki, Finland, ISBN 951-758-431-8

Katsoni, A.; Frontistis, Z.; Xekoukoulotakis, N.P.; Diamadopoulos, E. & Mantzavinos, D. (2008) Wet air oxidation of table olive processing wastewater: Determination of key operating parameters by factorial design. *Water Research*, Vol. 42, No. 14, pp. 3591-3600, ISSN 0043-1354

Keav, S.; Martin, A.; Barbier Jr. J. & Duprez, D. (2010) Deactivation and reactivation of noble metal catalysts tested in the catalytic wet air oxidation of phenol. *Catalysis Today,* Vol. 151, No. 1-2, pp. 143–147, ISSN 0920-5861

Khoufi, S.; Aloui, F. & Sayadi, S. (2006) Treatment of olive oil mill wastewater by combined process electro-Fenton reaction and anaerobic digestion. *Water Research,* Vol. 40, No. 10, pp. 2007-2016, ISSN 0043-1354

Khoufi, S.; Aloui, F. & Sayadi, S. (2009) Pilot scale hybrid process for olive mill wastewater treatment and reuse. *Chemical Engineering and Processing,* Vol. 48, No. 2, pp. 643–650, ISSN 0255-2701

Kment, S.; Kmentova, H.; Kluson, P.; Krysa, J.; Hubicka, Z.; Cirkva, V.; Gregora, I.; Solcova, O. & Jastrabik, L. (2010) Notes on the photo-induced characteristics of transition metal-doped and undoped titanium dioxide thin films. *Journal of Colloid and Interface Science,* Vol. 348, No. 1, pp. 198-205, ISSN 0021-9797

Kolaczkowski, S.T.; Plucinski, P.; Beltran, F.J.; Rivas, F.J. & McLurgh, D.B. (1999) Wet air oxidation: a review of process technologies and aspects in reactor design. *Chemical Engineering Journal,* Vol. 73, No. 2, pp. 143-160, ISSN 1385-8947

Kushwaha, J.P.; Srivastava, V.C. & Mall, I.D. (2010) Organics removal from dairy wastewater by electrochemical treatment and residue disposal. *Separation and Purification Technology,* Vol. 76, No. 2, pp. 198-205, ISSN 1383-5866

Kyriacou, A.; Lasaridi, K.E.; Kotsou, M.; Balis, C. & Pilidis, G. (2005) Combined bioremediation and advanced oxidation of green table olive processing wastewater. *Process Biochemistry,* Vol. 40, pp. 1401–1408, ISSN 0032-9592

Lafi, W.K.; Shannak, B.; Al-Shannag, M.; Al-Anber, Z. & Al-Hasan, M. (2009) Treatment of olive mill wastewater by combined advanced oxidation and biodegradation. *Separation and Purification Technology,* Vol. 70, No. 2, pp. 141–146, ISSN 1383-5866

Legrini, O.; Olivieros, E. & Braun, A.M. (1993) Photochemical processes for water treatment. *Chemical Reviews,* Vol. 93, No. 2, pp. 671-698, ISSN 0009-2665

Lehto, M.; Salo T.; Sorvala S.; Kemppainen R.; Vanhala P.; Sipilä I.; Puumala M. (2007) *Peruna- ja juureskuorimon jätteet ja jätevedet.* MTT Tampere: Tampereen Yliopistopaino, Juvenes Print, Finland, ISSN 1458-5073

Levec, J. & Pintar, A. (2007) Catalytic wet-air oxidation processes: A review. *Catalysis Today,* Vol. 124, No. 3-4, pp. 172-184, ISSN 0920-5861

Lixiong, L.; Chen, P. & Gloyna, E.F. (1991) Generalized kinetic model for the wet oxidation of organic compounds. *AIChE Journal,* Vol. 37, No. 11, pp. 1687-1697, ISSN 0001-1541

Lopez, A.; Pagano, M.; Volpe, A. & Di Pinto, A.C. (2004) Fenton`s pre-treatment of mature landfill leachate. *Chemosphere,* Vol. 54, No.7, pp. 1005-1010, ISSN 0045-6535

Lucas, M.S. & Peres, J.A. (2009b) Removal of COD from olive mill wastewater by Fenton's reagent: Kinetic study. *Journal of Hazardous Materials,* Vol. 168, No. 2-3, pp. 1253–1259, ISSN 0304-3894

Lucas, M.S.; Peres, J.A. & Puma, G.L. (2010) Treatment of winery wastewater by ozone-based advanced oxidation processes (O_3, O_3/UV and O_3/UV/H_2O_2) in a pilot-scale bubble column reactor and process economics. *Separation and Purification Technology,* Vol. 72, No. 3, pp. 235-241, ISSN 1383-5866

Lucas, M.S.; Peres, J.A.; Lan, B.Y. & Puma, G.L. (2009a) Ozonation kinetics of winery wastewater in a pilot-scale bubble column reactor. *Water Research,* Vol. 43, No. 6, pp. 1523-1532, ISSN 0043-1354

Luck, F. (1999) Wet air oxidation: past, present and future. *Catalysis Today,* Vol. 53, No. 1, pp. 81-91, ISSN 0920-5861

Ma, J. & Graham, N.J.D. (1997) Preliminary investigation of manganese-catalyzed ozonation for the destruction of atrazine. *Ozone; Science and Engineering,* Vol. 19, No. 3, pp. 227-240, ISSN 0191-9512

Mantzavinos, D. & Kalogerakis, N. (2005) Treatment of olive mill effluents Part I. Organic matter degradation by chemical and biological processes—an overview. *Environment International,* Vol. 31, No. 2, pp. 289–295, ISSN 0160-4120

Mantzavinos, D.; Hellenbrand, R.; Livingston, A.G. & Metcalfe, I.S. (1996a) Catalytic wet oxidation of p-coumaric acid: Partial oxidation intermediates, reaction pathways and catalyst leaching. *Applied Catalysis B: Environmental,* Vol. 7, No. 3-4, pp. 379-396, ISSN 0926-3373

Mantzavinos, D.; Hellenbrand, R.; Livingston, A.G. & Metcalfe, I.S. (1996b) Catalytic wet air oxidation of polyethylene glycol. *Applied Catalysis B: Environmental,* Vol. 11, No. 1, pp. 99-119, ISSN 0926-3373

Masten, S.J. & Davies, H.R. (1994) The use of ozonation to degrade organic contaminants in wastewater. *Environmental Science & Technology,* Vol. 28, No.4, pp. 180-185A, ISSN 0013-936X

Menendez-Flores, V.M.; Bahnemann, D.W. & Ohno, T. (2011) Visible light photocatalytic activities of S-doped TiO_2-Fe^{3+} in aqueous and gas phase. *Applied Catalysis B: Environmental,* Vol. 103, No. 1-2, pp. 99-108, ISSN 0926-3373

Metcalf & Eddy (2003) *Wastewater Engineering: treatment and reuse,* 4. ed, McGraw-Hill, Boston, USA, ISBN 0-07-041878-0

Minh, D.P.; Gallezot, P.; Azabou, S.; Sayadi, S. & Besson, B. (2008) Catalytic wet air oxidation of olive oil mill effluents 4. Treatment and detoxification of real effluents. *Applied Catalysis B: Environmental,* Vol. 84, No. 3-4, pp. 749–757, ISSN 0926-3373

Mishra, V.S.; Mahajani, V.V. & Joshi, J.B. (1995) Wet Air Oxidation. *Industrial & Engineering Chemistry Research,* Vol.34, No.1, pp. 2-48, ISSN 0888-5885

Munter R. (2001) Advanced oxidation processes - Current status and prospects. *Proceedings of the Estonian Academy of Sciences, Chemistry,* Vol. 50, No. 2, pp. 59-80, ISSN 1406-0124)

OECD guidelines for the testing of chemicals (1992) *Test no. 301: Ready biodegradability: 301 F. Manometric respirometry*. OECD, 1992

Oller, I.; Malato, S. & Sánchez-Pérez, J.A. (2010) Combination of advanced oxidation processes and biological treatments for wastewater decontamination - review. *Science of the Total Environment,* doi: 10.1016/j.scitotenv.2010.08.061 (article in press, No of pages 26), ISSN 0048-9697

Pala, A. & Erden, G. (2005) Decolorization of baker´s yeast industry effluent by Fenton oxidation. *Journal of Hazardous Materials,* Vol. 127, No. 1-3, pp. 141-148, ISSN 0304-3894

Pérez, M.; Torrades, F.; Peral, J.; Lizama, C.; Bravo, C.; Casas, S.; Freer, J. & Mansilla, H.D. (2001) Multivariate approach to photocatalytic degration of cellulose bleaching affluent. *Applied Catalysis B: Environmental,* Vol. 33, No. 2, pp. 89-96, ISSN 0926-3373

Rivas, F.J.; Beltrán, F.J. & Gimeno, O. (2000) Joint treatment of wastewater from table olive processing and urban wastewater. Integrated ozonation - aerobic oxidation. *Chemical Engineering & Technology,* Vol. 23, No. 2, pp. 177-181, ISSN 0930-7516

Rivas, F.J.; Beltrán, F.J.; Gimeno, O. & Alvarez, P. (2001) Chemical-biological treatment of table olive manufacturing wastewater. *Journal of Environmental Engineering,* Vol. 127, No. 7, pp. 611-619, ISSN 0733-9372

Rizzo, L.; Lofrano, G.; Grassi, M. & Belgiorno, V. (2008) Pre-treatment of olive mill wastewater by chitosan coagulation and advanced oxidation processes. *Separation and Purification Technology,* Vol. 63, No. 3, pp. 648-653, ISSN 1383-5866

Rokhina, E.V.; Repo, E. & Virkutyte, J. (2010) Comparative kinetic analysis of silent and ultrasound-assisted catalytic wet peroxide oxidation of phenol. *Ultrasonics Sonochemistry,* Vol. 17, No. 3, pp. 541–546, ISSN 1350-4177

Sangave, P.C.; Gogate, P.R. & Pandit, A.B. (2007) Combination of ozonation with conventional aerobic oxidation for distillery wastewater treatment. *Chemosphere,* Vol. 68, No. 1, pp. 32-41, ISSN 0045-6535

Sena, R.F.; Tambosi, J.L.; Genena, A.K.; Moreira, R.F.P.M.; Schröder, H.F.R. & José, H.J. (2009) Treatment of meat industry wastewater using dissolved air flotation and advanced oxidation processes monitored by GC-MS and LC-MS. *Chemical Engineering Journal,* Vol. 152, No. 1, pp. 151-157, ISSN 1385-8947

SES (1994) *The UV/Oxidation Handbook,* Solarchem Environmental Systems, Markham, Ontario, Canada

SFS 3020 (1979) *Determination of chemical oxygen demand (COD_{Cr}) in water. Oxidation with dichromate,* Finnish Standards Association SFS, Helsinki, Finland

SFS-EN 1484 (1997) *Water analysis. Guidelines for the determination of total organic carbon (TOC) and dissolved organic carbon (DOC),* Finnish Standards Association SFS, Helsinki, Finland

SFS-EN 1899-1 (1998) *Water quality. Determination of biochemical oxygen demand after n days (BODn). Part I: Dilution and seeding method with allylthiourea addition,* Finnish Standards Association SFS, Helsinki, Finland

Sreethawong, T. & Chavadej, S. (2008) Color removal of distillery wastewater by ozonation in the absence and presence of immobilized iron oxide catalyst. *Journal of Hazardous Materials,* Vol. 155, No. 3, pp. 486–493, ISSN 0304-3894

Sroka, E.; Kamfliski, W. & Bohdziewicz, J. (2004) Biological treatment of meat industry wastewater. *Desalination,* Vol. 162, No.10, pp. 85-91, ISSN 0011-9164

Staehelin, J. & Hoigne, J. (1982) Decomposition of ozone in water: rate of initiation of hydrogen ions and hydrogen peroxide. *Environmental Science and Technology,* Vol. 16, No. 10, pp. 676-681, ISSN 0013-936X

Szpyrkowicz, L.; Juzzolino, C. & Kaul, S.N. (2001) A comparative study on oxidation of disperse dyes by electrochemical process, ozone, hypochlorite and Fenton reagent. *Water Research,* Vol. 35, No. 9, pp. 2129-2136, ISSN 0043-1354

Verenich, S. (2003) *Wet oxidation of concentrated wastewaters: process combination and reaction kinetic modeling,* PhD thesis, Lappeenranta University of Technology, Lappeenranta, Finland, ISBN 951-764-739-5

Wang, P.H.; Yap, P.S. & Lim, T.T. (2011) C-N-S tridoped TiO_2 for photocatalytic degradation of tetracycline under visible light irradiation. *Applied Catalysis A – General,* Vol. 399, No. 1-2, pp. 252-261, ISSN 0926-860X

Yavuz, Y. (2007) EC and EF processes for the treatment of alcohol distillery wastewater. *Separation and Purification technology,* Vol. 53, No. 1, pp. 135-140, ISSN 1383-5866

Zayas, P.T.; Geissler, G. & Hernandez, F. (2007) Chemical oxygen demand reduction in coffee wastewater through chemical flocculation and advanced oxidation processes. *Journal of Environmental Sciences,* Vol. 19, No. 3, pp. 300-305, ISSN 1001-0742

Zinatizadeh, A.A.L.; Mohamed, A.R.; Najafpour, G.D.; Hasnanin Isa, M. & Nasrollahzadeh, H. (2006) Kinetic evaluation of palm oil mill effluent digestion in a high rate up-flow anaerobic sludge fixed film bioreactor. *Process Biochemistry,* Vol. 41, No. 5, pp. 1038-1046, ISSN 1359-5113

17

Corrosion in the Food Industry and Its Control

Benjamín Valdez Salas, Michael Schorr Wiener,
Margarita Stoytcheva, Roumen Zlatev and Monica Carrillo Beltran
Universidad Autonoma de Baja California.
Instituto de Ingeniería, Mexicali, Baja California,
Mexico

1. Introduction

From ancient times human beings have survived on a diet consisting on a relatively few species of plants and animals, domesticated and then cultivated and grown. Three cereals, wheat, rice and corn, supply the need of human energy, protein and vitamins requirements for the network of metabolic processes to maintain normal body function and temperature. In a prehistoric era, indigenous peoples all over the world were moving in inhospitable grounds obtaining their daily sustenance by hunting and gathering fruits, seeds and roots.

Actually, the three largest markets worldwide, according to their production extent, the number of consumers and their economic and social significance are the food, energy and water markets. Furthermore, their increasing scarcity and soaring prices lead to a global critical situation. The demand for increased food supply is related both, to population increase and personal and family income. Consequently, the food market is the largest one, including all the inhabitants of this planet, about seven billion, since everyone eats!

The organized food production and supply starts with the agricultural revolution, developed and implemented in the fertile valleys of the rivers Tigris-Euphrates in Mesopotamia and the Nile in ancient Egypt. Afterwards the food industry expanded, avoided widespread famine and ensured that sufficient food is supplied for all people to stay healthy. Current food research had been largely stimulated by rapidly growing world demand but technological advances in food processing, equipment and production plants have also contributed. A most significant aspect in the search of new nutritional food is the requirement for adequate protein in regions where meat and fish are not available. Additionally, advances in the food industry (FI) such as preservation, packaging and storage facilitate food delivery and minimized health hazards. Space flight conditions have stimulated the creation of space food which meets highly demanding standards for conservation and to be ready for easy digestion e.g. solid dehydrated food easily converted into liquid or paste food.

Techniques for preserving food from natural deterioration following harvest or slaughter dated to prehistoric times applying drying, salting, fermentation of milk and fruits and pickling of vegetables. Modern techniques include canning, freezing, dehydration, cooking under vacuum and addition of chemicals. The principal causes of food spoilage are growth of microorganism, enzyme action, oxidation and dehydration. The economic and social relevance of the FI is evident by the diverse international and national professional associations, R & D institutions, regulation and standards agencies and potent multinational industrial enterprises involved in all aspects of food science, engineering and technology (S.O. Jekayinfa et al, 2005). It includes authorities from government, industry and academia that address progress crucial for national and global prosperity. Lately, the threats of bioterrorism by poisoning food have pushed the FI to the front of efforts by safety and security organizations to prevent these hazards events.

Among others, aluminium, tin, cooper, titanium and mainly stainless steel (SS) are widely used in FI for the manufacture of processing, production, storage and transportation equipment and machinery. Modern science and technology have developed an extended range of materials with increased corrosion resistance, improved mechanical strength, easier forming and fabrication, weldability and health friendly features. The workhorses of food processing industry are UNS S30400 and S31600, but other austenitic and duplex SS are in useful service in food plants.

Food consists mainly of proteins, carbohydrates and fats. Processed foods contain diverse aqueous solutions, syrups and additives, used to improve food appearance, quality and preservation. They have a wide pH range and salt, water and vinegar content, that impact on food corrosivity.

Many cleaning and sanitation agents are employed to remove bacteria, scale, fouling and corrosive biological and mineral deposits. They include alkaline, acidic, strong or weak oxidizing and reducing chemicals to ensure a high hygiene level. This great variety of corrosive environments and aggressive chemical agents require the use of Corrosion Resistant Alloys (CRAs). SS are the obvious choice of the FI to prevent equipment damage and food contamination.

2. General corrosion considerations

FI is a complex network conformed by primary producers and the industries linked to them. The plants involved in the processes of products and packaging must preserve the capital invested and minimize operation costs. The physicochemical characteristics of processed foods permit them to achieve different corrosivity grades depending on the type of content (Valdez et. al., 2004, Jellesen, et. al., 2006).

Three types of foods are recognized according to their corrosivity:

- Non corrosive: milk, meat, fish, oil, fat, cereals
- Mild corrosivity, foods with pH 6 – 7 and less than 1% of salt: dairy products, fruit syrups, wine, carbonated sweet drinks, beer, soups, canned meat.
- High corrosivity, foods with pH 3 – 5, such as citric fruit juices, jams and acidic canned fruits or hot gravies, sauces and dressings, vegetables and fish pickled in brines with 1 – 3 % salt.
- Many cleaning, disinfection and sanitation agents are employed to remove bacteria, scale, fouling and corrosive biological and mineral deposits. They include alkaline, acidic, strong or weak oxidizing and reducing chemicals to ensure a high hygiene level.

Acetic acid, used as an acidifying agent for the production of pickles and canned vegetables is mildly corrosive to SS. Three types of cleaning and sanitation agents are applied in the FI:

1. Alkaline: such as caustic soda ($NaOH$), alkali phosphates (Na_3PO_4), sodium carbonate and bicarbonate (Na_2CO_3, $NaHCO_3$).
2. Acidic: phosphoric, citric and sulphamic acids.
3. Oxidizers: chlorine, nitric acid, ozone, hypochlorite, hydrogen peroxide (H_2O_2)

This great variety of corrosive environments and aggressive chemical agents require the use of Corrosion Resistant Alloys (CRAs). SS are the obvious choice of the food industry to prevent equipment damage and food contamination.

Food main constituents.- Foods are constituted by two essential groups of substances: those with nutritional values: proteins, carbohydrates and fats and the components imparting particular tastes: salt ($NaCl$) for saltiness, acids for sourness or acidity and sugars for sweetness. Few peculiars foods have a bitter taste e.g. almonds. A central component of food is water as in fruits, vegetables, meat, eggs, milk, etc.

Water.- The vegetables and ripe fruits contain a considerable amount of water between 70% and 90%. Since water is used as a medium for food preparation e.g. cooking , confectionery, hot and cold beverages, even cooked foods contain water. The water behaviour and its interaction with food depends on its physicochemical properties, its molecular and ionized structure and its chemical bonding. Water is considered as the universal solvent since all chemical substances have a finite solubility in water ranging from acids, bases, salts, sugars, alcohols, etc. Water is dissociate into ions: $H_2O \rightarrow H^+ + OH^-$ forming the hydronium ion H_3O^+ a bonding between an hydrogen ion H^+ and a water molecule, these are the ions involved in corrosion of metallic materials used for the construction of food processing equipment.

Acidic metallic salts undergo hydrolysis in water:

$$2FeCl_3 + 3H_2O \rightarrow 6HCl + Fe_2O_3, \qquad (1)$$

forming an acidic and corrosive environment.

Water is the medium for countless chemical and biological reactions. With the exception of petrochemical and combustion reaction, most chemical processes occur in aqueous systems, including vegetal and animal respiratory and metabolic functions. Food preparation involving dehydration, softening, takes places in water. The operation requires some energy to break hydrogen bond and form new bonds between the solute: food components and water: the solvent.

Salt.- Table salt ($NaCl$) is encountered in many natural environments: seawater, natural brines, enclosed seas, e.g The Dead Sea, Israel and the Salt Lake, Utah, USA, with a salt concentration of about 300 g/L; in aboveground and underground mines found as rock salt. An aqueous salt solution is decomposed by electrolysis into several useful products: metallic sodium, hydrogen, sodium hydroxide and chlorine, all important raw materials for the chemical processing industry (CPI). The meat packing, sausage making, fish curing and food conservation employ salt as a preservative or seasoning or for both purposes. Up to the 19th Century, during the invention of industrial food freezing and cannery; commercial and navy ships conserved meat in wood barrels full of solid, granular salt for their long voyages since putrefaction bacteria cannot live in salt. Salt is a hygroscopic substance, absorbing atmospheric moisture, forming a concentrated salt solution or slurry, corroding steels and deteriorating ceramic materials such a masonry bricks. Salt is part of mankind history and tradition. Modern and ancient words, in many languages, are derived from salt: salary, salad, salutary, salud (Spanish for health) . The Bible describes hospitable reception of visitors and travellers offering water, salt and bread (M. Schorr et al., 2011).

Acids.- Acids in water solution impart a sharp sour taste to food but also a corrosive action on steel. Therefore steel for food containers should be protected with corrosion-resistant material or fabricate from stainless steels or plastic materials. The word acid derives from ancient European language meaning sour. Acids and bases play a large part in industrial chemistry and particularly in chemical cleaning in food processing equipment. Water, the "universal" solvent, dissolves acids and bases. The acidity of foodstuffs is determined by measuring their pH. The majority of the vegetables and fruits consumed by human beings are acidic, with a pH between 5 and 2; in particular citrus fruits displaying pH values between 2 and 4. Two acids: acetic acid (vinegar CH_3-COOH) and phosphoric acid (H_3PO_4) are widely applied in the FI, the first for the conservation of vegetables and the second for acidifying beverages, drinks and juices.

Sugar.- Sugar is a sweet substance obtained from the juices of various plants, chiefly the sugarcane and the sugar beet but also from date palm and the maple tree. It is an article for human nutrition and a sweetener for other foods, especially beverages, confectionaries and canned fruits. It is a carbohydrate, called sucrose and the prime source of energy in the human diet, needed to maintain body temperature and activity. Most of the sugar is consumed in the form of white, granulated sugar, produced in refineries processing sugar cane or beets. Liquid concentrated sugar syrup is used for sweet food manufacture purposes, such as confectionary and candies. Sugar dissolves easily in water forming a molecular solution without electrical conductivity and therefore slightly corrosive since there is some amount of dissolved oxygen.

Environments in FI.- The pH values associated with foods ranges from 3 to 8 that corresponds to a mildly acidic water and has an acceptable corrosive behaviour for aluminium, stainless steel and tin. Table 1 shows the pH ranges for different foodstuffs branch in the FI. Nevertheless, the aggressive ions hypochlorite and chloride could be concentrated by evaporation of the cleaning solutions on specific localized areas of the metal surface, inducing pitting corrosion. The combination of both anions chloride and hypochlorite can lead to a more positive redox potential (E^0 = +0.9 V vs Satured Hydrogen Electrode (SHE)) and to synergize the pitting corrosion process of stainless steel.

Foodstuff	pH range
Vegetables	3.0 to 6.0
Fruits	2.0 to 5.0
Bakery	5.0 to 6.5
Meat	6.0 to 7.0
Fish	5.5 to 6.0
Dairy	5.0 to 6.5
Beverages	2.0 to 5.5

Table 1. Ranges of pH values for different foodstuffs

The cost of corrosion.- The food processing industry is one of the largest manufacturing industry in the United States of America, accounting for approximately 14 percent of the total U.S. manufacturing output. The total food sales including the supermarket and traditional food store, restaurant food and drinks, non-traditional food store and convenience stores sales were 1,638 billion US dollars in 2010 (Plunkett Research, 2010). Assuming that SS consumption and cost in the FI is entirely attributed to corrosion, total annual direct cost corrosion was $ 2.1 billion, according to National Association of Corrosion Engineers report for corrosion costs in 1999. This cost includes SS usage for beverage production, food machinery, cutlery and utensils, commercial and restaurant equipment and appliances.

3. Public health and hygiene

Food can be an important source of disease-causing organisms, the investigations on risk management in food handling, diet and health (Buchanan et. al., 2001, and Oddy et. al., 1985), revealed an association between human diseases and food contaminated by flies. Food can contain beneficial bacteria, but contamination with harmful microorganisms from external reservoirs must be prevented not only by removal of food residues and disinfection but also by the modification of surfaces specially designed to avoid sites for the development of bacterial colonies, such as accumulated corrosion products or degraded protective coatings. These surfaces must not introduce toxic substances or influence flavour. Materials that fulfil these requirements are glass, earthenware, some plastics, and metals like SS and aluminium alloys. Glass is sensitive to shock and fragments are sharp, while inexpensive plastics are not heat resistant and, in thin gage, they are permeable to gases. Metals do not have these disadvantages, but they must be corrosion resistance in food environments.

The modern FI combines the application of the latest production processes, the use of cleanable corrosion resistant SS equipment and computerized information with an efficient, environmentally sound approach to meet the needs of food products consumers everywhere. Cleanability is important in relation to taste, colour, odour and contamination of edible products such as milk, processed and canned foods and alcoholic beverages. This is of particular hygienic importance in food handling. Food processing involves operations by which raw foodstuff are suitable for human consumption, it includes the basic preparation of food, preservation and packaging techniques. Many innovations have resulted in new products such as concentrated fruit juice, freeze-dried coffee and instant food.

Sterilization is one of the most important aspects of hygiene and sodium hypochlorite is a disinfectant widely used to kill microorganisms through an oxidizing reaction that yields sodium chloride as final product.

$$NaClO \rightarrow NaCl + O^* \qquad (2)$$

Where O^* represents oxygen species capable to kill biological material by oxidation. Both sodium chloride and sodium hypochlorite can be easily removed from surfaces used in the FI by rinsing with water, but it is not permissible to leave any residue because residual chlorides and hypochlorites can be concentrated by evaporation and induce localized pitting corrosion

4. Microbiological induced corrosion in the food industry

Microorganisms are primitive unicellular organisms capable to live in colonies at aqueous media. They are diverse and for the interest of this chapter it is appropriate to mention;

bacteria, fungi and algae. These microorganisms are not corrosives by themselves, but due to the impossibility to produce their nutrients by photosynthesis, they must metabolize organic content present in the media, producing other substances that are returned to the solution. The nutrients needed include a source of energy such as glucose that supplies carbon and minerals containing phosphorous, sulfur and nitrogen for the cell structure of bacteria. The growth, development activity and death of bacteria are also influenced by the pH and concentration of some ions, which establish the best conditions. Bacteria are diverse and can live in extreme conditions such as high concentration of chlorides (halophilic), higher temperatures (thermophilic) or to realize different chemical transformations; reduce sulfates, reduce sulfides, use hydrogen or reduce nitrates, etc. Bacteria can participate in corrosion processes in different ways: changing the environment by replacing a substance by another, covering partially the metallic surface with their biofilms creating local corrosion cells or inducing the corrosion process by depolarization of hydrogen at the metal surface.

5. Stainless steel in the food industry

The corrosion resistance of SS is due to the presence on its surface of a protective, passive oxide film which is stable, tightly adherent and very thin: about 50 Å. When broken, the film regenerates itself by exposure to air or moisture. A broad summary of the utilization of various categories of SS in the FI is given in Table 2.

SS UNS * number	Chemical composition % w/w				Characteristics/Uses
	Cr	Ni	Mo	C_{max}	
Martensitic and Ferritic chromium steels					
S41000	11.5-13.5	---	---	0.15	Hardenable by heat treatment. Turbine blades, valve trim, freezer blades.
S41600	12-14	---	---	0.15	Easy machinable. Valve stems, plugs and gates.
S42000	12-14	---	---	0.35-0.45	Hardenable by heat treatment. Cutlery, cladding over steel.
S44000	16-18	---	---	0.6	Very hard. Pumps, plungers, gears, bearings.
S43000	16-18	---	---	0.2	Good corrosion resistance. Structural purposes
S43100	15-17	1.5-2.5		0.16-0.17	
Austenitic chromium-nickel steels					
S30200	17-19	8	---	---	Good corrosion resistance. General purpose.
S30400	18-20	8-12	---	0.08	Good corrosion resistance. General dairy equipment.
S31000	24-26	19-22	---	0.25	Heat resistance. High temperature equipment.
S31600	16-18	10-14	2-3	0.10	Superior corrosion resistance. The workhorse of the dairy industry

***UNS: Unified Numbering System**

Table 2. Stainless steel for use in the food industry

Selection of SS is an activity that requires the participation of specialized expertise in materials technology, coupled with a sound knowledge of the chemical and physical characteristics of the food products (S.H. Zhang and B. Monitz, 2006, R.M. Davidson et al., 1987). Ferritic low carbon SS UNS S44400 has the advantage to be immune to Stress Corrosion Cracking (SCC) while its resistance to localized pitting and crevice corrosion is comparable to UNS S31600 austenitic SS. The austenitic SS group is the most popular group used for the construction of infrastructure in the FI, mainly the UNS S30400 and S31600 due to their good corrosion resistance properties. The 400 series, e.g., UNS S41000, is specified for applications such as pump impellers, plungers, cutting blades, scrapes and bearings. Corrosion resistant properties of the different SS depend on the alloying elements such as chromium, nickel and molybdenum. The duplex SS is a group that contains similar amounts of ferrite and austenite in their microstructure are increasing their use in the FI in the last years. These steels have high strength and good corrosion resistance compared with the austenitic SS, but one concern is that duplex SS tend to form brittle intermetallic phases and the service temperature must be limited to a maximum of 315 C.

In the FI, the equipment is manufactured from SS with a finely polished surface that tends to have a better corrosion resistance regarding to non-polished ones, and is easy to clean and to keep clean. A smooth surface is less susceptible to an accumulation of deposits, fouling or biofouling, which often become focal points for localized corrosion. Seamless and welded austenitic SS sanitary tubing and pipelines intended for use in the FI should have special surface finishes, as the ASTM A270 standard specifies. The finishing process results in SS with varying smoothness, brightness and light reflectivity.

The standard designations for SS finishes are classified by mill forms. For instance, standard mechanical sheet finishes comprise unpolished or rolled finishes: No. 1; No.2D and No 2B (dull and bright respectively) and polished finishes, since called mirror finishes from No. 3 to No. 8, with a most reflective surface. No. 4 finish is a general-purpose polished finish commonly used for food equipment. In addition to the mechanically rolled and polished finishes, other types of finishes are achieved by chemical, electrochemical or thermal treatment and sometimes these techniques are applied for rehabilitation of damaged SS surfaces.

6. Cleaning and sanitation of stainless steel

SS are used extensively by the FI to ensure purity and quality of the products and cleanness and durability of the equipment. Hygiene and sanitation are basic requirements of this industry, and the metallic surfaces are cleaned, maintained or restored by manual or automated mechanical or chemical cleaning, the later referred to as ´cleaning in place´ (CIP), without disassembling the equipment.

The different foodstuffs have a considerable quantity of nutrients representing an ideal media for bacterial growth, which can induce corrosion processes on the metallic SS surfaces by microbiologically induced corrosion (MIC) or corrosion under acidic, salty, fatty fouling deposits or calcareous scaling from cooling water on heat transfer surfaces. The concentration of bacterial population present on a surface is the parameter to measure in order to establish the ability of such surfaces to be disinfected. Several studies have been published showing that the SS have very similar characteristics for disinfection regarding to aluminum, rubber, polyester, etc., (J.T. Hola and R.H. Thorpe, 1990). Cleaning is a mechano-chemical operation for sanitation and disinfection of the machinery used for the food processing, to prevent contamination of products and damage to the SS surfaces. The CIP system evolved from recirculation cleaning solutions in tubing, pipelines and equipment to an update automatic system with actuated valves, controls and timers. The easiness in which SS can be cleaned allows fast turnarounds of tankers, coolers, pumps, heat exchangers and pipes, particularly during high production cycles. The SS alloys resist corrosive attack from chemical specialties formulated for cleaning and sanitation, such as chlorine, alkalis, mixed acids, organic ammonium quaternaries, halogenated hydrocarbons and detergents. To obtain better luster finishes, soft clothes or sponges are used in order to avoid scratching highly reflective surfaces. After cleaning, the surfaces are rinsed with hot water, dried and exposed to air to rebuild the protective, passive oxide layer of SS. Mechanical conveyors constructed of SS for moving and transportation of food raw materials and processed products containers are easy to maintain clean and sanitized. Water or food steam cleaning and visual inspection can verify that all products residue has been removed and that surfaces do not become stained, corroded or abraded. Currently environmental concerns are resulting in some modifications of CIP systems to allow water reuse and to minimize waste discharges.

Materials scientists at the University of Birmingham, UK, made stainless steel surfaces resistant to bacteria in a project funded by the Engineering and Physical Sciences Research Council. By introducing silver or copper into the steel surface (rather than coating it on to the surface), researchers developed a technique that not only kills bacteria but the surface is very hard and resistant to wear and tear during cleaning. This type of modified steels could be used in the FI and kitchens (ASM News, 2011).

7. Food industry and corrosion

The corrosion processes occurs due to different factors or conditions that define the corrosive characteristics of the media in which the SS elements of the FI are exposed. To illustrate the corrosion characteristics of the FI, dairy, beverage and canned food products have been selected and are described as follows:

7.1 The dairy industry

Dairying has been an agricultural practice before the dawn of history. Milk and its derivatives are a vital part of the diet for human beings. The dairy industry (DI) collect, treat and distribute milk that is used as feedstock for the manufacturing of other products particularly cheese, butter, yogurt and creams. The corrosion control in DI plants depends on the nature of these products and the processes involved during their production. DI has a large capital investment in infrastructure, which is exposed to a mild corrosive environment. Nevertheless, a considerable effort is expended to reduce corrosion losses by selecting corrosion resistant alloys (CRA´S), in particular SS and by the implementation of preventive, corrective and preferably, predictive maintenance. At first, milk looks as a simple white liquid, but in fact, is a complex emulsion/suspension mixture of lactose, whey proteins, fat globules, minerals and vitamins among others, in aqueous solution. The fats are present as an emulsion of tiny globular particles, while casein, the most abundant protein combines with calcium ions and remains dispersed in the liquid phase as a colloidal suspension. Fermentation processes of milk are possible due to its content of sugar in the form of lactose, which used as a nutrient by bacterial cultures naturally presents in the milk. The bacteria *Lactococcus lactis* and *Lactococcus cremoris*, convert lactose into lactic acid during souring. The minerals presents are calcium and phosphorous and the vitamins are fat soluble (A and D) and water soluble (B_1, B_2, B_{12}, C and M), these last are unstable when milk is heated modifying the corrosive conditions of the media. The composition of different cow species milk is shown in Table 3.

Milk is fairly neutral in reactions, although lactic acid is present; the lactic acid content is increased by natural souring or by the artificial souring necessary for cheese and butter manufacture. This is perhaps the only constituent of milk which is responsible for metal attack. For a critical FI, such as the DI, the equipment and accessories must be constructed with a smooth, impervious, corrosion - resistant material which can be cleaned easily. Obviously the choice material that fulfills these requirements is the SS. It will be important to take care and to follow the adequate procedures in the use of SS alloys, in order to avoid affecting the odor, flavor and color of milk products.

COW:	FAT %	PROTEIN %	LACTOSE %	ASH %	TOTAL SOLIDS %
Brown Swiss	4.0	3.6	5.0	0.7	13.3
Ayrshire	4.1	3.6	4.7	0.7	13.1
Cebu	4.9	3.9	5.1	0.8	14.7
Jersey	5.5	3.9	4.9	0.7	15.0
Holstein	3.5	3.1	4.9	0.7	12.2
Guernsey	5.0	3.8	4.9	0.7	14.4

Table 3. Composition of milk obtained from different cow breeds.

The metallic alloys to be used in the DI operations where the milk and its products will be in contact with them, must have low toxicity and physiological tolerance, corrosion resistance to milk, products and other chemical food additives and cleaners, and availability of the desirable profiles in the market at acceptable costs. The main SS equipment installed in a modern dairy plant is listed in Table 4.

Milking machines	Homogenizers
Vacuum pumps	Pasteurizers
Centrifugal pumps	Tubular and plate heat exchangers
Agitators and Mixers	Vacuum evaporators
Milk coolers	Milk dryers
Bulk milk storage tanks	Spray milk dryers
Clarifiers	Conveyors
Cream separators	Vessels for cleaning solutions
Centrifuges	Piping and tubing equipment fittings
Bulk milk tankers	CIP spray devices

Table 4. Typical stainless steel dairy equipment

Electric energy is used for heating in small equipment, and for instrumentation and control of process machinery. Steam and hot water required for heating, pasteurization, sanitation and production of evaporated and dried milk is supplied by boilers. A wide variety of equipment for cooling, pasteurization, homogenization, transportation, etc, is fabricated from austenitic SS mainly USN S30400 and S31600. SS tubing and piping of different nominal diameter and wall thickness fitted with standard SS fittings are widely used for moving milk and its products throughout the plant.
The standards and specifications for design, construction, operation and sanitation of equipment for producing, handling and processing milk products have been formulated jointly by the International Association of Milk, Food and Environmental Sanitarians, the United States Public Health Service and the Dairy Industry Committee. These three associations Sanitary Standards are recognized and accepted as the official guides for the sanitary design of equipment for the DI. For the surfaces not in contact with the product the SS are preferred over other less expensive materials, nevertheless, if for economical reasons, other metals are used in the proximity to SS, precautions must be taken to avoid galvanic corrosion problems. Most of the SS are passivated and are less reactive than the majority of the engineering metals and can stimulate attack (D.E. Talbot and J.D. Talbot., 2007)

7.2 The beverage industry

A great variety of beverages includes beer and wine made by fermentation; distilled spirits requiring alcoholic fermentation and distillation, and soft drinks, carbonated and non carbonated, consisting of water flavored with sweeten natural or artificial sweetened syrup. The fabrication of beer requires barley, hops, yeast, and water as main raw materials. A series of by-products result from the processed yeast for health foods, animal feed from spent barley and the used hope is used as a fertilizer.
Beer is produced by fermentation of germinated barley and wine of the fermented juice of grape. Soft drinks are non-alcoholic beverage; they include mineral water or treated water containing sweetening agents, edible acid, flavors and sometimes fruit juices. Carbon dioxide gas gives the beverage its sparkle and tangy taste and prevents spoilage. The production of these beverages involves the use of great quantities of water in the cleaning, storage and bottling procedures. Many beverages are acidic and aggressive to carbon steel requiring corrosion –resistant SS equipment. In addition, the wet, damp and high humidity conditions contribute to plant corrosion and premature equipment failure. Application of SS will prevent the occurrence of these noxious corrosion events.
At the beginning, breweries were small places with a low production volume and their operations were carried out in wooden vessels and the first metallic components used were made form carbon steel and copper. Nowadays, many of the plants for beer, wine or sweetened beverages are large and most of the equipment is manufactured from UNS S30400 and include tanks, piping, plate coolers, pasteurizers and conveyors for bottle and can filling. Sometimes scaling occurs with chloride content in the hard, salt deposits. Remedial action involve acid treating of the water to remove dissolved carbonates and cleaning with mild acids such as phosphoric or citric to remove dissolved carbonate scaling. SS is the main material for fabrication equipment of the beverage industry infrastructure to avoid the formation corrosion products that could contaminate the beverage and affect its taste. The producing equipment include tanks, vessels, steamers pasteurizers, coolers,

blenders, filters, working tables, packaging and vending machines (J. Beddoes, 1999, H.S. Kathak, 2002).

From a point of view of general corrosion resistance aspects, beer and their raw materials are in the pH range from 4 to 5. The S30400 SS low carbon content and stabilized variants are resistant to this pH range under cold and fully aerated conditions and are used for fermenting vessels, grist hoppers, holding tanks and the piping accessories, fabricated with welded sheet or plate. Under conditions of hot or boiling liquids the environment turns more aggressive because of the higher temperature and the depletion of the oxygen content. This inconvenient is solved by the choice of a more strongly passivating SS; S31600 is used for the mash tuns, whirlpool separators and heat exchanger plates in the wort cooler. SS30400 is also satisfactory for casings used to hold the mineral lagging used as isolating element to conserve the heat in hot zones of these vessels.

The SCC is a common cause of corrosion failure in vessels handling hot liquids and it can occur due to external or internal sources. The external source situation appears when chloride ion is leached from the lagging by condensing water vapor and the SS surface has not an appropriate anti-chloride coating or interposed vapor sealing barriers of aluminum foil between successive layers of lagging. In the case of internal source, the chloride ions are contained in the water used for brewing or residues of disinfectants and from acids used in descaling. The chlorides become concentrated locally after thermal cycling or evaporation, which is a well known mechanism for SCC. The internal source of SCC can be controlled by deionizing the water and prohibiting the use of hypochlorite solutions as disinfectants.

In the wine industry, bisulfites are widely used to prevent non desirable biochemical process. The most corrosive agent found in wines is the sulfur dioxide that forms sulfurous acid. Sulfur dioxide is usually present in wine either bound to carbonyl or unsaturated compounds and/or phenol derivatives or free, as HSO_3^- and SO_2 (Araújo C., et. al., 2005, Ruiz-Capillas C., et. al., 2009). It is added to wine during its production as small amounts of sulphur dioxide (Preservative 220) or potassium metabisulphite (Preservative 224). Sulfur dioxide provides three important properties in winemaking:

1. SO_2 has antiseptic qualities.
2. It helps to protect wine from the deleterious effects of oxygen.
3. Destroy the enzyme that causes enzymatic browning in juice (similar to what happens to apples when they are sliced and exposed to the air). Without SO_2, wine would likely be brown or amber in color, smell oxidized (or have a sherry-like aroma), and probably be ruined by bacterial spoilage.

When sulfur dioxide is added to wine it is rapidly converted into bisulphite ion (HSO_3^-). Approximately half quickly combines with other wine constituents and is bound within complexes formed with aldehydes, ketones, phenolics, etc. The remaining sulfur dioxide/bisulfite is thus free in solution and it is this free portion that is readily available as an antioxidant. If oxygen is dissolved into a wine the sulfur dioxide levels will be reduced as they oxidize to sulfates (SO_4^{2-} and HSO_4^-). If SO_2 is greatly in excess, it can also produce a pungent aroma in white wines, considered by most to be a fault. The aroma is best described as that of a match that has just been struck. Many people have trouble smelling the sulfurous aroma of SO_2, but instead perceive it as an irritation of the membranes of the nose. High SO_2 can also render the palate of the wine harsh, metallic and frequently bitter.

In addition citric, tartaric, tannic, acetic and malic acids are also found in wines. SS is corrosion resistant to all of them in the concentration ranges in which they usually are found in wine fabrication steps (B.J. Connolly, 1971 in ASM).

7.3 Canned food products

The practice of preservation of vegetables and fruits by drying or salting can be traced to prehistory. Today, major processes for food preservations are cooling, freezing, dehydration, candying and canning. The main concern is for food quality, conservation and the prevention of spoilage. For cans fabrication is used tin one of the oldest available metals with a non toxic nature. The cans have been made from tin since 80 years ago when refrigeration of food was not easily available. According to their convenience the use of cans has been focused on food and beverages dispensers creating an intense competition between manufacturers not only for tinplate cans but also other materials, notably chromized steel, aluminum alloys and polymers.

Modern tinplate for cans is getting by continuous electrodeposition on steel strips, chemical conversion coating and finish polymeric coating. The manufacturing is made following three different types classified according to the methods of fabrication: a) The three piece can, b) The two piece draw/redraw (DRD), and c) The draw/wall ironed (DRW). In all of them the tinplate process is customized to economize and fulfill the requirements for corrosion resistance, filling and mechanical properties.

Aluminum cans are also widely used as containers for pressured beverages which are designed to have a minimum of material at the strength required for operation. The walls acquire rigidity by effect of the internal gas pressure from the content, but the bottom base support and the top are made from thicker metal and formed into an internal dome.

The processed fruits and vegetables are hermetically sealed and sterilized by heat and stored in glass jars and aluminum or coated steel cans. Sterilized canned food, include all types of vegetables under acidic or salty conditions and fruits in acidic, sweet syrup. The canning process consists of several stages: vegetable and fruit cleaning, filling the containers, closing and sealing, sterilizing the canned product, labeling and warehousing the finished goods. All these operation, carried out under excessive, humid conditions; require corrosion resistant equipment and machinery, currently fabricated from SS such as S30400 and S31600.

8. Corrosion control

Corrosion prevention and control require the application of appropriate methods and techniques from the early stages of design through the construction, erection and operation of the food processing plant.

Today, the main and fastest source of information on corrosion control of industrial equipment, plant and facilities is the Internet. Computer-based expert systems dealing with various aspects of corrosion prevention and control for many industries are listed in Roberge's Handbook. It is very important at the time to select an anti-corrosion material, ensure that we have investigated that the choice material is the most appropriate for the expected type of corrosive environment occurring in the plant (R. Nash, 2007).

Data on corrosion resistance of SS is displayed on NACE International's website (www.nace.org). Questions on corrosion prevention and control are answered in the NACE

Corrosion Network. SS for food processing industry is reported by the International Association of Food Industry Suppliers (www.iafis.org).

The Nickel Development Institute (www.nidi.org) presents info on the use of SS and conducts workshops on related subjects. The SS Appeal (SSA), which groups SS equipment suppliers, launched a website (www.stainlessappeal.com) to promote the application of SS. The International Association of Milk, Food and Environmental Sanitarians (www.iamfes.org) provides food safety professionals worldwide with a forum for exchange of information on sanitation practices. Many producers of SS and fabricators of SS processing equipment are listed in the Stainless Steel World Buyer's Guide published in the SS World Journal, the Netherlands.

The National Sanitation Foundation (NSF) in the USA is best known for its role in the development of standards for equipment, primarily in the food service area. The NSF's materials and finishes guide refers to three zones: the food zone, the splash zone and the non-food zone. The NSF stipulates that the food zone "surface materials shall be smooth, corrosion resistant, non toxic, stable and non absorbent under use conditions. They shall not impart odor, color or taste, nor contribute to the adulteration of the food. Exposed surfaces shall be easily cleaned".

Corrosion control in the FI involves three different methods to reduce corrosion: design considerations, materials selection and protective coatings.

8.1 Design considerations

The variety of structures: tanks, heat exchangers, cooling systems, hygiene control systems, conveyors, cans, fillers, etc., used in the FI, should be designed to provide functional qualities along to all the process steps.

It is necessary to avoid risks by underdesign or excessive costs due to overdesign considerations. Knowledge, technical information, and creative engineering principles must be used in an intelligent way to avoid corrosion. There are three important aspects in general to remember when designing a system for demanding service (P. Roberge, 2008, Avery et al., 1992):

1. Design for complete and free drainage
2. Eliminate or seal weld service
3. Make it easy to inspect.

Cathodic protection and corrosion inhibitors are also employed to protect some installations for water supply, cooling system and underground tanks and pipelines. Normally they are applied as a dual anti-corrosion system together with coatings, linings or paints.

8.2 Materials selection

Corrosion resistance is the main property to be considered in the choice of materials for such plant, but the final selection must be a compromise between technological and economic factors. It is sometimes more economical to use high-priced SS that will provide long and trouble-free service than to use a lower priced material that may require frequent maintenance or replacement.

Although the materials selection by trial and error frequently works, it does not always lead to an optimal or innovative decision. The evolution of computer and software has revolutionized the availability of data bases on materials characteristics and performance. These data bases are integrated in smart system as tools for the design operation improving the materials selection required for a specific environment. The methodology created by Michael Ashby integrates in this way the materials selection to the design process (M. Ashby, 2005). In this model the material has attributes such as density, strength, electrical properties, costs, etc., that must be compared with those of real materials to find the best match.

8.3 Protective coatings

The coating of surfaces is so ancient as well as rock paintings in caves dated to be thirty thousand years old. They were done on the surface of granitic rock; an igneous, hard rock with a coarse - or medium - grained texture, rich in quartz and feldspar. The ancient painters took advantage of the natural, exfoliated, rough surface, with diminutive cracks, voids and hollows for the mechanical anchorage of its mineral-based paint applied as a paste or as slurry (B. Valdez et al, 2008). Varnishes were used by the Egyptians, polychrome sculptures were realized by the Greek hundreds of years BC and Romans used coatings for protective and decorative purposes.

The organic coatings form an excellent physical barrier between possible corrosive environments and metallic surfaces protecting the material from degradation and corrosion. The organic coatings include paints, resins, lacquers and varnishes, while the inorganic comprise enamels, glass linings and chemical conversion coatings. They are used in the FI protect the interior of cans, water tank containers, and the exterior of installations made with carbon and galvanized steel, concrete or wood.

For best results in corrosion protection coatings are commonly applied as a system composed by several layers that are classified as follow:

- Primer. Is the first coating applied to the substrate and has adhesive affinity for it and to provide if it is necessary a better adhesion for the subsequent coat layer.
- Intermediate coating. Is added when multiple thin layers are required. It is also called secondary coating when applied as top or final coat. Sometimes the intermediate is used to provide thickness to the coating film.
- Topcoats. It is the final layer applied to extend the life of the previous coatings. This film is more dense and hydrophobic in order to avoid the penetration of moisture to the underlying coats. Commonly provides an aesthetic appearance to the surface.

9. References

Araújo C, Carvalho J, Mota D, Araújo C, Coelho N. (2005). Determination of sulphite and acetic acid in foods by gas permeation flow injection analysis. Food Chem. 92: 765-770.

Ashby, M.F., (2005) Materials Selection in Mechanical Design, 3rd Edition, Oxford, U.K., Elsevier.

ASM News, http://www.asminternational.org/portal/site/www/NewsItem, consulted on July 27, 2011.

Avery R.E., and Tuthill A.H., (1992) Guidelines for the Welded Fabrication of Nickel-Containing Stainless Steels for Corrosion Resistance Services. NiDI Report 11007. Toronto, Canada: Nickel Development Institute (NiDI).

B. Valdez and M. Schorr, (2004) Stainless steels for corrosion control in the food processing industry, Stainless Steel World 2004, KCI Publishing BV, pp. 419-422, USA.

B. Valdez, J. Cobo, M. Schorr, R. Zlatev y L. Cota, (2008) Characterization of Materials and Techniques used in Mexican Rock Paintings, Rock Art Research, Vol. 25, No. 2, 131-136.

B.J. Connolly, (2006) Australian. Corros. Eng., Vol. 15, No. 11, p. 23, 1971, ASM Handbook, Corrosion: Environments and Industries, Vol. 13, USA.

Buchanan R L et al., (2001) Moving beyond HACCP - Risk management and food safety objectives, in Symposium abstracts IAFP 88th Annual Meeting, Minneapolis, USA.

D.E. Talbot and J.D. Talbot., (2007) Corrosion Science and Technology, 2nd edition, USA, CRC Press, pp. 409 - 435.

H.S. Kathak, (1992) Corrosion of Austenitic Stainless Steel. Mitigation and Monitoring, Handbook of Case Histories in Failure Analysis, ASM International, edited by Khlefa A. Esaklul, Vol. I. pp. 194-197.

http://www.plunkettresearch.com/foodbeveragegrocerymarketresearch/industryandbusinessdata/statistics, consulted on July 21, 2011.

J. Beddoes, Introduction to Stainless Steel, 3rd ed. Materials Park, OH: ASM International, 1999.

J.T. Hola and R.H. Thorpe, (1990) Bacteria retention on cleaned surfaces, J. Appl. Bacteriol., Vol. 69, pp. 599-608.

Jellesen, M. S., Rasmussen, A. A. and Hilbert, L. R., (2006) A review of metal release in the food industry. Materials and Corrosion, Vol. 57, pp. 387–393.

M. Schorr, B. Valdez, J. Ocampo and A. Eliezer. (2011) Corrosion Control in the Desalination Industry, Desalination, Trends and Technologies, pp. 71-86 INTECH, India.

Oddy D J and Millar S (1985), Diet and health in modern Britain, pp. 148-147, London, 1985.

Pierre R. Roberge, (2000) Handbook of Corrosion Engineering, McGraw Hill, USA.

Pierre R. Roberge, (2008) Corrosion Engineering. Principles and Practice, McGraw Hill, USA.

R.M. Davidson, T. DeBold and M.J. Johnson, (1987) Corrosion of Stainless Steels, Corrosion Metals Handbook, Vol. 13, 9th ed., ASM International, pp. 547-565.

Rick Nash, (2007) Battling Corrosion in the Food Processing Industry, Pump & Systems, Food Processing Section, www.pump-zone.com, consulted on July 27, 2011.

Ruiz-Capillas C, Jiménez-Colmenero F. (2009). Application of flow injection analysis for determining sulphites in food and beverages: A review. Food Chem. 112: 487-493.

S.H. Zhang and B. Monitz, (2006) Corrosion in the Food and Beverage Industries, Corrosion: Environments and Industries, Handbook Vol. 13C, 1th ed., ASM International, pp. 803-809.

S.O. Jekayinfa, P.O. Okekunle, I.G. Amole, J.A. Oyelade, (2005) "Evaluation of corrosion cost in some selected food and agro-processing industries in Nigeria", Anti-Corrosion Methods and Materials, Vol. 52 Iss: 4, pp.214 - 218

18

X-Ray Microtomography for Food Quality Analysis

Janine Laverse, Pierangelo Frisullo,
Amalia Conte and Matteo Alessandro Del Nobile
University of Foggia, Agricultural Faculty, Department of Food Science,
Via Napoli,
Italia

1. Introduction

In an effort to understand the physical and rheological behavior as well as the mechanical and sensory attributes of foods, processing focus and emphasis have shifted to the microstructure level. Microstructure elements such as air bubbles or cells, starch granules, protein assemblies and food biopolymer matrices contribute greatly to the identity and quality of foods (Aguilera, 2005). The microstructure of food has an influence over the key attributes of a product as evaluated by consumers. Many of these properties are synergetic, therefore having multiple interactions, and are poorly understood as a result. Advances in the last decade in microscopy techniques, along with an improvement in computing capabilities, has made it possible to understand a food's structure; its relation to physical properties (so called structure-property relationships) and how to engineer and control these properties (Aguilera, 2005). Structure-property relationships can strongly affect the physiochemical, functional, technological and even nutritional properties of foods. For example, with regards to solid food foams like bread, extruded cereals, biscuits and cakes, the consumer appreciation of these products is strongly linked to the texture. For texture, sensory properties of solid food foams are related to both mechanical properties and cellular structure. In this context, determining the relationships between a given mechanical property and the cellular structure is thus of prime importance. It has also been found that the structural organization of the components of cheese, especially the protein network, affect the texture of cheese: in particular the stress at fracture, the modulus and work at fracture could be predicted very well from the size of the protein aggregates (Wium et al., 2003). Cheeses having a regular and close protein matrix with small and uniform (in size and shape) fat globules show a more elastic behavior than cheeses with open structure and numerous and irregular cavities (Buffa et al., 2001). The mechanical properties of cocoa butter are strongly dependent from its morphology at microscopic level and, in particular, from the polymorphic transformation of the fat crystals and the coexistence of different polymorphic forms (Brunello et al., 2003). Thorvaldsson et al. (1999) studied the influence of heating rate on rheology and structure of heat-treated pasta dough. They found that the fast-heated samples had pores smaller than the slowly heated one and that the pore dimension affects the energy required to cause a fracture. In particular, the energy required to determine a fracture in the samples having the smallest pores was more than for the samples having the highest pores. A study carried out on the effects of grind size on peanut butter texture demonstrated that an increase of that variable decreases sensory smoothness, spreadability and adhesiveness (Crippen et al., 1989). Langton et al. (1996) studied correlations existing between microstructure and texture of a particulate protein gel (spherical particles joined together to form strands). They found that the texture, as measured with destructive methods, was sensitive to pore size and particle size, whereas it was sensitive to the strand characteristics if measured with non-destructive methods. Martens and Thybo (2000) investigated the relationships among microstructure and quality attributes of potatoes. They found that volume fraction of raw starch, volume fraction of gelatinized starch and dry matter were positively correlated to reflection, graininess, mealiness, adhesiveness and chewiness and negatively correlated to moistness. From the evidence that microstructure affects food sensorial properties, an important consideration derives: foods having a similar microstructure also have a similar behavior (Kalab et al., 1995). All foods can be analyzed in terms of their chemical composition. This gives limited information about the structure, physical state or sensorial properties. The natural building blocks of foods can be considered as water, air, carbohydrates, proteins and fats. The way in which these are structured during processing ultimately determines the functionality of the food. For example, fat content in meat products is a very important compound influencing the palatability characteristics such as taste, juiciness and texture. In addition, the visual appearance of the fat could affect the consumers overall acceptability of product and therefore the choice when selecting meat product before buying (Helgesen et al., 1998). The design of a food product must account for all these relationships whilst maintaining the high standards expected by consumers. Since microstructure is determined both by nature and processing, food processing can be considered as the way to obtain the desired microstructure (and consequently the desired properties) from the available food components (Aguilera, 2000). As a consequence, knowledge of microstructure must precede the regulation of texture (Ding and Gunasekaran, 1998) and other food attributes. It is also possible to obtain microstructural information by studying mechanical and viscoelastic properties of foods. A food sample submitted to mechanical tests gives rise to a force-time curve from which several parameters related to microstructure can be extrapolated: hardness, cohesiveness, springiness, chewiness, gumminess, stickiness (Martinez et al., 2004). When submitted to a stress (under compression, tension or shear conditions), food samples suffer a strain. The elastic modulus or Young modulus of the analyzed sample can be obtained from the stress-strain curve (Del Nobile et al., 2003; Liu et al., 2003). The viscoelastic properties of a food can be expressed in terms of G', G'' and tan δ parameters. G' takes into account the elastic (solid-like) behavior of a material, G'' is a measure of the viscous (fluid-like) behavior of a material and tan δ represents the ratio between G'' and G'. These parameters can be evaluated by performing dynamic-mechanical and rheological tests (Kokelaar et al., 1996; Brunello et al., 2003; Wildmoser et al., 2004; Ross et al., 2004). Therefore, relationships between microstructural and mechanical properties can therefore be analyzed by means of directionally dependent morphological parameters.

2. Imaging techniques for microstructure studies

X-ray microtomography (μCT) is a miniaturized version of medical CT or CAT (computed axial tomography) scanning and given the enormous success of x-ray computed tomography (μCT) in medical applications and material science, it is not surprising that ir recent years much attention has been focused on extending this imaging technique to food science as a useful technique to aid in the study of food microstructure. The microstructure of food products determines to a large extent the physical, textural and sensory properties of these products. Developing a proper understanding of the microstructure, particularly the spatial distribution and interaction of food components, is a key tool in developing products with desired mechanical and organoleptic properties. Information about the 3-D microstructure of food products and ingredients can be obtained using various imaging techniques. To-date, commonly used techniques are bright-field, polarising and fluorescence light microscopy (LM), confocal scanning laser microscopy (CSLM), transmission electron microscopy (TEM) and scanning electron microscopy (SEM). Other techniques such as atomic force microscopy (AFM), ultrasound and magnetic resonance imaging (MRI) are used for specific food applications. LM requires the staining of the different chemical components of a food (proteins, fat droplets, proteins etc.), therefore it is a more suitable technique for the investigation of multicomponent or multiphase foods such as cereal-based foods (Autio and Salmenkallio-Marttila, 2001). LM, SEM and TEM can be used to highlight various aspects of particulate structures, e. g. in a study on micro-porous, particulate gels (Langton et al., 1996), LM was used to visualize pores, TEM was applied to evaluate particle size and SEM was used to detect how the particles were linked together, i.e. the three-dimensional structure. As these techniques require some sample preparation (freezing, dehydration, staining etc.) that may lead to artifacts (Kalab, 1984). On the other hand, CLSM is a more suitable alternative method of analysis of food microstructure, as it requires minimum sample preparation. CLSM has been used for examining the three-dimensional structure of the protein network of pasta samples (Fardet et al., 1998), doughs (Thorvaldsson et al., 1999) and for high-fat foods (Wendin et al., 2000) that cannot be prepared for conventional microscopy without the loss of fat (Autio and Laurikainen, 1997). Atomic force microscopy (AFM) and magnetic resonance imaging (MRI) have been recently introduced into food science as non-destructive techniques. The former is particularly suitable for studying surface roughness, especially in fresh foods (Kaláb et al., 1995) and the latter can be successfully applied for studying processing such as frying, foam drainage, fat crystallization and other operations in which a dynamic study of food structure needs (Kaláb et al., 1995). Takano et al. (2002) and Grenier et al. (2003) used the MRI technique to study qualitatively and quantitatively the local porosity in dough during proving, stage in which invasive analytical methods may cause the dough to collapse. The ultrasound imaging technique is used primarily to investigate the structural properties of meat. It has been used to distinguish crystalline fats from liquid fats (McClements and Povey, 1988) or to determine a food's composition (Chanamai and McClements, 1999). Although, these wide varieties of imaging techniques exist, they are mostly invasive, as they require sample preparation hence, formation of artifacts or are restricted to certain types of food products.

3. X-Ray microtomography – An overview

X-ray microtomography (μCT), on the other hand, is a non-invasive technique that has several advantages over other methods, including the ability to image low moisture materials. It uses the differences in X-ray attenuation arising, principally, from differences in density within the specimen. A series of X-ray projections are recorded at a number of angles around the specimen (usually over a range of either 180 or 360 degrees). In μCT, unlike medical CT, the specimen is usually rotated, rather than the X-ray source and detector. If the projections are taken through a single plane in the specimen, it is possible to reconstruct a cross sectional image of that plane. In most µCT scanners today, 2D images are recorded, making it possible to reconstruct a complete 3D map of X-ray attenuation. In such cases, because of the divergence of the X-ray beam, it is necessary to use a conebeam reconstruction algorithm. This generally gives only an approximate reconstruction, with errors increasing with distance from the central plane (normal to the rotation axis). By using a spiral locus (translating the specimen along the rotation axis as it rotates) an exact (barring artefacts) reconstruction is possible, but this requires complex reconstruction algorithms, which are currently impracticable for large data sets. In the ideal case, each voxel of data represents the X-ray linear attenuation coefficient (LAC) of the corresponding volume in the specimen only. This is related to the composition and density of the material within that volume. Thus µCT studies can be used both for pure geometric studies, where the LAC is used only to determine the presence or absence of a phase, and quantitative studies where the LAC is used to determine density or concentration. The latter generally requires a higher signal to noise ratio, requiring high dynamic range detection and long X-ray exposures.

A complete µCT analysis is normally made by acquiring a number of radiographs (typically about 1000) of the same sample under different viewing angles (one orientation for each radiograph). A series of 2D X-ray images are obtained as a sample is rotated. A final computed reconstruction step is required to produce a three-dimensional map of the linear attenuation coefficients in the material. This three-dimensional map indirectly gives a picture of the structure density. In µCT, the X-ray source and the detector are placed at the opposite sides of the sample. The spatial resolution of the attenuation map depends on the characteristics of both the detector and the number of X-ray projections. Differences in the linear attenuation coefficients within a material are responsible for the X-ray image contrast. The main contrast formation in µCT is due to absorption contrast. Manipulation and analyses of µCT data using special software also allows reconstruction of cross-sections at depth increments as low as 15 micrometer, and along any desired orientation of the plane of cut. A series of non-invasive µCT slices of the same sample in any direction can provide much more information than just one Scanning Electron Microscopy or optical imaging picture for example. The true 3-D shape of the cells can also be visualized from its 2-D slices (Trater et al, 2005). This technique has been successfully used to observe the stability of gas bubbles in dough during the bread making process (Whitworth and Alava, 2002), the microstructure of foams (Lim and Barigou, 2004.) and ice crystals within frozen foods (Mousavi et. al, 2005).

X-ray microtomography has also proven to be a very useful technique for the non-invasive visualization and measurement of the internal microstructure of cellular food products, such as porous rice kernels and whipped cream (van Dalen et al., 2003), aerated chocolate and muffins (Lim et al., 2004), bread (Falcone et al., 2004; 2005; Lassoued et al., 2007), cornflakes (Chaunier et al., 2007), dough (Mousavi et al., 2005), extruded starches (Babin et al., 2007), French fries (Miri et al., 2006) and biopolymer foams (Trater et al., 2005). X-ray microtomography is non-destructive and provides in-depth information on the microstructure of the food product being tested; therefore a better understanding of the physical structure of the product and from an engineering perspective, knowledge about the microstructure of foods can be used to identify the important processing parameters that affect the quality of a product. Processes are no longer designed from a macroscopic level; knowing the properties of foods on the micro scale determines the process specification.

Hence, X-ray microtomography is fast becoming a very useful tool to aid in the study of food microstructure and is an important development in imaging technology that has eliminated some of the drawbacks of traditional imaging and enabled noninvasive characterization of microstructure food in three dimensions (Flannery et al.1987; Sasov and Van Dyck, 1998).

As stated above, the fat content in meat products is a very important compound and nowadays, lots of meat products with different fat contents and different physical and chemical features (protein network, moisture content, ingredients, additives and so on) are being manufactured. Consumers of today, require some of these information e.g. total fat content, types fat, ingredients, additives etc. to be stated Therefore total fat content of meat products (e.g. salami, steak etc) is an important quantity used in numerous studies. Thus, reliable methods for the quantitative analysis of fat from this type of food products are of critical importance. There are several methods to analyze fat content quantitatively (Monin 1998), although the method (AOAC, 1995) that is commonly used is based on chemical analysis, it is quite expensive and time consuming. Furthermore this technique is destructive to the sample, as a result, the same sample cannot be measured more than once and sometimes uses harmful, flammable solvents with health and environmental hazards.

With regards to bread, characteristics such as cell wall thickness, cell size, and uniformity of cell size affect the texture of bread crumb (Kamman, 1970) and also the appearance, taste perception and stability of the final product (Autio and Laurikainen, 1997). Crumb elasticity can be predicted from its specific volume and is strongly affected by the amylose-rich regions joining partially gelatinized starch granules in the crumb cell walls (Scanlon and Liu, 2003). Although, microstructure parameters like size and number density of air cells and their contribution to mechanical properties of solid food foams have been studied before (Barrett and Peleg 1992; Barrett et al. 1994a; Van Hecke et al. 1995; Gao and Tan 1996), the underlying mechanism relating cellular structure to the mechanics of these products is still not well understood. The limitations of traditional imaging techniques like scanning electron microscopy (SEM) and optical microscopy, which are two dimensional (2-D) and destructive in nature and also provide poor contrast, make it difficult to characterize cellular structure accurately. For the last few years, X-ray tomography has been proven to be particularly well suited for the 3D investigation of cellular materials. These studies have shown that tomography images allow describing accurately in three dimensions the complexity of the morphology of cellular food products. In addition, this technique enables dynamic studies. From a mechanical point of view, structure-property relationships of heterogeneous materials are often addressed through theories incorporating more or less realistic microstructural information. In order to understand the relation between microstructure and mechanical structure, mathematical models can be developed. For these models a number of free parameters are needed *e.g.* initial and final moisture content, structural parameters and type of diffusion. Microstructural parameters can be derived from the 3-D structure visualized by μCT and mechanical structure parameters can be derived from dynamical mechanical analysis.

A μCT image is typically called a slice and corresponds to a certain thickness of the object being scanned. Therefore, whereas a typical digital image is composed of pixels (picture elements), a μCT slice image is composed of voxels (volume elements). An X-ray shadow image corresponds to a two-dimensional projection from the three- dimensional object. In the simplest case, it can be described as a parallel X-ray illumination. In this approximation, each point on the shadow image contains the integration of absorption information inside the three-dimensional object in the corresponding partial X-ray beam. The X-rays that are transmitted through the object are scattered and/or absorbed. The gray levels in a slice correspond to the X-ray attenuation, which reflects the proportion of X-rays absorbed or scattered as they pass through each voxel. X-ray attenuation is a function of X-ray energy and the density and atomic number of the material being imaged. Directing X-rays through the slice plane from multiple orientations and measuring their resultant decrease in intensity creates a μCT image. A specialized algorithm is used to reconstruct the distribution of X-ray attenuation in the slice plane. By acquiring a stacked, contiguous series of μCT images

4. Case studies introduction and objectives

The aim of the following case studies is to demonstrate the capability of X-ray microtomography as a useful technique to study fat distribution and the percentage of fat in meat products such as salami. It also aims to help understand the correlation between the physical aspect and microstructural properties of these food products. It also aims demonstrates how the application of X-ray microtomography technique can aid in the study of cellular food microstructure such as coffee, breadsticks and biscuits with different porosity, by imaging and visualising the internal structure.

4.1 Case study 1: Meat products such as salami

4.1.1 Materials and methods

Five different types of Italian salami, chosen to exhibit variability in terms of visible structure of fat, were used for this experiment: Milano, Ungherese, Modena, Norcinetto and Napoli. They were purchased locally and were tested on the same day of purchase. Three samples were prepared for each type of salami, each 28mm in diameter and a thickness of 18mm. Each sample used for x-ray microtomography (μCT) analysis was wrapped with parafilm to avoid dispersion of moisture; the parafilm does not interfere with the x-rays. The same samples used for μCT analysis were also used for chemical analysis of fat composition.

4.1.2 Results and discussion

In order to verify the results of the μCT analysis, the percentage fat content obtained from the chemical analysis for each sample was compared to the μCT analysis results. Furthermore, the additional information gained from the μCT analysis i.e. the geometric parameters mentioned provided the required information to characterize and to see if there is any correlation between the microstructures of the different types of salami.

μCT technique validation

The percentage dark areas for each x-ray microtomographical image was calculated as a representation of the percentage total fat content within the sample. This value was directly derived from the geometric parameter, Percent object volume (POV). The percentage fat content was also calculated by chemical analysis on the same samples to verify the results obtained by μCT analysis. There were no statistical differences among the results obtained by chemical analysis and the POV values therefore suggesting that the POV calculated by μCT analysis for each sample is therefore a true representation of the percentage total fat content present within the sample. Therefore x-ray microtomography has proven to be a useful technique to quantitatively analyse fat content in meat products.

Application of μCT technique

Table 1 shows the average values obtained for the following six parameters using the CTAn software (Skyscan): Percent object volume (POV), Object surface/volume ratio (OSVR), Fragmentation index (FI), Structure thickness (ST), Structure separation (SS) and Degree of anisotropy (DA) and the results of the statistical analysis carried out as reported below.

	POV	OSVR	FI	ST	SS	DA
milano	22.70a±1.45	1,42c±0.08	-0,041a±0.04	4,23ab±0.19	8,48b±0.69	0,15a±0.05
modena	27.65b±4.64	0,92a±0.10	-0,004a±0.30	5,10c±0.41	8,23bc±0.35	0,21a±0.04
napoli	33.10a±1.32	1,28b±0.07	-0,195a±0.21	4,15a±0.25	10,88c±1.64	0,21a±0.03
norcinetto	32.32c±1.80	0,84a±0.02	-0,287a±0.18	4,94c±0.18	10,43c±1.25	0,21a±0.03
ungherese	21.34c±0.98	1,23b±0.04	-0,309a±0.05	4,65bc±0.19	6,53a±0.31	0,15a±0.06

* All parameters obtained were submitted to one-way analysis of variance (ANOVA) and Duncan's test ($p<0.05$) through the statistic package Statistica for Windows (Statsoft, Tulsa, USA).

Table 1. Values for the geometric parameters for the salami samples.

Where FI is the index of connectivity and is a measure of relative convexity or concavity of the total solid surface, based on the principle that concavity indicates connectivity, and convexity indicates isolated disconnected structures (Lim et al., 2004). A lower FI signifies better-connected solid lattices and has a negative index while on the other hand a higher FI indicates a more disconnected solid structure and has a positive index. As the fat in salami is considered generally to be of a concave structure, it can be noted from the table that the FI is negative for all samples and there are no statistically significant differences among the samples for this geometric parameter. The degree of anisotropy (DA) is a measure of the 3D structural symmetry, i.e. in this case it indicates the presence or absence of preferential alignment of the fat along a particular direction (Lim et al., 2004). A value of 0 would correspond to total isotropy, whereas a value of 1 would indicate total anisotropy. According to the results obtained for DA (see table 1), the fat present in all samples have a fairly good degree of isotropy and there are no statistical differences among the samples. POV is the percentage of the total fat content present in the sample as proven above. It can be seen that the POV for the Ungherese salami and the Norcinetto salami are statistically equal and are also the highest. There are also no statistical differences in the POV values between the Milano salami and the Napoli salami. OSVR indicates the fat globule size distribution within the sample, the higher the value, the more finely distributed is the fat present in the sample. It can be seen from the table that the Milano salami has the highest value hence its fat content is more finely distributed. Whilst, the Modena salami and the Norcinetto salami are statistically equal i.e. having approximately the same type of fat globule distribution, their low values indicate that the fat globules are more largely distributed with respect to the Milano salami. The results also show that the Napoli salami and the Ungherse salami are statistically equal, hence they have a similar structure with respect to the fat globule size distribution. ST is the average thickness of fat present and SS is the average distant between the fat globules in the samples. The results for ST from table 1 show that Modena, Norcinetto and Ungherese salami are statistically equal, whilst for SS,

Napoli and Norcinetto are statistically equal and also Milano and Modena. As reported above, in order to group the samples for each geometric parameter, a cluster analysis was run but only the results for ST are reported in this paper, as it could be considered the main explicative parameter for salami microstructure. As can be inferred from the results (not reported), the statistical analysis classified the 5 samples in 2 different groups. A group included Milano and Napoli salami and the other one all other samples. This second group was not homogeneous and could be divided in two different sub-groups: one for Ungherese and another one for Modena and Norcinetto salami. Also from table 1 a correlation of the parameters is not evident therefore cluster analysis was performed on all the geometric parameters in order to group the salami in terms of similarity (results not reported) The results of the cluster analysis showed that the Milano and Napoli salami are similar, this similarity occurs only between the microstructural parameters ST and POV (statistical figures not reported), therefore suggesting that the geometric parameters, structure thickness and percentage object volume are important parameters to consider when defining the microstructure of salami. Due to the lack of literature for this work, further investigation will have to be carried out to help achieve a better understanding of the relationships between the microstructure, mechanical structure and sensory properties of salami.

4.2 Case study 2: Porous foods such as biscuits and breadsticks

4.2.1 Materials and methods

Two different porous food products were used for this experiment, Italian sweet biscuits and breadsticks. Three types of Italian commercial sweet biscuits: 'Campagnole', 'Abbracci' and 'Frollini' were used for the first part of this experiment. Two types of brands were used for each types of biscuits: 'Campagnole' produced by Mulino Bianco and Coop, 'Abbracci' produced by Mulino Bianco and Coop and 'Frollini' produced by Ottimini Divella and Doemi. 'Campagnole' biscuits are made from rice flour and milk, 'Abbracci' are made from cream and chocolate and 'Frollini' are traditional made biscuits. A total of 30 samples were used for this experiment, 10 samples for each type of biscuit, i.e. 5 from each brand. Their dimensions were measured out of twenty to thirty random samples and were as following: 'Marie' thickness: 0.60 ± 0.07 (cm) and diameter 6.7 ± 0.03 (cm), 'Petit Beurre' thickness 0.61 ± 0.01 (cm), great and small axis: 6.23 ± 0.07 (cm), 5.45 ± 0.06 (cm), respectively. Biscuit diameter was measured as the average value of two orthogonal diameters. Thickness was measured with a digital micrometer. For each measurement the average of two to three readings was recorded. For the second part of the experiment, three different brands of breadsticks (Grissini) with the same typology were used: 'Mulino Bianco', 'Coop' and 'Bon'.

4.2.2 Results and discussion

Biscuits

The three-dimensional geometrical parameters were calculated using the CTAn software (Skyscan). Table 2 shows the average values obtained for the following six parameters using the CTAn software (Skyscan): Percent object volume (POV), Object surface/volume ratio (OSVR), Fragmentation index (FI), Structure thickness (ST), Structure separation (SS) and Degree of anisotropy (DA) and the results of the statistical analysis carried out as reported below.

	POV	OSVR	FI	ST	SS	DA
Abbarcci MB	37.91±4.36ab	0.88±0.17a	-0.08±0.03°	5.41±0.48ab	6.09±0.37a	0.36±0.02a
Abbarcci C	39.76±5.18ab	0.85±0.15a	-0.13±0.15a	5.89±1.52ab	6.03±1.17a	0.38±0.02a
Campagnole C	38.02±3.33ab	0.92±0.10ab	-0.12±0.10a	5.13±0.40ab	5.86±0.66a	0.37±0.06a
Campagnole MB	31.23±5.04a	1.18±0.16b	-0.02±0.06a	4.52±0.82a	5.74±0.36a	0.39±0.06a
Traditional D	42.91±5.64b	0.74±0.19b	-0.04±0.11a	6.72±1.34b	5.80±0.39a	0.35±0.09a
Traditional OD	41.03±2.54ab	0.77±0.07a	0.03±0.03a	6.23±0.89ab	5.63±0.16a	0.33±0.01a

* All parameters obtained were submitted to one-way analysis of variance (ANOVA) and Duncan's test ($p<0.05$) through the statistic package Statistica for Windows (Statsoft, Tulsa, USA).

Table 2.Values of the geometric parameters for all the biscuit the samples.

Where FI is the index of connectivity and is a measure of relative convexity or concavity of the total solid surface, based on the principle that concavity indicates connectivity, and convexity indicates isolated disconnected structures (Lim et al., 2004). A lower FI signifies better-connected solid lattices and has a negative index while on the other hand a higher FI indicates a more disconnected solid structure and has a positive index. From the results obtained for FI it can be said that the air cells observed in the samples are of a concave structure, as the FI is negative for all samples. It can also be noted that there are no statistically significant differences among the samples for this geometric parameter. The degree of anisotropy (DA) is a measure of the 3D structural symmetry, i.e. in this case it indicates the presence or absence of preferential alignment of the air cells along a particular direction (Lim et al, 2004.). A value of 0 would correspond to total isotropy, whereas a value of 1 would indicate total anisotropy. According to the results obtained for DA (see table 2), the air cells present in all samples have a fairly good degree of isotropy and there are no statistical differences among the samples. POV is the percentage of the total air cells present in the sample. It can be seen that the POV for both the Frollini samples although statistically there are little or no differences among the samples. OSVR indicates the cell size distribution within the sample, the higher the value, the more finely distributed are the air cell present in the sample. It can be seen from the table that the Campagnole MB has the highest value hence its air cells are more finely distributed, this can also be observed from the images obtained. Whilst, both the Frollini samples have the lowest OSVR samples hence their air cells are more largely distributed. ST is the average size of the cells in the samples and SS is the average distant between the air cells in the samples. The results for ST from the table show that there are larger air cells present in both the Frollini sample. From table 2 a correlation of the parameters is not evident therefore cluster analysis was performed on all the geometric parameters in order to group the salami in terms of similarity (results not shown); the results showed that there are three groups for terms of similarity. The Campagnole MB in terms of the geometric parameters cannot be grouped with any of the other samples, whilst, Abbracci C and Campagnole C are similar in terms of their geometric parameters and Frollini D, Frollini OD and Abbracci MB are also similar. Further investigation can be carried out to help achieve a better understanding of the relationships between the geometrical parameters, mechanical structure parameters and sensory properties.

Breadsticks

The three-dimensional geometrical parameters were calculated using the CTAn software (Skyscan). Table 3 shows the average values obtained for the following six parameters using the CTAn software (Skyscan): Percent object volume (POV), Object surface/volume ratio (OSVR), Fragmentation index (FI), Structure thickness (ST), Structure separation (SS) and Degree of anisotropy (DA) and the results of the statistical analysis carried out as reported below.

	POV	OSVR	FI	ST	SS	DA
Grissini MB	52.39±4.29a	0.58±0.05a	-0.38±0.12a	10.17±0.46ab	4.38±0.15a	0.34±0.02a
Grissini C	57.66±3.25b	0.67±0.67a	-0.30±0.06a	9.65±1.02a	4.48±0.16a	0.42±0.02a
Grissini B	50.26±4.81a	0.89±0.89b	-0.58±0.18b	13.13±1.59b	4.37±0.13a	0.55±0.02a

* All parameters obtained were submitted to one-way analysis of variance (ANOVA) and Duncan's test ($p<0.05$) through the statistic package Statistica for Windows (Statsoft, Tulsa, USA).

Table 3. Values of the geometric parameters for all the breadstick samples.

Where FI is the index of connectivity and is a measure of relative convexity or concavity of the total solid surface, based on the principle that concavity indicates connectivity, and convexity indicates isolated disconnected structures (Lim et al., 2004). A lower FI signifies better-connected solid lattices and has a negative index while on the other hand a higher FI indicates a more disconnected solid structure and has a positive index. From the results obtained for FI it can be said that the air cells observed in the samples are of a concave structure, as the FI is negative for all samples. It can also be noted that the FI for the Grissini B sample is statistically higher. The degree of anisotropy (DA) is a measure of the 3D structural symmetry, i.e. in this case it indicates the presence or absence of preferential alignment of the air cells along a particular direction (Lim et al, 2004.). A value of 0 would correspond to total isotropy, whereas a value of 1 would indicate total anisotropy. According to the results obtained for DA (see table 3), the air cells present in all samples are isotropic in nature and there are no statistical differences among the samples. POV is the percentage of the total air cells present in the sample. The Grissini C sample has the highest POV, hence a greater amount of air cells are present in this sample, whilst for Grissini MB and Grissini B the Frollini, statistically there are little or no differences among these samples. OSVR indicates the cell size distribution within the sample, the higher the value, the more finely distributed are the air cell present in the sample. It can be seen from the table that the Grissini B has the highest value hence its air cells are more finely distributed, this can also be observed from the images obtained. ST is the average size of the cells in the samples and SS is the average distant between the air cells in the samples. There are no statistical differences among the samples for ST and SS. From table 3 a correlation of the parameters is not evident therefore cluster analysis was performed on all the geometric parameters in order to group the salami in terms of similarity (results not shown); results obtained show that there are two groups in terms of similarity. The Grissini C in terms of the geometric parameters cannot be grouped with any of the other samples, whilst, Grissini MB and Grissini B are similar in terms of their geometric parameters. As for the case of the biscuits, further investigation can be carried out to help achieve a better understanding of the relationships between the geometrical parameters, mechanical structure parameters and sensory properties.

4.3 Case study 3: Emulsions

4.3.1 Materials and methods

Four different types of commercially produced mayonnaises, chosen for their variability of fat contents, were used for this experiment: 'kraft' 'calvé', 'kraft legeresse' and 'calvé-mayò'.

They were purchased locally and all tests, microstructural and rheological, were carried out on the same day. Three 10ml samples were prepared for each type of mayonnaise; with regards to μCT analysis, each sample was placed in a cylindrical tube.

4.3.2 Results and discussion

Application of μCT technique

Table 4 shows the average values obtained for the following seven parameters using the CTAn software (Skyscan): Percent object volume (POV), Object surface/volume ratio (OSVR), Fragmentation index (FI), Structure thickness (ST), Structure separation (SS), structure modelling index (SMI) and Degree of anisotropy (DA) and the results of the statistical analysis carried out as reported below.

	*POV	*OSVR	Av FI	Av SMI	*Av ST	*Av SS	Av DA
Calvé	$65.28^{a} \pm 1.45$	$0.48^{a} \pm 0.04$	0.05 ± 0.08	2.06 ± 0.69	$5.54^{a} \pm 0.19$	$5.18^{a} \pm 0.05$	0.12 ± 0.03
Calvé-mayò	$24.12^{b} \pm 1.32$	$0.51^{a} \pm 0.30$	0.17 ± 0.10	2.49 ± 1.64	$4.76^{b} \pm 0.07$	$8.18^{b} \pm 0.25$	0.22 ± 0.08
kraft	$69.08^{a} \pm 1.80$	$0.60^{b} \pm 0.21$	0.21 ± 0.07	2.57 ± 1.25	$2.43^{c} \pm 0.18$	$4.69^{a} \pm 0.02$	0.17 ± 0.98
kraft legeresse	$21.69^{b} \pm 0.04$	$0.61^{b} \pm 0.18$	0.15 ± 0.04	2.64 ± 0.31	$2.45^{c} \pm 0.05$	$7.80^{b} \pm 0.28$	0.24 ± 0.31

*All parameters obtained were submitted to one-way analysis of variance (ANOVA) and Duncan's test ($p<0.05$) through the statistic package Statistica for Windows (Statsoft, Tulsa, USA).

Table 4. Values for the geometric parameters for the mayonnaise samples.

Where FI is the index of connectivity and is a measure of relative convexity or concavity of the total solid surface, based on the principle that concavity indicates connectivity, and convexity indicates isolated disconnected structures (Lim et al., 2004). A lower FI signifies better-connected solid lattices and has a negative index, on the other hand a higher FI indicates a more disconnected solid structure and has a positive index. As it can be noted from the results the FI is positive for all samples and there are no statistically significant differences among the sample, therefore the fat structures in these samples have more or less disconnected solid lattices and therefore convex in structure. The degree of anisotropy (DA) is a measure of the 3D structural symmetry, i.e. in this case it indicates the presence or absence of preferential alignment of the fat present along a particular direction (Lim et al., 2004). A value of 0 would correspond to total isotropy, whereas a value of 1 would indicate total anisotropy. According to the results obtained for DA (see table 4), the fat present in all samples are isotropic and there are no statistical differences among the samples, this in accordance with Langton et al., (1998) whom found out that the fat droplets in mayonnaise samples had no orientation and could be regarded as isotropic. POV is the percentage of the total fat content present in the sample as proven above. It can be seen that the 'calvé' and 'kraft' mayonnaise samples have the highest POV values that are also statistically equal; therefore these two samples are similar in terms of fat content. On the other hand, the 'calvé-mayò' and 'kraft legeresse' samples have the lowest values and are also statistically equal. OSVR indicates the oil droplet/fat distribution within the sample. The higher the value, the more finely distributed is the oil droplet present in the sample. It can be observed from the table that the 'kraft legeresse' and the 'kraft' samples have the highest OSVR values that are also statistically equal; therefore they have a much finer dispersion of oil droplets compared to the other two samples. On the other hand, the samples, 'calvé' and 'calvè-mayò', have the lowest OSVR values and are also statistically equal i.e. having approximately the same type of fat distribution. Their low values indicate that the oil droplets present in these samples are more largely distributed with respect to the other two samples. ST is the average thickness of the fat structure present; this parameter calculates a volume-based thickness of the structure in three dimensions independent of an assumed structure type. It can be noted from table 4 that the samples, 'calvé' and 'calvè-mayò' have the highest ST values, therefore they have a similar fat thickness. While on the other hand the samples 'kraft legeresse' and the 'kraft' samples have low ST values that are statistically equal. SS is the average distant between the fat structures in the samples, it can be noted from this table that 'calvé' and 'kraft' have low SS values that are statistically equal. On the other hand the samples, 'kraft legeresse' and 'calvé-mayò' light have high SS values this could be due to the fact that the fat content in these samples are very low hence the average distant between each fat globule is greater with respect to the samples; 'calvé' and 'kraft'. No significant differences were found for SMI index in the different mayonnaise samples. SMI parameter, that is a topological index, gives an estimate of the characteristic shape of a structure in terms of plates and cylinders composing the 3D structure (Hildebrand & Rüegsegger, 1997) and is calculated using a differential analysis of triangulated surface of the structure under examination. The SMI assumes integer values of 0, 3 and 4 for ideal plates, cylinder and spheres, respectively. The SMI values calculated for all the samples ranged from 2.06 to 2.64, these values are fairly close to 3 therefore suggesting that the characteristic shape of the oil droplets present in these commercially produced mayonnaise are cylindrical in nature.

Rheological analysis

It was noted that for the rheological data, G′ and G″ values vs. the oscillatory frequency all samples exhibited a viscoelastic behaviour with storage modulus (G′) greater than the loss modulus (G″) (Subramanian & Gunasekaran, 1997; Rao & Steffe, 1992). As reported in literature (Ma & Barbosa-Canovas, 1995) emulsions with a greater fat content show higher values of G′. This was found for all samples except for the samples 'calvè-mayò' and 'kraft legeresse' that exhibited a higher storage modulus than sample 'calvè' in spite of their lower fat content. This could be due to the fact that the G′ reduction was counterbalanced by the higher carbohydrate content of the 'calvè-mayò' and 'kraft legeresse' samples that structure the emulsions (Munoz & Sherman, 1990; Peressini et al., 1998).

It can also be stated from the results obtained that all the mayonnaise samples exhibited solid-like gel behaviour with rheological spectra resembling that of weak gel (Ross-Murphy, 1988; Richardson et al., 1989). Typical weak gel characteristics were observed: G′ was greater than G″ throughout the frequency range, and the moduli showed a slight dependence on frequency.

The experimental data on the frequency sweep tests were correlated using the following power law according to Bohlin's theory of flow as a cooperative phenomenon (Bohlin, 1980):

$$G' = A \cdot w \cdot \exp\left(1 / z\right) \tag{1}$$

On the basis of this theory, emulsions are modeled as a network of rheological units that interact for establishing system structure. The coordination number z gives the level of these interactions and the coefficient A their magnitude. The z and A values were obtained for all samples. These parameters give an idea of the emulsion stability. From the results obtained (not shown) the 'calvè', 'kraft' and 'kraft legeresse' samples showed statistically similar values of the z parameter and higher than that the 'calvè-mayò' sample. Therefore, the calvè-mayo light with higher z values has a more complex structure and contains more microstructural interactions with respect to the other samples. Regarding the A parameter all samples differ significantly. Moreover, the highest and the lowest A values were recorded for the 'kraft' and 'calvè' samples, respectively, therefore the 'kraft' 'sample was more stable (Peressini et al., 1998) with respect to the calvè sample.

Correlation

Table 5 shows the results of the correlation among the microstructural parameters and rheological properties of the mayonnaise samples.

	Av POV	Av OSVR	Av FI	Av SMI	Av ST	Av SS	Av DA	A	z
Av POV	1.00								
Av OSVR	-0.38	1.00							
Av FI	-0.21	0.83	1.00						
Av SMI	-0.51	0.98	0.87	1.00					
Av ST	-0.58	-0.22	0.09	0.00	1.00				
Av SS	-0.68	0.94	0.74	0.97	0.06	1.00			
Av DA	0.86	-0.71	-0.68	-0.84	-0.50	-0.90	1.00		
A	-0.10	0.68	0.97	0.75	0.19	0.59	-0.59	1.00	
z	0.54	0.07	-0.30	-0.15	-0.97	-0.17	0.58	-0.41	1.00

* The highlighted parameters indicate a strong correlation among the parameters for this the EXCEL statistic package for Windows was used.

Table 5. Results of the correlation among the microstructural parameters and rheological properties of the mayonnaise samples.

It can be noted from this table that a correlation exists among some of the following microstructural properties of the mayonnaise samples, OSVR, ST, SMI and FI. With regards to the microstructural and rheological relationship, there is a strong correlation among the coefficient A and FI and the coordination number z and ST, suggesting that the average thickness, i.e. the volume based thickness, of the fat structure present affects the level of microstructural interactions and the connectivity of the fat surfaces affects the magnitude of these interactions. The correlation thus confirms that as the fat structures in these mayonnaise samples have more or less disconnected solid lattices and being convex in structure there is an amplified level of microstructural interactions and the level of these interactions is correlated to the thickness of the fat structure, hence the results obtained demonstrate that there is a link between some microstructural features and rheological properties. As stated above, structure-property relationships can strongly affect the physiochemical, functional, technological and even nutritional properties of foods. For example, with regards to food emulsions like ice cream, yoghurt, spreads and dressings such as mayonnaise the consumer appreciation of these products is strongly linked to the texture. Therefore, this preliminary investigation allows us to carry out further study on the main process variables that affect the characteristics of mayonnaise that are involved in consumer satisfaction to promote full product acceptability.

4.4 Case study 4: Coffee

4.4.1 Materials and methods

Coffea arabica var. S-795 from India, crop 2008-2009, has been used for this experiment. The green coffee beans were roasted with a Probat model BR74 (220°C, 100 g starting sample) and sampled during roasting process at 2, 3, 4 and 5 minutes. The roasting degree has been determined gravimetrically and expressed in terms of total weight loss (6.8, 10.7, 13.9, and 21.1 % w/w, respectively) it corresponded to very light, light, medium and dark, respectively. Five coffee beans at each roasting time were chosen for the experiment. Each coffee bean was weighed on a digital precision balance (± 0.1g) (Gibetini Eurpoe, Italy) to calculate the density of the sample. For x-ray microtomographical analysis, each sample was imaged under the same conditions, using the Skyscan 1172 high-resolution desktop X-ray microtomography system (Skyscan, Belgium).

4.4.2 Results and discussion

Figures 1a, 1b and 1c show as an example the set of flat cross sections that were obtained for these samples after binarisation of the images using the reconstruction software CTAn software (Skyscan), from these images the structure/air matrix is clearly visible. The three-dimensional geometrical parameters were calculated using the CTAn software (Skyscan).

From the images of above, it can be noted that the structure and the number of holes in the samples changed and increased with increasing roasting time.

Following data analysis, figure 2 shows the graph for the average distribution of the total volume of the holes present for the coffee bean samples at the five different roasting times.

The values of the total volume represented in the graph does not take into account the large central hole of the coffee bean, as this structure is only air it is therefore excluded in this micro-structural study. It can be noted from this graph that as the roasting time increases, the total volume of holes increases, with a significant peak at roasting time 3 due to the rupture of bonds in the internal structure of the coffee beans, this is in concurrence with Schenker et.al (2000). Statistical analysis, performed on the results of this graph confirmed that there are statistical differences for the total volume between roasting times t0 and t3 and roasting times t3 and t5, although there are no statistical differences between the total volumes at roasting time t4 and t5. On the other hand, the percentage hole volume, i.e. the geometric parameter POV, was calculated for each sample as a representation of the percentage total hole content within the sample. From the values obtained for POV (results not shown), it can be noted also in this case that the POV values increase with increasing roasting time. Statistical analysis, performed on the results of confirmed that there are statistical differences among all the POV values at the different roasting times. This result is in accordance with the results for the total volume, as an increase in total volume of holes results in an increase of POV.

Figure 3 shows the graph for the total number of pores present in the samples of coffee examined at different roasting times excluding the large central hole as before.

The trend of the graph shows that at roasting time 2 there is a significant peak due to the increase in temperature and therefore a change of the internal structure, this can also be noted from the images obtained. After roasting time 2 the increase of the total number of holes is gradual and statistical analysis performed on the results confirmed that although there is a statistical difference between the total number of holes at roasting times t2 and t5, there are no statistical differences among the values for roasting time t3 and t4. The average bean densities for the different roasting times are shown in figure 3. There is significant decrease in bean density due to its increase in pore volume and simultaneous weight loss. This is in accordance with Dutra et. al (2001) whom found also a significant decrease in bean density with increase in roasting time.

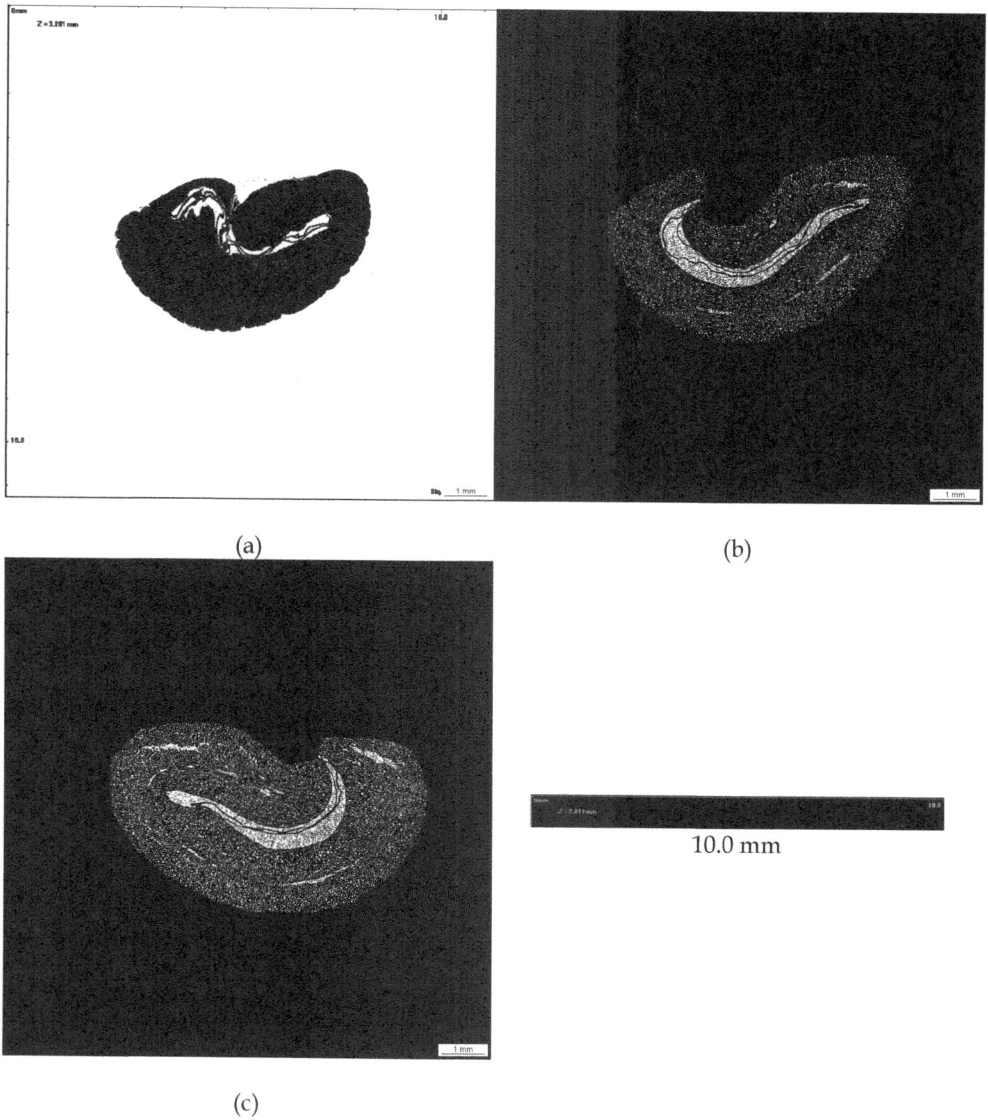

Fig. 1. Examples of the binary tomograph illustrating the separation of the air and structure phases: a) green coffee bean; b) roasted coffee bean at 2 minutes; c) roasted coffee bean at 5 minutes.

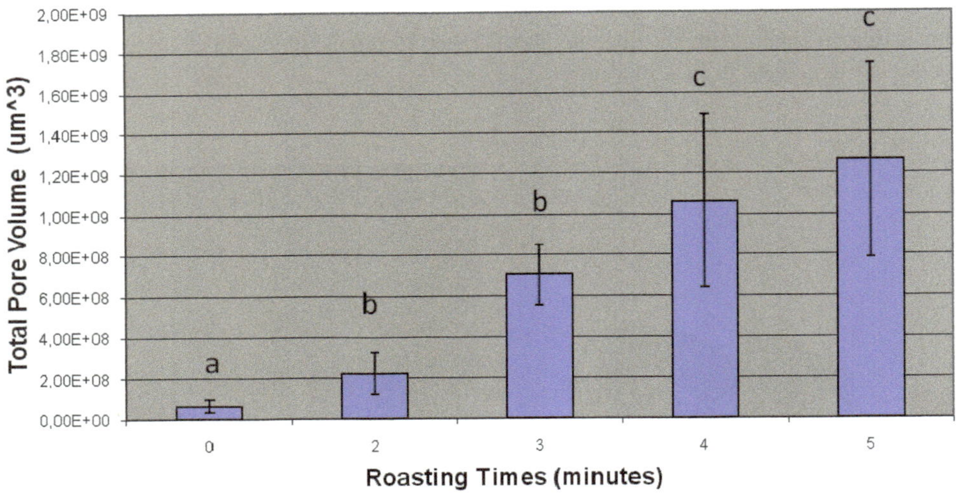

Fig. 2. The average distribution for the total volume of the holes.

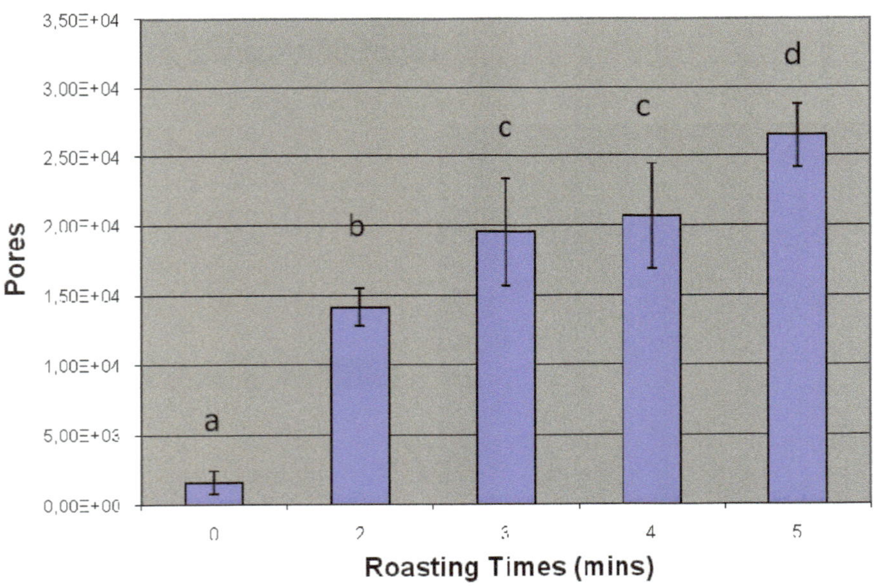

Fig. 3. The total number of pores.

5. Conclusion

X-ray microtomography has proven to be a very useful technique to image the 3-D microstructure of food products. X-ray microtomography can be complementary to other microscopic techniques used for food research. With X-ray microtomography, a full 3-D image of large samples can be obtained with a voxel resolution of about 0.5 μm and image analysis of the full 3-D microstructure, measuring the size, shape, networking/connectivity and distribution of various phases is possible. These measurements represent the full 3-D microstructure, which is not always possible by 2-D image analysis using statistical techniques.

As stated above, the fat content in meat products is a very important compound and nowadays, lots of meat products with different fat contents and different physical and chemical features (protein network, moisture content, ingredients, additives and so on) are being manufactured. Consumers of today, require some of these information e.g. total fat content, types fat, ingredients, additives etc. to be stated. Therefore total fat content of meat products (e.g. salami, steak etc) is an important quantity used in numerous studies. X-ray microtomography has proven to be a very useful and sophisticated tool for the quantification analysis of fat in meat and meat products. Combined visualisation of the microstructure using X-ray microtomography and other microscopic techniques, extraction of quantitative data obtained by image analysis, and modelling of the microstructure based on characteristics of the structuring elements should point to the optimal food product. The consumer appreciation of solid food foams like bread, extruded cereals, biscuits and cakes is strongly linked to the texture. The control of the sensory properties of such products, which is still a challenge, requires a better understanding of relationships between composition, cellular structure formation mechanisms and the final texture. For texture, sensory properties of solid food foams are related to both mechanical properties and cellular structure. In this context, determining the relationships between a given mechanical property and the cellular structure is thus of prime importance. Since cellular cereal products can be considered, from morphological and topological points of view, like metallic or polymeric foams, it is tempting to address this problem by referring to Gibson & Ashby's model. Such scaling laws are shown to be efficient to assess the effect of the relative density on mechanical properties like Young's modulus or strength of extruded starchy materials or bread. However, for the same density, mechanical properties and sensory properties are also sensitive to microstructural dispersions.

With regards to cellular cereal products, X-ray microtomography to date has proven to be a very useful technique for the non-invasive 3-D visualization and quantitative analysis of the microstructure of cellular cereal products. Further work can be carried out to study the inner cellular structure of the cereal matrix or to assess the integrity of moisture barriers applied on cereal product. The obtained quantitative information can be used as input for simulation models for moisture diffusion. The technique has significant benefits for the design, analysis, and processing science of certain food products. Such an advance in cellular food measurement will undoubtedly open up new horizons for the development of mathematical and computational models that link product microstructure to product mechanical properties and rheology.

With regards to emulsions such as mayonnaise, the fat distribution could be observed and quantified in the microtomographic images, as well as fat droplets size. With μCT, a full 3-D image of large samples can be obtained with high resolution and image analysis of the full 3-D microstructure, measuring the size, shape, networking/connectivity and distribution of various phases is possible. These measurements represent the full 3-D microstructure, which is not always possible by 2-D image analysis using statistical techniques. The correlation of the μCT analysis and rheological analysis identified the microstructural-rheological structure relationships. It showed that the average volume base thickness of the fat structure present is correlated to the level of microstructural interactions and due to fact that the fat structures in mayonnaise having more or less disconnected solid lattices and being convex in structure this allows for there to be an amplified level of these microstructural interactions. The identification of the microstructural-rheological structure relationships is of importance as these relationships strongly affect the physiochemical, functional, technological and even nutritional properties of foods.

With regards to coffee beans, the case study demonstrated that with x-ray microtomography a full 3-D image of the bean samples could be obtained with high resolution. Image analysis of the full 3-D microstructure does allow the measuring of size, shape, total volume distribution, porosity and density. These measurements calculated from the 3-D microstructure are not always achievable from 2-D image analysis even by using statistical techniques. Therefore, as proven also in this work X-ray microtomography is a useful and sophisticated tool to provide detail information on the microstructure of porous foods such as coffee beans at different roasting degree: from green to very dark. In fact x-ray microtomography was able to quantify the structural alteration of the microstructure caused by the high internal pressure generated by the large amount of gases released as a consequence of the thermal treatment.

6. References

Abecassis, J. C.; Faure, J. & Feillet, P. (1989). Improvement of cooking quality of maize pasta products by heat treatment. *Journal of the Science of Food and Agriculture*, Vol.47, No.4, pp. 475-485.

Aguilera, J.M. (2000). Microstructure and food product engineering. *Food Technology*, Vol.54, No.11, pp. 56-65.

Aguilera, J.M. (2005). Why food microstructure. *Journal of Food Engineering*. Vol.67, No.1-2 (March 2005), pp. 3-11.

Ahmad, R. & Magan, N. (2002). Microfloral contamination and hydrolytic enzymes differences between monsooned and nonmonsooned coffees. *Letters in Appl Microbiol*, Vol.34, No.4, pp. 279–282

Amorim, H.V. (1976). Some physical aspects of brazilian green coffee beans and the quality of the beverage. *Turrialba* , Vol.26, No.1, pp. 24-27

AOAC, 1995. Official Methods of Analysis, 16th ed. AOAC International, Washington, DC.

Autio, K. & Laurikainen, T. (1997). Relationships between flour/dough microstructure and dough handling and baking properties. Review. Trends in Food Science & Technology, Vol.8, No.6, pp. 181-185.

Babin, P.; Della Valle, G.; Dendievel, R.; Lourdin, D. & Salvo, L. (2007). X-ray tomography study of the cellular structure of extruded starches and its relations with expansion phenomenon and foam meachanical properties. Carbohydrate Pol, Vol.68, pp. 329-40

Barrett, A.H & Peleg, M. (1992). Extrudate cell structure-texture relationships. J. Food Science, Vol.57, No.5 (Semptember 1992), pp. 1253–1257

Barrett, A.H.; Cardello, A.V.; Lesher, L.L. & Taub, I.A. (1994)a. Cellularity, mechanical failure and texture perception of corn meal extrudates. Journal of Texture Studies, Vol.25, No.1 8March 1994), pp. 77–95

Bellido, G.G.; Scanlon, M.G.; Page, J.H. & Hallgrimsson B. (2006). The bubble size distribution in wheat flour dough. *Food Research International*, Vol.39, No.10, pp. 1058–1066

Bohlin, L. (1980). A theory of flow as a cooperative phenomenon. Journal of Colloid and Interface Science, Vol.74, No.2 (April 1980), pp. 423-434

Brunello, N.; McGauley, S.E. & Marangoni, A. (2003). Mechanical properties of cocoa butter in relation to its crystallization behavior and microstructure. Lebensm.-Wiss. u.-Technol., Vol.36, No.5 (August 2003), pp. 525-532

Buffa, M.N.; Trujillo, A.J.; Pavia, M. & Guamis, B. (2001). Changes in textural, microstructural, and colour characteristics during ripening of cheeses made from raw, pasteurized or high-pressure-treated goats' milk. International dairy Journal, Vol.11, No.1, pp. 927-934

Chanamai, R. & McClements, D.J. (1999). Ultrasonic determination of chicken composition. Journal of Agricolture and Food Chemistry, Vol.47, No.11, pp. 4686-4692

Chaunier, L.; Della Valle, G. & Lourdin, D. (2007). Relationships between texture, mechanical properties and structure of cornflakes. Food Research International, Vol.40, pp. 493-503

Cowin, SC. (1985). The relationship between the elasticity tensor and the fabric tensor. Mechanics of Materials Vol.4, No.4, pp. 137-47

Crippen, K.L.; Hamann, D.D. & Young, C.T. (1989). Journal of Textural Studies, Vol.20, No.1, pp. 29-41

Del Nobile, M.A.; Martoriello, T.,+; Mocci, G. & La Notte, E. (2003). Modeling the starch retrogradation kinetic of durum wheat bread. Journal of Food Engineering, Vol.50, No.2 8semptember 2003), pp. 123-128

Dentan, E. (1977). Structure fine du grain de café vert. Proceed. 8th ASIC Colloquium; Paris. pp. 59-64

Dentan, E. & Illy, A. (1985). Étude microscopique de grains de café matures, immatures et immatures fermentés Arabica Santos. Proceed 11th ASIC Colloquium; Paris. pp. 341-368

Depree, J. A. & Savage, G.P. (2001). Physical and flavour stability of mayonnaise. Trends in Food Science and Technology, Vol.12, No.5-6, pp 157–163

Dutra, E.R.; Oliveira, L.S.; Franca, A.S.; Ferraz, V.P. & Afonso, R.J.C.F. (2001). A preliminary study on the feasibility of using the composition of coffee roasting exhaust gas for the determination of the degree of roast. Journal of Food Engineering, Vol.47, No.3, pp. 241–246

Liu, Z.; Chuah, C.S.L. & Scanlon, M.G. (2003). Compressive elastic modulus and its relationship to the structure of a hydrated starch foam. Acta Materialia, Vol.51, No.2 (January 2003), pp. 365-371

Ding, K. & Gunasekaran, S. (1998). Three-dimensional image reconstruction procedure for food afficrostructure evaluation. Artificial Intelligence Review, Vol.12, pp. 245-262

Falcone, P.M.; Baiano, A., Zanini, F., Mancini, L., Tromba, G. & Montanari, F. (2004). A novel approach to the study of bread porous structure: Phase-contrast x-ray micro- tomography. Journal of Food Science, Vol.69, No.1, pp. 38-43

Falcone, P.M.; Baiano, A.; Zanini, F.; Mancini, L.; Tromba, G. & Dreossi, D. (2005). Three-dimensional quantitative analysis of bread crumb by X-ray microtomohraphy. Journal of Food Sciences, Vol.70, No.3, pp. 265-72

Fardet, A.; Baldwin, P.M.; Bertrand, D.; Bouchet, B.; Gallant, D.J. & Barry, J.L. (1998). Textural image analysis of pasta protein networks to determine influence of technological processes. Cereal Chemistry, Vol.75, No.5, pp. 699-704

Flannery, B.P.; Deckman, H.W.; Roberge, W.G. an& D'Amico, K.L. (1987). Three-dimensional x-ray microtomography. Science Vol.237, 1439–1444

Frisullo, P.; Laverse, J.; Marino, R. & Del Nobile M.A. (2009). X-ray computer tomography to study processed meat micro-structure. Journal of Food Engineering, Vol.94, No.3-4 (October 2009) pp. 283-289

Gao, X. & Tan, J. (1996). Analysis of expanded-food texture by image processing. Part II. Mechanical properties. Journal of Food Process Engineering, Vol.19, No.4 (November 1996), pp. 445–456

Grenier, A.; Lucas, T.; Collewet, G. & Le Bail, A. (2003). Assessment by MRI of local porosity in dough during proving. Theoretical considerations and experimental validation using a spin-echo sequence. Magnetic Resonance Imaging, Vol.21, No.9 (November 2003), pp. 1071-1086

Gutiérrez, C.; Ortolá, M.D.; Chiralt, A. & Fito, P. (1993). Análisis por meb de la porosidad del café tostado. Proceed. 15th ASIC Colloquium; Paris, pp. 661-671

Hahn, M.; Vogel, M.; Pompesius-Kempa, M. & Delling, G. (1992). Trabecular bone pattern factor - a new parameter for simple quantification of bone microarchitecture. Bone, Vol.13, pp. 397-330

Harringan, T.P. & Mann, R.W. (1984). Characterisation od microstructural anisotropy in orthotropic materials using a second rank tensor. Journal of Mater Science, Vol.19, No.3, pp. 761-767

Helgesen, H.; Solheim, R. & Næs, T. (1998). Consumer purchase probability of dry fermented lamb sausages. Food Quality and Preference, Vol.9, No.5, pp. 295-301

Hildebrand, T. & Ruegsegger, P. (1997). A new method for the model independent assessment of thickness in three dimensional images, Journal of Microscopy, Vol.185, No.1, pp. 67-75

Illy, A. & Viani, R. (1995). Espresso coffee. 1st ed. London: Academic Press Ltd, pp. 253

Inglis, D. & Pietruszczak, S. (2003). Characterization of anisotropy in porous media by means of linear intercept measurements. International Journal of Solids and Structures, Vol.40, No.5, pp. 1243-1264

Kaláb, M. (1984). Artifacts in conventional scanning electron microscopy of some milk products. Food Microstructure, Vol.3, No.2, pp. 95-111

Kaláb, M.; Allan-Wojtas, P. & Miller, S.S. (1995). Microscopy and other imaging technique in food structure analysis. Trends in Food Science, and Technology, Vol.6, No.6, pp. 177- 186

Kamman, P.W. (1970). Factors affecting the grain and texture of white bread. Baker's Dig., Vol.44, No.2, pp. 34-38

Kanatani, K.I. (1984). Distribution of directional data and fabric tensors. International Journal of Engineering Science, Vol.22, No.2, pp. 149-164

Kokelaar, J.J.; Van Vliet, T. & Prins, A. (1996). Strain hardening properties and extensibility of flour and gluten doughs in relation to breadmaking performance. Journal of Cereal Science, Vol.24, No.3, pp. 199-214

Langton, M.; Aström, A. & Hermansson, A. (1996). Texture as a reflection of microstructure. Food Quality and Preference, Vol.7, No.3/4, pp. 185-191

Lassoued, N.; Babin, P.; Della Valle, G.; Devaux, M.F. & Réguerre, A.L. (2007). Granulometry of bread crumb grain: Contributions of 2D and 3D image analysis at different scale. Food Research International, Vol.40, pp. 1087-97

Léonard, A.; Blacher, S.; Nimmol, C. & Devahastin, S. (2008). Effect of far-infrared radiation assisted drying on microstructure of banana slices: An illustrative use of X-ray microtomography in microstructural evaluation of a food product. Journal of Food Engineering, Vol.85, pp. 154–162

Lim, K.S. & Barigou, M. (2004). X-ray micro-tomography of cellular food products. Food Research International, Vol.37, No.10, pp. 1001-1012

Liu, H. Xu.; X.M. & Guo Sh.D. (2006) Rheological, texture and sensory properties of low-fat mayonnaise with different fat mimetics. LWT, Vol.40, No.6. pp. 946-954

Ma, L. & Barbosa-Canovas, G.V. (1995). Rheological Characterization of Mayonnaise. Part II: Flow and Viscoelastic Properties at Different Oil and Xanthan Gum Concentrations. Journal of Food Engineering, Vol.25, No.3, pp. 409-425

Maire, E.; Fazekas, A.; Salvo, L.; Dendievel, R.; Youssef, S.; Cloetens, P. & Letang, J.M. (2003). X-ray tomography applied to the characterization of cellular materials. Related finite element modeling problems. Components Science and Technology, Vol.63, pp. 2431-2443

Martin, M.J.; Pablos, F.A. & Gonzàlez, G. (1997). Discrimination between arabica and robusta green coffee varieties according to their chemical composition. Talanta, Vol.46, pp. 1259–1264

Martens, H.J. & Thybo, A.K. (2000). An integrated microstructural, sensory and instrumental approach to describe potato texture. Lebensm.-Wiss. u.-Technology, Vol.33, pp. 471-482

Martinez, O.; Salmerón, J.; Guillén, M.D. & Casas, C. (2004). Texture profile analysis of meat products treated with commercial liquid smoke flavourings. Food Control, Vol.15, No.6 (September 2004), pp. 457-461

Massini, R.; Nicoli, M.C.; Cassarà, A. & Lerici, C.R. (1990). Study on physical and physicochemical changes of coffee beans during roasting. Note 1. Italian Journal of Food Science, Vol.2, pp. 123-130

McClements, D.J. & Povey, M.J.W. (1988). Comparison of pulsed NMR and ultrasonic velocity measurements for determining solid fat contents. International Journal of Food Science and Technology, Vol.23, pp. 159-170

McDonald, S.A., Mummery, P.M., Johnson, G., & Withers, P.J. (2006). Characterization of the three-dimensional structure of a metallic foam during compressive deformation. Journal of Microscopy, Vol.223, No.2 (August 2006), pp. 150-158

Mendoza, F., Verboven, P., Mebatsion, H.M., & Kerckhofs, G. (2007). Three-dimensional pore space quantification of apple tissue using X-ray computed microtomography. Planta, Vol.226, No.3, pp. 559–570

Miri, T.; Bakalis, S.; Bhima, S.D. & Fryer, P.J. (2006). Use of X-ray Micro-CT to characterize structure phenomena during frying. Available from http://iufost.edpsciences.org

Monin, G. (1998). Recent methods for predicting quality of whole meat. Meat Science, Vol.49, No.1, pp. 231-243

Mousavi, R.; Miri, T.; Cox, P.W. & Fryer, P.J. (2005). A novel technique for ice crystal visualization in frozen solids using X-ray micro-computed tomography. Journal of Food Sciences, Vol.70, No.7, pp. 437-42

Munoz, J, & Sherman, P. (1990). Dynamic viscoelastic properties of some commercial salad dressings. Journal of Texture Studies, Vol.24, No.4, pp. 411-426

Murthy, P.S. & Manonmani, H.K. (2009). Physico-chemical, antioxidant and antimicrobial properties of Indian monsooned coffee. European Food Research Technology, Vol.229, pp. 645–650

Odgaard, A. (1997). Three-dimensional methods for quantification of cancellous bone architecture. Bone. Vol.20, No.4, pp. 315-328

Pareyt, B.; Talhaoui, F.; Kerckhofs, G.; Brijs, K.; Goesaert, H.; Wevers, M. & Delcour J.A. (2009). The role of sugar and fat in sugar-snap cookies: Structural and textural properties. Journal of Food Engineering, Vol.90, No.3, pp. 400-408

Peressini, D.; Sensidoni, A. & de Cindio, B. (1998). Rheological Characterization of Traditional and Light Mayonnaises. Journal of Food Engineering, Vol.35, No.4 (March 1998), pp. 409-417

Puhlmann, R.; Sobek, E. & Bartsch, G. (1986). Strukturveränderungen der Kaffeebohne im Röstprozess. Lebensmittelindustrie, Vol.33, No.6, pp. 278-279

Radke, R. (1975). Das Problem der CO2-Desorption von Röstkaffee unter dem Gesichtspunkt einer neuen Packstoffentwicklung. Proceed. 7th ASIC Colloquium; Paris, pp. 323-333

Rao, M.A. & Steffe, J.F. (1992). Viscoelastic Properties of Foods. Elsevier Applied Science, New York, NY.

Richardson, R.K.; Morris, E.R.; Ross-Murphy, S.B.; Taylor, L.J. & Dea I.C.M. (1989). Characterisation of the perceived texture of the thickened systems by dynamic viscosity measurements. Food Hydrocolloids, Vol.3, pp. 175-191

Ross-Murphy, S. B. (1988). Small deformation measurements. In Food Structure: Its Creation and Evaluation, (pp. 387-400). J. M. V. Blanshard, and J. R. Mitchell, ed. Butterworths, London, UK

Ross, K.A.; Pyrak-Nolte, L.J. & Campanella, O.H. (2004). Food Research International, Vol.37, No.6 (July 2004), pp. 567-577

Sahoo, P.K.; Soltani, S.; Wong, A.K.C. & Chen, Y.C. (1988). A survey of thresholding techniques. Computer Vision, Graphics, and Image Processing, Vol.41, No.2, pp. 233-260

Saleeb, F.Z. (1975). Adsorption of carbon dioxide on roast and ground coffees. Proceed. 7th ASIC Colloquium; Paris, pp. 335-339

Sasov, A. & Van Dyck, D. (1998). Desktop X-ray microscopy and microtomography. Journal of Microscopy, Vol.191, No.2 (August 1998), pp. 151–158

Schatzki, T.F. & Fine, T.A. (1988). Analysis of radiograms of wheat kernels for quality control. Cereal Chemistry, Vol.65, No.3, pp. 233-239

Schenker, S.; Handschin, S.; Frey, B.; Perren, R. & Escher, F. (2000). Pore Structure of Coffee Beans Affected by Roasting Conditions. Journal of Food Science, Vol.65, No.3, pp. 452-457

Subramanian, R, & Gunasekaran, S. (1997). Small amplitude oscillatory shear studies on Mozzarella cheese. Part I. Region of linear viscoelasticity. Journal of Texture Studies, Vol.28, pp. 633-642

Takano, H.; Ishida, N.; Koizumi, M. & Kano, H. (2002). Imaging of the fermentation process of bread dough and the grain structure of baked breads by magnetic resonance imaging. Journal of Food Science, Vol.67, No.1, pp. 244-250

Terzia, S.; Salvoa, L.; Suerya, M. & Bollerb, E. (2009). In situ X-ray microtomography characterization of the entrapped liquid formed during partial remelting of a cold-rolled Al–8 wt.% Cu alloy. Scriptia Materialia, Vol.60, No.8, pp. 671-674

Thorvaldsson, K.; Stading, M.; Nilsson, K.; Kidman, S. & Langton, M. (1999). Rheology and structure of heat-treated pasta dough: influence of water content and heating rate. Lebensm.-Wiss. u.-Technology, Vol.32, No.3 pp. 154-161

Toledo, R. T. (1999). Fundamentals of Food Process Engineering, 2nd edition, ASPEN, pp. 234

Tollner, E.W.; Hung, Y.C.; Upchurch, B.L. & Prussia, S.E. (1992). Relating x-ray absorption to density and water content in apples. Transactions of the ASAE. Vol.35, No.6, pp. 1921-1928

Torrente, Y.; Gavina, M.; Belicchi, M.; Fiori, F.; Komlev, V.; Bresolin, N. & Rustichelli, F. (2006). High-resolution X-ray microtomography for three-dimensional visualization of human stem cell muscle homing. FEBS Letters, Vol.580, No.24, pp. 5759-5764

Trater, A.M.; Alavi, S. & Rizvi, S.S.H. (2005). Use of non-invasive X-ray microtomography for characterizing microstructure of extruded biopolymer foams. Food Research International, Vol.38, No.6 (July 2005), pp. 709–719

Van Dalen, G.; Blonk, H.; Van Aalst, H. & Hendriks, C.L. (2003). 3D imaging of foods using x-ray microtomography. G.I.T. Imaging & Microscopy, pp. 18-21

Van der Burg, W.J.; Aartse, J.W.; Van Zwol, R.A.; Jalink, H. & Bino, R.J. (1994). Predicting tomato seedling morphology by x-ray analysis of seeds. Journal of the American Society for Horticultural Science, Vol.119, No.2, pp. 258-263

Van Hecke, E.; Allaf, K. & Bouvier, J.M. (1995). Texture and structure of crispy-puffed food products. I: Mechanical properties in bending. Journal of Texture Studies, Vol.26, No.1, pp. 11–25

Wendin, K.; Langton, M.; Caous, L. & Hall, G. (2000). Dynamic analyses of sensory and microstructural properties of cream cheese. Food Chemistry, Vol.71, No.3, pp. 363-378

Whitworth, M.B. & Alava, J.M. (1999). The imaging and measurement of bubble in bread doughs. In . Bubbles in Foods, G. M. Campbell, C. Webb, S.S. Pandiella & K. Niranjan Editors, pp. 221-231, St. Paul, MN: Eagan Press

Wildmoser, H.; Scheiwiller, J. & Windhab, E.J. (2004). Impact of disperse microstructure on rheology and quality aspects of ice cream. Lebensm.-Wiss. u.-Technology, Vol.37, No.8, pp. 881-891

Wilson, A.J. (1997). Preliminary investigations of oil biosynthesis in the coffee cherry. Proceed. 17th ASIC Colloquium; Paris, pp. 92

Wium; H.; Pedersen, P.S. & Qvist, K.B. (2003). Effect of coagulation conditions on the microstructure and the large deformation properties of fat-free Feta cheese made from ultrafiltered milk. Food Hydrocolloids, Vol.17, No.3, pp. 287-296

19

Content of Total Polyphenols in Biodynamic Organic Acid Lime (*Citrus latifolia* Tanaka) Clarified by Microfiltration

Daniela Soares Viana, Lucia Maria Jaeger de Carvalho,
Ediane Maria Gomes Ribeiro and Gisela Maria Dellamora Ortiz
Federal University of Rio de Janeiro,
Brazil

1. Introduction

The main citrus producing countries according to FAO (2008) are China, Brazil and the United States, followed by Mexico, Spain and Italy. However, by species, Brazil is the first world producer of oranges and lemons by conventional tillage, followed by the United States, while China stands out in the production of tangerines.

In Brazil, the acid lime cv. *Tahiti* (*Citrus latifolia* Tanaka), also known as *Tahiti* lemon, is regarded as one of the commercially most important citric fruits, occupying over 30,000 hectares of cultivated area (Fig.1) (Barros, 1986).

The acid lime is used raw or in the form of juice, especially as an ingredient in the composition of blends to enhance the flavor of beverages (FAO, 2008).

Fig. 1. Leaves and fruits of *Citrus latifolia* Tanaka, cv. Tahiti (Bom Jesus Farm, São Paulo, July, 2009).

According to the Secretary of Agricultural Protection (SAP, 1999), the organic system adopts technologies that optimize natural resources, including ecological, biodynamic, natural, sustainable, regenerative, organic and agroecological crops.

So far, the most accurate and reliable system that allows the consumer to determine if a food has been produced by organic or biodynamic organic cultivation is the quality seal provided by national and international entities, such as the Biodynamic Institute (IBD), an organization established in 1991 that develops certification of organic and biodynamic products (Darolt, 2003).

The organic quality seal is an indication that the food was produced and processed in accordance with the guidelines and standards of organic farming, resulting in a positive agronomic quality when compared to conventionally grown food, along with safety and reliability for the consumers that the products have not been subjected to ionizing radiation and have no transgenic or genetically modified organisms (GMOs) in their composition. Therefore, cultivation (without pesticides, fertilizers or chemicals) plus the quality certificate of the Biodynamic Institute (IBD) characterize biodynamic agriculture (IBD, 2009; Darolt, 2003).

The citrus flavor and aroma are pleasant, and in its composition are essential nutrients and micronutrients such as vitamins and minerals, and fiber (Pellegrini et al., 2003; Morton et al., 2000). Vitamin C and phenolic compounds present in acid lime have the ability to capture free radicals, having an important role in its antioxidant activity (Dhuique-Mayer, 2005; Harbone & Williams, 2000).

Polyphenols are a heterogeneous group, composed of several classes of substances with antioxidant properties that are found in different foods and drinks. Typical compounds that possess antioxidant activity include phenols, phenolic acids and their derivatives, carotenoids, flavonoids (anthocyanins and anthoxantins), tocopherols, phospholipids, amino acids, phytic acid, ascorbic acid and sterols (Roesler et al., 2008).

The first studies on determination of total phenolics were published by Folin & Ciocalteau (1927), Swain & Hillis (1959), Singleton & Rossi (1965), and Fantozzi & Montedoro (1978). Different methods can be employed to determine the total content of phenolic compounds in foods and plants. The most utilized procedures employ the *Folin-Ciocalteau* reagent, in which the mixture of phosphowolframic and phosphomolybdic acids, in alkaline media, is reduced by oxidizing phenolic compounds, giving rise to blue wolframium (W_8O_{23}) and molybdenum (Mo_8O_{23}) oxides (Mirsaeedghazi et al, 2010; Mezadri et al, 2008; Kuskoski et al., 2006).

Some citric fruits such as acerola, orange and lemon contain high concentrations of ascorbic acid, while certain flavonoids such as anthocyanins, flavonols and flavones predominate in strawberry, mulberry and açaí (Cordenunsi et al., 2005; Pozo-Insfran et al., 2004; Assis et al., 2001).

Jardini & Mancini-Filho (2007) evaluated antioxidant activity of different extracts from the pulp and seeds of rum, and found that the fruit showed high antioxidant activity due to the presence of phenolic and reducing compounds. Water extracts from the pulp and seeds were the most effective regarding antioxidant activity.

Santos et al. (2008) showed that anthocyanins and total phenolics had a positive and significant correlation with the high antioxidant capacity of açaí pulps.

Beneficial health effects, especially a significant decrease on the incidence of chronic and degenerative deseases, have been attributed to ascorbic acid and phenolic constituents of fruits, vegetables, cereals, tea and wine. Epidemiological, clinical and *in vitro* studies indicate relationships between dietary polyphenols and certain biological effects, such as antioxidant, anti-inflammatory, antimicrobial and anticancer activities (Rodrigo et al, 2011; Beer et al, 2005).

The physiological role of phenolic compounds can be more easily understood if one takes into account that *in vitro* the antioxidant capacity of these substances is related not only to their chemical structures, but also depends upon the type and polarity of the solvent used for the extraction, the isolation process and purity of the compounds, as well as on the substrate to be protected by the antioxidant, and if an aqueous or lipophylic assay system is to be used (Giada, 2005).

Ultrafiltration (UF) and microfiltration (MF) processes have been used by food industries for the clarification, concentration, and cold sterilization of liquids, producing food without chemical additives, with low power consumption, providing alternative products with nutritional quality and agreeable flavor (Carvalho et al., 2008; Koroknai et al., 2008; Cassano et al., 2004; Girard, & Fukumoto, 2000).

Investigation of the nutritional and sensory post-process quality is required to fulfill the needs of the consumer market for clarified juices, because several substances with nutritional value such as vitamins and minerals, among others, cannot be recovered in the clarified juice or even be retained in the concentrated pulp, or on the surface or inside the membrane pores after UF and MF of juices, with or without prior enzymatic treatment (Koroknai et al., 2008; Cianci et al., 2005; Carvalho et al., 1998).

As well as any other industrial food processing technology, membrane separation processes have several advantages and disadvantages (Habert et al., 2003). The most important drawback is "fouling", a key event in UF and MF processes that can affect economic and commercial viability of a system, since it reduces productivity and useful life of the membrane (Nilsson, 1990). Advantages of membrane processes are increased yields of clarified juices, possibility of operation in one step, lower operational time, operation at room temperature (avoiding food degradation), simple and fast cleaning system, as well as preservation of nutritional and sensory characteristics of foods (Castro et al., 2007; Vaillant et al., 2001).

The aim of this study was to determine the total polyphenol content in biodynamic organic acid lime juice clarified by membrane processes of microfiltration.

2. Material and methods

2.1 Materials

A total of 113 kg of biodynamic organic acid lime (*Citrus latifolia* Tanaka), cultivar *Tahiti*, 2009 crop, (certified by IBD/IFOAM - Federación Internacional de Movimientos de Agricultura Orgánica) and 58.40 kg of conventionally cultivated acid lime, 2009 crop, were supplied by Bom Jesus Farm, located in Santa Rita do Passa Quatro, São Paulo, Brazil.

After selection, whole, non-injured fruits were weighed and sanitized by immersion in chlorinated water (100 ppm for 10 min.) and washed. The juice was extracted in a semi-industrial multiprocessor (Konik, Class 700 model, São Paulo, Brazil), sieved, packed in PVC containers and stored at – 15 °C, until processing and analyses.

2.2 Water permeability and juice recovery (%)

Before each MF process, the membrane water permeability was measured at transmembrane pressures of 0.5, 1.0 and 2.0 Bar, respectively, to verify its recovery, according to the following formula:

$$\text{Flux}\left(\text{L.m}^2.\text{h}\right) = \frac{(\text{L}/\text{h}) \times \text{P}(\text{Bar})}{0.05\ \text{m}^2}$$

Where: 0.05 m^2 = total membrane area; P = applied transmembrane pressure
The water permeability (L.m^2.h) was measured with a chronometer, at 10 minutes intervals. Cleanup procedures (alkaline and chlorine-alkaline) were performed before and after each process, aiming to recover water permeability of the membrane.
Membrane alkaline cleaning was performed by adding 80 mL of a NaOH solution (Tedia, Rio de Janeiro, Brazil) (1N/pH 11) to 10 L of distilled water at 40 °C, and re-circulating the solution in the system for 30 minutes. The solution was then discarded, using 80 L of distilled water. Chlorine-alkaline-cleaning was performed by using 40 mL of NaOH (1N/pH 11) and 130 mL of NaClO (Tedia, Rio de Janeiro, Brazil) (6% sodium hypochloride) diluted in 10 L of distilled water at 40 °C, with recirculation in the system for 30 minutes, followed by washing with 80 L distilled water. After each cleanup, pH was checked.

2.3 Membrane processes

2.3.1 Pressure optimization

The juices were clarified in a PROTOSEP IV system (Koch Membrane Systems Inc., Massachusetts, USA) using a 0.3 µm (mean pore diameter) polyethersulfone membrane.
For optimization of the best transmembrane pressure to be applied, 8.0 L of conventional acid lime juice was used, operating at 0.5, 1.0 and 2.0 Bar, for 60 minutes.
For flux calculation, the volume of the clarified juice was measured each 5 minutes. The best fluxes were obtained at 0.5 Bar TMP, which was used for all the processes, in triplicate.

2.4 Total polyphenol content

The whole and clarified acid lime juices were previously lyophilized in a Labconco Lyophilizer (model 75223, Kansas City, Missouri, USA) at - 40 °C, under vacum (< 200 x 10^{-3} mBar), for 18 hours.
The Folin-Ciocalteau method (Folin & Ciocalteau, 1927; Singleton & Rossi, 1965) was used for quantification of the total polyphenol in the juices. After reaction with the Folin-Ciocalteau reagent (2N) (Sigma Aldrich, Steinhem, Germany), absorbance was measured at 760 nm (λmax = 765 nm) using a UV- vis spectrophotometer (Shimadzu, model 1240, Kyoto, Japan). The amount of polyphenols was calculated from a standard curve of gallic acid (MP Biomedicals, Ilkirch, France). The results were expressed as mg of gallic acid equivalents (GAE)/mL of juice sample (Andrade et al., 2007).

2.5 Statistical analysis

Statistical analysis was performed using one-way ANOVA and mean values were compared applying Duncan's multiple range tests using Minitab Microsoft Excel. Trends were considered significant when means of compared sets differed at $P < 0.05$.

3. Results and discussion

3.1 Conventional lime juices (CLJ) and biodynamic organic lime juices (BLJ) yields

The CLJ and BLJ yielded 52.60 and 44.65%, respectively, in accordance with the minimum yield of 42% recommended by Swisher & Swisher (2000) for citric juices. On the other hand, CLJ yield was significantly higher than that of BLJ ($P < 0.05$), possibly due to the higher

percentage of peels in the BLJ samples. However, a yield of 47% was previously reported by Pedrão et al. (1999) for conventional lime juice, cv. *Tahiti*. Other factors may affect juice yield such as crops with high concentrations of nitrogen and potassium, leading to increased peel thickness, as well as relative humidity, and ambient temperature (Chitarra et al., 1990).
The maturity stage of the fruit also influences the yields of acid lime juice that have been observed, ranging from 55.60% in green fruit and 59.40% in ripe fruit (Ziena, 2000).

3.2 Water permeability and permeate fluxes

Mean water permeability flux was 1,241.95 L.m^2.h (± 0.48) in the three processes using the 0.3 µm membrane.
In order to determine the best operating transmembrane pressure for all MF processes, 8.0 L of CLJ were initially used in the feed, resulting in 6.0 L (75%) of clarified juice. The mean fluxes obtained at transmembrane pressures of 0.5, 1.0 and 2.0 bar were 49; 47 and 35 L.m^2.h, respectively (Fig.2). A slight difference was observed between the fluxes at the applied transmembrane pressures during the first twenty minutes of the process, but after this time there was a significant flux decline for processes operated at 1.0 and 2.0 Bar, demonstrating that concentration polarization was the result of polarization on the membrane surface and/or fouling, providing a decrease in the clarified juice flux over time (60 minutes) in all cases, which was more pronounced at 2.0 Bar. For both UF and MF processes, flux decline is high even when the water permeability of the membrane is completely recovered by washing (Oliveira et al., 2006).

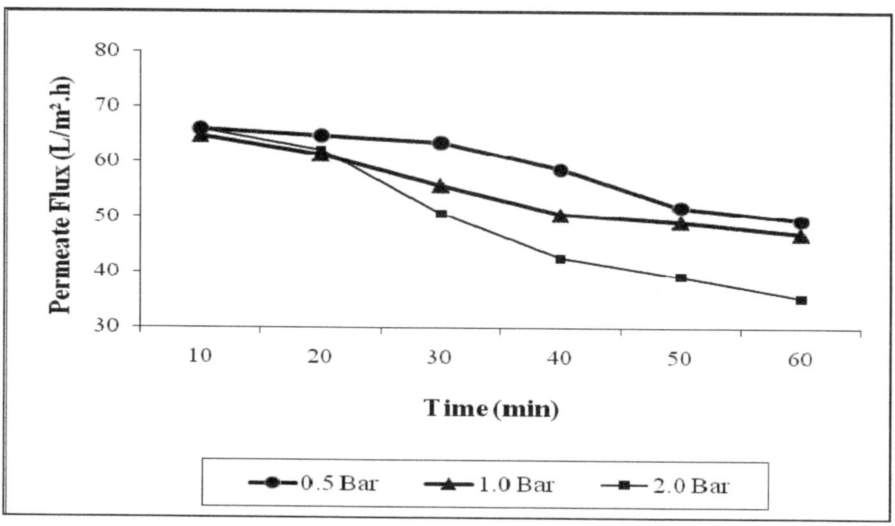

Fig. 2. Effect of process mechanism on the permeate flux at 0.5, 1.0 and 2.0 Bar transmembrane pressures with conventional lime juice.

Thus a 0.5 Bar transmembrane pressure for 60 minutes was adopted to process BLJ by MF, resulting in a mean volume of 6.0 L of clarified juice from a 8.15 L feed and yields of 73, 76 and 54% of clarified juice in the first, second and third processes, respectively (Fig. 3). In the first and third process, after the first 20 minutes there was a decrease in the flux that declined gradually declined until the end of the process, probably due to fouling.

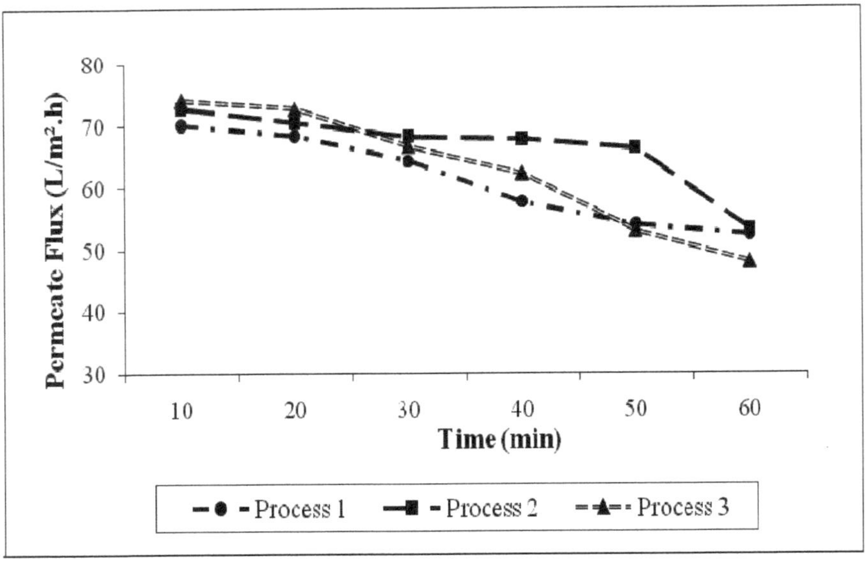

Fig. 3. Effect of process mechanism on the permeate flux at 0.5 Bar with biodynamic lime juice.

MF processes have been employed to clarify apple (Onsekizoglu et al., 2010), orange and kiwi juices (Galaverna et al., 2008), resulting in the same behavior. The processes presented mean fluxes of 75 L.m^2.h, with a progressive and constant decrease until the end (50 L.m^2.h). Kozák et al. (2006) and De Paula et al. (2004) using a 0.3 µm tubular membrane (polyesthersulphone) at 0.5 Bar transmembrane pressure in the clarification of strawberry and passion fruit juices, observed higher mean flux reductions of clarified juice (53% and 56%, respectively). Concentration polarization causes the initial decrease in permeate flux and "fouling" results from accumulation of material on the membrane surface reducing flux along the process time.

A different behavior was observed for clarification of pomegranate juice by MF, using hydrophilic mixed cellulose esters flat membranes (Plate and frame system - 0.22 µm), where the flux decreased rapidly in the early stage (10 minutes - 5.0 L.m^2.h), remaining constant until the end (Mirsaeedghazi et al., 2010).

Moreover, Yasan et al. (2007) clarified pasteurized apple juice, with prior enzymatic treatment, using a 0.2 µm polyethersulfone flat membrane (Plate and Frame system) obtaining higher than 60 L.m^2.h fluxes at 2.0 Bar.

Laorko et al. (2010) used 0.1 and 0.2 µm polysulphone hollow fiber membranes at 1.0 Bar in the clarification of hydrolyzed pineapple juice, obtaining fluxes of 24.2 and 22.0 L.m^2.h., respectively. However, it was noticed that the systems used by the aforementioned authors, especially those using flat and sheet membranes, promote faster fouling formation.

Carvalho et al. (2008) obtained a mean flux of 31.37 L.m^2.h in clarification by MF of non hydrolyzed pineapple juice, using a 0.3 µm tubular polyethersulfone membrane at 3.0 Bar, while for the UF of the same juice, the mean flux was 17.39 L.m^2.h, at 6.0 Bar.

The clarified juices in this study presented light green color and a limpid, translucent and very attractive aspect, as expected. MF processes have been employed to clarify apple

(Onsekizoglu et al., 2010), orange (Galaverna et al., 2008) and kiwi juices (Cassano, Marchio and Drioli, 2007) resulting in similar characteristics.

3.3 Total polyphenol contents

The total polyphenol contents of the whole and clarified CL and BL juices were 304 mg/100 g, 336 mg/100 g, 242 mg/100 g and 263 mg/100 g, respectively (Fig. 4). It must be emphasized that the gallic acid calibration curve employed to calculate total polyphenols expressed as gallic acid equivalents (GAE) (mg GA/100 g sample) showed a determination coefficient (R^2) of 0.9967, proving a positive correlation (R = 0.9983) (Fig. 5).

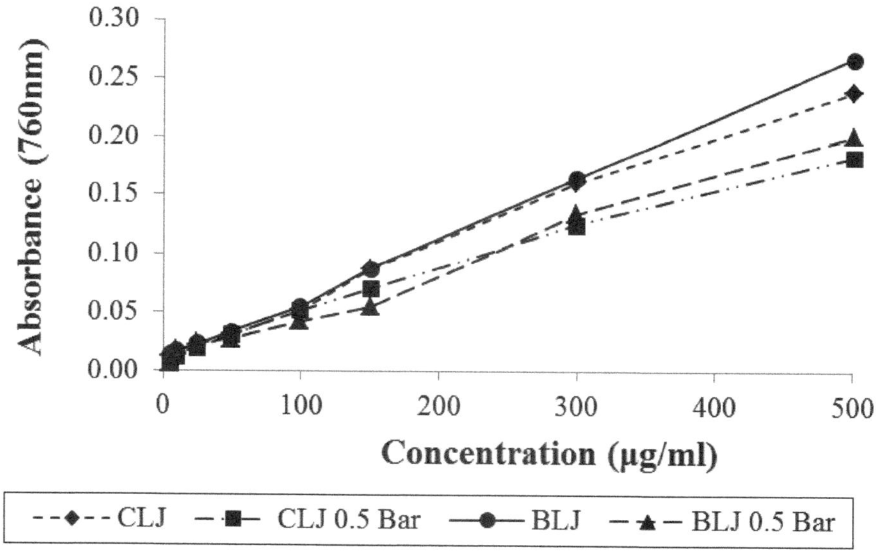

Fig. 4. Total polyphenol contents in conventional lime (CLJ) and biodynamic lime (BLJ) whole and clarified juices at 0.5 Bar.

Total polyphenols in fruits and juices of fruits of different origins (tropical and exotic) have been investigated for their antioxidant activity. Kuskoski et al. (2005 and 2006) found contents of 897.60, 229.60, 580.10, 544.90, 136.80 and 132.10 mg GA/100 g, respectively, in extracts of baguaçu and jambolão, acerola, mango, açaí and strawberry pulps. Lower total polyphenol contents were found for jambolão, açaí and strawberry extracts (whole juices) than found in the present study for the whole and clarified acid lime juices. Cavalcante et al. (2006) found total polyphenol contents of 119 mg GA/100 g, in the cashew juice, while Mondello et al. (2000) obtained 217 mg GA/100 g in orange juice, 145 mg GA/100 g in caqui, and 134 mg GA/100 g in pineapple, and Gorinstein et al. (1999) found 164 mg GA/100 g in mango.

However, much higher levels were reported by Vargas et al. (2008) in red grapes (508.4 mg GA/100 g) and white grapes (487.3 mg GA/100 g) recognized as rich in anthocyanins. Rapisarda et al. (2008) also found higher levels of polyphenols in orange juices (507.01 mg GA/100 g).

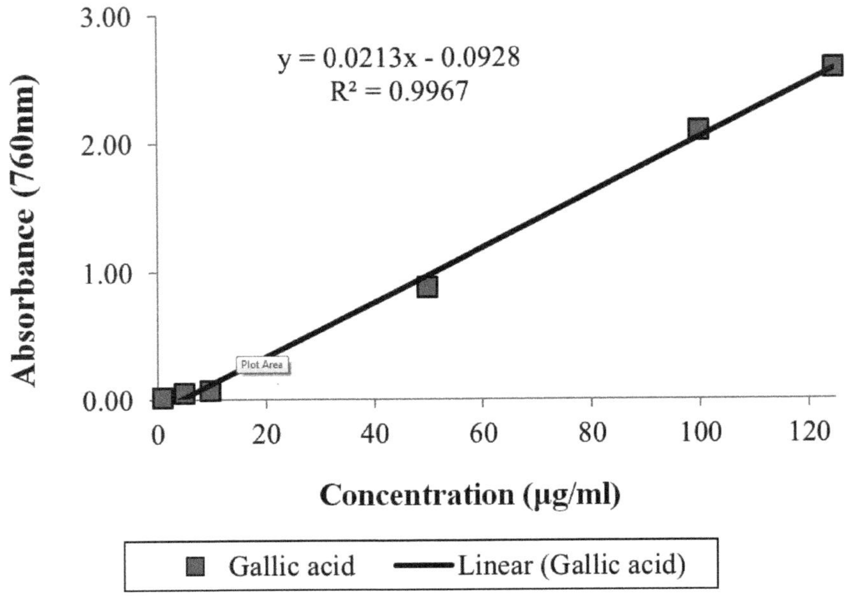

Fig. 5. Calibration curve of gallic acid.

4. Conclusions

The MF process recovered 78-79% of total polyphenols in the whole and clarified juices. Although the antioxidant activity and total polyphenols have been preserved in acid lime juices of both crops, after the MF process, the use of raw materials initially richer in total polyphenols than acid lime is recommended in order to assure a higher content of these compounds in the clarified juice. Studies are being conducted by this research group aiming to compare the different methods for determination of antioxidant activity in whole and clarified juices, since to this moment no official standard method is available.

5. Acknowledgments

We wish to thank the Fundação Carlos Chagas de Amparo à Pesquisa do Estado do Rio de Janeiro/FAPERJ for financial support to the project and the granting of scientific initiation scholarships during its development. The authors are also grateful to CNPq for financial support and the Instituto Nacional de Ciência e Tecnologia para a Inovação Farmacêutica (INCT_if) for support.

6. References

Andrade, C.A., Costa, C.K., Bora, K., Miguel, M.D., Miguel, O.G., & Kerber, V.A. (2007). Determination of phenolic content and antioxidant activity evaluation of *Acacia podalyriifolia* A. Cunn. Ex G. Don., Leguminosae-mimosoideae. *Brazilian Journal of Pharmacognosy*, v. 17, n. 2, 231-235.

Assis, S.A.; Lima, D.C.; Oliveira, O.M.M.F. (2001). Activity of pectinmethylesterase, pectin content and vitamin C in acerola fruit at various stages of fruit development. *Food Chemistry*, v. 74, p. 133-137.

Barros, C.B. (1986). Óleos essenciais cítricos do Brasil. 2 ed. Revisão Atualizada, *Fundação Cargill*, Campinas, 45p.

Beer, D.; Joubert, E.; Gelderblom, W.C.A.; Manley, M. (2005). Antioxidant activity of South African red and white cultivar wines and selected phenolic compound: I vitro inhibition of microsomal lipid peroxidation. *Food Chemistry*, v. 90, p. 569-577.

Carvalho, L.M.J., Silva, C.A.B., Pierucci, & A.P.T.R. (1998). Clarification of pineapple juice (*Ananas comosus* L. Merril) by ultrafiltration and microfiltration: physicochemical evaluation of clarified juices, soft drink formulation, and sensorial evaluation. *Journal of Agricultural and Food Chemistry*, v. 46, n. 6, p. 2185–2189.

Carvalho, L.M.J., Castro, I.M., & Silva, C.A.B. (2008). A study of retention of sugars in the process of clarification of pineapple juice (*Ananas comosus*, L.Merril) y micro and ultrafiltration. *Journal of Food Engineering*, v. 87, p. 447 - 454.

Cassano, A., Jiao, B., & Drioli, E. (2004). Production of concentrated kiwifruit juice by integrated membrane process. *Food Research International*, v. 37, p. 139-148,

Cassano, A., Marchio, M., & Drioli, E. (2007). Clarification of blood orange juice by ultrafiltration: analyses of operating parameters, membrane fouling and juice quality. *Desalination*, v. 212, p. 15-27.

Castro, T.R.; Abreu, F.A.P.; Carioca, J.O.B. (2007). Obtaining clarified cashew juice (*Anacardium occidentali, L*) using membrane separation processes. *Revista Ciência Agronômica*, v. 38, n. 2, p. 164-168.

Cavalcante, A.A.C.M., Leite, A.S., Salvador, M., & Rubensam, G., & Henriques, J.A.P. (2006). Phenolic compounds, carotenoids and vitamin C in antioxidant activity of cashew juice and cajuína. *I CONEPI*.

CHITARRA, M. I. F., CHITARRA, A. B. (1990). *Post-harvest fruits abd vegetable: physiology and handling*. Lavras, MG: FAEP, 230p.

Cianci, F.C., Silva, L.F.M., Cabral, L.M.C., & Matta, V.M. (2005). Clarification and concentration of cashew juice by membrane processes. *Ciência e Tecnologia de Alimentos*, v. 12, p. 579-583.

Cordenunsi, B.R.; Genovese, M.I.; Nascimento, J.R.O.; Hassimotto, N.M.A.; Santos, R.J.; Lajolo, F.M. (2005). Effects of temperature on the chemical composition and antioxidant activity of three strawberry cultivars. *Food Chemistry*, v. 91, p. 113-121.

Darolt, M.R. (2003). Comparison between the quality of organic and conventional food. In: *Organic Food: production, technology and certification. Stringheta & Muniz*. Editora UFV, Viçosa, p. 289-312.

De Paula, M.J., Lissi, C.V., Monteiro, B.F., & Venildes, H.R. (2004). Improvement in the efficiency of clarification of passion fruit juice by a combination of microfiltration and enzymatic processes. *Ciência e Tecnologia de Alimentos*, v. 22, n. 2, p. 311-324.

Dhuique-Mayer, C.; Caris-Veyrat, C.; Ollitrault, P.; Curk, F.; Amiot, M.J. (2005). Varietal and interspecific influence on micronutrient contents in citrus from the Mediterranean area. *Journal of Agricultural and Food Chemistry*, v. 53, p. 2140-2145.

Fantozzi, P.; Montedoro, G. (1978). Dosage des composés phénoliques dans les drupes d'olives récoltés à différents stades de maturation. *Industries Alimentaires* v. 17, p. 1335–1339.

FAO - FOOD AGRICULTURE ORGANIZATION (2008). FAOSTAT *Statistics database*, Agriculture, Rome, Italy, p. 52-58.

Folin, C., & Ciocalteau, V. (1927). Tyrosine and tryptophan determination in proteins. *Journal of Biology Chemistry*, v. 73, p. 627-650.

Galaverna, G., Silvestro, G., Cassano, A., Sforza, S., Dossena, A., Drioli, E., & Marchelli, R. (2008). New integrated membrane process for the production of concentrated blood orange juice: effect on bioactive compounds and antioxidant activity. *Food Chemistry*, v. 106, p. 1021–1030.

Giada, M.L.R. (2005). Evaluation of the antioxidant capacity of phenolic compounds from the cotyledon of sunflower seeds (*Helianthus annuus* L.) rajada. Doctoral Thesis (Food Science) – Food and Experimental Nutrition Departament, Universidade de São Paulo (USP), São Paulo, Brazil. 206 p.

Girard, B., & Fukumoto, L.R. (2000). Apple juice clarification using microfiltration and ultrafiltration polymeric membranes. *Lebensmittel-Wissenscheft und Technologie*, v. 32, n. 5, p. 290-298.

Gorinstein, S., Zemser, M., Haruenkit, R., Chuthakorn, R., Grauer, F., Martín-Belloso, O., Trakhtenberg, S. (1999). Comparative content of total polyphenols and dietary fiber in tropical fruits and persimmon. *Journal of Nutritional Biochemistry*, v. 10, p. 367-371.

Habert, A.C.; Borges, C.P.; Nobrega, R. (2003). Membrane separation processes. *Chemical Engineering Program*. COPPE/UFRJ.

Harborne, J.B.; Williams, C.A. (2000). Advances in flavonoid research since 1992. *Phytochemistry*, v. 52, p. 481-504.

IBD - INSTITUTO BIODINÂMICO (2009). *Guidelines for the organic quality standard.*. 17th Edition, Doc. 812, Botucatu, São Paulo.

IFOAM - FEDERACIÓN INTERNACIONAL DE MOVIMIENTOS DE AGRICULTURA ORGÂNICA (2002) – IFOAM *Basic Standards for organic production and processing.*

Jardini, F.A. ; Mancini-Filho, J. Evaluation of antioxidant activity in different extracts of the pulp and seeds of the pomagranate (*Punica granatum*, L.) (2007). *Brazilian Journal of Pharmaceutical Sciences*, v. 43, n.1, p. 138-147.

Koroknai, B., Csanádi, Z., Gubcza, L., & Bélafi-Bakó, K. (2008). Preservation of antioxidant capacity and flux enhancement in concentration of red fruit juices by membrane processes. *Desalination*, v. 228, p. 295-301.

Kozák, A., Rektor, A., Vatai, G. (2006). Integrated largescale membrane process for producing concentrated strawberry fruit juices. *Desalination*, v. 200, p. 540–542.

Kuskoski, E.M., Asuero, A.G., Troncoso, A.M., Mancini-Filho, J., & Fett, R. (2005). Application of several chemical methods to determine antioxidant activity in pulp. *Ciência e Tecnologia de Alimentos*, v. 25, n. 4, p. 726-732.

Kuskoski, E.M., Asuero, A.G., Morales, M.T., & Fett, R. (2006). Tropical and wild fruits and frozen pulps: antioxidant activity, polyphenols and anthocyanins. *Ciência Rural*, Santa Maria v. 36, p. 1283-1287.

Laorko, A.; Li, Z.; Tongchitpakdee, S.; Chantachum, S.; Youravong, W. (2010). Effect of membrane property and operating conditions on phytochemical properties and permeate flux during clarification of pineapple juice. *Journal of Food Engineering*, v 100, p. 514-521.

Mezadri, T.; Villaño, D.; Fernández-Pachón, M.S.; García-Parrilla, M.C.; Troncoso, A.M. (2008). Antioxidant compounds and antioxidant activity in acerola (*Malpighia emarginata* DC.) fruits and derivates. *Journal of Food Composition and Analysis*, v. 21, p. 282-290.

Mirsaeedghazi, H.; Mousavi, S.M.; Djomeh, Z.E.; Rezaei, K.; Aroujalian, A.; Navidbakhsh, M. (2011). Comparison between ultrafiltration and microfiltration in the clarification of pomegranate juice. *Journal of Food Processing Engineering*. Online ISSN: 1745-4530

Mondello, L., Cotroneo, A., Errante, G., Dugo, G., & Dugo, P. (2000). Determination of anthocyanins in blood Orange juices by HPLC analysis. *Journal of Pharmaceutical and Biomedical Analysis*, v. 23, p. 191-195.

Morton, L.W., Cacetta, R.A., Puddey, I.B., Croft, K.D. (2000). Chemistry and biological effects of dietary phenolic compounds: Relevance to cardiovascular disease. *Clinical and Experimental Pharmacology and Phisiology*, v. 27, p. 152-159.

Nilsson, L.S. (1990). Protein fouling of UF membrane: Causes and consequences. *Journal of Membrane Science*, v. 56, p. 13-28.

Oliveira, J. C; Setti-Perdigão P.; Siqueira, K. A. G.; Santos A. C.; Miguel, M. A. L. (2006). Microbiology of orange juice. *Ciência e Tecnologia de Alimentos*, Campinas, v. 26, p. 241-5.

Onsekizoglu, P., Savas Bahceci, K., & Jale Acar, M. (2010). Clarification and the concentration of apple juice using membrane processes: A comparative quality assessment. *Journal of Membrane Science*, v. 26, p. 376-381.

Pedrão, M.R.; Beleia, A.; Modesta, R.C.D.; Prudencio-Ferreira, S.H. (1999). Physical and chenical stability and sensory Tahiti natural lemon juice and sweteneed, frozen. *Ciência e Tecnologia de Alimentos*, v. 19, n.2, p. 156-162.

Pellegrini, N.; Serafini, M.; Colombi, B.; Del Rio, D. (2003). Total antioxidant capacity of plant foods beverages and oils consumed en Italy assessed by three different in vitro assays. *Journal of Nutrition*, v. 133, p. 2812-2819.

Pozo-Insfran, D.D.; Brenes, C.H.; Talcott, S.T. (2004). Phytochemical composition and pigment stability of açaí (*Euterpe oleraceae* Mart.). *Journal of Agricultural and Food Chemistry*, v. 52, n. 6, p. 1539-1545.

Rapisarda, P., Bianco, M.L., Pannuzzo, P., & Timpanaro, N. (2008). Effect of cold storage on vitamin C, phenolics and antioxidant activity of Five Orange genotypes (*Citrus sinensis* L. Osbeck). *Postharvest Biology and Technology*, v. 49, p. 348-354.

Rodrigo, R., Miranda, A., Vergara, L. (2011). Modulation of endogenous antioxidant system by wine polyphenols in human disease. *Clinica Chimica Acta*, v. 412, p. 410–424.

Roesler, R., Catharino, R.R.; Malta, L.G.; Eberlin, M.N.; Pastore, G. (2008). Antioxidant activity of Caryocar brasiliense (pequi) and characterization of components by electrospray ionization mass spectrometry. *Food Chemistry*, v. 110, p. 711–717.

Santos, G.M.; Maia, G.A.; Sousa, P.H.M.; Costa, J.M.C.; Figueiredo, R.W.; Prado, G.M. (2008). Correlation betwen antioxidant activity and bioactive compounds of commercial açaí pulp (*Euterpe oleracea* Mart). *Latin Americal Files Nutrition*, v. 58, n. 2, p. 187-192.

SAP - SECRETARY OF AGRICULTURAL PROTECTION. *Normative Instruction number 7* of May 17th, (1999). Official Daily of State of May 19th, 1999, Chapter I, 11p. Available on: www.agricultura.gov.br.

Singleton, V.L.; Rossi, J.A. (1965). Colorimetry of total phenolics with phosphomolybdic-phosphotungstic acid reagents. *American Journal of Enology and Viticulture*, p. 144-158.

Swain, T.; Hillis, W.E. (1959). The phenolic constituents of *Punnus domestica*. I. Quantitative analysis of phenolic constituents. *Journal of the Science of Food and Agriculture*, v. 10, p. 63-68.

Vaillant, F.; Millan, A.; Dornier, M.; Decloux, M.; Reynes, M. (2001). Strategy for economical optimisation of the clarification of pulpy fruit juices using cross flow microfiltration. *Journal of Food Engineering*, v. 48, p. 83-90.

Vargas, P.N., Hoelzel, S.C., & Rosa, C.S. (2008). Determination of total polyphenols content and antioxidant activity in commercial grape juices. *Alimentos e Nutrição*, v. 19, n. 1, p. 11-15.

Yasan, H.; Zhijuan, J.; Shunxin, L. (2007). Effective clarification of apple juice using membrane filtration without enzyme and pasteurization pretreatment. *Separation and Purification Technology*, v. 57, p. 366-373.

Ziena, H.M.S. (2000). Quality attributes of Bearss Seedless lime (*Citrus latifolia* Tanaka) juice during storage. *Food Chemistry*, v.71, p.167-172.

20

Computer-Based On-Line Assessment of Sterilizing Value and Heat Distribution in Retort for Canning Process

Montip Chamchong, Vilasinee Sangsom and Nuttakorn Charoeamkitti
Department of Food Engineering, Faculty of Engineering at Kamphaengsaen
Kasetsart University, Nakhon Pathom
Thailand

1. Introduction

Heating process for food is of importance to the consumers since it is considered to be one of food preservation techniques. Under these techniques food can be stored or edible within a long period of time. One of them which require heat treatment is sterilization process. Thermal sterilization of prepackaged canned foods in retort has been the most widely used during the twentieth century. Typically this method consists of heating food containers in pressurized retorts at specified temperatures for prescribed lengths of times (Teixeira and Tucker, 1997).

The process time for canned food is indicated based on the sufficient achievement of bacterial inactivation in each container in order to comply with public health standards or food safety. In addition it will minimize the probability of food spoilage. The traditional methods for thermal process calculations or validation such as Ball and Stumbo methods were developed and widely used ever since. However they required the off-line input of tables and consequently series of calculation steps which might be resulting in too-long or too short heating process. At present there are a lot of commercial software available which could be used either on-line or off-line analysis for sufficient heat treatment or process lethality (F_o) such as CAN-CALC©, and CALSoft™ etc. Balaban (1996 cited by Teixeira et al., 1999) described that CAN-CALC software needed to get f_h (heating rate factor) and j_h (heating lag factor) from heat penetration test prior to be able to predict internal center product temperatures in response to any dynamic boundary temperature for products of any shape and size as shown in figure 1 and 2. Therefore if assumed that the selected can was at the slowest heating point of the retort, simulated system F_o for food products that heated by any combination of conduction or convection heat transfer also could be obtained. However, the software performance was emphasizing with its capability to deal with process deviation such as steam shutting off and back on. The CALSoft™ software (Anonymous, 2011) was designed specifically for conducting heat penetration and temperature distribution testing, evaluating the collected data, calculating a thermal process or vent schedule/come-up time, and evaluating process deviations. It was supposed to use with CALPlex™ data logger and claimed for the most widely used commercial thermal processing software.

Accomplished F_o of the coldest point in canned food can be expressed as

$$F_o = \int_0^{t_h} 10^{(T-T_{ref})/z}\ dt \tag{1}$$

It is calculated in the unit of minutes at reference temperature (T_{ref}) ; T is the temperature at coldest point in the container; T_{ref} is a standardized reference temperature, usually 121.1°C or 250°F ; z is temperature dependency factor from microbial thermal death time curve and expressed as temperature change required for a ten-fold change of destruction time. Usually it is 10 °C for Botulinum cook. Equation (1) can be usually evaluated by numerical integration as general method or computed from the sum of the different sterilizing value (ΔF_i) accomplished after small time intervals (Δt) as temperature change throughout the process (Teixeira and Tucker, 1997).

$$F_o = \sum_{i=1}^{n} \Delta F_i = \sum_{i=1}^{n} 10^{(T_i - T_{ref})/z} \Delta t \tag{2}$$

When comparing to all other methods to calculate F_o, the general method has been accepted to be the most accurate. However, the disadvantage of this method is used to be the clumsiness since it has to obtain the lethal rate in every time step. The smaller time step, the more accurate it will be. But when using computer as a tool to perform all these calculation, F_o determination become rapid and simple.

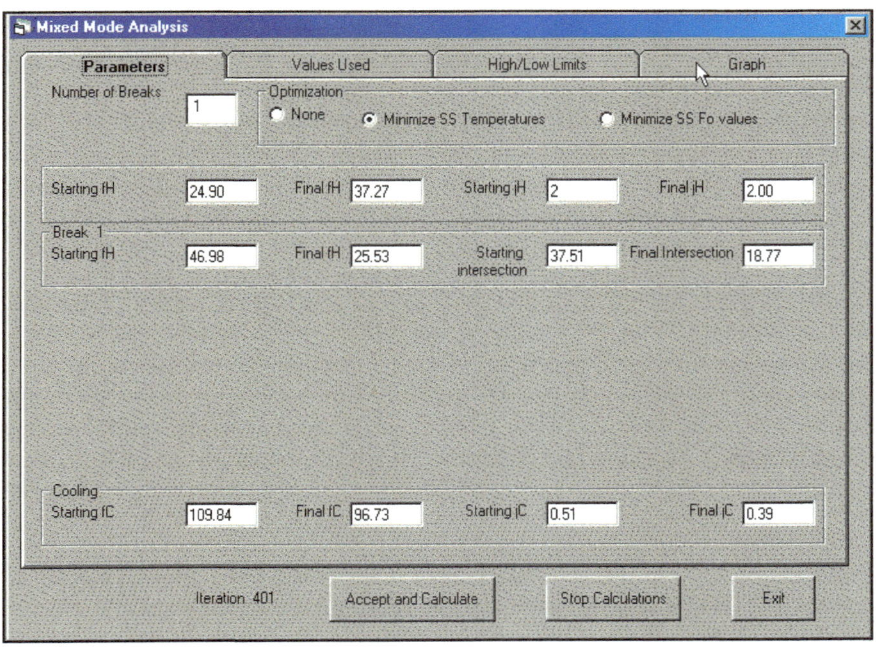

Fig. 1. Parameters input in CAN-CALC software before simulating for system F_o (Balaban, 2004)

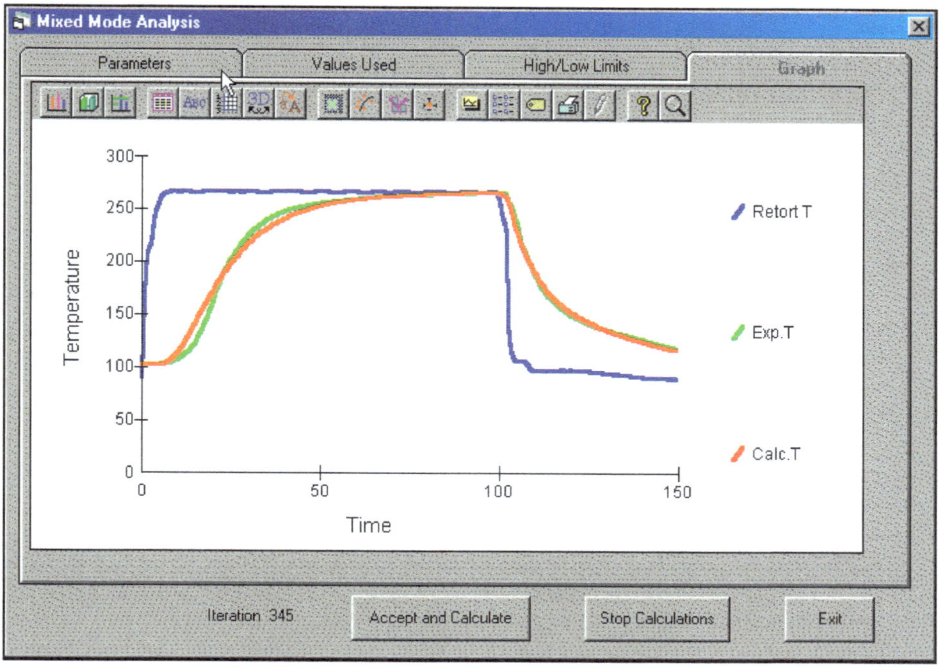

Fig. 2. Graphical display of calculated or predicted and experimented temperature at coldest point in CAN-CALC software (Balaban, 2004)

Many researchers (Lappo and Povey ,1986; Ryniecki and Jayas, 1993) had employed the accumulated process lethality to design system process control for batch steam retort. A number of thermocouples were connected to the cans. The mean temperature at the center of those cans was used for calculating process lethality in real time. Datta et al. (1986) used the numerical solution of 2 dimensional heat transfers in a finite cylinder as a part of the decision-making software in a computer-based retort control system. Actual retort temperature was read directly from sensors located in the retort and it was continually updated with each iteration of the numerical solution. Heating was continued until the accumulated lethality was reached some designated target value and the process would always end with the desired level of sterilization. However their solution of the model has some limitations since purely conduction-heated canned food was simulated for. Later many research works (Bichier et al.,1995 ; Teixeira, 1992) had been done without these limitations.

Visual Basics computer simulation package for thermal process calculation was developed by Chen and Ramaswamy (2007). This graphical user interface (GUI) program was designed for training and testing of artificial neural artwork models and for study of process design or other research purposes. It is applicable to different retort thermal processing with different types of food such as solids, liquids and liquids containing particles in containers of different shapes and size. Temperature in container was solved by using finite difference and a numerical integration method was used for calculating process lethality and quality retention.

There have been several attempts to develop control approaches for thermal process operation in food canning. Traditionally it consists of maintaining specified operating conditions that have been predetermined from product or process heat penetration test. The first control strategy was to employ real-time heat penetration data acquisition for intelligent on-line control of thermally processed foods. It was the most effective way to handle process deviation. However prior to start thermal operation, a number of product containers are instrumented with temperature probes then filling and seaming. Connection is made between these containers to data logger through the lead wires. Computer thus have the real-time accessing the data from data logger and perform calculation for accomplished sterilizing value at the coldest spot of container. The calculated accomplished sterilizing value is continually compared with the target value required at the end of heating. This strategy provides very accurate calculation of process lethality and is able to handle the process deviation without operator intervention and without any unnecessary degree of over-processing. The most valuable feature of this control strategy is that it is nearly foolproof since any thing that might have gone wrong earlier in the product preparation is revealed and accounted for. However, the obvious disadvantage for this type of control strategy would be cost prohibitive (Simpson et al., 2007).

Another retort control strategy that many researchers had worked on is about on-line correction of process deviation which is integrating of the real-time data acquisition for retort temperature, on-line correction factor and mathematical heat-transfer model of can temperature (Teixeira and Manson, 1982; Datta et al., 1986; Teixeira and Tucker, 1997). However, the strategy that will be the future trend is microcontroller-based retort control system or simply on-line temperature measurement of retort to lap top computer. When the calculated accomplished lethality reaches the specified target value, computer will automatic shut off or turn on valves (Simpson el al., 2007). Awuah *et al.* (2007) discussed that Can-Calc process simulation software also was tested for its performance and further integrated into a computer-based on-line control system by Noronha et al. (1995) and Teixeira et al. (1999).

As a whole, the purpose of software design and hardware control was based on the fact that foods should not be overheated since it leads to detrimental effect to food quality as well as the waste of energy and water. Thus heat should be minimally applied or applied as necessary as it needs. In order to get the above mentioned process, it is essential to have proper machine or devices associated with analysis method to assess the process efficacy involving heat treatment for any one of canned products and heat distribution in sterilizing device.

However, in Thailand most of the hardware and software available now are imported. They are designed basically on either the post assessment (after completely heating foods) or undergo heating. Up to now more efforts have been carried out for developing the intelligent on-line retort control system which is capable of rapid evaluation, on-line correction and printed documentation. The development of local devices or software for such purposes is still rarely found in Thailand. Thus the objectives of the research are to develop visual basic computer software to integrate the on-line data acquisition. The assessment of sterilizing value or process lethality (F_o) as well as heat distribution in retort was performed while heating. The software also can be used as an education tool for thermal processing study.

2. Materials and method

2.1 On-line data acquisition and sterilizing value (F_o) assessment

Quick Basics program was designed and developed to obtain the interfacing data from PCL-812PG card (multifunction data acquisition card) together with PCLD-889 boards (amplifier/multiplexer board with signal conditioning and cold junction sensing circuit) as in figure 3. Up to 8 thermocouples could be instrumented to the loaded cans and hard-wired through a retort (figure 4). They were used for sensing the analog inputs of temperatures from different locations in the retort and then they were transformed to digital temperature data via the interface PCL-812PG A/D card.

Thus time-temperature history data from tested cans and some for temperature in the retort were recorded and displayed graphically in every 4.5 second in the developed program as GUI software coded by Visual Basic 6.0. Prior to the test, all the temperature reading probes were calibrated from the temperature range of ambient to 140°C by comparing to the reading of reliable portable digital thermometer measuring hot oil.

The designed computer program is able to access the recorded Quick-Basic data file which provide real-time of time-temperature history in cans and retort, and calculate them for lethal rate in every 4.5 second by Simpson's rule of numerical integration to obtain accomplished F_0 dynamically during sterilizing. In order to evaluate the accuracy of this program, time-temperature history data was also tested with F-ADDING which is a computer program for calculating F_0 coded by Rouweler (2000). The minimum accomplished F_0 among all of them from each probe is obtained as system F_0 and it is compared simultaneously with the target F_0 needed to end the process in the minimum of process time. The flowchart of the algorithm for on-line F_0 assessment is shown in figure 5. This approach was accepted as unquestionably the most effective and most certainly the very safest on-line correction when process deviation occurs (Teixeira and Tucker, 1997; Simpson et al., 2007) since thermocouples were used to on-line measure temperature not only retort but also the cans.

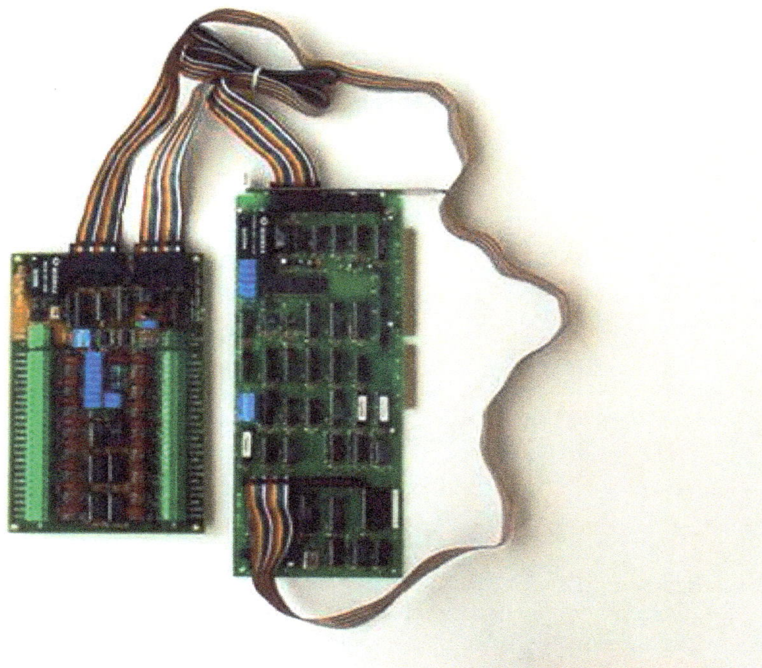

Fig. 3. Enhanced A/D multi-lab card & programmable gain amplifier/multiplexer board

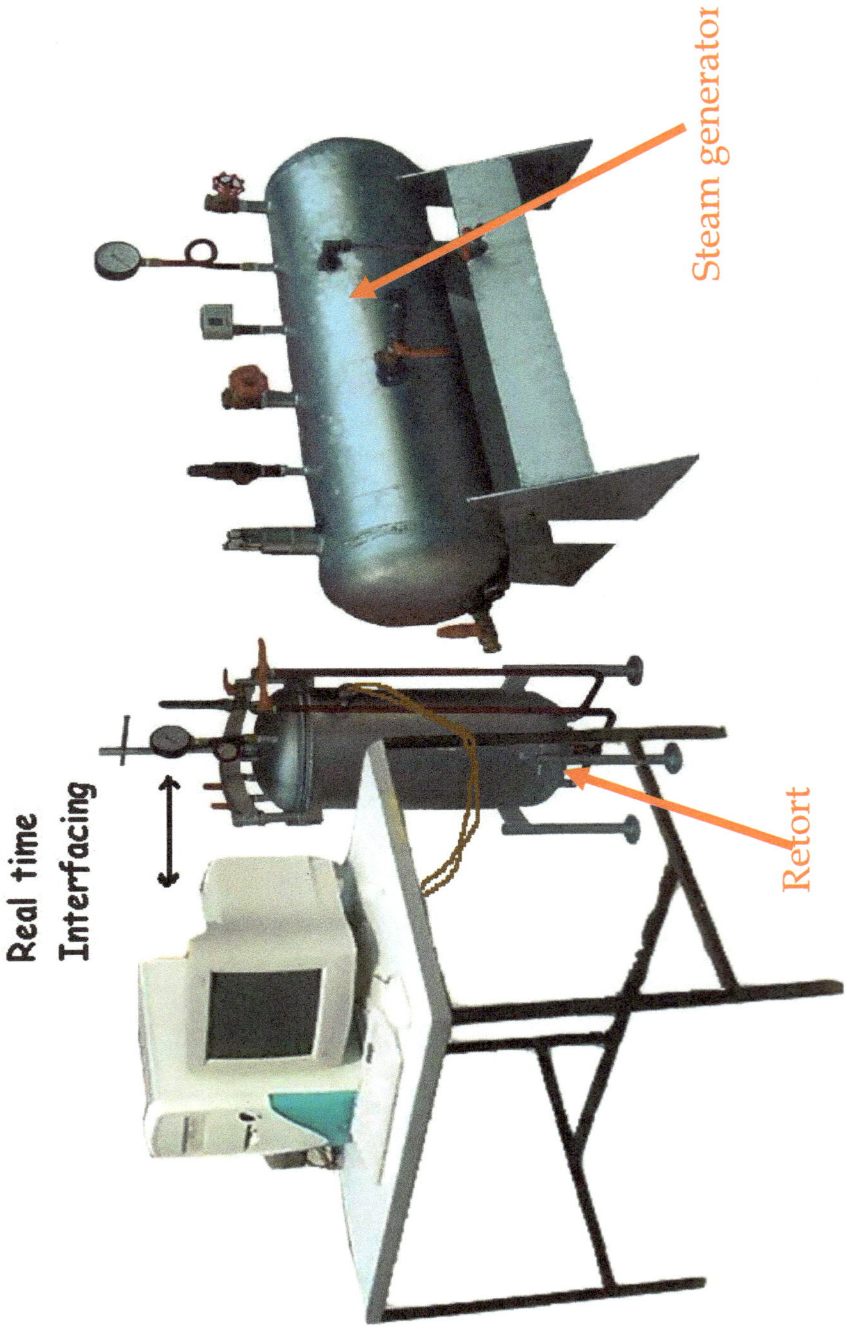

Fig. 4. Computer-based on-line assessment systems

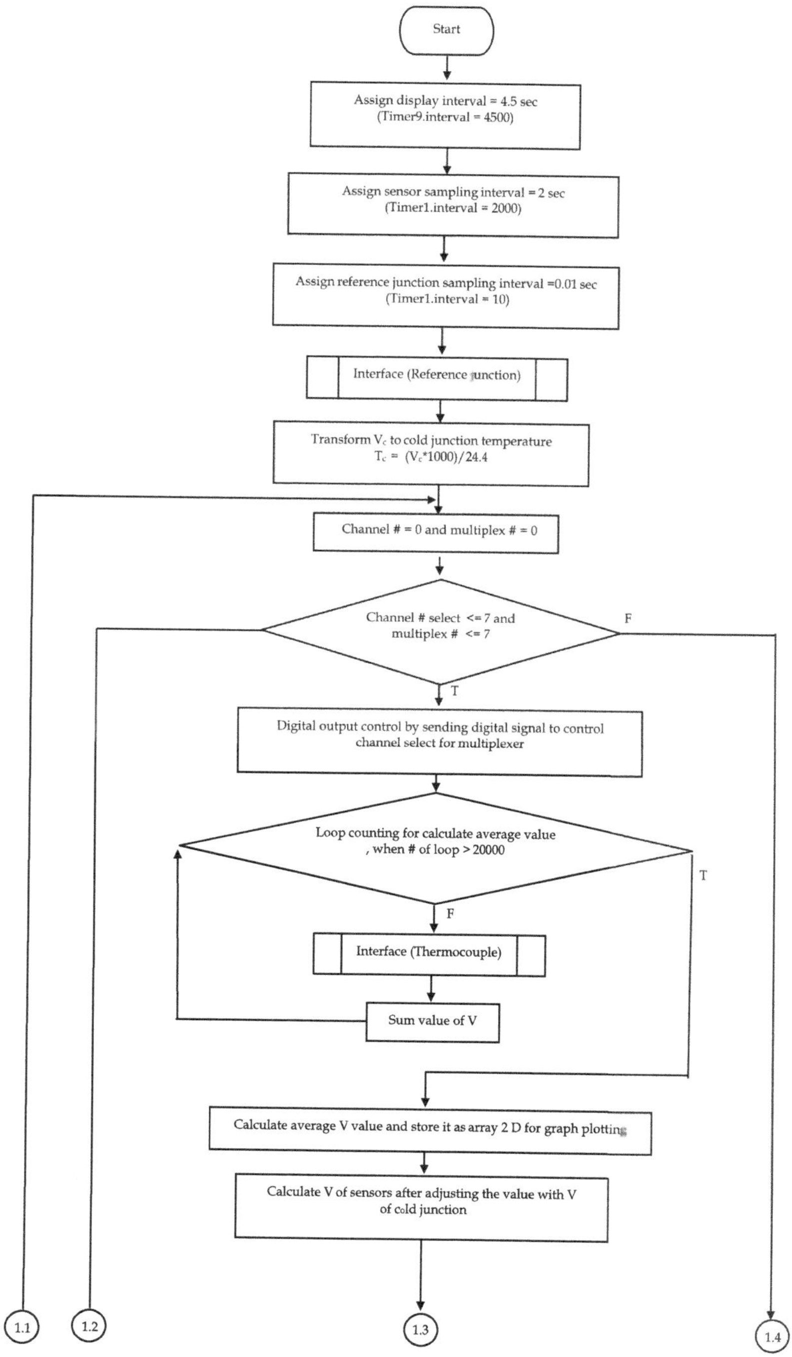

Fig. 5. Program algorithm for dynamic accomplished F_0 determination

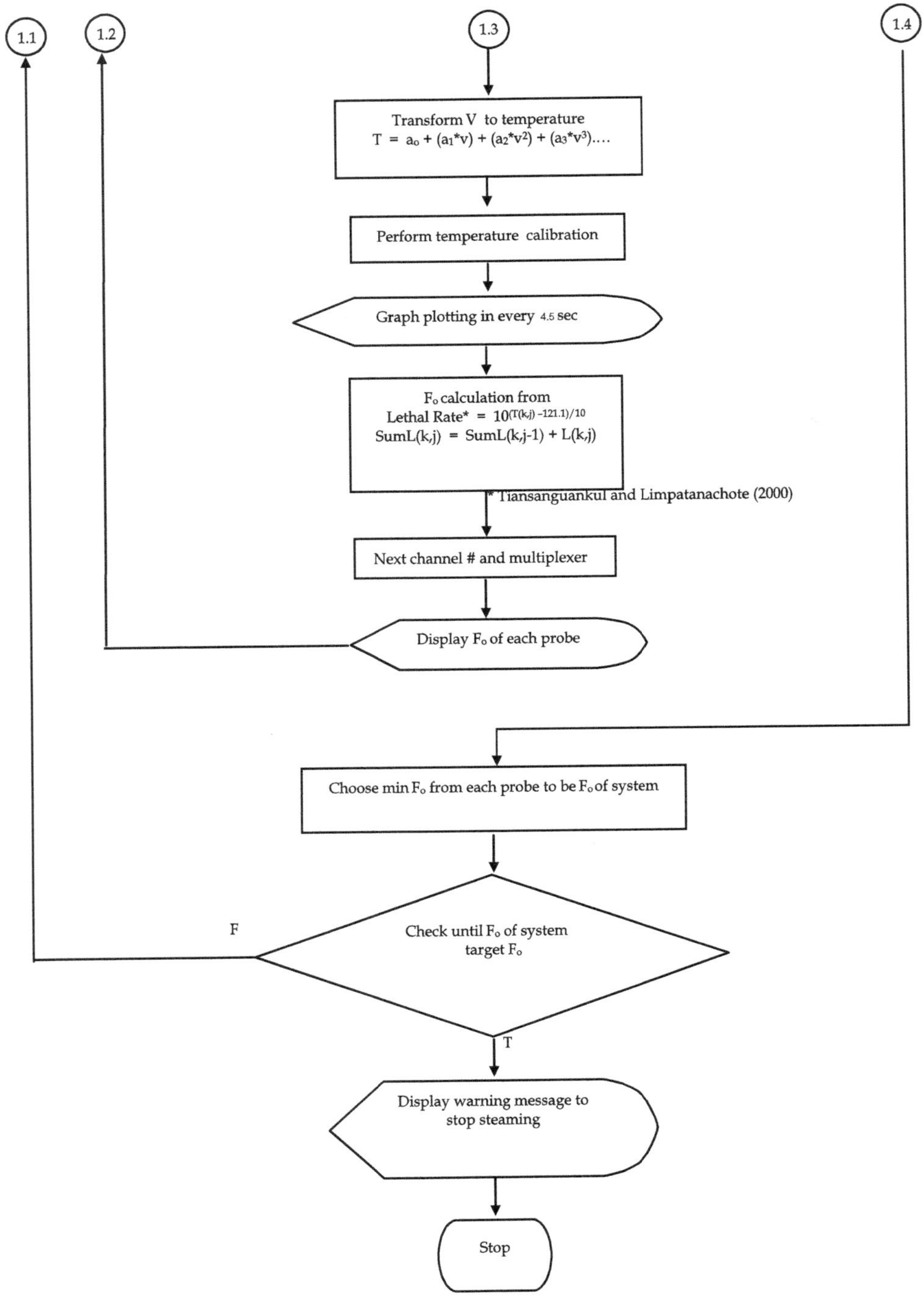

Fig. 5. Program algorithm for dynamic accomplished F_o determination (continue)

2.2 Heat distribution performance in retort

A small vertical retort with diameter of 38.8 cm and electric boiler were constructed for the test as shown in figure 4. The interfacing devices was assembled - interface cards, thermocouples, connectors, computer and peripheral equipments- vertical retort and electric boiler. One probe of thermocouples (probe # 1) were connected to the end tip of mercury thermometer in the retort and one (probe # 8) connected to the center of can which was hot filled with distilled water and then seamed. The rest of them (6 probes) were distributed appropriately inside the retort as shown in figure 6. The on-line graphical display of temperature from 8 thermocouple probes was shown while heating. In addition, the probes # which provides minimum and maximum temperature, as well as maximum temperature difference were indicated through out the heating process. The sterilization temperature at 110 and 121°C were chosen to investigate the heat distribution in the retort.

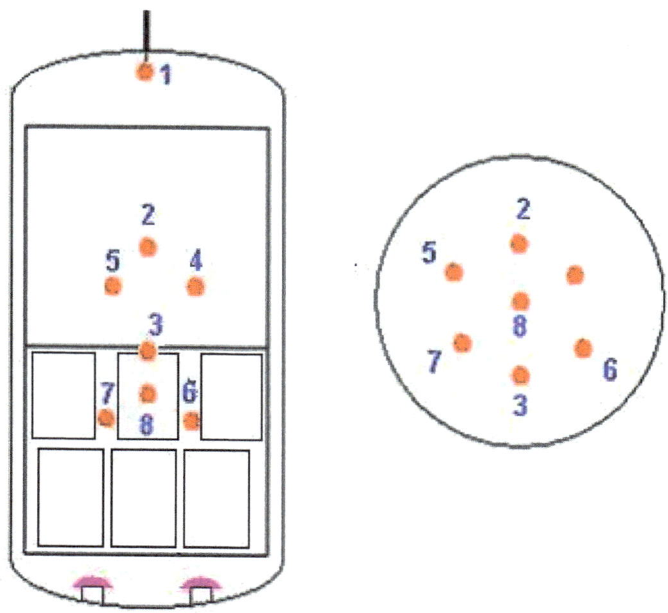

Fig. 6. Positions of thermocouple probes in heat distribution test

2.3 Process design and minimum heat accumulation in canned products

Heat accumulation in canned products during sterilizing could be investigated either from their heat penetration profiles or accomplished F_0 values. Thus three probes of thermocouples were connected to the cans and located those (3 cans) in the basket since these 3 locations tend to be the cold points of system - probes # 3, 4, and 5 attached to the cans located at the positions 1, 3, and 5 in the basket respectively (figure 7) and probe # 6 exposed directly to the temperature of the heating medium in the retort.

The cans were hot filled with concentrated pineapple juice then seamed and put into the basket at specific locations in the retort as mentioned above. The retort was full loaded with the rest of the cans. Specify target sterilizing value was chosen according to product characteristics (table 1) in GUI window. The information about target sterilizing value could be added to the file by pressing updated button. Subsequently sterilization was commenced by removing air in retort by replacing it with steam. Start button in F_o determination GUI window was pressed to begin recording time-temperature data via the interfacing devices until the minimum value of accomplished F_o (system F_o) reached specified target F_o. Thus process schedule was recorded automatically and displayed graphically. Even though concentrated pineapple juice is acid food (pH > 4.0), mild sterilizing is usually sufficient and therefore could be applied. Specified target F_o would be chosen as $F_{121.1}^{10}$ = 0.6-0.8 minutes (Rouweler, 2000) in this case. However in order to demonstrate the process design with this educational tool, the experiments were carried out with the holding temperature during sterilization selected to be at 110°C and 120°C.

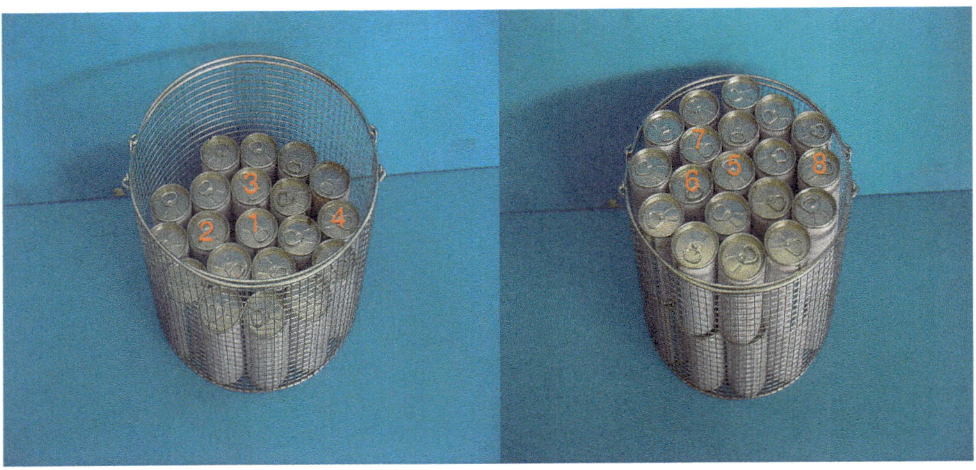

Fig. 7. Location of cans which connected to thermocouples in the bask

Product	Can size	pH	°Brix	Sterilizing Temp.(°C)
Concentrated pineapple juice	200 x 202	4.06	60.4	110
				120

Table 1. The characteristics of tested product.

2.4 Coldest spot in container

To validate the capability of program when it was used to find the coldest spot in a container, the interfacing devices were assembled as before. One of the cans was hard-wired with 2 thermocouple probes: probe# 4 at 1/4 of central axis from bottom of can, and probe# 5 at half of central axis as shown in figure 8 while probe# 6 was for measuring temperature of medium in the retort. This can was filled up with baby corns in saline solution which was solids in liquid type of canned food, and then seamed. The position to put this can in the retort was the slowest heating point of this equipment. After the retort was full loaded with cans, sterilization process was carried out at 121°C for some certain period of time as before, minimum accomplished F_o of which could be obtained.

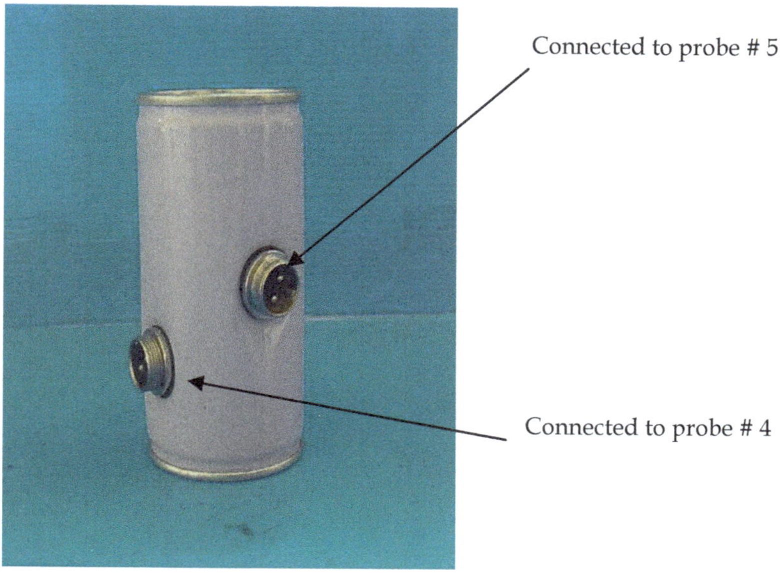

Fig. 8. The connecting points to validate the coldest spot in a can

2.5 Portable educational tools for computer-based off-line assessment of sterilizing unit

The objective of this part was to design the computer program for assessing process lethality from interfacing data system via USB-A/D board. Thus the driver and interfacing program for data logger, National Instrument USB-9211A, 4 Channel 24 bit (figure 9) was installed to a note book computer. The commercial interfacing software was sensing voltage signal through 1 to 4 type-T thermocouples and transforming to digital data stored in a note book computer (figure 10).

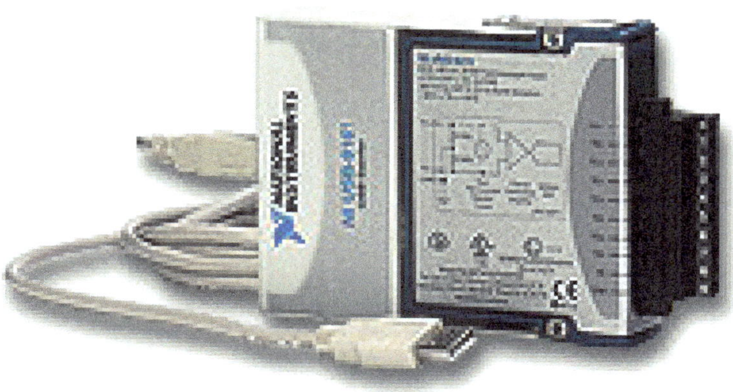

Fig. 9. USB A/D board (NI USB-9211A 4 CH, 24 BIT)

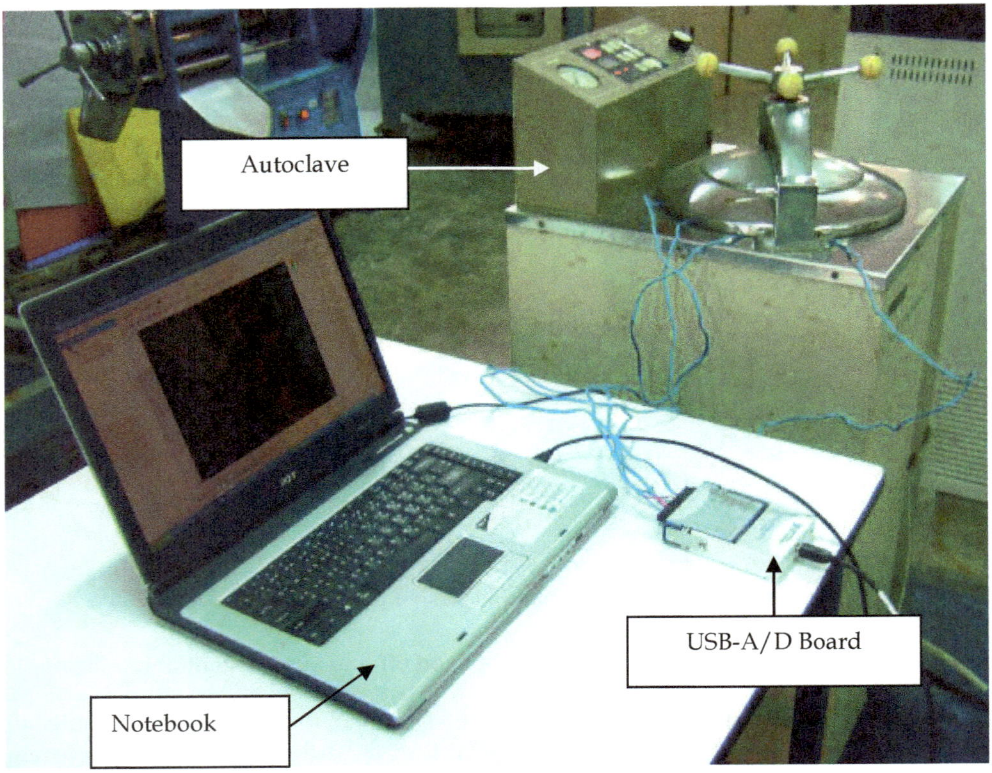

Fig. 10. Hardware installed for the off-line assessment of sterilizing unit

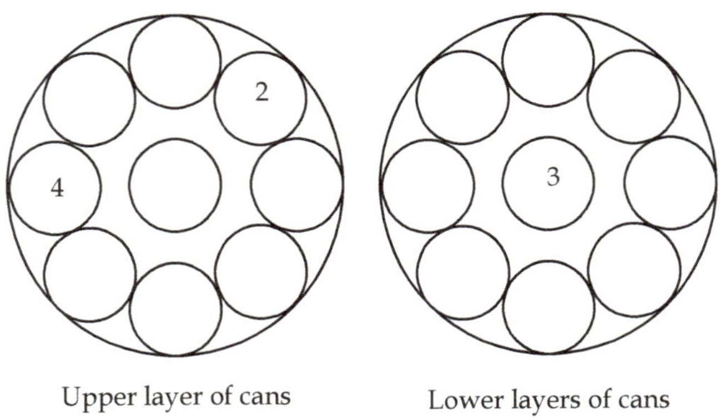

Fig. 11. The locations of cans connected to thermocouple probes # 2-4 in the autoclave

Tab water samples in 300*407 size cans were prepared to be full loaded in vertical commercial autoclave (HA-240MII/-300MII, Japan). Three cans were hard-wired to the thermocouple probes # 2-4 at the coldest point of cans (figure 11) and located at possibly slowest heating point in the autoclave. Therefore one of the can attached to probe # 3 was located in the basket at the center of bottom layer as the most probably slowest heating point. However one thermocouple (probe # 1) was exposed in the autoclave indicating medium temperature measurement during sterilizing. Temperature of 121 °C for 15 min was chosen for demonstrating sterilizing condition. Temperatures from 4 channels was recorded and stored in text file (*.txt) for every 2 seconds after running autoclave until finishing cooling process. QuickCalFo (Chamchong et al., 2008) was software designed to perform the off-line process lethality assessment by using Visual Basic 6.0 program. Input data of temperature and time during sterilization was retrieved from stored text file (figure 12) while target F_o for each product with specific can size was pre-entered and saved into the program or selected from the list of available data before starting analysis for system F_o. The result could display the temperature and time record in a spread sheet as well as heat penetrating curves and lethal rate profiles. F_o values from each temperature-time profile were calculated by Simpson's rule general method and the minimum value was shown in the combo box nearby as system F_o or accomplished F_o. The accuracy of F_o calculation from this software was validated as before by comparing it with that obtained from F-ADDING program coded by Rouweler (2000).

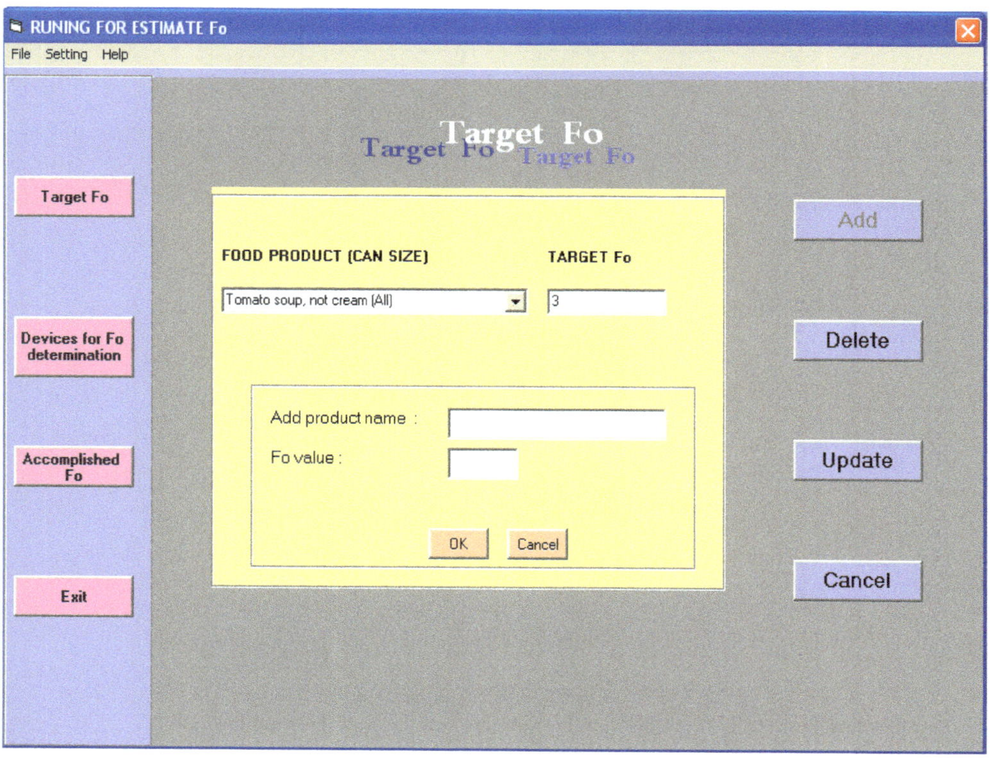

Fig. 12. Visual Basics form to specified target sterilizing value (F_o)

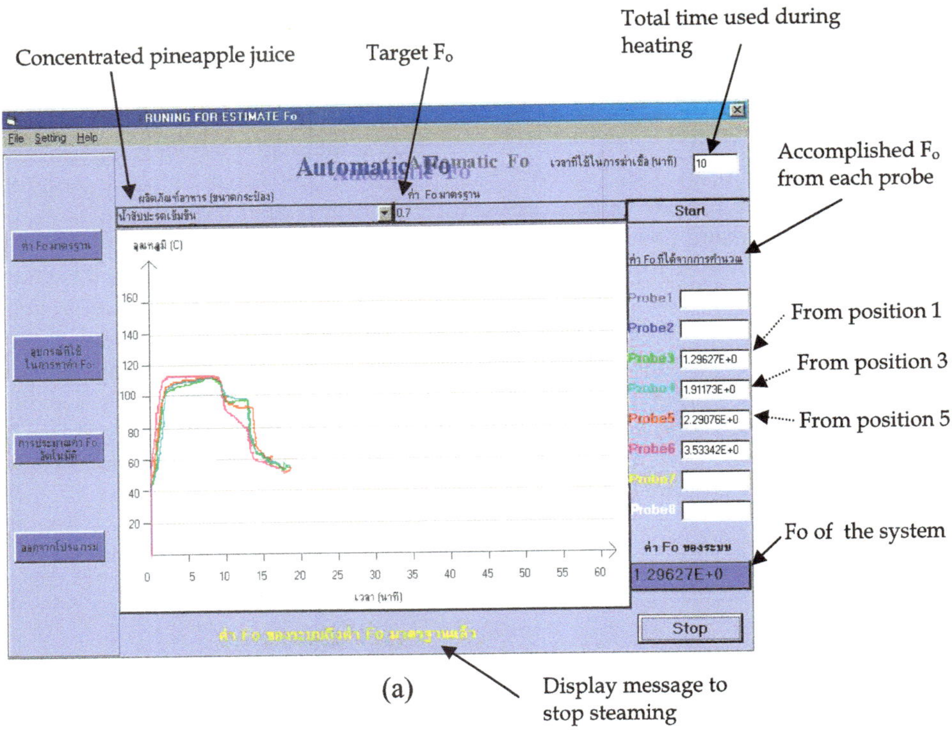

(a)

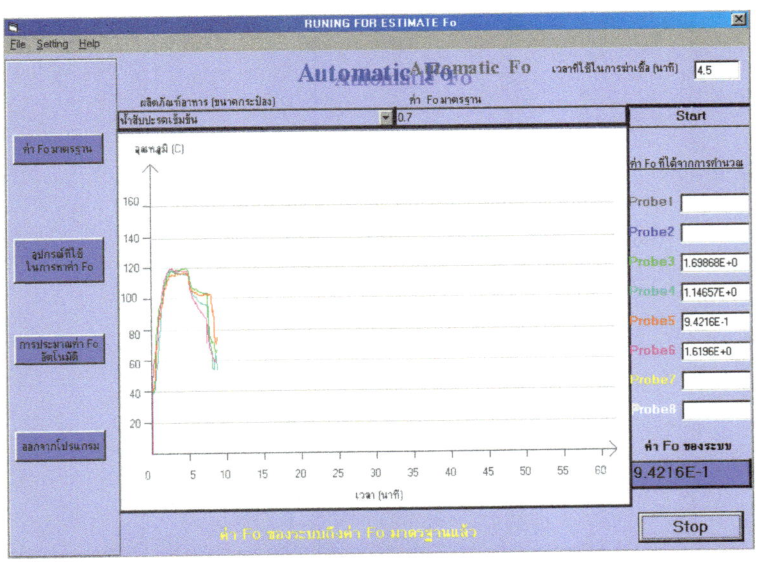

(b)

Fig. 13. Real-time heat penetration curve and F_o of concentrated pineapple juice when sterilizing at (a) 110 °C (b) 120°C

3. Results and discussion

3.1 Computer-based on-line assessment of sterilizing value

The software package for process design was divided into 3 parts: (1) the main window of the GUI to receive the input parameter which is target sterilizing value. The user can choose this value from pull down combo box or add/delete and update to have more choices for later use (figure 12). (2) Graphical window of temperature and time profiles with 8 corresponded text boxes to display accomplished sterilizing values from maximum 8 probes (figure 13). There is one text box at the bottom to display system sterilizing value which is the minimum value among all of accomplished sterilizing values from each probe. System sterilizing value increases while the process is underway heating and cooling and ultimately reaches the designated target sterilizing value. Program then displays text message at the bottom of GUI for the operator to stop steaming and total process time during heating is shown in upper right corner text box. The temperature and time record can be used for process design or as documentation in quality assurance system. (3) The spread sheet of temperature-time recorded from 8 thermocouple probes which could display minimum and maximum temperature, as well as maximum temperature difference (max-min) at each time interval through out the heating process (figure 14).

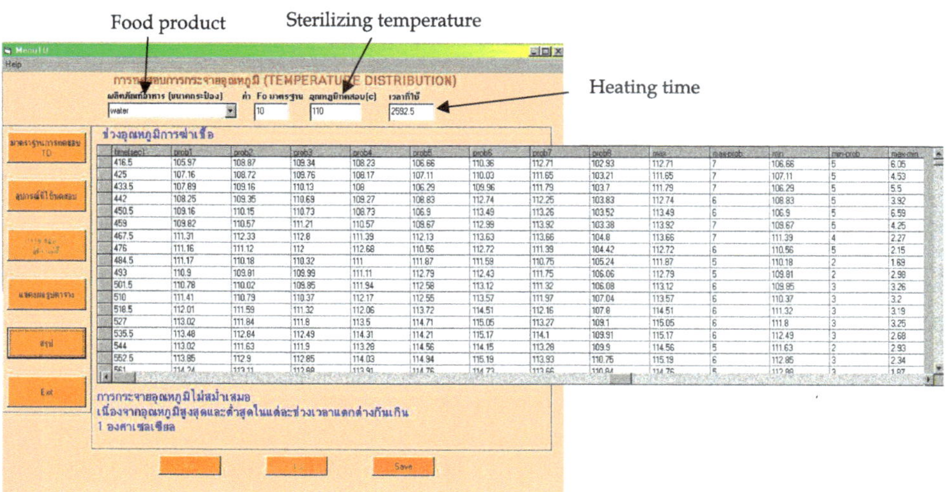

Fig. 14. Real-time heat distributions in retort, record from 8 thermocouple probes

3.2 Heat distribution in retort

Practically heat distribution in a retort should be carried out before performing the assessment of sterilizing value of the process in order to validate the slowest heating point. Therefore heat distribution in a retort (as in part 3 from mentioned above) was observed from on-line temperature record obtained from different locations of this equipment. For sterilizing at 110°C in a small retort unit distributed heat could be indicated by temperature values at positions 1-8 in the retort corresponding to probe # 1-8 (figure 6). In addition minimum heating reading from thermocouple probe which was connected to the can located at the slowest heating point was able to be quantified as the accomplished F_0 for the system. Then it could be used as an indicator for stopping steaming. Therefore the display was able to assure the minimum heat distribution occurring while heating. The coldest point of the system should be coming from the can which had the thermocouple connected with and was located at position 8 or at the upper layer of cans and at the center of the basket. To get enough heat treatment for the products, heat distribution test must be carried out once the machine was installed or process/product was modified.

For the same retort, it was found that when holding temperature of sterilization was moderate at 110°C heat distribution was more uniform than that at 121°C. Temperature difference from max and min at any holding time or temperature deviation was between 1.8 to 3.1°C (1.6-2.8%) when sterilizing temperature was at 110°C but it was between 5-14°C (4.1-11.6%) at 121°C sterilizing temperature. This was possible since heating at 121°C required higher heating rate. However more of stagnant point or dead legs would appear. According to the steam flow pattern in this retort, the probe located at positions 3 (on top of the can which was at the center of basket) and 6 (upper layer and in between cans) were found to be the stagnation points and shown minimum convective heat transfer in each run at higher sterilizing temperature. However the minimum heating point for lower sterilizing temperature was found to get changed to be either position 5 or 2 (top of the retort) at the early stage of holding temperature in sterilizing period and then changed to location 3 for the rest of holding time. This was possible because the more amount of steam used during heating at 121°C, the narrower the stagnation area would be. In addition, when lower amount of steam used at 110 °C sterilizing temperature, the probes at position 5 and 2 in the top layer of cans initially would get contacted with steam slower than any other locations. After heating at this temperature for a while, heating was up to the top of retort. Temperature at position 5 and 2 would be no longer the minimum.

However the exit point of coming steaming was from the bottom. Whenever steam valve was not fully opened the thermocouples in the lower layer would be affected or heated first. Therefore accomplished F_0 obtained from the can at the center upper layer of the basket was suitable to be system F_0 because the temperature from the probe nearby (probe 3) had demonstrated minimum heat received.

3.3 Process design or schedule and minimum heat accumulation in canned products

The process design or schedule of acid food like concentrated pineapple juice was shown in table 2. The obtained process time of sterilizing acid food at 110 and 120°C was 10 and 4.5 minutes excluding cooling period respectively. At higher sterilizing temperature (120°C), it was shorter than that at lower temperature (110°C) since both were calculated based on the same specified target F_0 (0.7 minutes) or having the same area under the heat penetration curve before stopping steaming. Although the specified target sterilizing value for such a product was chosen to be 0.7 minutes but F_0 of the system obtained was 1.29 and 0.94 minutes for sterilizing at 110 and 120°C, respectively. This was because the calculation was including come-up time, holding and cooling period. Thus a little over-process could occur in each run since there was slow removal of heat during cooling. Improved and proper design of cooling system in the retort would provide better product quality in terms of organoleptic properties.

Heat accumulation in canned products can be observed from all accomplished F_0 values obtained from different locations in retort. For sterilizing at 110°C in figure 13 (a), accumulated heat which could be indicated by accomplished F_0 at positions 1, 3, and 5 were

1.29, 1.91 and 2.29 minutes, respectively, while the one outside the can was 3.53 minutes. Thus this was assuring that the coldest point was from the position 1 or at the bottom layer of cans and at the center of the basket. To get enough heat treatment for the products, heat distribution test must be carried out once the machine was installed or process/product was modified.

Sterilizing period	Time used during sterilizing at 110°C (min)	Time used during sterilizing at 120°C (min)
Come-up time	2.5	2.5
Holding period	7.5	2.0
Cooling period	7.0	1.5
Total process time	10.0	4.5
Calculated system F_o	1.29	0.94

Remark: Calculated system F_o is including come-up time, holding and cooling period.
Total process time does not include cooling period.

Table 2. Process design or schedule obtained from heat penetration curve of concentrated pineapple juice when target F_o = 0.7 minutes

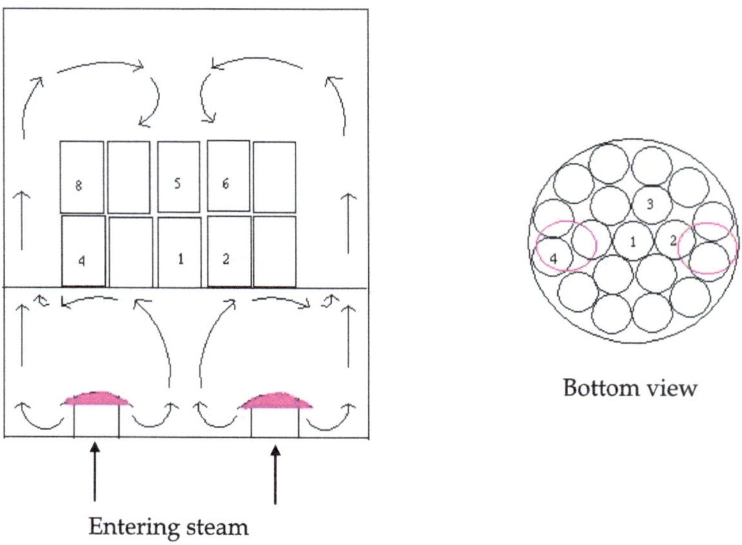

Fig. 15. Configuration of steam flow pattern in retort

From heat penetration curve, F_o of the system represents the minimum accumulated heat at coldest point of can in the retort. As in figure 13 (a) and (b), system F_o was obtained from different probes or positions of cans i.e. probe # 3 (position 1) from figure 13 (a) and probe # 5 (position 5) from figure 13 (b), the F_o of which were 1.29 and 0.94 minutes respectively. According to the steam flow pattern of this retort shown in figure 15, the cans located at positions 1 and 5 were supposed to be at the stagnation points and had minimum convective heat accumulation in each run. However the minimum accumulated heating point was found to get changed from position 1 located at the bottom center of the basket to position 5 located at the upper center of the basket while sterilizing at higher temperature. This was possible because the more amount of steam used during heating at 120°C, the narrower the stagnation area would be. In addition, while cooling, the can at position 5 (top layer of cans) would get contacted with blowing air into retort during balancing pressure right after stop steaming. Therefore system F_0 would be obtained from the can at the center upper layer of the basket due to smaller heat accumulation. However cold water could significantly enhance heat removal when it was leveled up to that position in retort.

3.4 Coldest spot in a container

The sterilizing values at 2 different points could be used as indicated tool to validate the coldest spot in a container while undergo sterilizing process. As shown in figure 8 the can was instrumented with 2 thermocouple probes and hard-wired through a retort. Probe# 4 was at 1/4 of central axis from bottom of can, and probe# 5 was at the half of central axis while probe# 6 was for measuring temperature in the retort. It was found that temperature rising and dropping from probes #4 and 5 were almost identical and hard to distinguish. In addition, during sterilizing there were 3 main different periods of operations - come-up time, holding at sterilizing temperature and cooling periods. Heating rate at different period through the can and different spot in the can could be varied. Thus practically coldest spot in the can may not be the same point at all the time. However minimum accumulated heat in the can need to be known since it was used as critical point to evaluate for enough heat treatment of the process. Thus minimum sterilizing value was useful and reliable to represent the minimum accumulated heating point in the can. From table 3, the sterilizing values corresponding to probes # 4, 5 and 6 were 3.69, 2.85 and 9.54 minutes respectively. According to the meaning of accomplished sterilizing value of the process, the coldest spot was found from the minimum value (2.85). So for this type of food, baby corns in saline solution, heat transfer was the slowest at half of central axis of container.

Probe #	Connecting point	Sterilizing value at 121°C (min)
4	1/4 of central axis from bottom of can	3.69
5	the half of central axis of can	2.85
6	Exposed to heating medium in retort	9.54

Table 3. Validation of coldest spot in the can by sterilizing value

3.5 Portable educational tools for computer-based off-line assessment of sterilizing unit

The temperature and time logging on-line was displayed as in figure 16 but they were retrieved off-line as a text file shown in figure 17. Then F_0 was calculated from these data which were coming from thermocouple probes connected to 3 cans and one exposed to heating medium in the autoclave. From QuickCalFo as shown in figure 18, it showed that the slowest heating point in the retort was from thermocouple probe # 3 which was attached to the can located at the center bottom layer of the basket in the autoclave. Thus the accomplished sterilizing value or system F_0 was chosen from the minimum of 12.81 minutes among all from 4 thermocouple probes. In analysis F_0 frame box at the bottom right corner, the message after

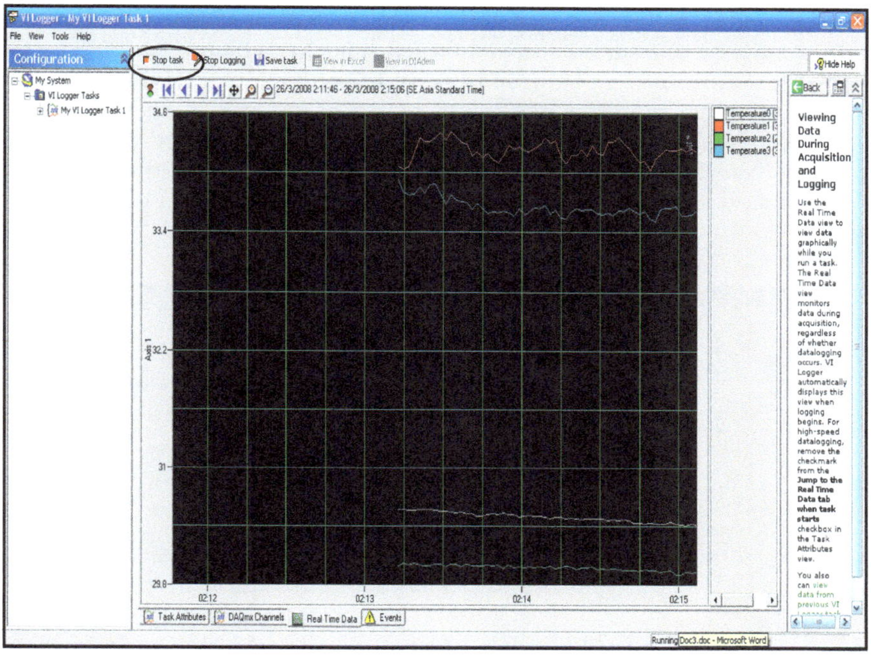

Fig. 16. Temperature and time display in real time recording via USB-A/D board

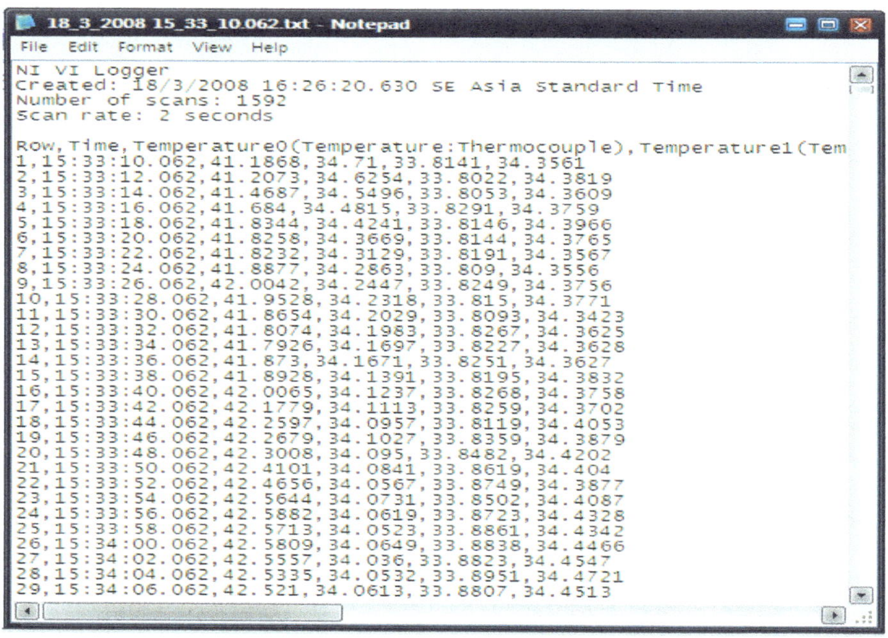
18_3_2008 15_33_10.062.txt - Notepad

File Edit Format View Help

```
NI VI Logger
Created: 18/3/2008 16:26:20.630 SE Asia Standard Time
Number of scans: 1592
Scan rate: 2 seconds

Row,Time,Temperature0(Temperature:Thermocouple),Temperature1(Tem
1,15:33:10.062,41.1868,34.71,33.8141,34.3561
2,15:33:12.062,41.2073,34.6254,33.8022,34.3819
3,15:33:14.062,41.4687,34.5496,33.8053,34.3609
4,15:33:16.062,41.684,34.4815,33.8291,34.3759
5,15:33:18.062,41.8344,34.4241,33.8146,34.3966
6,15:33:20.062,41.8258,34.3669,33.8144,34.3765
7,15:33:22.062,41.8232,34.3129,33.8191,34.3567
8,15:33:24.062,41.8877,34.2863,33.809,34.3556
9,15:33:26.062,42.0042,34.2447,33.8249,34.3756
10,15:33:28.062,41.9528,34.2318,33.815,34.3771
11,15:33:30.062,41.8654,34.2029,33.8093,34.3423
12,15:33:32.062,41.8074,34.1983,33.8267,34.3625
13,15:33:34.062,41.7926,34.1697,33.8227,34.3628
14,15:33:36.062,41.873,34.1671,33.8251,34.3627
15,15:33:38.062,41.8928,34.1391,33.8195,34.3832
16,15:33:40.062,42.0065,34.1237,33.8268,34.3758
17,15:33:42.062,42.1779,34.1113,33.8259,34.3702
18,15:33:44.062,42.2597,34.0957,33.8119,34.4053
19,15:33:46.062,42.2679,34.1027,33.8359,34.3879
20,15:33:48.062,42.3008,34.095,33.8482,34.4202
21,15:33:50.062,42.4101,34.0841,33.8619,34.404
22,15:33:52.062,42.4656,34.0567,33.8749,34.3877
23,15:33:54.062,42.5644,34.0731,33.8502,34.4087
24,15:33:56.062,42.5882,34.0619,33.8723,34.4328
25,15:33:58.062,42.5713,34.0523,33.8861,34.4342
26,15:34:00.062,42.5809,34.0649,33.8838,34.4466
27,15:34:02.062,42.5557,34.036,33.8823,34.4547
28,15:34:04.062,42.5335,34.0532,33.8951,34.4721
29,15:34:06.062,42.521,34.0613,33.8807,34.4513
```

Fig. 17. Recorded data obtain via USB-A/D board

analyzing indicated whether the product had enough heat treatment or not. To display temperature and time as a spread sheet on the left side of this figure, time interval of data was selected first at the bottom between in every 1, 2 or 5 minutes. Then the temperature and time only at this specific interval was shown in the spread sheet of GUI. Heat penetration and lethal rate profiles from all 4 probes were also displayed graphically. The x and y range of these 2 graphs were adjusted automatically according to process temperature and time span used. In addition, this visual basic form could be printed out for food safety documents.

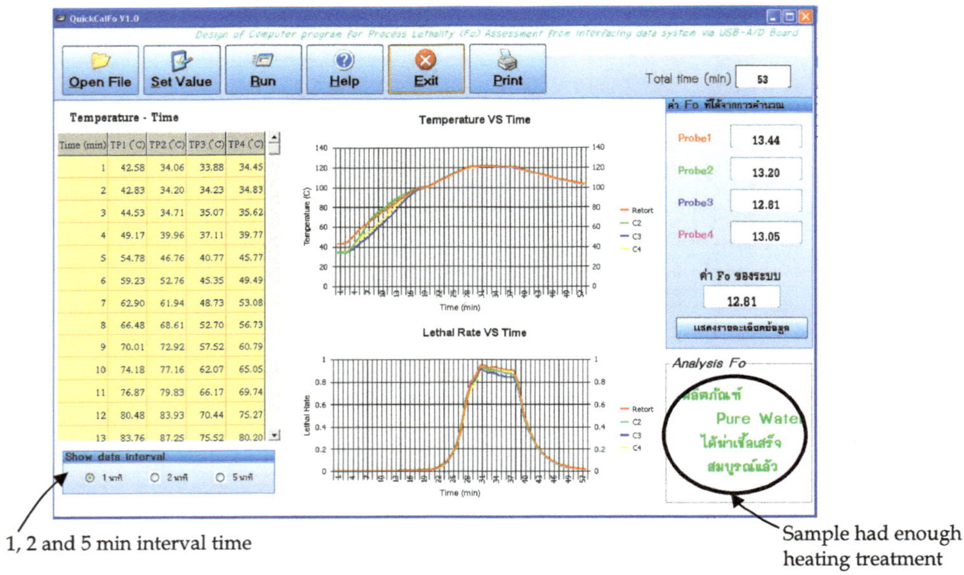

Fig. 18. The result displayed by QuickCalFo

4. Conclusion

A computer program for on-line data acquisition and accomplished F_0 assessment was developed by MS Visual Basic 6.0 language. This computer-based on-line device was able to evaluate the coldest point of the can in the retort and calculate process lethality or system F_0 dynamically while sterilizing. Too much over or under process was avoidable for process design or schedule with the integration of such a device for on-line accomplished F_0 determination during preprocessing. The setup of hardware and software for computer-based on-line assessment of sterilizing unit would be needed for the cases of new products, processes or equipments.

Non-uniform heat distribution in retort always exists. The designed program was able to perform heat distribution evaluation by recording and displaying maximum/minimum temperature deviation at different locations in retort during holding temperature in sterilization process. Lower sterilization temperature at 110°C had lower temperature deviation (1.6-2.8%) in the retort during the holding temperature comparing to 4.1-11.6% at 121 ^{o}C. Thus lower temperature tended to lower the deviation.

Coldest spot in a container during sterilization process can be verified by employing sterilizing value (F_o). The minimum of these values is corresponding to the point in the can which has the lowest accumulated heat. The coldest spot in the can of baby corns in saline solution while undergo sterilizing was confirmed to be at half of central axis of can by committing minimum value of F_o.

Portable or handy educational tools for sterilizing process was necessary to be available since most of the processes may routinely carry out for the same process schedule. The result could be used to assure food safety control.

5. Acknowledgement

This study was supported by faculty of Engineering at Kamphaengsaen, Kasetsart University, Nakhon Pathom 73140 Thailand.

6. References

Anonymous. (January 2011). TechniCAL. Consulting Services and Products for Food Processing Industry. 6 January 2011, Available from http://www.tcal.com/CALSoftSoftware.php

Awuah, G. B., H. S. Ramaswamy and A. Economides. 2007. Thermal Processing and Quality : Principles and Overview. *Chemical Engineering and Processing*, 46 , pp. 584-602.

Balaban, M. O. (March 2004). Software for Food Engineering. Thermal Processing. 15 January 2008, Available from http://mob.ifas.ufl.edu/SOFTWARE.HTM

Bichier, J. G., A. A. Teixeira, M. O. Balaban and T. L. Heyliger. 1995. Thermal Process Simulation of Canned Foods under Mechanical Agitation. *Journal of Food Process Engineering*, Vol. 18, No.1, pp. 17-40.

Chamchong, M., Sukcham, K and T. Katsarapong. 2008. Design of Computer Program for Process Lethality (F_o) Assessment from Interfacing Data System via USB-A/D Board. Engineering Project. Department of Food Engineering. Faculty of Engineering at Kamphaengsaen. Kasetsart University. Nakhon Pathom, Thailand.

Chen, C. R. and H. S. Ramaswamy. 2007. Visual Basics Computer Simulation Package for Thermal Process Calculations. Chemical Engineering and Processing, 46 ,pp. 603-613.

Datta, A. K., A. A. Teixeira and J. E. Manson. 1986. Computer-based Retort Control Logic for On-line Correction of Process Deviations. Journal of Food Science, 51, pp. 480-483.

Lappo, B. P. and M. J. W. Povey. 1986. Microprocessor Control system for Thermal Sterilization Operations. *Journal of Food Engineering*, 5, pp. 31-53.

Noronha, J., M. Hendricks, A. van Loey and P. Tobback. 1995. New Semi-empirical Approach to Handle Time-variable Boundary Conditions during Sterilization of Non-conductive Heating Foods. *Journal of Food Engineering*, 24, pp. 249-268.

Rouweler, J. 2000. Main Unit Operations in Food Processing. Material handout in 10th International Course on Food Processing. International Agricultural Centre. Wageningen, The Netherlands.

Ryniecki, A. and D. S. Jayas. 1993. Automatic Determination of Model Parameters for computer Control of Canned Food Sterilization. *Journal of Food Engineering*, Vol. 19, No. 1, pp. 75-94.

Simpson, R., A. A. Teixeira and S. Almonacid. 2007. Advances with Intelligent On-line Retort Control and Automation in Thermal Processing of Canned Foods. *Food Control*, 18, pp. 821-833.

Tiansanguankul, J. and P. Limpatanachote. 2000. The Application of Computer Interfacing Technique for F_0 Determination in Canned Food. 212499 Food Engineering Project. Department of Food Engineering. Faculty of Engineering at Kamphaengsaen. Kasetsart University, Nakhon Pathom.

Teixeira, A. A. 1992. Thermal Process Calculation. Chapter 11, pp. 563-619. *In* D. R. Heldman and D.B. Lund (eds). Food

Engineering Handbook. Marcel Dekker, Inc, New York.

Teixeira, A. A. and J. E. Manson. 1982. Computer Control of Batch Retort Operation with On-line Correction of Process Deviations. *Food Technology*, Vol. 36, No. 4, pp. 85-90.

Teixeira, A. A. and G. S. Tucker. 1997. On-line Retort Control in Thermal Sterilization of Canned Foods. *Food Control*, Vol. 8, No. 1, pp. 13-20.

Teixeira, A. A., M. O. Balaban, S. P. M. Germer, M. S. Sadahira, R. O. Teixeira-Nero and A. A. Vitali. 1999. Heat Transfer Model Performance in Simulation of Process Deviations. *Journal of Food Science*, 64, pp. 488-493.

21

Ice-Temperature Storage Technology of Fruits and Vegetables

Liu Bin, Cai Bihao and Shen Jiang
Tianjin Key Lab of Refrigeration Technology, Tianjin University of Commerce, P.R.China

1. Introduction

The ice-temperature storage technology belongs to the no-freezing storage technology, in which the food is stored at a temperature from 0°C to the freezing point. The fundamental thought of ice temperature storage technology is that the stored product is a living organism. After the cooling disposition on certain conditions, it can reach to an approximate "hibernate" state. Preserved at this state, the product can be maintained in good quality and its life can reach to the longest level since its metabolic rate is the smallest and so is the energy consumption.

The ice-temperature storage technology was originated from Japan. Their found that the freeze point of the snake, frog and meat is below 0°C for the antifreeze substance such as the sugar, the protein and alcohol and they can be kept as a living organism. The results pointed out that the temperature limit of life and death is below zero centigrade and at which temperature, the cell can be kept a living state. The effect of the ice-temperature storage technology on the fresh food mainly includes the followings.

1. The ice-temperature storage can inhibit respiration, delay the respiratory peak and reduce the loss of nutrients. The exhaled volume of CO2 in the ice-temperature (-0.8°C) is less than that stored in normal temperature about 30% to 60%.
2. The ice temperature storage won't destroy the cells, but improve the quality of fruits and vegetables in the vicinity of freezing temperature. In order to prevent from forming ice, fruits will secrete large amounts of antifreeze (its main ingredients are glucose, amino acids, aspartic acid etc.) to reduce the freezing point, or decompose the starch into sugar. These physiological changes improve the quality of fruits and vegetables in different degrees.
3. The ice temperature can inhibit the growth of microbial efficiently. Under the condition of the ice temperature, water molecules in fruits or vegetables are arranged in an orderly state, which reduce the content of free water available for the microbial. In the short-period storage, the ice temperature can inhibit multiplication of microorganisms, better than the frozen temperature. In the long-term storage, the ice temperature and the frozen temperature maintain the same level of a reproduction rate of bacteria, fungi and low-temperature bacteria, which is much smaller than that of the cold temperature.
4. Since ice temperature can inhibit chemical reaction strongly, the food quality in ice temperature is better than that of the normal cold storage. The ice temperature also can inhibit lipid oxidation, non-enzymatic and other chemical reaction.

Based on these advantages of the ice temperature storage, the study of the ice temperature technology on fruits and vegetables was carried out in our lab. The obtained result shows that the key for the ice temperature preservation is to realize a hibernation state of the product. The "hibernation" process is a cooling process during which the product can reduce its activity ability and energy consumption through the self adaption, starting from the change of components within cells, and at the same time ensuring their own living life characteristics. It is a typical phenomenon of natural adaption. However, the traditional precooling process ignores that and the storage effect couldn't reach to the best level.

Consequently, for the ice temperature storage technology, the prime technical key points are determination of the freezing point, the cooling process and the stability of storage environment. According to these key points, studies of the influence of freezing point, processing and storage environment were carried out.

2. Determination of freezing point

The freezing point is the storage critical point of all the cryogenic stored products and plays a decisive role in corresponding storage process. Therefore, the research of the freezing point has not only formed the basis of refrigeration field, but also been a hot researching topic. The moisture in food is not pure water, but liquor including organic and inorganic substances. So the ice crystals won't produce until the temperature of food reduce to subzero. According to LaWuEr second law, the reduction of liquor freezing point is proportional to the solute concentration. Freezing point will decrease 1.86°C with increasing 1 molar concentration. Due to the difference of food types, the stored conditions after harvest and the concentration of muscle plasma etc, different food has different freezing point. The freezing point of general food is from -0.5°C to -2.5°C.

There are two common ways to measure the freezing point, one is the traditional freezing method, and the other is DSC testing method. In this study the first method was adopted, for that we can clearly see the cooling process, the corresponding supercooling points and phase change during the measuring time in the method. Figure 1 shows the test setup.

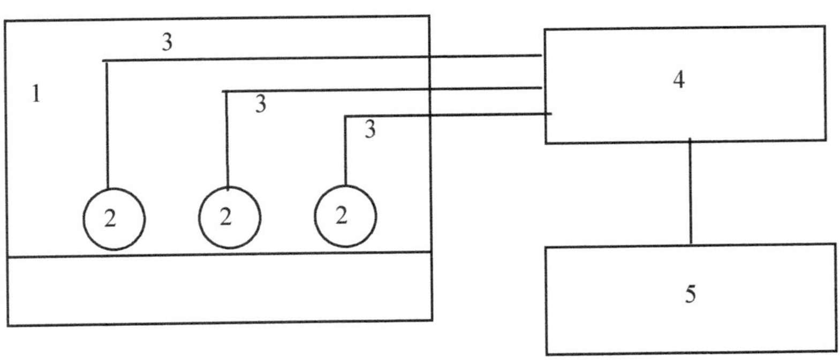

1. alternating thermostatic box, 2. tested items
3. Thermocouple, 4. Data collector, 5. computer

Fig. 1. Diagram of the setup testing the freezing point

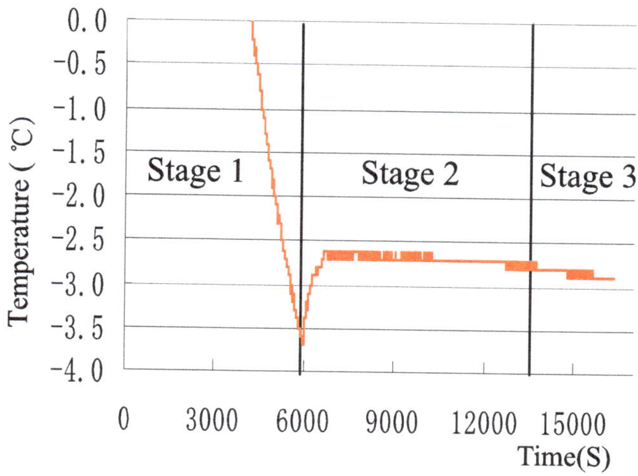

Fig. 2. Curve of temperature change during freezing

From the Figure 2, the temperature profile can be divided into three stages:
The first stage: the food's temperature is reduced from the initial temperature to the freezing point, the heat releasing from food is sensible heat. Compared with the total heat, this value is small, so the cooling speed is fast and the freezing curve is relatively steep.
The second stage: the food's temperature is almost kept at a constant temperature, which value is from 0°C to -5°C. At this stage, most of the water in food is frozen and release a lot of latent heat which is about 50~60 times as much as sensible heat. It is in the second stage that most of the heat in food freezing process is released. So there is a flat segment in the curve.
The third stage: the food's temperature reduce from the phase change temperature to the final temperature, the heat released at this time is partly due to ice cooling ,and partly because of the residual small amount of water getting frozen. The freezing curve in this stage is also relatively steep.

2.1 Influence of reducing sugar on freezing point

The freezing point shown in Fig. 3 has an increasing trend with the increase of mass fraction of reducing sugar, which is not apparent. But it doesn't match traditional recognition. Usually the freezing point is believed to drop with the increase of mass fraction of reducing sugar. The tendency of pear's freezing point just reflects the relationship in Fig. 4.The reason maybe that kiwi has a large amount of acid which influence the effect of sugar on freezing point. This is just reflected in figure 7 and figure 8.

2.2 Influence of total sugar on freezing point

From the trend line of relationship between freezing point and total sugar in Fig. 5 and Fig. 6, we can find that freezing point increases with the increase of mass fraction of total sugar, but this relationship is weak.

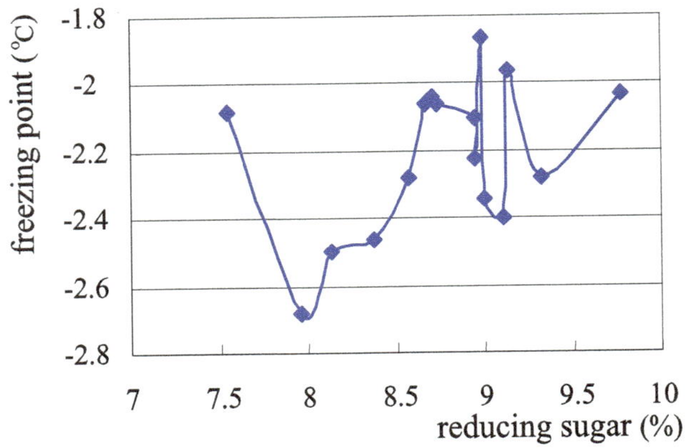

Fig. 3. freezing point of Kiwi vs. reducing sugar

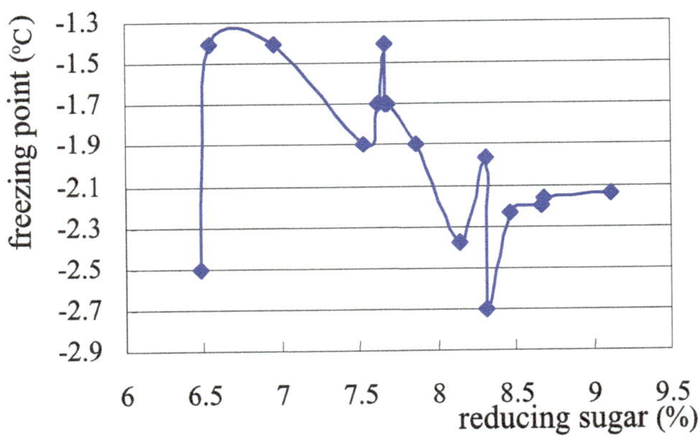

Fig. 4. freezing point of pear vs. reducing sugar

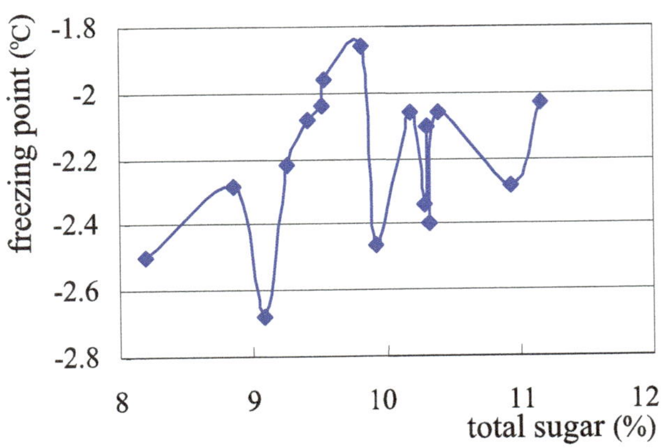

Fig. 5. freezing point of kiwi vs. total sugar

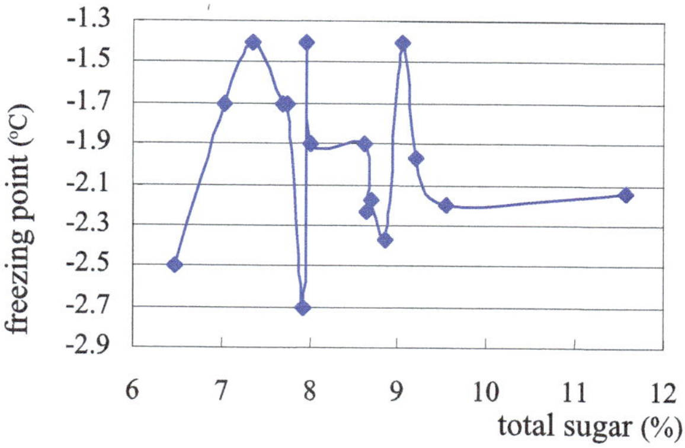

Fig. 6. freezing point of pear vs. total sugar

2.3 Influence of acid on freezing point

From Fig.7 and Fig.8 it can be found that the relationship between freezing point and acid is not apparent, but the freezing point has a downward tendency with the increase of acid in the case of smaller acidity.

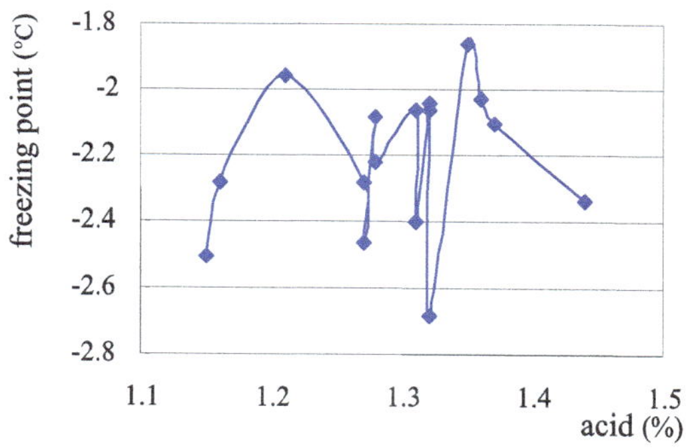

Fig. 7. freezing point of kiwi vs. acid

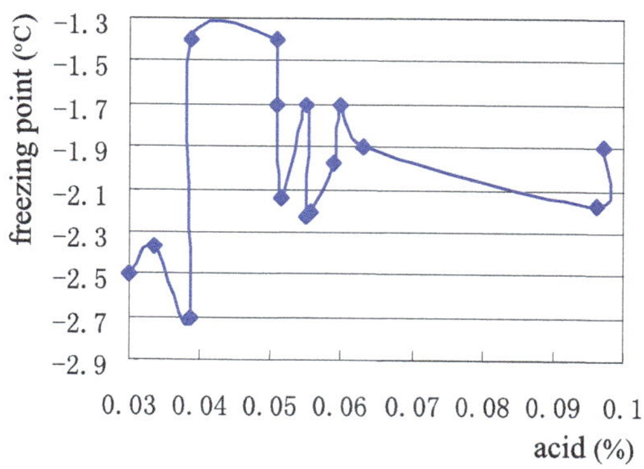

Fig. 8. freezing point of pear vs. acid

2.4 Influence of soluble solid on freezing point

Seen from Fig.9 and Fid.10, the freezing point has a downward tendency with the increase of mass fraction of soluble solid, but the decline rate of kiwi's freezing point is smaller than that of the pear.

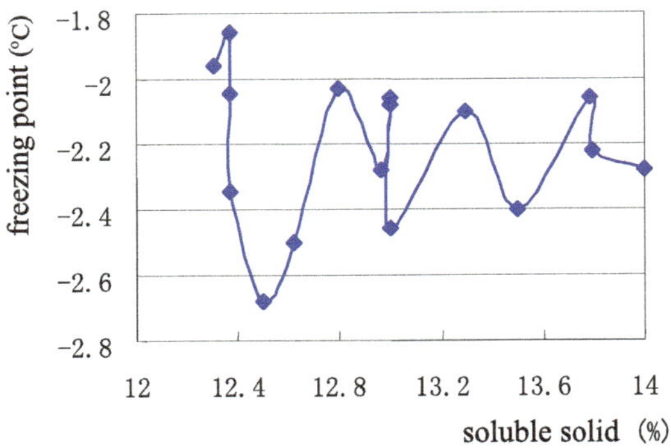

Fig. 9. freezing point of kiwi vs. soluble solid

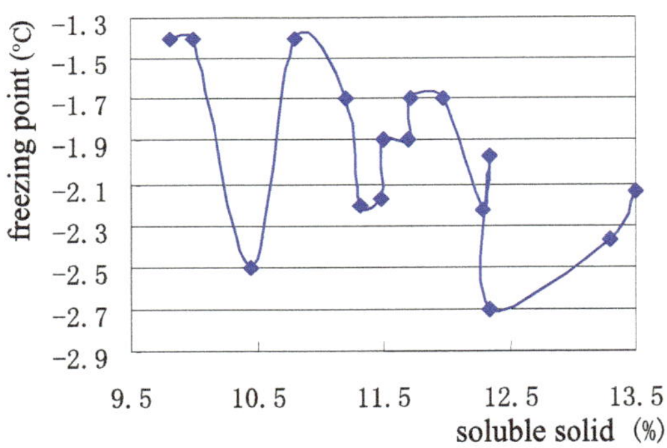

Fig. 10. freezing point of pear vs. soluble solid

From the curve relationship between above four kinds of nutrition and freezing point, it can be found that there is a strong relationship between freezing point and the nutritional proportions for a defined fruit. The experimental results show that freezing point of fruits is related with multiple nutritional proportions, among which the interaction effect may influence the changes in freezing point. For example, the acid has a reverse effect on sugar. In general, the freezing point will drop accordingly as sugar content increases. However, this effect becomes weak in the role of acid, which can be verified by the relationship between the freezing point and reducing sugar and total sugar shown in Fig. 3 and Fig.5.

2.5 Regression of freezing point and mass fraction

Based on the experimental data, the linear regression of the relationship between freezing point and the mass fraction of each component were carried out shown in Eq.(1) and Eq.(2) as follows.

Kiwi:

$$t = 0.117198 \times x_{\text{reducing sugar}} + 0.055817 \times x_{\text{total sugar}} + 0.095504 \times x_{\text{acidity}} - 0.07142 \times x_{\text{soluble solid}} - 2.97913 \quad (1)$$

Pear:

$$t = -0.39643 \times x_{\text{reducing sugar}} + 0.213383 \times x_{\text{total sugar}} + 5.746773 \times x_{\text{acidity}} - 0.14007 \times x_{\text{soluble solid}} + 0.669404 \quad (2)$$

By the checking, calculation, the freezing point of pear by Eq.(1) is calculated, the maximum error is 15%and the minimum error is 1.7%,while calculated by Eq.(2),the maximum error of pear is 25% and the minimum error is 1%. The predicted results coincide better with experimental ones.

3. Influence of precooling process

As living organizations, the fruit and vegetable are affected by severe environmental changes in precooling process. During the precooling process, there are physiological changes in the inside of the fruit and vegetable to adapt to the environment, which can be reflected from two aspects. One is the change of physical structure, such as the body stress, epidermal pore and so on; another one is biochemistry changes in vivo like respiratory intensity and nutrition. In order to investigate the effect of the precooling process, the preceooling experiments of three kinds of fruits were carried out, and the results were compared with the traditional one.

3.1 Material preparation and test method

In this experiment, the experimental storage of different precooling processes on three fruits was carried out, including kiwi, pear and peach.

3.1.1 Precooling process of kiwi

The kiwi was produced in Shengxi Province, after harvested, they was transported to lab in 24 hours.

Traditional storage process of kiwi: kiwi is precooled rapidly from 29°C to 3°C in 3 hours and be stored in the storehouse of which the ambient temperature is 3°C and the relative humidity is 85%.

The variable cooling rate precooling process of kiwi: the kiwi was precooled rapidly from 29°C to 5°C in 2.5 hours, then change the precooling rate, the kiwi was cooled slowly to 3°C in 24 hours ,then to 0°C in 10 hours ,then to -0.8°C in 10 hours, and at last to the storage temperature -1.3°C in 10 hours, stored in the ice-temperature storehouse at -1.3°C and its relative humidity is 85%.

$$29^{\circ}\mathrm{C} \xrightarrow{2.5\,hours} 5^{\circ}\mathrm{C} \xrightarrow{24\,hours} 3^{\circ}\mathrm{C} \xrightarrow{10\,hours} 0^{\circ}\mathrm{C} \xrightarrow{10\,hours} -0.8^{\circ}\mathrm{C} \xrightarrow{10\,hours} -1.3^{\circ}\mathrm{C}$$

3.1.2 Precooling process of pear

The pear was produced in Xingjian Province, and after harvest they were transported to lab in 48 hours.

Traditional storage process of pear: pear was precooled rapidly from 29°C to 3°C in 3 hours and was stored in the storehouse of which the ambient temperature is 3°C and the relative humidity is 85%.

The variable-rate pre-cooling process of kiwi: the kiwi was precooled rapidly from 29°C to 5°C in 3 hours, then cooled the kiwi slowly to 3°C in 24 hours, then to 0°C in 10 hours, then to -1°C in 10 hours, and at last to the storage temperature -1.7°C in 10 hours, after that stored the kiwi in the freezing storehouse at -1.3°Cand its relative humidity is 85%.

$$29°C \xrightarrow{3hours} 5°C \xrightarrow{24hours} 3°C \xrightarrow{10hours} 0°C \xrightarrow{10hours} -1°C \xrightarrow{10hours} -1.7°C$$

3.1.3 Precooling process of peach

The peach was produced in Beijin City, and after harvested they were transported into lab in 24 hours.

Traditional storage process of peach, A: the peach was precooled rapidly from 28°C to 2.5°C in 3.5 hours and be stored in the storehouse of which the ambient temperature is 3°C and the relative humidity is 85%.

In order to verify the influence of the precooling at different cooling rate on the storage of peach, two precooling experiments were carried out, after the precooling, the peaches were stored in the storehouse at -0.7°C with a relative humidity of 85%.

(1)the pre-cooling process of peach at slow cooling rate, B: the peach was precooled from 28°C to 5°C in 3.5 hours, then cooled the peach slowly to 3°C in 3.5 hours, then to 1.7°C in 24 hours ,then to 1°C in 24 hours, then to 0.5°C in 24 hours ,then to 0°C in 24 hours, then to -0.3°C in 24 hours, and at last to the storage temperature 0.7°C in 24 hours, after that stored the peach in the freezing storehouse at -1.3°C and its relative humidity is 85%.

slow cooling rate，B:

$$28°C \xrightarrow{3.5h} 5°C \xrightarrow{35h} 3°C \xrightarrow{24h} 1.7°C \xrightarrow{24h} 1.0°C \xrightarrow{24h} 0.5°C \xrightarrow{24h} 0°C$$
$$\xrightarrow{24h} -0.3°C \xrightarrow{24h} -0.7°C$$

(2) the precooling process of peach at fast cooling rate, C: the peach was precooled from 28°C to 5°C in 3.5 hours, then cooled the peach slowly to 2.5°C in 24 hours, then to the storage temperature 0.7°C in 24 hours. After that, stored the peach in the freezing storehouse at -0.7°C and its relative humidity is 85%.

②rapid cooling rate, C:

$$28°C \xrightarrow{3.5h} 0.5°C \xrightarrow{24h} 2.5°C \xrightarrow{24h} 0.5°C \xrightarrow{24h} -0.7°C$$

3.1.4 Measurement of respiration rate and nutrients

During the experiments, according to different types of fruit, different fruit has different components in fruits were measured, including the respiration rate, the water-soluble vitamin C, the total sugar, the acidity. The respiration rate is determined by gas flow processes, the vitamin C is measured by GB 6195-86 (Chinese), the total sugar is measured GB 6194-86 (Chinese) and the acidity is measured by GB 12293-90 (Chinese). Except the measurement of respiration, six samples were measured and the average value of 6 measurements was used in the analysis.

3.2 Experimental results and discussion

3.2.1 Experimental results and analysis of kiwi

Fig.11 and Fig. 12 show curves of the respiration rate of kiwi and the change of water-soluble vitamin C over storage time in preservation process.

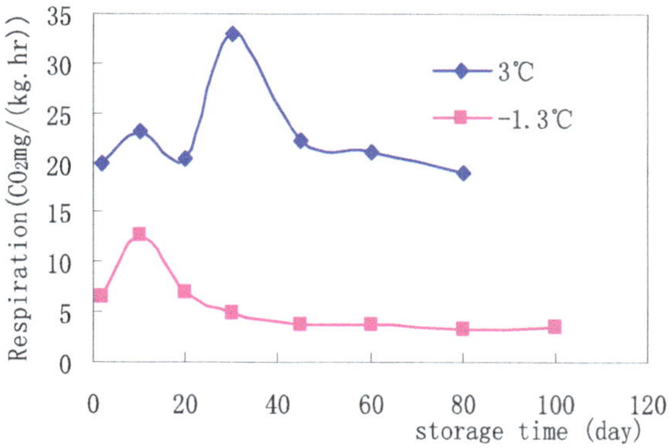

Fig. 11. Curves of kiwi respiration v.s. storage time

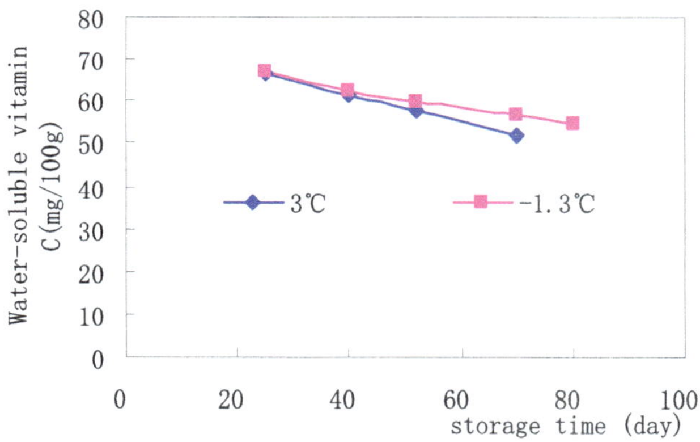

Fig.12 Curves of kiwi water-soluble vitamin C v.s. storage time

From Fig.11, it can be found that the kiwi has two reparatory peaks under the traditional storage condition, while under the condition of the ice temperature, there is only one reparatory peak, and on both storage conditions, the time to show the first reparatory peak is almost the same. The main reason is that when kiwi was put into the storehouse, the stimulation of the low temperature strengthened the respiration rate to protect itself in cold environment. So there is a minor reparatory peak in both the storage conditions. As for the second reparatory peak under the traditional condition, the reason is that the kiwi belongs to breathing-type fruit, the reparatory peak is delayed under the traditional storage condition. But as the the fruit precooled at variable cooling rate and stored in ice temperature, it dosen't has the second reparatory peak .The main reason for that is in the pre-cooling process with different cooling rate, the kiwi has adapted the surrounding well and is stored in the vicinity of ice temperature.

Compared the respiration rate of the two storage conditions, it can be found that the respiratory rate decline rapidly in the vicinity of freezing point, which doesn't match the law of temperature coefficient. From the two figures, the temperature difference between these two storage conditions is only 4.3°C, but the ratio of respiration rate is about 5, which does not meet with the common views.

3.2.2 Experimental results and analysis of pear

Fig. 13 to Fig. 15 show the curves of the respiration rate, the total sugar and the acid of pear over storage time.

From Fig.13, we can found that the pear under traditional storage condition has a reparatory peak after 70 days, which means that the pear has begun dying to death. As for the pear precooled at variable cooling rate, it has a smaller reparatory peak at the beginning of storage (ten days) at -1.7°C and at other time the respiration maintains at a smaller value, which prolong storage time in a certain degree. The main reason for this smaller reparatory peak of the pear with different cooling rate is that when pears are just stored in the storehouse at -1.7°C, they show a conditional stimulation reaction.

From the nutrition showed in Fig.14 and Fig.15, the pear precooled at variable cooling rate has higher sugar content and acidity, so that it can ensure a better quality.

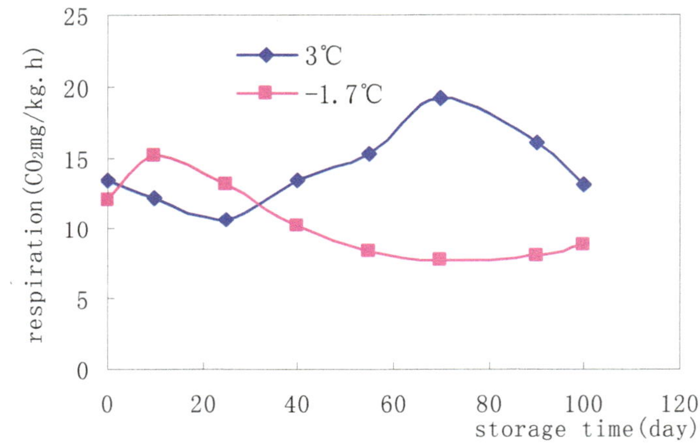

Fig. 13. Curves of pear respiration v.s. storage time

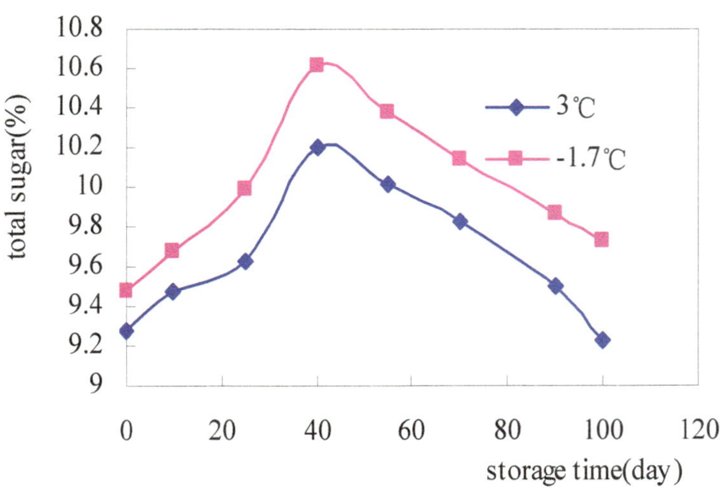

Fig. 14. Curves of pear total sugar v.s. storage time

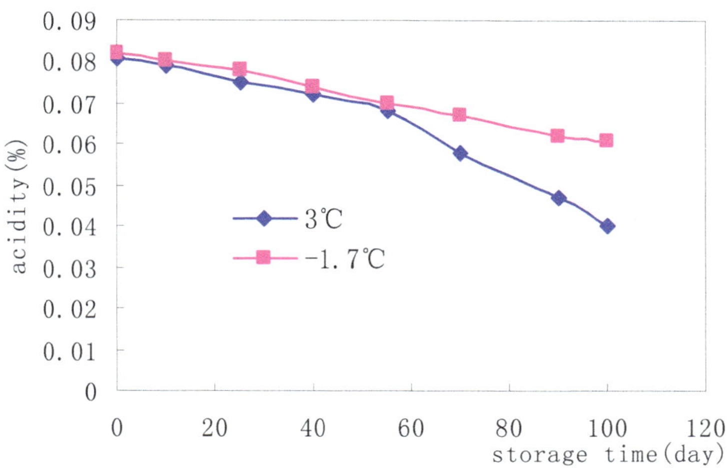

Fig. 15. Curves of pear acidity v.s. storage time

3.3 Experimental results and analysis of peach

Fig.16 to Fig.18 show the curves of the respiration rate, the total sugar and the acid of peach over storage time.

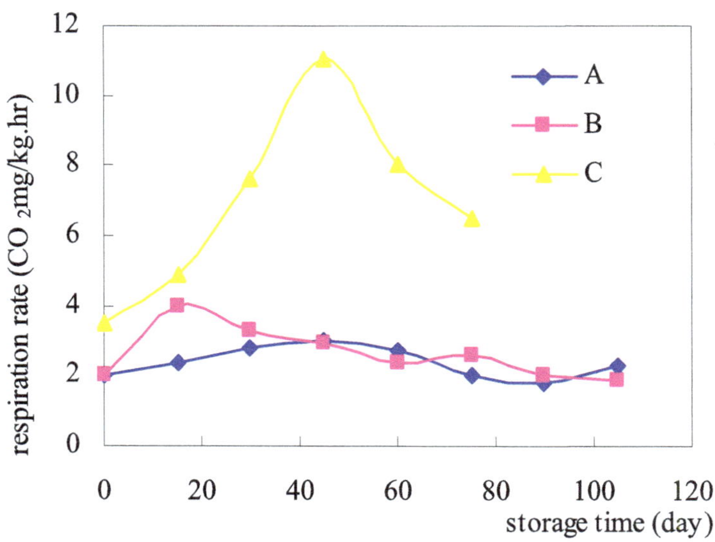

Fig. 16. Curves of peach respiration v.s. storage time

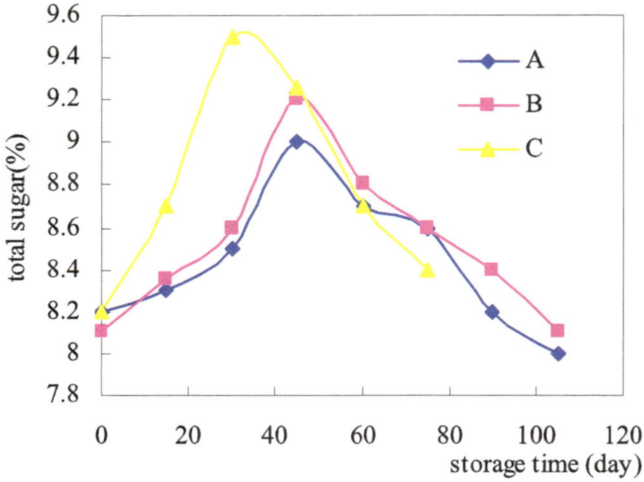

Fig. 17. Curves of peach total sugar v.s storage time

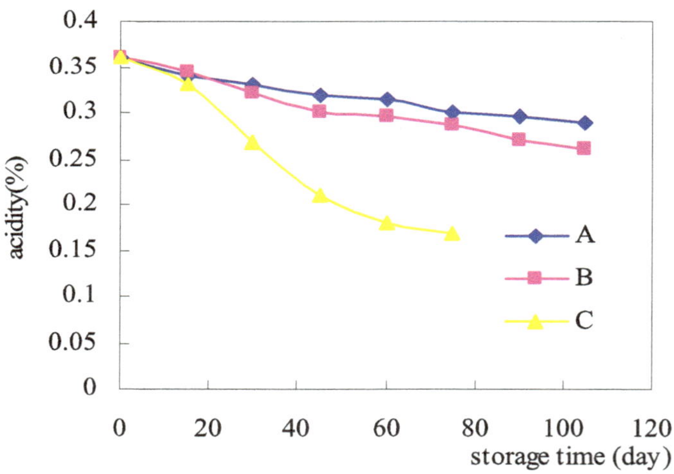

Fig. 18. Curves of peach acidity v.s storage time

As it can be seen from Fig.16, the peach has the first reparatory peak after 45 days in traditional storage while the ones precooled at variable cooling rate does not have. Compared with the two processes with different cooling rate, the peach cooled by the slower cooling rate shown a smaller reparatory peak, but which doesn't influence the respiration rate in the whole storage time. And from Fig.17, it can be found that the total sugar always has a peak value in the storage period. And the peak value of the peach cooled by variable cooling rate was smaller than that of the peach cooled in a traditional method. Fig. 18 shows that the acidity of peach precooled at variable rate is smaller than that of the one cooled in traditional method. However, after stored 100 days, the peach cooled by variable cooling rate still had better appearance and hardness shown in Fig.19, there is a serious decline in sugar and the flavor become worse.

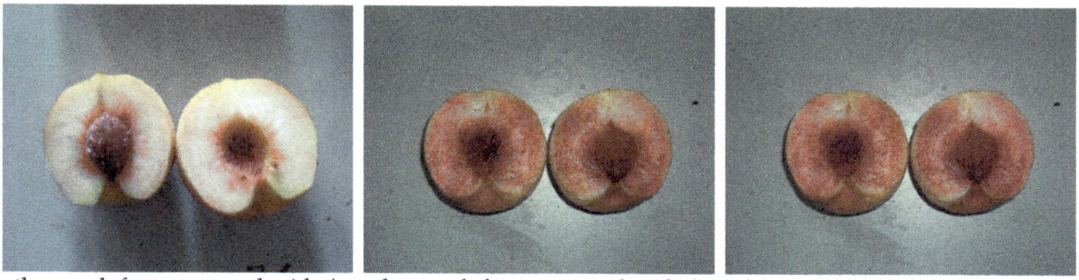

the peach feature stored with A processing after 75 days | the peach feature stored with A processing after 105 days | the peach feature stored with C processing after 75 days

Fig. 19. the peach feature at different storage time

Compared with the color of the peach shown in Fig.19, we can find that the peach cooled by the variable cooling rate has a better appearance than that of the peach treated by the traditional method. After 75 days in storage, the stored peach in the ice temperature has little difference with the new peach in color, hardness, etc., while the peach with traditional treatment has reached to the end of shelf period. And after 105 days, the peach at ice temperature reach to the end of shelf period with low sugar concentration shown in Fig.17.

Experimental results show that precooling process with variable temperature can make the fruit to adapt the low temperature environment and delay the emergence of the respiratory peak. Compared to the conventional low temperature storage, the fruit stored under the ice temperature has smaller respiratory rate, which guarantee the quality of stored fruit effectively and extend the storage time by 30%.

4. Effect of ice temperature storage

As discussed in 2, the cooling rate and heating rate of fruits before and after the ice temperature storage finally influence the storage period、shelf life and nutrition.

Under normal circumstances, the organization with rapid cooling stress will take over the depth of the cooling method to keep warm, and the one with slow cooling stress will take over the way of ice frost in the cells [5-10]. Slowly and staging cooling the organization can reduce the critical lethal temperature, thus making some of the critical lethal temperature higher than the organization of plant in the zero and can be preservation for a long time at 0°C or below zero, extend their storage period. Different heating rate after storage at ice-temperature has different effects on the shelf life of fruits. The reason is the act of fruit tissue self-adaption. For the study of the reactions of kiwi and pear under the processing conditions, the experimental studies were carried out to a better understanding the mechanism of ice-temperature storage.

According to the difference of respiratory type, two kinds of fruits were chosen to do the ice-temperature storage research, one is pear (breathing -jump variant), the other is kiwi(non-breathing-jump variant).After certain days, the storage temperature was increased in different warming ways. The results showed that compared with the comparison group, ice-temperature storage can maintain the nutrients in fruits better; as to the carbohydrate material, slow cooling is helpful to maintain the sugar during storage period. Ice-temperature storage by slow cooling can make the fruits have higher acidity while the fast cooling one make the fruits have more soluble solids.

4.1 Experimental materials and equipments

In this study, the mostly used equipments included two equipments. One is the DSC (Differential Scanning Calorimetry) used to measure the ice point and the thermal action of experimental materials in cooling or heating, but at here the results were not shown. The other equipment is two setups of ice-temperature with two parts. One part is used to cooling or heating the stored fruit, the other part is used to store the fruit. The two equipments are illustrated respectively on Figure 1 and 2.

The thermal dynamic process is a tool and the fruit quality is the goal. The total sugar, the total acid and the soluble solid of Kiwi and Pear fruit are measured. The measurement method are according to Chinese National Standard GB 6194-86, GB 12293-90 and GB12295-90.

Fig. 20. Diagram of DSC

Fig. 21. Experimental setup of Ice temperature

In experiments, two fruits were selected. One is the pear produced in Xinjiang Province, which was harvested in September 2006 and stored with a temperature of 5°C. The other is the Kiwi produced in Shanxi Province, which is harvested in October 2006 and stored with a temperature of 6°C. In Jan 6, 2007, the two fruits were transported to Tianjin University of Commerce and was stored in two cold room with a temperature of 5°C. The two fruits were transfered to the ice-temperature lab (the Jan 12, 2007), where the temperature of Kiwi was 5.8°C and 4.6°C, and the temperature of pear was 5.9°C and 4.3°C. In Table 1, the processes of cooling of the two fruits were shown. In order to compare to the experimental results, before the cooling, the two fruits were cooled to 5°C and 3.5°C.

The thermal treatment consists of three steps, where the first step is a cooling process in order to bring the fruit to the ice temperature, the second stage is a storage period (one month) at ice temperature and the last step is the heating step in order to warm the fruits to a temperature of 15°C.

Type of fruit	Processing mode (name)	Initial Temperature	Final Temperature	Begin of cooling Jan 16th, 2007. 16 : 00	End of cooling Jan 17th, 2007 10:00
Kiwi	fast cooling (KF)	5.0°C	0.0°C		
	slow cooling (KS)	3.5°C	0.0°C		
Pear	fast cooling (PF)	5.0°C	0.0°C		
	slow cooling (KS)	3.5°C	0.0°C		

Table 1. The step 1 of thermal cooling processes of experimental materials

After the two fruits were cooled down, they were stored one month in the ice temperature rooms with a set temperature of -1°C and -0.2°C. During the third step, the fruit were moved out of the ice temperature rooms, they were heated as shown in Table 2.

Heating mode	fruit	Cooling mode (name)
Referenced group (5.7°C)	Kiwi	5.7°C
	pear	5.7°C
Slow heating :	Kiwi	Fast cooling (KF-S)
		Slow cooling (KS-S)
	Pear	Fast cooling (PF-S)
		Slow cooling (PS-S)
Fast heating :	Kiwi	Fast cooling (KF-F)
		Slow cooling (KS-F)
	pear	Fast cooling (PF-F)
		Slow cooling (PS-F)

*Temperature rise : slow (from -1°C to 15°C in five days) and fast (from -0.2°C to 15°C in three days)

Table 2. The step 3 of thermal heating processes of experimental material

4.2 Results and discussions

During the step 2, one month storing time, the total sugar, the acidity and the soluble solid were measured each two week, and each measurement is an average of six samples. The results were shown in Figure 22 to Figure 24.

From Fig. 22 it can be easily found compared with the referenced group, t the total sugar of Kiwi and pear is higher, which means that a certain thermal process is needed before the cold storage and also means that the method of the ice-temperature is better than the method of cold storage. As for the different cooling processes, the difference is not obvious. But for different fruits, the cooling process plays an important role.

After 30 days in the ice-temperature storage, the total sugar of Kiwi stored in No.1 setup with -1°C by slow cooling is the highest, while the total sugar of pear stored in No.2 setup with -0.2°C by slow cooling is the highest.

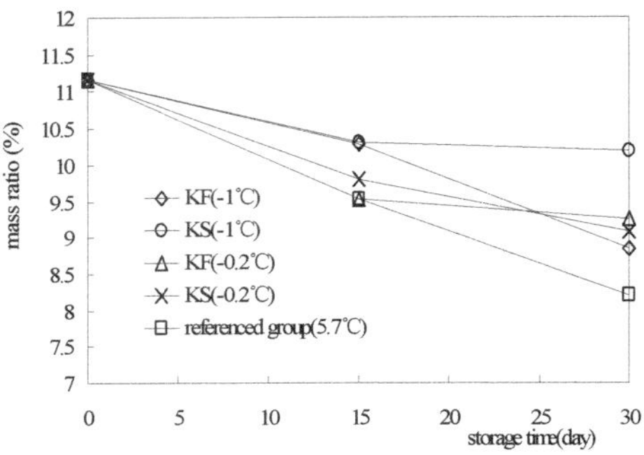

(a) total sugar of kiwi fruit

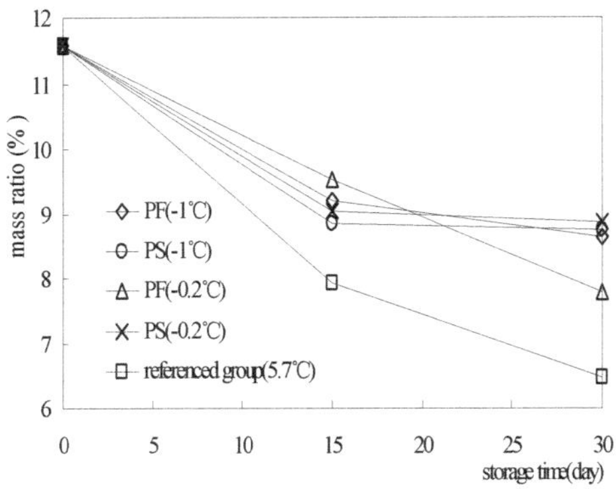

(b) total sugar of bergamot pear

Fig. 22. Curves of total sugar vs. storage time

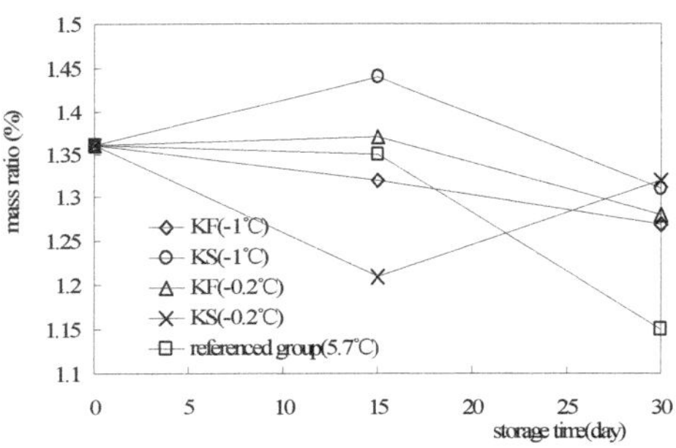

(a) total acidity of kiwi fruit

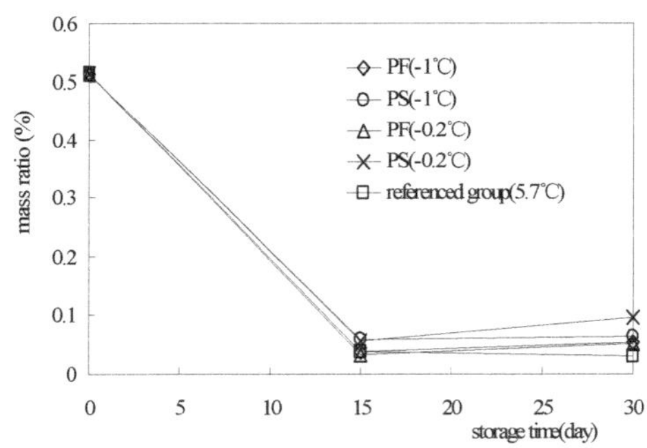

(b) total acidity of bergamot pear

Fig. 23. Curve of total acidity vs. storage time

From Fig.23-a, it can be found that the acidity of Kiwi first increased with the storage time, and after some days, it reduced. But in Fig. 23-b, the acidity of pear reduced with the time, and after some days, it rose a little. Comparing the different thermal processes in Figure 4, the acidity in the referenced group without any process is the lowest, and the acidity in the group by slow cooing is higher that that of fast cooling. Fig.5 shows the curves of soluble solid vs. time.

In Figure24-a, it can be found that the soluble solid of Kiwi increased with the storage time, but in Fig.24-b it can be found that the soluble solid of Pear increased with the storage time. While we can find that the two fruits processed by fast cooling all have a higher soluble solid. Table.3 shows the content change after different heating model. Table.3 shows the content change after different heating model.

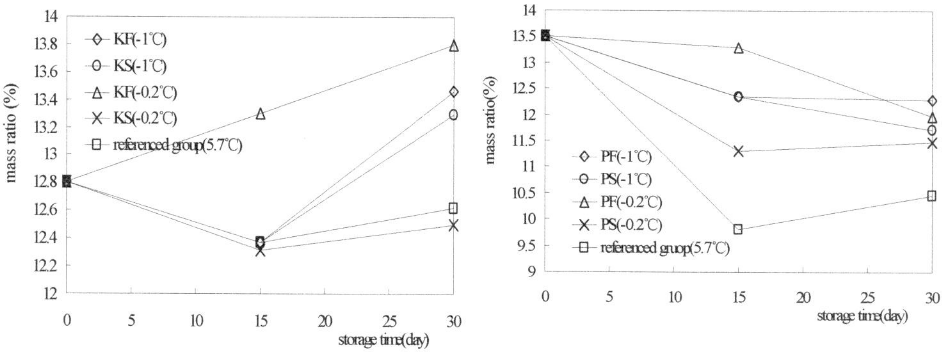

(a) soluble solid of kiwi fruit (b) soluble solid of bergamot pear

Fig. 24. Curve of soluble solid vs. time

Heating model	Total sugar		Acidity		Soluble solid	
	Before heating	After heating	Before heating	After heating	Before heating	After heating
Kiwi referenced(5.7°C)	8.2	10.94	1.15	1.16	12.62	14
Pear referenced (5.7°C)	6.47	7.35	0.03	0.051	10.4569	10
KF-S	8.85	9.92	1.27	1.27	12.96	13
KS-S	10.19	10.33	1.31	1.31	13.79	13.5
PF-S	8.64	7.04	0.055	0.055	12.28	11.2
PS-S	7.75	8.62	0.06	0.063	11.72	11.7
KF-F	9.26	9.41	1.28	1.28	13.8	13
KS-F	9.08	10.4	1.32	1.32	12.5	13
PF-F	7.68	7.96	0.051	0.051	11.97	10.8
PS-F	8.69	8	0.096	0.097	11.49	11.5

Table 3: content change after heating

From Table 3, it can be found that the acidity is not affected by the heating model, but the total sugar and the soluble solid are affected strongly.

The above results were obtained by experiments with little theoretical analysis. In our opinions, the fresh fruits have an ability to adapt to the change of environment. When the fresh fruit is processed by different thermal processing, they will change their inner contents to make them comfortable in a new environment and to prolong their life time liking human being. In future, more works should be paid on thermal action of fresh fruits when they are undertaken thermal processions, such as heat and mass transfer of cells.

5. Conclusion

Based on the characteristic that fruit and vegetable are living organisms, the ice temperature storage technology of fruit and vegetable enable them to adjust to the cryogenic environment by regulating the precooling process in order to prolong their life period. The ice temperature storage can greatly reduce the metabolic rate, resulting in obvious deviation of breathing index from the value at common temperature. It is a kind of cryogenic physical storage technology of fruit and vegetable which will has an significant development in the future.

6. References

[1] Shen Jiang, Wang Xiaodong, Wang Suying,,et al, 2009, Experimental Study on Application of Hyo-on Technology, Journal of Refrigeration (Chinese), vol 30 (4), pp 40-45.

[2] Shen Jiang, Qi Han-fei, Liu Xing-hua, 2011,The Influence of Heat Treatment on Peach Close to Freezing Point Storage Characteristic, Food Research And Development (Chinese), vol32 (2), pp 159-162.

[3] Bin Liu， Shen Jiang, Yali Guo, et al., 2009, Experimental Studies of Kiwi and Pear Fruit in Ice-Temperature Storage, Defect and Diffusion Forum, Vols. 283-286, pp 59-64.

[4] Shen Jiang, Li Chao, Miao Hui, et al., 2011, Experiment on Different Storage Temperature Impact on Litchi's Quality Change, Procedia Environmental Sciences, Volu 11, Part B, , pp 719-725.

[5] Shen Jiang ,He Xiaonan, Wang Suying,et al, 2010, Experiment on Influence of Hyo-on Storage on Amino Acid and Other Contents of Muskmelon, Journal of Refrigeration (Chinese), vol31 (3) .pp42-45.

[6] Guo Xiao-wei, Dong Shi-yuan, Liu Zun-ying, et al, 2010, Effect of controlled freezing-point on shelf life of oysters, FOOD SCIENCE AND TECHNOLOGY (Chinese), vol35 (11), pp86-90

[7] Dong Chang-lin,Li Jiang-kuo,Zhang Peng, et al,2011, Effect of freezing point storage on browning and physio-quality in persimmon, Journal of Fruit Science (Chinese), vol 28 (2), pp263-267.

Permissions

The contributors of this book come from diverse backgrounds, making this book a truly international effort. This book will bring forth new frontiers with its revolutionizing research information and detailed analysis of the nascent developments around the world.

We would like to thank Benjamin Valdez, Michael Schorr and Roumen Zlatev, for lending their expertise to make the book truly unique. They have played a crucial role in the development of this book. Without their invaluable contribution this book wouldn't have been possible. They have made vital efforts to compile up to date information on the varied aspects of this subject to make this book a valuable addition to the collection of many professionals and students.

This book was conceptualized with the vision of imparting up-to-date information and advanced data in this field. To ensure the same, a matchless editorial board was set up. Every individual on the board went through rigorous rounds of assessment to prove their worth. After which they invested a large part of their time researching and compiling the most relevant data for our readers. Conferences and sessions were held from time to time between the editorial board and the contributing authors to present the data in the most comprehensible form. The editorial team has worked tirelessly to provide valuable and valid information to help people across the globe.

Every chapter published in this book has been scrutinized by our experts. Their significance has been extensively debated. The topics covered herein carry significant findings which will fuel the growth of the discipline. They may even be implemented as practical applications or may be referred to as a beginning point for another development. Chapters in this book were first published by InTech; hereby published with permission under the Creative Commons Attribution License or equivalent.

The editorial board has been involved in producing this book since its inception. They have spent rigorous hours researching and exploring the diverse topics which have resulted in the successful publishing of this book. They have passed on their knowledge of decades through this book. To expedite this challenging task, the publisher supported the team at every step. A small team of assistant editors was also appointed to further simplify the editing procedure and attain best results for the readers.

Our editorial team has been hand-picked from every corner of the world. Their multi-ethnicity adds dynamic inputs to the discussions which result in innovative outcomes. These outcomes are then further discussed with the researchers and contributors who give their valuable feedback and opinion regarding the same. The feedback is then collaborated with the researches and they are edited in a comprehensive manner to aid the understanding of the subject.

Apart from the editorial board, the designing team has also invested a significant amount of their time in understanding the subject and creating the most relevant covers. They scrutinized every image to scout for the most suitable representation of the subject and create an appropriate cover for the book.

The publishing team has been involved in this book since its early stages. They were actively engaged in every process, be it collecting the data, connecting with the contributors or procuring relevant information. The team has been an ardent support to the editorial, designing and production team. Their endless efforts to recruit the best for this project, has resulted in the accomplishment of this book. They are a veteran in the field of academics and their pool of knowledge is as vast as their experience in printing. Their expertise and guidance has proved useful at every step. Their uncompromising quality standards have made this book an exceptional effort. Their encouragement from time to time has been an inspiration for everyone.

The publisher and the editorial board hope that this book will prove to be a valuable piece of knowledge for researchers, students, practitioners and scholars across the globe.

List of Contributors

Jafar Milani and Gisoo Maleki
Sari Agricultural Sciences and Natural Resources University, Iran

Ítalo Herbert Lucena Cavalcante and João Marcos de Sousa Miranda
Federal University of Piauí, Brazil

Lourival Ferreira Cavalcante
Federal University of Paraíba, Brazil

Antonio Baldo Geraldo Martins
São Paulo State University, Brazil

Seiichi Ishikawa, Bunko Cho, Shin Li, Yuji Okumura, Yoshikazu Iida, Teiji Tanizaki and Masayuki Higuchi
The University of Kitakyushu, Japan

D. M. Cabezas and M. C. Tomás
Centro de Investigación y Desarrollo en Criotecnología de Alimentos (CIDCA) - CCT La Plata - CONICET - Facultad de Ciencias Exactas (FCE), Universidad Nacional de La Plata (FCE - UNLP), La Plata, Argentina

R. Madoery
Cátedra de Química Orgánica - Facultad de Ciencias Agrarias, Universidad Nacional de Córdoba (FCA - UNC), Córdoba, Argentina

B. W. K. Diehl
Spectral Service GmbH Laboratorium für Auftragsanalytik, Cologne, Germany

María Eugenia Hidalgo and Estela M. Alvarez
Facultad de Ciencias Bioquímicas y Farmacéuticas, Universidad Nacional de Rosario, Argentina

Bibiana D. Riquelme
Facultad de Ciencias Bioquímicas y Farmacéuticas, Universidad Nacional de Rosario, Argentina
Instituto de Física Rosario (IFIR), CONICET-UNR, (2000), Rosario, Argentina

Jorge R. Wagner
Departamento de Ciencia y Tecnología, Universidad Nacional de Quilmes, Buenos Aires, Argentina

Patricia H. Risso
Facultad de Ciencias Bioquímicas y Farmacéuticas, Universidad Nacional de Rosario, Argentina
Instituto de Física Rosario (IFIR), CONICET-UNR, (2000), Rosario, Argentina
Facultad de Ciencias Veterinarias, Universidad Nacional de Rosario, Rosario, Argentina

Maricica Stoica, Petru Alexe, Rodica Dinica and Geta Cârâc
Dunarea de Jos University of Galati/ Department of Chemistry Physics and Environment, Romania

Lai Teng Ling
University of Malaya, Malaysia

U.D. Palanisamy
Monash University Sunway Campus, Malaysia

María J. Yebra, Vicente Monedero, Gaspar Pérez-Martínez and Jesús Rodríguez-Díaz
Departamento de Biotecnología de los Alimentos, IATA- CSIC, Spain

Nándor Nemestóthy, Tamás Bányai, Katalin Bélafi-Bakó, and László Gubicza
University of Pannonia, Research Institute on Bioengineering, Membrane Technology and Energetics, Argentina

László Bartha
Department of MOL Hydrocarbon and Coal Processing, Hungary

Francesca Bosco and Chiara Mollea
Politecnico di Torino, Italy

Carmen Sieiro, Belén García-Fraga, Jacobo López-Seijas and Abigaíl F. da Silva
Department of Functional Biology and Health Sciences, University of Vigo, Spain

Tomás G. Villa
Department of Microbiology and Parasitology, University of Santiago de Compostela, Spain

Karla Bigetti Guergoletto
Universidade Estadual de Londrina, Brazil

Margarita Stoytcheva, Roumen Zlatev and Benjamin Valdez
Universidad Autónoma de Baja California, Instituto de Ingeniería, México

Marcela Ovalle
Universidad Autónoma de Yucatán, Facultad de Química, México

Zdravka Velkova
Plovdiv University of Food Technology, Bulgaria

Velizar Gochev
Plovdiv University "Paisii Hilendarski", Bulgaria

Deborah Markowicz Bastos, Érica Monaro, Érica Siguemoto and Mariana Séfora
Nutrition Department, School of Public Health, São Paulo University, Brazil

Claudia Muro and María del Carmen Díaz
Institute Technological of Toluca, México

Francisco Riera
University of Oviedo, Spain

Anne Heponiemi
Department of Chemistry, University of Oulu, Oulu, Finland

Ulla Lassi
Department of Chemistry, University of Oulu, Oulu, Finland
Kokkola University Consortium Chydenius, Kokkola, Finland

Benjamín Valdez Salas, Michael Schorr Wiener and Monica Carrillo Beltran
Universidad Autonoma de Baja California, Instituto de Ingeniería, Mexicali, Baja California, Mexico

Janine Laverse, Pierangelo Frisullo, Amalia Conte and Matteo Alessandro Del Nobile
University of Foggia, Agricultural Faculty, Department of Food Science, Via Napoli, Italia

Daniela Soares Viana, Lucia Maria Jaeger de Carvalho, Ediane Maria Gomes Ribeiro and Gisela Maria Dellamora Ortiz
Federal University of Rio de Janeiro, Brazil

Montip Chamchong, Vilasinee Sangsom and Nuttakorn Charoeamkitti
Department of Food Engineering, Faculty of Engineering at Kamphaengsaen, Kasetsart University, Nakhon Pathom, Thailand

Liu Bin, Cai Bihao and Shen Jiang
Tianjin Key Lab of Refrigeration Technology, Tianjin University of Commerce, P.R. China